Hazardous Materials for First Responders

Fifth Edition

Alex Abrams, Leslie Miller, and Libby Snyder
Lead Senior Editors

Leslie Miller
Technical Writer

Cindy Brakhage, Tony Peters, and Rikka Strong
Senior Editors

Simone Rowe and David Schaap
Lead Instructional Developers

P9-CRU-829

IFSTA

Validated by the International Fire Service Training Association

Published by Fire Protection Publications • Oklahoma State University

RECYCLABLE

Cover photo courtesy of Dennis Walus. • Title page photo courtesy of Barry Lindley.

The International Fire Service Training Association (IFSTA) was established in 1934 as a *nonprofit educational association of fire fighting personnel who are dedicated to upgrading fire fighting techniques and safety through training*. To carry out the mission of IFSTA, Fire Protection Publications was established as an entity of Oklahoma State University. Fire Protection Publications' primary function is to publish and distribute training materials as proposed, developed, and validated by IFSTA. As a secondary function, Fire Protection Publications researches, acquires, produces, and markets high-quality learning and teaching aids consistent with IFSTA's mission.

IFSTA holds two meetings each year: the Winter Meeting in January and the Annual Validation Conference in July. During these meetings, committees of technical experts review draft materials and ensure that the professional qualifications of the National Fire Protection Association® standards are met. These conferences bring together individuals from several related and allied fields, such as:

- Key fire department executives, training officers, and personnel
- Educators from colleges and universities
- Representatives from governmental agencies
- Delegates of firefighter associations and industrial organizations

Committee members are not paid nor are they reimbursed for their expenses by IFSTA or Fire Protection Publications. They participate because of a commitment to the fire service and its future through training. Being on a committee is prestigious in the fire service community, and committee members are acknowledged leaders in their fields. This unique feature provides a close relationship between IFSTA and the fire service community.

IFSTA manuals have been adopted as the official teaching texts of many states and provinces of North America as well as numerous U.S. and Canadian government agencies. Besides the NFPA® requirements, IFSTA manuals are also written to meet the Fire and Emergency Services Higher Education (FESHE) course requirements. A number of the manuals have been translated into other languages to provide training for fire and emergency service personnel in Canada, Mexico, and outside of North America.

ISBN 978-0-87939-613-8   Library of Congress Control Number: 2017931043

Fifth Edition, First Printing, March 2017   *Printed in the United States of America*

10   9   8   7   6   5   4

If you need additional information concerning the International Fire Service Training Association (IFSTA) or Fire Protection Publications, contact:

Customer Service, Fire Protection Publications, Oklahoma State University
930 North Willis, Stillwater, OK 74078-8045
800-654-4055   Fax: 405-744-8204

For assistance with training materials, to recommend material for inclusion in an IFSTA manual, or to ask questions or comment on manual content, contact:

Editorial Department, Fire Protection Publications, Oklahoma State University
930 North Willis, Stillwater, OK 74078-8045
405-744-4111   Fax: 405-744-4112   E-mail: editors@osufpp.org

Chapter Summary

Table of Contents

List of Tables

Acknowledgements

The fifth edition of **Hazardous Materials for First Responders** is designed to meet the requirements of NFPA 1072, NFPA 1072, *Standard for Hazardous Materials/Weapons of Mass Destruction Emergency Response Personnel Professional Qualifications (2017)*.

Acknowledgement and special thanks are extended to the members of the IFSTA validating committee who contributed their time, wisdom, and knowledge to the development of this manual.

Special thanks go to committee member, Brian D. White, for his assistance writing Chapters 14 and 15. Thanks are also due Barry Lindley for contributing Appendix E, Global Harmonized System of Classification and Labeling of Chemicals (GHS).

IFSTA Hazardous Materials for First Responders
Fifth Edition Validation Committee

Chair
Scott D. Kerwood
Fire Chief
Hutto Fire Rescue
Hutto, TX

Vice Chair and Secretary
Rich Mahaney
Manager/Instructor
Mahaney Loss Prevention Services

Committee Members

Tyler Bones
Chief
Fairbanks North Star Borough Hazardous
   Materials Response Team
Fairbanks, AK

Dennis Clinton
Missouri HAZ-MAT Consultants, LLC
Ozark, Missouri

David Coates
Hazardous Materials Coordinator
South Carolina Fire Academy
Columbia, South Carolina

Brent Cowx
Program Coordinator, Hazardous Materials,
   Technical Rescue
Justice Institute of British Columbia
(Retd. Vancouver Fire Rescue Services)

Bryn Crandell
Training Developer
DoD Fire Academy
San Angelo, TX

Michael Fortini
Los Angeles Fire Department
Camarillo, CA

Doug Goodings
Continuing Education Coordinator
Blue River College
Missouri

CJ Haberkorn
Assistant Chief, Shift Commander
Denver Fire Department
Denver, CO

Butch Hayes
Firefighter/Hazmat Technician
Houston Fire Department
Conroe, TX

Steve Hergenreter
IAFF Haz Mat Training
Fort Dodge, IA

Robert Kronenberger
City of Middletown, CT Fire Department
Middletown, CT

Barry Lindley
Specialized Professional Services, Inc
Charleston, WV

IFSTA Hazardous Materials for First Responders
Fifth Edition Validation Committee
Committee Members (cont.)

Thomas Miller
Instructor III
Sissonville Volunteer Fire Department,
    National Volunteer Fire Council
Charleston, WV

Carlos Rodriguez
Fire Captain
City of Wichita Fire Department
Wichita, KS

Walter G. M. Schneider III, Ph.D., P.E., CBO, MCP
Agency Director
Centre Region Code Administration
State College, Pennsylvania

Department Chief
Bellefonte Fire Department
Bellefonte, Pennsylvania

Fred Terryn
Fire Program Manager
U.S. Air Force Fire Emergency Services
Tyndall AFB, FL

Brian D. White
US Department of Justice
Washington DC

Much appreciation is given to the following individuals and organizations for contributing information, photographs, and technical assistance instrumental in the development of this manual:

- Rich Mahaney, for providing so many photos used throughout the manual, including in tables and skill sheets. Rich's photos have enriched IFSTA's hazmat manuals for several editions.

- Barry Lindley, for fixing the little details, providing pictures, and giving the editors valuable chemistry lessons. He also answered innumerable questions about product and container behavior.

- Carlos Rodriguez, for sharing his technical expertise, providing exceptional feedback, and troubleshooting. As well, Carlos kept us on task with reminders to make this text usable for the target audience.

- The entire NFPA 472/1072 technical committee for answering questions about the standards' intent, as well as random technical questions.

- Dennis Walus for letting us use his fabulous cover photo.

Thanks also to:

Boca Raton Fire Rescue

Canadian Centre for Occupational Health and Safety

CBRN Responder Training Facility, Fort Leonard
    Wood

CDC Public Health Image Library

FEMA News Photos

Fort Leonard Wood Fire Department

Hutto Fire Rescue
    Rob Bocanegra
    John Gibson

Joe Mayberry
Derek Rogers
Michael Pal Parks
Ivan Valenzuela

International Association of Fire Fighters

Mohave Museum of History and Arts

Moore Memorial Library, Texas City, TX

Moore (OK) Fire Department

MSA

New South Wales Fire Brigades

Oklahoma State Fire Service Training

Oklahoma Highway Patrol Bomb Squad

Owasso (OK) Fire Department

Round Rock Fire Department

Valentin Diaz

Dalton Everett

Andrew Heustis

Stillwater (OK) Fire Department
  Wes Dotter
  Ty Lewis

Texas Commission on Fire Protection

David Alexander

Sherry Arasim

Lucas M. Atwell

Jocelyn Augustino

Andrea Booher

Ben Brody

Brian Canady

Deborah Carter

Tom Clawson

Gary Coppage

Charles Csavossy

John Deyman

Ray Elder

Mark D. Faram

Matthew Flynn

Brent Gaspard

Robert R. Hargreaves, Jr.

Win Henderson

Greg Henshall

Joan Hepler

William Hester

Chiaki Iramina

Steve Irby

Ron Jeffers

Scott Kerwood

Bradley A. Lail

J.A. Lee, II

David Lewis

Phil Linder

Todd Lopez

Chris E. Mickal

Ron Moore

Warren Peace

Todd Pendleton

Christopher D. Reed

Michael Rieger

Liz Roll

Antonio Rosas

Walter Schneider

Alissa Schuning

Brian A. Tuthill

August Vernon

Doug Weeks

Brian White

Kirk Worley

Sean Worrell

Charlie Wright

Wayne Yoder

Thanks also go to the agencies and organizations producing various resources used throughout this manual:

Health Canada

Los Alamos National Laboratory

Sandia National Laboratories

Transport Canada

Union Pacific Railroad

Oklahoma Highway Patrol Bomb Squad

U.S. Air Force

U.S. Army

U.S. Centers for Disease Control and Prevention

U.S. Coast Guard

U.0. Department of Defense

U.S. Department of Energy

U.S. Department of Homeland Security

U.S. Department of Justice

U.S. Department of Transportation; Pipeline, Hazardous Materials, and Safety Administration

U.S. Drug Enforcement Agency

U.S. Environmental Protection Agency

U.S. Federal Bureau of Investigation

U.S. Federal Emergency Management Agency

U.S. Fire Administration

U.S. Marines

U.S. National Institute for Occupational Safety and Health

U.S. National Nuclear Security Administration

U.S. Navy

U.S. Nuclear Regulatory Commission

U.S. Occupational Safety and Health Administration

Last, but certainly not least, gratitude is extended to the following members of the Fire Protection Publications staff whose contributions made the final publication of this manual possible.

Hazardous Materials for First Responders, Fifth Edition, Project Team

Lead Senior Editors
Alex Abrams, Senior Editor
Leslie A. Miller, Senior Editor
Libby Snyder, Senior Editor

Director of Fire Protection Publications
Craig Hannan

Curriculum Manager
Lori Raborg
Leslie A. Miller
Colby Cagle

Editorial Manager
Clint Clausing

Production Manager
Ann Moffat

Editor(s)
Cindy Brakhage, Senior Editor
Tony Peters, Senior Editor
Rikka Strong, Senior Editor

Illustrator and Layout Designer
Errick Braggs, Senior Graphic Designer

Lead Instructional Developers
Simone Rowe, Curriculum Developer
David Schaap, Curriculum Developer

Photographer(s)
Jeff Fortney, Senior Editor
Leslie A. Miller, Senior Editor
Alex Abrams, Senior Editor

Editorial Staff
Tara Gladden, Editorial Assistant

Indexer
Nancy Kopper

Dedication

This manual is dedicated to the men and women who hold devotion to duty above personal risk, who count on sincerity of service above personal comfort and convenience, who strive unceasingly to find better and safer ways of protecting lives, homes, and property of their fellow citizens from the ravages of fire, medical emergencies, and other disasters

...The Firefighters of All Nations.

The IFSTA Executive Board at the time of validation of the **Hazardous Materials for First Responders, Fifth Edition** was as follows:

IFSTA Executive Board

Executive Board Chair
Bradd Clark
Fire Chief
Ocala Fire Department
Ocala, FL

Vice Chair
Mary Cameli
Fire Chief
City of Mesa Fire Department
Mesa, AZ

IFSTA Executive Director
Mike Wieder
Associate Director
Fire Protection Publications at OSU
Stillwater, OK

Board Members

Steve Ashbrock
Fire Chief
Madeira & Indian Hill Fire Department
Cincinnati, OH

Steve Austin
Project Manager
Cumberland Valley Volunteer Firemen's Association
Newark, DE

Dr. Larry Collins
Associate Dean
Eastern Kentucky University
Safety, Security, & Emergency Department
Richmond, KY

Chief Dennis Compton
Mesa & Phoenix, Arizona
Chairman of the National Fallen Firefighters Foundation Board of Directors

John Hoglund
Director Emeritus
Maryland Fire & Rescue Institute
New Carrollton, MD

Tonya Hoover
State Fire Marshal
CA Department of Forestry & Fire Protection
Sacramento, CA

Dr. Scott Kerwood
Fire Chief
Hutto Fire Rescue
Hutto, TX

Wes Kitchel
Assistant Chief, Retired
Sonoma County Fire & Emergency Services
    Department
Santa Rosa, CA

Brett Lacey
Fire Marshal
Colorado Springs Fire Department
Colorado Springs, CO

Robert Moore
Division Director
Texas A&M Engineering Extension Services
College Station, TX

Dr. Lori Moore-Merrell
Assistant to the General President
International Association of Fire Fighters
Washington, DC

Jeff Morrissette
State Fire Administrator
State of Connecticut
Commission on Fire Prevention and Control
Windsor Locks, CT

Josh Stefancic
Division Chief
Largo Fire Rescue
Largo, FL

Paul Valentine
Senior Fire Consultant
Global Risk Consultants Group
Chicago, IL

Steven Westermann
Fire Chief
Central Jackson County Fire Protection District
Blue Springs, MO

Kenneth Willette
Segment Director/Division Manager
National Fire Protection Association
Quincy, MA

Introduction

Introduction Contents

Introduction

Hazardous materials are found in every jurisdiction, community, workplace, and modern household. These substances possess a wide variety of harmful characteristics. Some hazardous materials can be quite deadly or destructive, and terrorists and other criminals may use them to deliberately cause harm. Because hazardous materials tend to complicate the emergency incidents in which they are involved, first responders (personnel who are likely to arrive first at an incident scene) must be alert to the presence of hazardous materials at incidents and take proper precautions. First responders must recognize and understand the hazards presented by various types of hazardous materials. Therefore, they must possess the skills necessary to address incidents involving them in a safe and effective manner.

Purpose and Scope

This book is written for emergency first responders who are mandated by law and/or called upon by necessity to prepare for and respond to hazardous materials and weapons of mass destruction (WMD) incidents. These first responders include the following individuals:

- Firefighters
- Law enforcement officers/personnel
- Emergency medical services personnel
- Military responders
- Industrial and transportation emergency response members
- Public works employees
- Utility workers
- Members of private industry
- Other emergency response professionals

The purpose of this book is to provide these first responders with the information they need to take appropriate initial actions at WMD incidents and hazardous materials spills or releases. Its scope is limited to giving detailed information about initial — and primarily defensive operations. More advanced procedures require hazardous materials technicians who have specialized training.

This book is designed to train responders to the Awareness and Operations level certification requirements of NFPA 1072, *Standard for Hazardous Materials/Weapons of Mass Destruction Emergency Response Personnel Professional Qualifications, 2017 Edition*. It addresses the first responders' responsibilities to recognize the presence of hazardous materials and WMDs, secure the area, provide personnel protection, and request the assistance of trained technicians and law enforcement personnel when necessary. Additionally, it addresses the control of hazardous materials releases and WMDs using operations for which the responders have been trained.

The following related regulations/standards are referenced in this book as applicable:

- NFPA 472, *Standard for Competence of Responders to Hazardous Materials/ Weapons of Mass Destruction Incidents (2013 edition)*, for the Awareness and Operations Levels (core competencies plus mission-specific competencies).

- OSHA regulations in Title 29 *Code of Federal Regulations* (*CFR*) 1910.120, *Hazardous Waste Operations and Emergency Response* (*HAZWOPER*), paragraph (q), for first responders at the Awareness and Operational Levels

Book Organization

To meet the competencies of NFPA 1072, this book is divided into the following parts:

Part 1. The first three chapters address NFPA 1072's Awareness level requirements.

Part 2. Chapters 4-8 address NFPA 1072's Operations level requirements for everything *except* personal protective equipment and decontamination.

Part 3. Chapters 9 and 10 address NFPA 1072's Operations and Operations Mission-Specific requirements for personal protective equipment and decontamination.

Part 4. Chapters 11-15 address strictly Operations Mission-Specific requirements.

Part 1. Awareness (NFPA 1072 Chapter 4)

Chapter 1 – Introduction to Hazardous Materials

Chapter 2 – Analyzing the Incident: Recognizing and Identifying the Presence of Hazardous Materials

Chapter 3 – Implementing the Response: Awareness Level Actions at Hazmat Incidents

Part 2. Operations (NFPA 1072 Chapter 5)

Chapter 4 – Analyzing the Incident: Identifying Potential Hazards

Chapter 5 – Analyzing the Incident: Identifying Containers and Predicting Behavior

Chapter 6 – Planning the Response: Identifying Response Options

Chapter 7 – Implementing and Evaluating the Action Plan: Incident Management

Chapter 8 – Implementing the Response: Terrorist Attacks, Criminal Activities, and Disasters

Part 3. Operations and Mission-Specific (NFPA 1072 Chapters 5 and 6, PPE and Decon Requirements)

Chapter 9 – Implementing the Response: Personal Protective Equipment

Chapter 10 – Implementing the Response: Decontamination

Part 4. Mission-Specific (NFPA 1072 Chapter 6)

Chapter 11 – Implementing the Response: Mission-Specific Detection, Monitoring, and Sampling

Chapter 12 – Implementing the Response: Mission-Specific Victim Rescue and Recovery

Chapter 13 – Implementing the Response: Mission-Specific Product Control

Chapter 14 – Implementing the Response: Mission-Specific Evidence Preservation and Public Safety Sampling

Chapter 15 – Implementing the Response: Mission-Specific Illicit Laboratories

Terminology

This manual is written with a global, international audience in mind. For this reason, it often uses general descriptive language in place of regional- or agency-specific terminology (often referred to as *jargon*). Additionally, in order to keep sentences uncluttered and easy to read, the word *state* is used to represent both state and provincial level governments (or their equivalent). This usage is applied to this manual for the purposes of brevity and is not intended to address or show preference for only one nation's method of identifying regional governments within its borders.

The glossary at the end of the manual will assist the reader in understanding words that may not have their roots in the fire and emergency services. The sources for the definitions of fire-and-emergency-services-related terms will be the *NFPA Dictionary of Terms* and the IFSTA **Fire Service Orientation and Terminology** manual. Additionally, when reading this text, remember the following points:

1. The terms *emergency*, *incident*, and *hazmat incident* are often used interchangeably, with the understanding that the types of incidents addressed by this book are emergencies.

2. NFPA and OSHA have different terms for persons trained to the Awareness Level. NFPA 1072 refers to these individuals as *personnel* whereas OSHA's 29 *CFR* 1910.120 uses the term *responders*. When the term *first responder* is used in this manual, it generally refers to both Awareness- and Operations-Level responders as defined by OSHA. The authority having jurisdiction (AHJ) is responsible for defining the actions allowed by persons trained to the Awareness Level, depending on the standard to which they are trained.

3. There are many different ways to refer to hazardous materials. You may see *hazmat*, *haz mat*, *dangerous goods*, or *hazardous materials*. This manual will use the abbreviation *hazmat*.

4. Weapons of mass destruction (WMDs) as addressed in this manual are considered to be hazardous materials. When the term *hazmat* is used throughout this manual in a general sense, it should be understood that WMDs are included.

Key Information

In this book, various types of information are given in shaded boxes marked by symbols or icons. See the following definitions:

Case Study

A case study analyzes an event. It can describe its development, action taken, investigation results, and lessons learned.

Safety Alert

Safety alert boxes are used to highlight information that is important for safety reasons. (In the text, the title of safety alerts will change to reflect the content.)

Information

Information boxes give facts that are complete in themselves but belong with the text discussion. It is information that needs more emphasis or separation. (In the text, the title of information boxes will change to reflect the content.)

What This Means To You

These boxes take information presented in the text and synthesize it into an example of how the information is relevant to (or will be applied by) the intended audience, essentially answering the question, "What does this mean to you?"

Flash Point — Minimum temperature at which a liquid gives off enough vapors to form an ignitable mixture with air near the surface of the liquid.

A **key term** is designed to emphasize key concepts, technical terms, or ideas that first responders need to know. They are listed at the beginning of each chapter, highlighted in bold **red** font, and the definition is placed in the margin for easy reference.

Three key signal words are found in the book: **WARNING!**, **CAUTION**, and **NOTE**. Definitions and examples of each are as follows:

- **WARNING** indicates information that could result in death or serious injury to first responders. See the following example:

- **CAUTION** indicates important information or data that first responders need to be aware of in order to perform their duties safely. See the following example:

```
////////////////////////
```
WARNING
When damaged or stressed by heat or flames,
pressure containers may explode! Keep your
distance!

```
////////////////////////
```
CAUTION
Immediately remove any clothing saturated with
a cryogenic material.

- **NOTE** indicates important operational information that helps explain why a particular recommendation is given or describes optional methods for certain procedures. See the following example:

NOTE: *Vapor* is a gaseous form of a substance that is normally in a solid or liquid state at room temperature and pressure. It is formed by evaporation from a liquid or sublimation from a solid.

Metric Conversions

Throughout this manual, U.S. units of measure are converted to metric units for the convenience of our international readers. However, please be advised that we use the Canadian metric system. It is very similar to the Standard International system, but may have some variation.

We adhere to the following guidelines for metric conversions in this manual:

- Metric conversions are approximated unless the number is used in mathematical equations.

- Centimeters are not used because they are not part of the Canadian metric standard.

- Exact conversions are used when an exact number is necessary such as in construction measurements or hydraulic calculations.

- Set values such as hose diameter, ladder length, and nozzle size use their Canadian counterpart naming conventions and are not mathematically calculated. For example, 1½ inch hose is referred to as 38 mm hose.

The following two tables provide detailed information on IFSTA's conversion conventions. The first table includes examples of our conversion factors for a number of measurements used in the fire service. The second shows examples of exact conversions beside the approximated measurements you will see in this manual.

U.S. to Canadian Measurement Conversion

Measurements	Customary (U.S.)	Metric (Canada)	Conversion Factor
Length/Distance	Inch (in) Foot (ft) [3 or less feet] Foot (ft) [3 or more feet] Mile (mi)	Millimeter (mm) Millimeter (mm) Meter (m) Kilometer (km)	1 in = 25 mm 1 ft = 300 mm 1 ft = 0.3 m 1 mi = 1.6 km
Area	Square Foot (ft^2) Square Mile (mi^2)	Square Meter (m^2) Square Kilometer (km^2)	1 ft^2 = 0.09 m^2 1 mi^2 = 2.6 km^2
Mass/Weight	Dry Ounce (oz) Pound (lb) Ton (T)	gram Kilogram (kg) Ton (T)	1 oz = 28 g 1 lb = 0.5 kg 1 T = 0.9 T
Volume	Cubic Foot (ft^3) Fluid Ounce (fl oz) Quart (qt) Gallon (gal)	Cubic Meter (m^3) Milliliter (mL) Liter (L) Liter (L)	1 ft^3 = 0.03 m^3 1 fl oz = 30 mL 1 qt = 1 L 1 gal = 4 L
Flow	Gallons per Minute (gpm) Cubic Foot per Minute (ft^3/min)	Liters per Minute (L/min) Cubic Meter per Minute (m^3/min)	1 gpm = 4 L/min 1 ft^3/min = 0.03 m^3/min
Flow per Area	Gallons per Minute per Square Foot (gpm/ft^2)	Liters per Square Meters Minute ($L/(m^2.min)$)	1 gpm/ft^2 = 40 $L/(m^2.min)$
Pressure	Pounds per Square Inch (psi) Pounds per Square Foot (psf) Inches of Mercury (in Hg)	Kilopascal (kPa) Kilopascal (kPa) Kilopascal (kPa)	1 psi = 7 kPa 1 psf = .05 kPa 1 in Hg = 3.4 kPa
Speed/Velocity	Miles per Hour (mph) Feet per Second (ft/sec)	Kilometers per Hour (km/h) Meter per Second (m/s)	1 mph = 1.6 km/h 1 ft/sec = 0.3 m/s
Heat	British Thermal Unit (Btu)	Kilojoule (kJ)	1 Btu = 1 kJ
Heat Flow	British Thermal Unit per Minute (BTU/min)	watt (W)	1 Btu/min = 18 W
Density	Pound per Cubic Foot (lb/ft^3)	Kilogram per Cubic Meter (kg/m^3)	1 lb/ft^3 = 16 kg/m^3
Force	Pound-Force (lbf)	Newton (N)	1 lbf = 0.5 N
Torque	Pound-Force Foot (lbf ft)	Newton Meter (N.m)	1 lbf ft = 1.4 N.m
Dynamic Viscosity	Pound per Foot-Second (lb/ft.s)	Pascal Second (Pa.s)	1 lb/ft.s = 1.5 Pa.s
Surface Tension	Pound per Foot (lb/ft)	Newton per Meter (N/m)	1 lb/ft = 15 N/m

Conversion and Approximation Examples

Measurement	U.S. Unit	Conversion Factor	Exact S.I. Unit	Rounded S.I. Unit
Length/Distance	10 in	1 in = 25 mm	250 mm	250 mm
	25 in	1 in = 25 mm	625 mm	625 mm
	2 ft	1 in = 25 mm	600 mm	600 mm
	17 ft	1 ft = 0.3 m	5.1 m	5 m
	3 mi	1 mi = 1.6 km	4.8 km	5 km
	10 mi	1 mi = 1.6 km	16 km	16 km
Area	36 ft²	1 ft² = 0.09 m²	3.24 m²	3 m²
	300 ft²	1 ft² = 0.09 m²	27 m²	30 m²
	5 mi²	1 mi² = 2.6 km²	13 km²	13 km²
	14 mi²	1 mi² = 2.6 km²	36.4 km²	35 km²
Mass/Weight	16 oz	1 oz = 28 g	448 g	450 g
	20 oz	1 oz = 28 g	560 g	560 g
	3.75 lb	1 lb = 0.5 kg	1.875 kg	2 kg
	2,000 lb	1 lb = 0.5 kg	1 000 kg	1 000 kg
	1 T	1 T = 0.9 T	900 kg	900 kg
	2.5 T	1 T = 0.9 T	2.25 T	2 T
Volume	55 ft³	1 ft³ = 0.03 m³	1.65 m³	1.5 m³
	2,000 ft³	1 ft³ = 0.03 m³	60 m³	60 m³
	8 fl oz	1 fl oz = 30 mL	240 mL	240 mL
	20 fl oz	1 fl oz = 30 mL	600 mL	600 mL
	10 qt	1 qt = 1 L	10 L	10 L
	22 gal	1 gal = 4 L	88 L	90 L
	500 gal	1 gal = 4 L	2 000 L	2 000 L
Flow	100 gpm	1 gpm = 4 L/min	400 L/min	400 L/min
	500 gpm	1 gpm = 4 L/min	2 000 L/min	2 000 L/min
	16 ft³/min	1 ft³/min = 0.03 m³/min	0.48 m³/min	0.5 m³/min
	200 ft³/min	1 ft³/min = 0.03 m³/min	6 m³/min	6 m³/min
Flow per Area	50 gpm/ft²	1 gpm/ft² = 40 L/(m².min)	2 000 L/(m².min)	2 000 L/(m².min)
	326 gpm/ft²	1 gpm/ft² = 40 L/(m².min)	13 040 L/(m².min)	13 000L/(m².min)
Pressure	100 psi	1 psi = 7 kPa	700 kPa	700 kPa
	175 psi	1 psi = 7 kPa	1225 kPa	1 200 kPa
	526 psf	1 psf = 0.05 kPa	26.3 kPa	25 kPa
	12,000 psf	1 psf = 0.05 kPa	600 kPa	600 kPa
	5 psi in Hg	1 psi = 3.4 kPa	17 kPa	17 kPa
	20 psi in Hg	1 psi = 3.4 kPa	68 kPa	70 kPa
Speed/Velocity	20 mph	1 mph = 1.6 km/h	32 km/h	30 km/h
	35 mph	1 mph = 1.6 km/h	56 km/h	55 km/h
	10 ft/sec	1 ft/sec = 0.3 m/s	3 m/s	3 m/s
	50 ft/sec	1 ft/sec = 0.3 m/s	15 m/s	15 m/s
Heat	1200 Btu	1 Btu = 1 kJ	1 200 kJ	1 200 kJ
Heat Flow	5 BTU/min	1 Btu/min = 18 W	90 W	90 W
	400 BTU/min	1 Btu/min = 18 W	7 200 W	7 200 W
Density	5 lb/ft³	1 lb/ft³ = 16 kg/m³	80 kg/m³	80 kg/m³
	48 lb/ft³	1 lb/ft³ = 16 kg/m³	768 kg/m³	770 kg/m³
Force	10 lbf	1 lbf = 0.5 N	5 N	5 N
	1,500 lbf	1 lbf = 0.5 N	750 N	750 N
Torque	100	1 lbf ft = 1.4 N.m	140 N.m	140 N.m
	500	1 lbf ft = 1.4 N.m	700 N.m	700 N.m
Dynamic Viscosity	20 lb/ft.s	1 lb/ft.s = 1.5 Pa.s	30 Pa.s	30 Pa.s
	35 lb/ft.s	1 lb/ft.s = 1.5 Pa.s	52.5 Pa.s	50 Pa.s
Surface Tension	6.5 lb/ft	1 lb/ft = 15 N/m	97.5 N/m	100 N/m
	10 lb/ft	1 lb/ft = 15 N/m	150 N/m	150 N/m

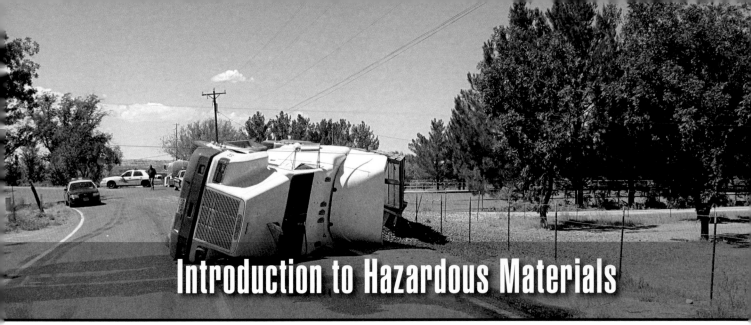

Introduction to Hazardous Materials

Photo courtesy of Rich Mahaney.

Chapter Contents

chapter 1

Key Terms

NFPA Job Performance Requirements

This chapter provides information that addresses the following job performance requirements of NFPA 1072, *Standard for Hazardous Materials/Weapons of Mass Destruction Emergency Response Personnel Professional Qualifications (2017).*

4.2.1

5.3.1

5.5.1

Introduction to Hazardous Materials

Learning Objectives

After reading this chapter, students will be able to:

1. Define a hazardous materials incident. (4.2.1)

2. Describe roles and responsibilities of first responders in hazardous materials incidents.

3. Recognize ways that hazardous materials harm people. (4.2.1, 5.3.1, 5.5.1)

4. List hazardous materials regulations, definitions, and statistics.

Chapter 1
Introduction to Hazardous Materials

This chapter:

- Defines a hazardous materials incident
- Explores emergency responders' roles and responsibilities at hazardous materials incidents
- Explains how hazardous materials can hurt you
- Provides information about hazardous materials definitions, regulations, and statistics

What Is a Hazardous Materials Incident?

Chemical substances, materials, and products are stored, manufactured, used, and transported throughout the world every day. While these products have necessary and beneficial uses, many present considerable risks to the public and to the environment if they are uncontrolled or uncontained. Emergency incidents involving these products may differ from other emergencies in significant ways, and first responders must be trained to respond safely and effectively.

Substances that possess harmful characteristics are called **hazardous materials** (or hazmat) in the United States and **dangerous goods** in Canada and other countries. When particularly dangerous hazardous materials, including chemical, biological, radiological, nuclear, or explosive (CBRNE) materials, are used as weapons, they are sometimes referred to as **weapons of mass destruction** (WMD) because they have the potential to cause mass casualties and damage.

NOTE: For simplicity, the term *hazmat* will be used in this manual in place of *hazardous materials/WMD*, except when WMDs are addressed directly.

A hazardous materials/WMD incident is an emergency involving a substance that poses an unreasonable risk to people, the environment, and/or property. It may involve a substance (product or chemical) that has been (or may be) released from a container or a substance that is on fire. The following are potential causes of hazmat incidents:

- Human error
- Mechanical breakdowns/malfunctions
- Container failures **(Figure 1.1, p. 12)**
- Transportation accidents
- Deliberate acts
  - Chemical suicides
  - WMD incidents

Hazardous Material — Any substance or material that poses an unreasonable risk to health, safety, property, and/or the environment if it is not properly controlled during handling, storage, manufacture, processing, packaging, use, disposal, or transportation.

Dangerous Goods — (1) Any product, substance, or organism included by its nature or by regulation in any of the nine United Nations classifications of hazardous materials. (2) Alternate term used in Canada and other countries for hazardous materials. (3) Term used in the U.S. and Canada for hazardous materials aboard aircraft.

Weapon of Mass Destruction (WMD) — Any weapon or device that is intended or has the capability to cause death or serious bodily injury to a significant number of people through the release, dissemination, or impact of toxic or poisonous chemicals or their precursors, a disease organism, or radiation or radioactivity; may include chemical, biological, radiological, nuclear, or explosive (CBRNE) type weapons.

Figure 1.1 Containers holding hazardous materials may fail, creating incidents that require cleanup. *Photo courtesy of Barry Lindley.*

Hazmat incidents are often more complex than other types of emergency incidents. Often, hazardous materials/WMDs are involved in fires, explosions, and criminal or terrorist activities, complicating the emergency response. For example, hazardous materials may:

- Present a variety of dangers, sometimes in small quantities.

- Be extremely difficult to contain and/or control.

- Require specialized equipment, procedures, and personal protective equipment (PPE) to **mitigate** safely.

- Be difficult to detect, requiring sophisticated monitoring and detection equipment to identify and predict their severity.

Mitigate — (1) To cause to become less harsh or hostile; to make less severe, intense or painful; to alleviate. (2) Third of three steps (locate, isolate, mitigate) in one method of sizing up an emergency situation.

⚠️ CAUTION

Hazardous materials incidents are not always clearly defined before first responders arrive. You must be constantly alert to the presence of hazardous materials and their possible effect on the incident. Whether involved or not, the mere presence of hazardous materials may change the incident's dynamics.

Case Study

On April 17, 2013, a fire started in a fertilizer storage and distribution facility in West, Texas. Twenty-two minutes after the fire broke out, an estimated 20-30 tons of ammonium nitrate exploded, leveling the facility and a nearby neighborhood. The blast, which registered with the force of a small earthquake, killed twelve first responders and two civilians who had volunteered to help fight the fire. It also injured hundreds.

The ammonium nitrate was stored in bulk granular form in wooden bins rather than a fireproof structure that would protect it from ignition sources. The facility did not have a sprinkler system to extinguish fires. Few safety systems were in place to prevent a fire and explosion from occurring.

The Texas State Fire Marshal's investigation into the incident found that the local volunteer fire department did not have a prefire or preincident plan for the facility, nor standard operating procedures (SOPs) or standard operating guidelines (SOGs) for emergency operations. While the strategies and tactics implemented at the incident were appropriate for most of the incidents to which the department typically responded (residential structural fires), they were inappropriate for a large commercial structure with hazardous materials.

The West, Texas, explosion is an example of how quickly a hazardous materials incident can turn deadly. Many first responders do not understand the hazards and risks presented by the products in their communities. In August, 2013, the U.S. Federal Government released a chemical advisory on the safe storage, handling, and management of ammonium nitrate. The emergency response section emphasizes evacuation and defensive strategies for managing these incidents and encouraged all emergency response agencies to preplan for the hazardous materials in their jurisdictions.

Roles and Responsibilities at Hazardous Materials Incidents

First responders must possess the skills necessary to address incidents involving hazmat in a safe and effective manner. You must understand the role you play in these incidents and understand your limitations, realizing when you cannot proceed any further. In part, this role is established in government laws that set forth the training requirements and response limitations imposed on personnel responding to these emergencies (see Information Box).

In addition to governmental regulations, the **National Fire Protection Association (NFPA)** has several consensus standards that apply to personnel who respond to hazmat emergencies. The requirements in these standards are recommendations, not laws or regulations, unless they are adopted as such by the **authority having jurisdiction (AHJ)**. However, because they are a national standard, they can be used as a basis for accepted practice. The NFPA's hazardous materials requirements are detailed in the following standards:

- NFPA 1072, *Standard for Hazardous Materials/Weapons of Mass Destruction Emergency Response Personnel Professional Qualifications*

- NFPA 472, *Standard for Competence of Responders to Hazardous Materials/ Weapons of Mass Destruction Incidents*

- NFPA 473, *Standard for Competencies for EMS Personnel Responding to Hazardous Materials/Weapons of Mass Destruction Incidents.*

NOTE: Individuals who train to meet NFPA 1072 will meet or exceed Occupational Safety and Health Administration (OSHA) requirements for Awareness, Operations, and Technician.

National Fire Protection Association (NFPA) — U.S. nonprofit educational and technical association devoted to protecting life and property from fire by developing fire protection standards and educating the public. Located in Quincy, Massachusetts.

Authority Having Jurisdiction (AHJ) — An organization, office, or individual responsible for enforcing the requirements of a code or standard, or approving equipment, materials, an installation, or a procedure.

North American Regulations Governing Emergency Response to Hazmat/WMD Incidents

The United States (U.S.) Occupational Safety and Health Administration (OSHA) and the U.S. Environmental Protection Agency (EPA) require that responders to hazardous materials incidents meet specific training standards. The OSHA versions of these legislative mandates are outlined in paragraph (q) of Title 29 (Labor) *Code of Federal Regulations* (*CFR*) 1910.120, Hazardous Waste Operations and Emergency Response (HAZWOPER). The training requirements found in 29 *CFR* 1910.120 (q) are included by reference in the EPA regulations in Title 40 (Protection of Environment) *CFR* 311, Worker Protection. This EPA regulation provides protection to those responders not covered by an OSHA-approved State Occupational Health and Safety Plan. See **Appendix B**, OSHA Plan States, for a list of state-plan and non-state-plan states.

If you are a first responder to hazmat incidents in the U.S., by law your employer must meet the requirements set forth in the HAZWOPER regulation (29 *CFR* 1910.120). If you belong to a volunteer fire and emergency services organization, you will also have to meet these regulations. Under 40 *CFR* 311, volunteers are considered employees. If your AHJ has formally adopted the applicable NFPA standards as law, your employer is required to meet them as well.

In Canada, the Ministry of Labour (in most provinces) or the Workers Compensation Board (WCB) in British Columbia are the regulatory bodies governing response to hazmat incidents and the training requirements for first responders. These provincial bodies also require employers to provide standard operating procedures (SOPs) or standard operating guidelines (SOGs) to protect their employees. Canadian firefighters and most emergency responders are trained to the same NFPA standards as their U.S. counterparts. While Canada does not have the definitive equivalent of OSHA 29 *CFR* 1910.120, the minimum acceptable level of training for first responders is NFPA 472.

If you are a first responder to hazmat incidents in Canada, your employer must provide you with SOPs/SOGs and the training required by your province. If you are a firefighter, you must be trained in accordance with the requirements in NFPA 472.

Mexico has developed and implemented a variety of national laws dealing with the handling and regulation of hazardous materials. However, national laws do not currently apply to the training of emergency hazmat first responders. Local jurisdictions may have their own training standards.

Awareness Level — Lowest level of training established by the National Fire Protection Association for personnel at hazardous materials incidents.

Operations Level — Level of training established by the National Fire Protection Association allowing first responders to take defensive actions at hazardous materials incidents.

Operations Mission-Specific Level — Level of training established by the National Fire Protection Association allowing first responders to take additional defensive tasks and limited offensive actions at hazardous materials incidents.

NFPA 1072 and 472 identify three training levels that are addressed in this manual **(Figure 1.2)**:

- **Awareness** — Awareness level personnel are typically at the incident when it occurs; for example, industrial personnel or utility workers who witness an accident. They perform limited defensive actions, such as calling for help, evacuating the hazard area, and securing the scene.

- **Operations** — Operations responders are dispatched to the scene in order to mitigate the incident. These responders may include firefighters, law enforcement, industrial response personnel, or others. They are allowed to perform defensive actions, but, with some exceptions, they are not expected to come into direct contact with the hazardous material.

- **Operations Mission-Specific** — Operations level responders may be trained beyond the set of core competencies to perform additional defensive tasks and limited offensive actions. These may include using specialized equipment and performing tasks where they might come into contact with the hazardous material.

The NFPA also identifies response personnel who perform more complex operations at hazmat incidents, including the following:

- **Hazardous Materials Technician** — Performs offensive tasks, including controlling releases at hazmat incidents and may supervise the activities of Operations level responders performing Mission-Specific tasks.

- **Hazardous Materials Technician With Specialty** — Provides additional expertise in areas, such as radiation, monitoring and detection equipment, or certain container types.

- **Hazardous Materials Incident Commander** — Manages the incident by making command decisions to utilize resources and determine strategies and tactics to mitigate the emergency.

- **Hazardous Materials Officer** — Manages the hazmat personnel and operations under the direction of the Incident Commander.

- **Hazardous Materials Safety Officer** — Ensures that recognized safe practices are followed at hazmat incidents.

- **Specialists** — Provide expertise in specialized areas, such as chemicals, processes, containers, and special operations, typically in an advisory capacity.

At complex hazmat incidents, first responders may be called upon to interact with responders with many different levels of training, representing many different agencies and interests. Depending on circumstances, the incident may expand from initial response personnel to include the following:

- Local or regional hazmat team(s)
- Private industry specialists and/or response teams
- Resources from state and national agencies

These diverse entities play important roles at the incident, and first responders should not be surprised to see additional and potentially unfamiliar resources called to the scene because of the expertise and equipment they can provide. More details about other response personnel at hazmat incidents will be provided in Chapter 6.

Hazmat Training Levels

Awareness

Operations

Mission-Specific

Figure 1.2 NFPA 1072 and 472 identify three training levels – Awareness, Operations, and Operations Mission-Specific.

APIE Process

When developing the NFPA 472 standard, the NFPA's Technical Committee on Hazardous Materials Response Personnel devised a simple, 4-step response model that can guide responders' actions at hazardous materials incidents. Known by the acronym, *APIE*, these steps form a consistent problem-solving process that can be used at any incident, regardless of size or complexity **(Figure 1.3)**:

Step 1: **Analyze the incident** — During this phase of the problem-solving process, personnel and responders attempt to understand the current situation. For example, first responders will attempt to identify the hazmat involved, what kind of containers are present, the quantity of materials released, the number of exposures, potential hazards, and other relevant information needed to plan a safe and effective response.

Step 2: **Plan the initial response** — During this phase, responders use the information gathered during the analysis phase to determine what actions need to be taken to mitigate the incident. For example, the **Incident Commander (IC)** will develop the Incident Action Plan and assign tasks to first responders.

Step 3: **Implement the response** — During this phase, responders perform the tasks determined in the planning stage. When implementing the response, responders direct actions to mitigate the incident.

Step 4: **Evaluate progress** — During this phase, which continues throughout the incident until termination, responders monitor progress to see whether the response plan is working. For example, first responders should report if their actions are completed successfully or if they notice changing conditions.

Responders with different levels of training have different responsibilities in each of these steps. Because Awareness level personnel and Operations level responders have limited responsibilities at hazmat incidents, not all aspects of APIE are addressed in the Awareness and Operations levels. As responsibilities increase, so do the components of APIE. These responsibilities will be addressed throughout this manual, often in the context of APIE.

NOTE: Each fire and emergency services organization should have written procedures describing appropriate actions consistent with the level of training. A sample of such a guideline is found in **Appendix C**, Sample Written Guideline.

Incident Commander (IC) — Person in charge of the Incident Command System and responsible for the management of all incident operations during an emergency.

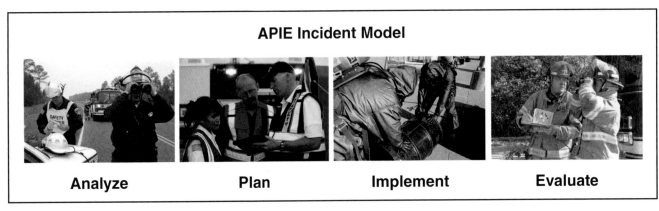

APIE Incident Model

Analyze   Plan   Implement   Evaluate

Figure 1.3 The acronym, *APIE*, can help responders devise a 4-step response to any hazardous materials incident.

Awareness Level Personnel

Personnel who are trained and certified to the Awareness level are individuals who, in the course of their normal duties, may be the first to arrive at or witness a hazmat incident **(Figure 1.4)**. Individuals trained to the Awareness level are expected to assume the following responsibilities when faced with an incident involving hazardous materials:

- Recognize that hazmat is present or potentially present at the incident

- Protect themselves and others from hazards

- Communicate information to an appropriate authority and call for appropriate assistance

- Establish scene control by isolating the hazardous area and denying entry

Figure 1.4 Personnel with Awareness level training might be the first persons at a hazmat incident.

Analyzing the Incident

Awareness personnel must always exercise **situational awareness**. Whether they are operating in public or at a fixed facility, they must be aware of where they are and what is around them. There are many clues that can help Awareness personnel recognize that an incident may involve hazardous materials. Some clues may include the location, a container's shape, transportation or facility markings, metering devices, or other sensory input, such as odor, taste, or appearance. More information about the clues to the presence of hazardous materials will be provided in Chapter 2.

Planning the Response

Awareness personnel are not responsible for planning the response to a hazmat incident. However, **standard operating procedures (SOPs)**, standard operating guidelines (SOGs), and/or predetermined procedures may provide Awareness personnel with the initial actions they should take in the event of an emergency. For example, Awareness personnel may be authorized to use the *Emergency Response Guidebook* to determine isolation distances (see Chapter 3).

Situational Awareness — Perception of the surrounding environment and the ability to anticipate future events.

Standard Operating Procedures (SOPs) — Standard methods or rules in which an organization or fire department operates to carry out a routine function. Usually these procedures are written in a policies and procedures handbook and all firefighters should be well versed in their content.

Implementing the Response

Awareness personnel serve an important role at hazmat incidents, and their initial actions can affect the course of the incident for better or worse. Awareness personnel are expected to do the following at hazmat incidents **(Figure 1.5)**:

- Communicate information to an appropriate authority and call for appropriate assistance.
- Initiate protective actions to protect themselves and others from hazards.
- Isolate the hazardous area and deny entry.

Evaluating Progress

As with planning the response, Awareness personnel are not expected to evaluate the incident response. However, if personnel have pertinent information about the incident or its status, that information should be relayed to an appropriate authority.

Awareness Level Responsibilities

| Transmit | Protect | Isolate |

Figure 1.5 After arriving at a hazmat incident, Awareness personnel should transmit information and call for assistance, initiate protective actions, and isolate the hazardous area and deny entry.

Operations Level Responders

Responders trained and certified to the Operations level are individuals who respond to hazmat releases (or potential releases) as part of their normal duties. Operations responders are expected to protect individuals, the environment, and property from the effects of the release in a primarily defensive manner (see Information Box for exceptions).

Responsibilities of the first responder at the Operations level include the Awareness level responsibilities. Additionally, first responders at the Operational level must be able to perform the following actions:

- Identify the potential hazards involved in an incident if possible.
- Identify response options.
- Implement the planned response to mitigate or control a release from a safe distance by performing assigned tasks to lessen the harmful incident and keep it from spreading.
- Evaluate the progress of the actions taken to ensure that response objectives are safely met.

Offensive Tasks Allowed by U.S. OSHA and Canada

U.S. OSHA and the Canadian government recognize that first responders at the Operations level who have appropriate training (including demonstration of competencies and certification by employers), appropriate protective clothing, and adequate/appropriate resources can perform offensive operations involving flammable liquid and gas fire control of the following materials:

- Gasoline
- Diesel
- Natural gas
- Liquefied petroleum gas (LPG)

Analyzing the Incident

At the Operations level, responders are expected to identify potential hazards at incidents, including (but not limited to) the following:

- Type of container involved
- Hazardous material involved
- Hazards presented by the material
- Potential behavior of the material

Operations responders are expected to analyze the surrounding conditions and determine the location and amount of any release, if possible **(Figure 1.6)**. Once they have an understanding of the material involved and the hazards present, they can begin to plan an appropriate response based on their training, SOPs, and/or predetermined procedures.

Figure 1.6 Operations responders might not have all the facts when they arrive at a scene, but they should be trained to determine where the hazardous material has been released and perhaps the amount of it. *Photo courtesy of Barry Lindley.*

Planning the Response

Operations responders must be able to identify response options for hazmat incidents. While they may not be responsible for planning the actual response, they must understand the tasks that they may be asked to perform and why. In order to protect themselves, operations responders must be aware of safety precautions, the suitability of the personal protective equipment available, and emergency decontamination needs. Response options at hazmat incidents will be addressed in Chapter 5.

Implementing the Response

Operations responders are expected to establish the Incident Management System, establish scene control, and implement protective actions such as evacuation. They must be able to:

- Follow safety procedures.
- Use personal protective equipment in the proper manner.
- Avoid hazards, and complete their assignments.
- Perform emergency decontamination.
- Identify and preserve potential evidence if the incident is a suspected crime.

Evaluating Progress

At the Operations level, responders are expected to evaluate the progress of their assigned tasks. Responders must report this information to their supervisor or other appropriate authority so the action plan can be adjusted if necessary.

Operations Mission-Specific Level

Operations personnel may be trained to perform mission-specific tasks, such as using specialized personal protective equipment or performing technical decontamination **(Figure 1.7)**. These tasks also fit into the APIE process and will be addressed in later chapters. The mission-specific tasks with specialized competencies are:

- Personal protective equipment
- Mass decontamination
- Technical decontamination
- Evidence preservation and sampling
- Product control
- Air monitoring and sampling
- Victim rescue and recovery
- Response to illicit laboratory incidents

How Can Hazmat Hurt You?

Hazardous materials/WMD can hurt you in a variety of ways. They may affect your health if they contact or get into your body, or they may cause harm by their behavior and/or physical properties, for example, if they burn or ex-

Figure 1.7 Operations Mission-Specific training may include using specialized personal protective equipment.

plode. To safely mitigate hazmat incidents, you must understand the variety of hazardous materials you may encounter, the potential health effects of the materials, and the physical **hazards** associated with them. Knowing some of these basic concepts will help prevent or reduce injury, loss of life, and environmental/property losses.

The following sections will address:

- Acute and chronic effects
- Routes of entry through which hazardous materials can contact and enter your body
- Specific mechanisms of harm caused by various hazardous materials

Acute vs Chronic

Many hazardous materials have potential health effects. Exposures to hazardous materials may be **acute** (single exposure or several repeated exposures to a substance within a short time period) or **chronic** (long-term, reoccurring) **(Figure 1.8, p. 22)**. Health effects can also be acute or chronic. **Acute health effects** are short-term effects that appear within hours or days, such as vomiting or diarrhea. **Chronic health effects** are long-term effects that may take years to appear, such as cancer.

Some harmful substances do not hurt the body right away. Delayed effects can occur hours or days later. For example, phosgene can cause serious health effects that may not become evident until many hours or days after exposure. In other cases, it may take years before a health issue arises (see Information Box, p. 22).

Hazard — Condition, substance, or device that can directly cause injury or loss; the source of a risk.

Acute — Characterized by sharpness or severity; having rapid onset and a relatively short duration.

Chronic — Marked by long duration; recurring over a period of time.

Acute Health Effects — Health effects that occur or develop rapidly after exposure to a hazardous substance.

Chronic Health Effects — Long-term health effects resulting from exposure to a hazardous substance.

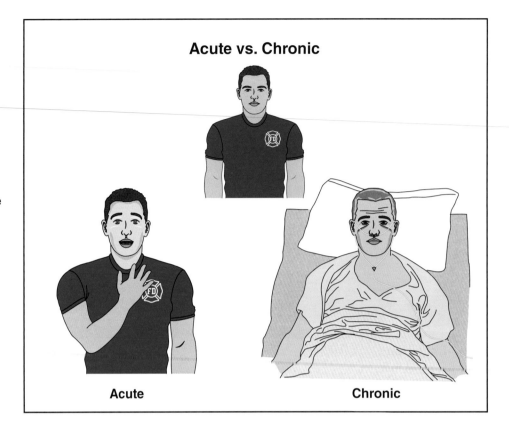

Figure 1.8 Effects of a hazardous material can be acute (short-term) or chronic (long-term).

Delay in Knowing Health Effects

In some cases, it may take many years for a chemical, agent, or substance to cause a disease such as cancer. Because of this delay (sometimes called *latency period*), it can be difficult to establish a direct chain of cause and effect between an exposure to a particular substance and the resulting disease.

The history of asbestos demonstrates how long it can take before enough evidence is gathered to produce action in the form of government intervention. Asbestos was first used in the U.S. in the early 1900s to insulate steam engines. In the 1940s, asbestos began to be used extensively. In particular, it was used in U.S. Navy shipyards to insulate the country's growing fleet of warships during World War II.

While some articles documenting the harmful effects of asbestos were published as early as the 1930s, it was not until the 1960s (15- to 40-year delay) that studies began to show an unquestionably clear relationship between the inhalation of asbestos fibers and the development of lung cancer, asbestosis, and mesothelioma in groups such as U.S. Navy shipyard workers. As a result of these studies and a growing public awareness of the hazard, the U.S. government began regulating asbestos in the 1970s.

Many substances (acetaldehyde, chloroform, progesterone, and polychlorinated biphenyls [PCBs]) are listed by the U.S. Department of Health and Human Services as reasonably anticipated to be carcinogens or suspected carcinogens because the body of evidence concerning their chronic effects is still being gathered and evaluated. Saccharin, for example, was listed as a suspected carcinogen for nearly 20 years before it was removed from the list in 2000 because there was little evidence that it caused cancer. In the same year, diesel exhaust particulate was added to the list.

Our understanding of the health effects associated with chemical products and substances is often changing, and new products are continually being developed. First responders should keep in mind that chronic health effects of substances may not be known for many years, and what is considered safe today, may not be tomorrow.

Routes of Entry

The following are the main **routes of entry** (also called *routes of exposure*) through which hazardous materials can enter the body and cause harm **(Figure 1.9, p. 24)**:

Routes of Entry — Pathways by which hazardous materials get into (or affect) the human body.

- **Inhalation** — Breathing hazardous materials in through your nose or mouth. Hazardous vapors, smoke, gases, liquid aerosols, fumes, and suspended dusts may be inhaled into your body. When a hazardous material presents an inhalation threat, respiratory protection is required. Inhalation is the most common exposure route.

- **Ingestion** — Eating or swallowing hazardous materials through your mouth. Taking a pill is a simple example of deliberate ingestion. However, poor hygiene after handling a hazardous material can lead to accidental ingestion. Other examples include:

  — Chemical residue on your hands can be transferred to your food and then ingested while eating. (Hand washing is important to prevent accidental ingestion of hazardous materials.)

  — Particles can become trapped in your mucous membranes and ingested after being cleared from your respiratory tract.

- **Absorption** — Process of taking in materials through your skin or eyes. Some materials pass easily through your mucous membranes or areas where your skin is the thinnest, allowing the least resistance to penetration. Your eyes, nose, mouth, wrists, neck, ears, hands, groin, and underarms are areas where this can occur. Many poisons are easily absorbed into your body in this manner. Others can enter your body easily if you unknowingly touch a contaminated finger to your eye.

- **Injection** — Process of taking in materials through a puncture of your skin. Protection from injection must be a consideration when dealing with any sort of contaminated (or potentially contaminated) objects easily capable of cutting or puncturing your skin. Such items include:

  — Broken glass

  — Nails

  — Sharp edges

  — Tools like utility knives

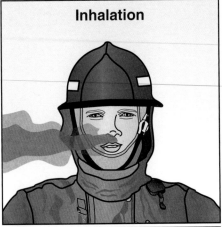

Inhalation

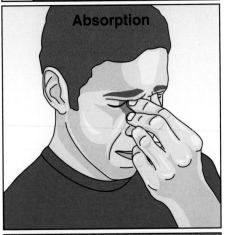

Ingestion

Absorption

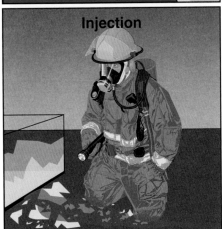

Injection

Figure 1.9 Inhalation, ingestion, absorption, and injection are the primary routes by which hazardous substances enter the body.

Be Aware!

Some chemicals may have multiple routes of entry **(Figure 1.10)**. For example, toluene (a solvent) can cause moderate irritation to the skin through absorption and skin contact, but it can also cause dizziness, lack of coordination, coma, and even respiratory failure when inhaled in sufficient concentrations. Other chemicals with multiple routes of entry include methyl ethyl ketone (MEK), benzene, and other solvents.

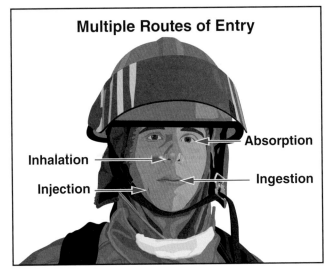

Figure 1.10 Many chemicals have multiple routes of entry. First responders must ensure that all resources are protected against exposure.

Three Mechanisms of Harm

The hazards presented by hazardous materials and hazardous materials incidents may vary from chemical hazards (such as toxicity) to physical hazards (such as flammability). Some hazards, such as electrical hazards, may be unrelated to that hazmat itself. The following sections will explore hazmat incidents' three primary mechanisms of harm **(Figure 1.11)**:

- Energy release
- Corrosivity
- Toxicity

Mechanisms of Harm

Energy Release	Corrosivity	Toxicity

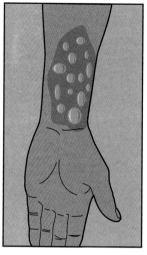

Figure 1.11 Energy release, corrosivity, and toxicity are considered the primary mechanisms of harm that hazardous materials and hazmat incidents present.

TRACEM

Another way to categorize the types of harm caused by hazardous materials uses the acronym, TRACEM. TRACEM stands for: **t**hermal, **r**adiological, **a**sphyxiation, **c**hemical, **e**tiological, and **m**echanical hazards. **Table 1.1** shows the relationship between TRACEM and the three mechanisms of harm. TRACEM hazards are summarized as follows:

- **Thermal hazards** — Thermal hazards are related to temperature extremes.

- **Radiological hazards** — Depending on many variables, exposure to radiological materials may cause mild to serious health effects.

- **Asphyxiation hazards** — Asphyxiants are substances that affect the oxygenation of the body and generally lead to suffocation.

- **Chemical hazards** — Exposure to hazardous chemicals may produce a wide range of adverse health effects.

- **Etiological/biological hazards** — Microorganisms such as viruses or bacteria (or their toxins) that may cause severe, disabling disease or illness.

- **Mechanical hazards** — Mechanical hazards cause trauma through direct contact with an object, most commonly via striking or friction.

Table 1.1 TRACEM Comparison	
TRACEM	**Three Mechanisms of Harm**
Thermal hazards	Energy release
Radiological hazards	Energy release
Asphyxiation hazards	Toxicity
Chemical hazards	Corrosivity/toxicity
Etiological/biological hazards	Toxicity
Mechanical hazards	Energy release

Energy Release

Energy release at hazmat incidents presents the greatest threat. Many hazardous materials will release energy due to their chemical or physical properties and the way they are shipped and/or stored. At hazmat incidents, you should always be aware of the following **(Figure 1.12)**:

Heat — Form of energy associated with the motion of atoms or molecules in solids or liquids that is transferred from one body to another as a result of a temperature difference between the bodies, such as from the sun to the earth. To signify its intensity, it is measured in degrees of temperature.

Mechanical Energy — Energy possessed by objects due to their position or motion, the sum of potential and kinetic energy.

Pressure — Force per unit area exerted by a liquid or gas measured in pounds per square inch (psi) or kilopascals (kPa).

- **Heat** — **Heat** hazards (also called *thermal hazards*) are common at hazmat incidents. Hazardous materials can cause temperature extremes such as with elevated-temperature materials or exothermic reactions (sudden release of heat energy). Fires and explosions involving flammable liquids and explosive materials can cause burns. Environmental factors such as hot weather can cause heat illnesses. Conversely, a lack of heat can also cause harm. For example, cryogenic liquids and liquefied gases are so cold that if contacted, they can cause damage and injuries. Cold atmospheric temperatures can complicate many hazmat operations, for example, decontamination.

- **Mechanical Energy** — **Mechanical energy** is possessed by objects due to their position or motion. At hazardous materials incidents, emergency responders can be injured by flying or falling objects, such as the failure of a pressurized container, an explosive detonation, a shifting container, or the reactivity of the hazardous material itself. Friction injuries may occur as a result of skin or other portions of the body rubbing against an abrasive surface, such as protective clothing, causing raw skin (abrasions), blisters, and burns.

- **Pressure** — Hazardous materials stored under **pressure** can release violently if their containers are damaged or mishandled. When released, these materials expand rapidly, spreading whatever chemical hazards they possess quickly over potentially large areas.

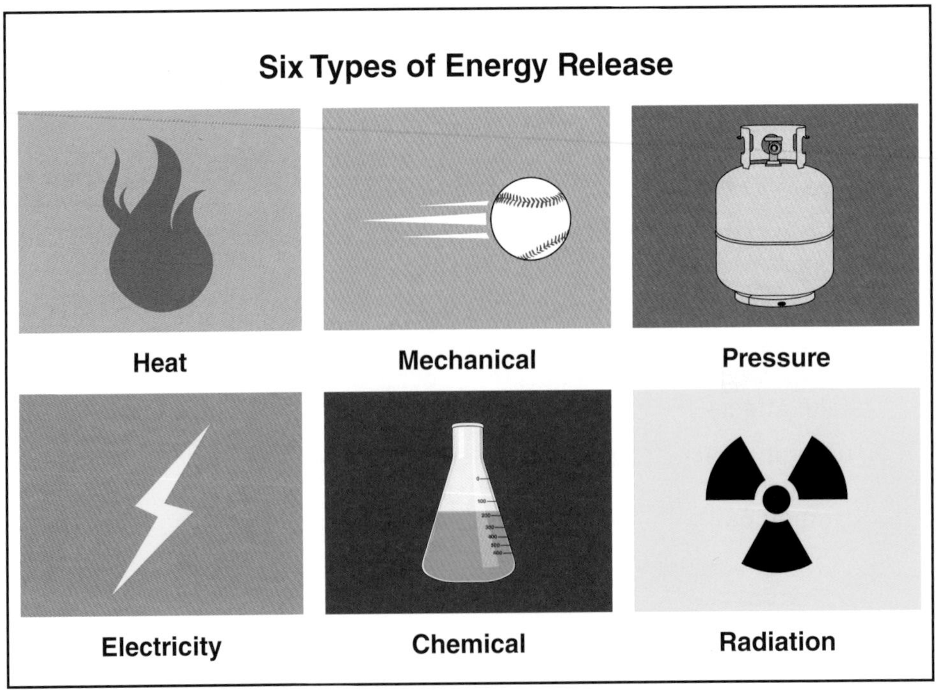

Six Types of Energy Release

| Heat | Mechanical | Pressure |
| Electricity | Chemical | Radiation |

Figure 1.12 Responders at hazmat incidents should be aware of heat, mechanical energy, pressure, electricity, chemical, and radiation.

- **Electricity** — Electrical hazards and **electricity** may be present at hazmat incidents in sources, including utilities, energized containers, and electrical equipment, such as portable generators and power tools.

- **Chemical** — When hazardous materials undergo chemical reactions, they release **chemical energy**. For example, some materials react violently when exposed to water. However, not all chemical reactions result in flames or explosions. Some may release heat, some may use heat, and some may create new hazardous materials with hazards differing from the original material(s).

- **Radiation** — **Radiation** is energy that is emitted as particles or waves. The potential for radiation exposure exists at incidents at medical centers, certain industrial operations, nuclear power plants, and research facilities. There is also the potential for exposure during terrorist attacks.

NOTE: More information about the six types of energy release will be provided in subsequent chapters.

Corrosivity

Corrosives are chemicals that destroy or burn living tissues and have destructive effects by virtue of their corrosivity (ability to cause corrosion, particularly to metals) **(Figure 1.13)**. Corrosive materials can hurt you if they contact your skin or body, and they may also damage tools and equipment. With the exception of liquid and gas fuels, corrosives comprise the largest usage class (by volume) in industry.

NOTE: More information about corrosives will be provided in Chapters 2 and 4.

Toxicity

Chemicals or biological substances that cause sickness, illness, or injury by doing damage on the molecular scale when in contact with the body are considered **toxic (Figure 1.14)**. Biological microorganisms, such as **viruses**

Electricity — Form of energy resulting from the presence and flow of charged particles.

Chemical Energy — Potential energy stored in the internal structure of a material that may be released during a chemical reaction or transformation.

Radiation — Energy from a radioactive source emitted in the form of waves or particles, as a result of the decay of an atomic nucleus; process known as *radioactivity*.

Corrosive — Capable of causing damage by gradually eroding, rusting, or destroying a material.

Toxic — Poisonous.

Virus — Simplest type of microorganism that can only replicate itself in the living cells of its hosts. Viruses are unaffected by antibiotics.

Figure 1.13 Corrosives cause damage to metals and skin.

Figure 1.14 Smallpox virus is a virus that can cause harm. *Courtesy of the CDC Public Health Image Library.*

Bacteria — Microscopic, single-celled organisms.

Toxin — Substance that has the property of being poisonous.

or **bacteria**, may cause severe, disabling disease or illness. **Toxins** can cause damage to organs or other parts of the body. Many toxins have fast-acting, acute toxic effects while others may have chronic effects that are not manifested for many years.

NOTE: More information about toxins will be provided in Chapters 2 and 4.

Hazmat Regulations, Definitions, and Statistics

This section explains the roles of different North American government agencies in regulating hazardous materials, as well as their definitions for such materials. Laws and regulations are often viewed negatively as too restrictive or limiting. However, in 1986, fire service professional organizations representing both labor and management testified before the U.S. Congress and requested inclusion of emergency responders in the provisions of the Superfund Amendment and Reauthorization Act (SARA). Emergency responders requested this inclusion based on a history of harmful and deadly incidents that affected the emergency response community.

Why Do We Have All These Regulations?

Texas City, TX (April 16, 1947)

The greatest industrial disaster in U.S. history occurred in April of 1947 when a French ship, the Grandcamp, ignited during loading in Texas City, Texas. The ship was already loaded with tobacco, twine, cotton, other commodities, and 2,300 tons (2 100 T) of ammonium nitrate fertilizer when workers were completing the loading process in Hold 4 of the ship. The hold contained more than 800 tons (720 T) of ammonium nitrate fertilizer and was being loaded with more when a fire ignited (thought to be caused by lax safety practices, including workers smoking tobacco products). The crew attempted to fight the fire but was driven from the cargo hold rapidly by thick smoke. The crew sounded a dock alarm at about 08:30.

By 08:45, the Texas City Fire Department was on scene and had deployed hoselines. By 09:00, extensive flames were coming from the ship. At 09:12, the ship disintegrated, killing all 27 firefighters of the Texas City Department and 34 vessel crew members and causing additional explosions and fires at nearby refineries

Figure 1.15 More than 550 people were killed and over 3,000 injured in Texas City, Texas, when a ship carrying ammonium nitrate fertilizer exploded at a dock in 1947. Moore Memorial Library 1701 9th Avenue North, Texas City, TX 77590.

and on adjacent ships. The initial blast sent the Grandcamps 3,200-pound (1 600 kg) anchor flying over 1.6 miles (2.5 km) inland. The initial explosion and subsequent events killed over 550 people and injured well over 3,000 **(Figure 1.15)**.

As a result of this incident, the U.S. Coast Guard (USCG) Board of Investigation recommended establishment of a federal office to do the following:

- Collect, evaluate, and disseminate information on fire prevention and extinguishment on board merchant vessels.

- Prepare and publish a fire prevention and fire extinguishment manual for use on board merchant vessels.

- Establish and operate a fire fighting school for training of key operating personnel of merchant ship operators and stevedores.

- Conduct other related marine safety activities.

SS *Torrey Canyon* tank ship, England and France (Mp4arch 18, 1967)

On March 18, 1967, one of the world's first oil supertankers (Torrey Canyon) ran aground off the southern coast of England. The mishap, caused by a series of navigational miscalculations, resulted in the release of approximately 31,000,000 gallons (125 000 000 L) of Kuwaiti oil from the Torrey Canyon. The spill caused extensive damage to marine life and shorelines of France and England. Additionally, since no one had ever planned for an event such as this one, mitigation techniques only worsened the damage.

Attempts to mitigate the spill used over 10,000 tons (9 000 T) of dispersing agents that became more toxic than the oil itself. Additionally, the wreckage and oil remaining in the ship were bombed and napalmed in attempts to burn the oil. Aviation gas was dumped into the spill in an attempt to get it to burn more completely.

As a result of the incident, many countries instituted national plans for dealing with spills of large quantities of oil off their shores. The U.S. passed legislation providing for such planning and designated the USCG as the agency responsible for protecting U.S. coastlines from oil spills and other hazardous materials emergencies. The USCG also placed dedicated Strike Teams in service.

Kingman, AZ (July 5, 1973)

Two workers had been preparing a railcar for unloading at the Kingman Doxol Gas plant. A leak was detected during the process, and as workers attempted to stop it, a fire ignited, seriously burning one of the workers and killing the other. The Kingman Fire Department was dispatched at 13:57, and first arriving units were on scene at 14:00. Fire department personnel initially deployed handlines and worked to secure water supply for deck-gun operations from the closest hydrant (approximately 1,200 feet [360 m] away).

Less than 20 minutes after the fire started, the railcar tank shell failed, releasing the propane contents in what has come to be known as a BLEVE (boiling liquid expanding vapor explosion). The BLEVE instantly killed four firefighters and burned seven others so badly that they succumbed to their injuries in the following days. One fire department member who had climbed into a truck to talk on the radio suffered severe burns but survived.

As a result of the incident, the Department of Transportation required all railcars in flammable gas service at the time to be retrofitted with thermal protection that would shield the tank during similar fire conditions. As a direct result of this incident, today's flammable gas railcars are thermally protected against 100 minutes of exposure to a 1,600°F fire (870°C) and at least 30 minutes of 2,200°F (1 200°C) impingement.

Love Canal, Niagara Falls (1978)

The Love Canal saga began nearly 100 years before the environmental nightmare came to the world's attention. In the 1890s, industrialist William T. Love planned a canal around Niagara Falls. The canal would allow marine traffic around the falls, provide a water source for inexpensive hydroelectric power, and create a distinct boundary for a model planned industrial community. The U.S. economy entered a sharp decline shortly after the project began, and the development ceased. The open canal was publicly auctioned, and by 1920, it was in use as a local landfill and swimming area. The canal's evacuated dimensions were 3,000-foot long, 60-foot wide, 40-foot deep (900 m long, 18 m wide, 12 m deep) with several 25-foot (7.5 m) deep trenches built near the canal.

In 1942, Hooker Chemical Company (later purchased by Occidental Chemical) purchased the site and used it as a waste disposal site for its Niagara Falls Plant. Between 1942 and 1954, Hooker dumped an estimated 22,000+ tons (20 000+ T) of chemical waste into the canal.

In 1952, the local school board began to pressure Hooker to sell a small portion of the filled canal to the board for the site of a new school because current facilities were overtaxed with a huge population surge. Hooker initially declined, but in 1953, he agreed to provide the district with the property for $1 provided the school district took the entire canal site and relieved Hooker of any liability for the site. Hooker was allowed to continue dumping on the site until construction began.

The deed-transfer paperwork carried specific warnings to the school board about not disturbing parts of the site. The school board built a school on the site, and the surrounding area was developed.

In 1978, a series of news reports identified local health issues and the presence of high levels of toxic materials, including dioxins and polychlorinated biphenyls (PCBs) **(Figure 1.16)**. Reporters uncovered high rates of illness, birth defects, neurological and respiratory disorders, and other anomalies when they surveyed the affected neighborhoods. The land itself proved to be contaminated with corroded barrels visible in yards and noxious substances coating many surfaces. The findings led to the evacuation of the entire area, the declaration of a federal emergency, and the development of the Comprehensive Environmental Response, Compensation, and Liability Act (CERCLA).

Figure 1.16 The resulting health problems from Love Canal led to the eventual evacuation of the area and the passage of the Comprehensive Environmental Response, Compensation, and Liability Act (CERCLA). Courtesy of U.S. Environmental Protection Agency.

Shreveport, LA (September 17, 1984)

On September 17, 1984, the Shreveport (LA) Fire Department responded to an anhydrous ammonia leak at Dixie Cold Storage. While working to control the leak in the warehouse, a spark ignited the ammonia. The explosion and flash fire severely burned two firefighters who were working in the area in chemical protective clothing. One firefighter died from his injuries a few days later. As a direct result of the incident, the NFPA standards on chemical protective clothing now address the hazards of flash fires and require garments to be constructed of materials that will not contribute to injuries in similar situations.

Kansas City, MO (November 29, 1988)

Kansas City (MO) Fire Department Engine Companies 41 and 30 responded to a report of a pickup truck fire at a highway construction site in the early morning hours of November 29, 1988. Security guards who phoned in the call reported explosives were stored on site and that information was communicated to the responding units. Arriving units found multiple fires on the site including two smoldering trailers that were loaded with explosives. The trailers' contents (which were not required to be marked or labeled) contained a total of nearly 50,000 pounds (25 000 kg) of ammonium nitrate, fuel-oil-mixture-based explosives. Other explosives stored on site were labeled. Shortly after arrival, the smoldering trailers detonated **(Figure 1.17)**. The first blast killed the six firefighters on scene instantly and destroyed the two pumping apparatus. The explo-

Figure 1.17 U.S. Department of Transportation regulations regarding placards were changed after an unmarked trailer containing 50,000 pounds (22 680 kg) of ammonium nitrate exploded in Kansas City in 1988, killing six firefighters. *Courtesy of Ray Elder.*

sion left an 80-foot (25 m) diameter and 8-foot (2.5 m) deep crater. Nearly 10 years later, U.S. Department of Transportation regulations regarding when placards can be removed from vehicles were changed as a direct result of the incident.

U.S. Regulations and Definitions

In the U.S., the four main agencies involved in the regulation of hazardous materials and/or wastes at the federal level are as follows:

- **Department of Transportation (DOT)** — The DOT issues transportation regulations in Title 49 (Transportation) *Code of Federal Regulations* (*CFR*). These regulations are sometimes referred to as the Hazardous Materials Regulations or HMR. They address the transportation of hazardous materials in all modes: air, highway, pipeline, rail, and water.

- **Environmental Protection Agency (EPA)** — The EPA is responsible for researching and setting national standards for a variety of environmental programs. The EPA issues legislation to protect the environment in Title 40 *CFR*.

- **Department of Labor (DOL)** — The Occupational Safety and Health Administration (OSHA), part of DOL, issues legislation relating to worker safety under Title 29 *CFR*. OSHA legislation of interest to first responders includes the HAZWOPER regulation (29 *CFR* 1910.120), the Hazard Communication regulation (29 *CFR* 1910.1200) and the Process Safety Management of Highly Hazardous Chemicals regulation (29 *CFR* 1910.119). The Hazard Communication Standard (HCS) is designed to ensure that information about chemical hazards and associated protective measures is disseminated to workers and employers. The Process Safety Management (PSM) of Highly Hazardous Chemicals (HHCs) standard is intended to prevent or minimize the consequences of a catastrophic release of toxic, reactive, flammable, or explosive HHCs from a process.

- **Nuclear Regulatory Commission (NRC)** — The NRC regulates U.S. commercial nuclear power plants and the civilian use of nuclear materials as well as the possession, use, storage, and transfer of radioactive materials through Title 10 (Energy) *CFR* 20, Standards for Protection Against Radiation. The NRC's primary mission is to protect the public's health and safety and the environment from the effects of radiation from nuclear reactors, materials, and waste facilities.

Table 1.2, p. 32, lists the main U.S. agencies, their spheres of responsibility and important legislation. It also details the hazardous materials terms and definitions associated with the legislation.

Several other U.S. agencies are involved with hazardous materials:

- **Department of Energy (DOE)** — Manages the national nuclear research and defense programs, including the storage of high-level nuclear waste. The DOE oversees the National Nuclear Security Administration (NNSA), which provides the main capability for responding to nuclear or radiological incidents within the U.S. and abroad.

- **Department of Homeland Security (DHS)** — Has three primary missions: (1) prevent terrorist attacks within the U.S., (2) reduce America's vulnerability to terrorism, and (3) minimize the damage from potential attacks and natural disasters. DHS assumes primary responsibility for ensuring that emergency response professionals are prepared for any situation in the event of a terrorist attack, natural disaster, or other large-scale emergency.

Agency	Sphere of Responsibility	Important Legislation	Hazardous Material Terms/Definitions
Department of Transportation (DOT) Research and Special Programs Administration (RSPA)	Transportation Safety	Title 49 (Transportation) *CFR* 100-185 Hazardous Materials Regulations (HMR)	***Hazardous Material:*** A substance or material (including hazardous wastes, marine pollutants, and elevated temperature materials) that has been determined by the U.S. Secretary of Transportation to be capable of posing an unreasonable risk to health, safety, and property when transported in commerce and which has been so designated.*
Environmental Protection Agency (EPA)	Public Health and the Environment	Title 40 (Protection of Environment) *CFR* 302.4 Designation of Hazardous Substances	***Hazardous Substance:*** A chemical that if released into the environment above a certain amount must be reported and, depending on the threat to the environment, federal involvement in handling the incident can be authorized.
		40 *CFR* 355 Superfund Amendments and Reauthorization Act (SARA)	***Extremely Hazardous Substance:*** Any chemical that must be reported to the appropriate authorities if released above the threshold reporting quantity.** ***Toxic Chemical:*** One whose total emission or release must be reported annually by owners and operators of certain facilities that manufacture, process, or otherwise use a listed toxic chemical.***
		40 *CFR* 261 Resource Conservation and Recovery Act (RCRA)	***Hazardous Wastes:*** Chemicals that are regulated under the Resource Conservation and Recovery Act (40 *CFR* 261.33 provides a list of hazardous wastes).
Department of Labor (DOL) Occupational Safety and Health Administration (OSHA)	Worker Safety	29 (Labor) *CFR* 1910.1200 Hazard Communications	***Hazardous Chemical:*** Any chemical that would be a risk to employees if exposed in the workplace (Hazardous chemicals cover a broader group of chemicals than the other chemical lists).
		29 *CFR* 1910.120 Hazardous Waste Operations and Emergency Response (HAZWOPER)	***Hazardous Substance:*** Every chemical regulated by the U.S. DOT and EPA.
		29 *CFR* 1910.119 Process Safety Management of Highly Hazardous Chemicals	***Highly Hazardous Chemicals:*** Those chemicals that possess toxic, reactive, flammable, or explosive properties (A list of these chemicals is published in Appendix A of 29 *CFR* 1910.119.)

Continued

Table 1.2 (concluded)

Agency	Sphere of Responsibility	Important Legislation	Hazardous Material Terms/Definitions
Consumer Product Safety Commission (CPSC)	Hazardous Household Products (chemical products intended for consumers)	Title 16 (Commercial Practices) *CFR* 1500 Hazardous Substances and Articles Federal Hazardous Substances Act (FHSA)	***Hazardous Substance:*** Any substance or mixture of substances that is toxic; corrosive; an irritant; a strong sensitizer; flammable or combustible; or generates pressure through decomposition, heat, or other means and if such substance or mixture of substances may cause substantial personal injury or substantial illness during or as a proximate result of any customary or reasonably foreseeable handling or use, including reasonably foreseeable ingestion by children. ***Also:*** Any radioactive substance if, with respect to such substance as used in a particular class of article or as packaged, the Commission determines by regulation that the substance is sufficiently hazardous to require labeling in accordance with the Act in order to protect the public health.****
Nuclear Regulatory Commission (NRC)	Radioactive Materials (use, storage, and transfer)	Title 10 (Energy) *CFR* 20 Standards for Protection Against Radiation	

* DOT uses the term hazardous materials to cover 9 hazard classes, some of which have subcategories called divisions. DOT includes in its regulations hazardous substances and hazardous wastes, both of which are regulated by the U.S. EPA if their inherent properties would not otherwise be covered.

** Each substance has a threshold reporting quantity. The list of extremely hazardous substances is identified in Title III of SARA of 1986 (see 40 CFR 355).

*** The list of toxic chemicals is provided in Title III of SARA (see 40 CFR 355). The EPA regulates these materials because of public health and safety concerns. While regulatory authority is granted under the Resource Conservation and Recovery Act, the DOT regulates the transport of these materials.

**** The complete definition of hazardous substance as found in the FHSA contains five parts (A–E) and includes such items as toys and other articles intended for use by children. Only parts A and C are cited here in their entirety.

Details:

— Responsibilities include providing a coordinated, comprehensive federal response to any large-scale crisis and mounting a swift and effective recovery effort.

— The Federal Emergency Management Agency (FEMA) and U.S. Coast Guard (USCG) are just a few of the agencies that were relocated within DHS following the terrorist attacks on September 11, 2001 **(Figure 1.18, p. 34)**.

• **Consumer Product Safety Commission (CPSC)** — Oversees and enforces compliance with the Federal Hazardous Substances Act (FHSA), which requires that certain hazardous household products (hazardous substances) carry cautionary labeling to alert consumers to the potential hazards that those products present and inform them of the measures needed to protect themselves from those hazards.

Figure 1.18 Hazmat/WMD incidents may be more complex and difficult to mitigate than other types of emergencies. *Courtesy of U.S. Air Force photo by Staff Sgt. Gary Coppage.*

- **Department of Defense Explosives Safety Board (DDESB), Department of Defense (DOD)** — Provides oversight of the development, manufacture, testing, maintenance, demilitarization, handling, transportation, and storage of explosives, including chemical agents on DOD facilities worldwide.

- **Bureau of Alcohol, Tobacco, Firearms and Explosives (ATF), Department of Treasury** — Enforces the federal laws and regulations relating to alcohol, tobacco products, firearms, explosives, and arson.

- **Department of Justice (DOJ)** — Assigns primary responsibility for operational response to threats or acts of terrorism within the U.S. and its territories to the Federal Bureau of Investigation (FBI). The FBI then operates as the on-scene manager for the federal government. It is ultimately the lead agency on terrorist incident scenes. The FBI performs the following duties related to hazmat incidents:

  — Investigates the theft of hazardous materials

  — Collects evidence for crimes

  — Prosecutes criminal violations of federal hazardous materials laws and regulations

Canadian Regulations and Definitions

In Canada, the four main agencies that are involved in the regulation of hazardous materials and/or wastes at the national level are:

- Transport Canada (TC)

- Environment Canada

- Health Canada

- Canadian Nuclear Safety Commission (CNSC)

Table 1.3, p. 36, lists the main Canadian agencies, their spheres of responsibility, important legislation, and the hazardous materials terms and definitions associated with the legislation. Table 1.4, p. 37, provides a brief summary of the regulatory programs administered by other agencies of the Canadian government relating to chemical substances.

Mexican Regulations and Definitions

The three main agencies that are involved in the regulation of hazardous materials and/or wastes at the national level in Mexico are:

- Secretaría de Comunicaciones y Transportes (SCT) — Ministry of Communications and Transport
- Secretaría de Medio Ambiente y Recursos Naturales (SEMARNAT) — Ministry of Environment and Natural Resources
- Secretaría del Trabajo y Previsión Social (STPS) — Ministry of Labor and Social Welfare

Table 1.5, p. 38, lists the Mexican agencies, their spheres of responsibility, and important legislation. It also covers the hazardous material terms and definitions with which first responders need to be familiar.

What This Means To You

Do not get caught up in the specific definitions of different hazmat terms. To you, as an emergency responder, they are all hazardous, and that means potentially dangerous. However, since you may hear other terms being used, be aware that the location where you find a hazardous material and how it is being used may determine what it is called for government purposes.

For example, when xylene is being transported in the U.S., the DOT regulates it, and it would be called a hazardous material. (If it were being transported in Canada, it would be called a dangerous good.) In the industry (or place of employment) where it is being used or manufactured, it becomes subject to the OSHA requirements protecting employees who work with it, and it would be considered a hazardous chemical. If it were marketed to consumers for purchase and use, it would fall subject to the Consumer Product Safety Commission, and it would be called a hazardous substance.

If at any point xylene was accidentally discharged from its packaging into the environment, it would become a hazardous substance as regulated by the EPA. When xylene completes its useful life in a plant or workplace and must be disposed of (in any manner), it becomes a hazardous waste and would be subject to both the EPA and DOT regulations (during transport). Additionally, if xylene was used to kill or injure a large number of people in a terrorist attack, it might be called a weapon of mass destruction by federal law enforcement authorities such as the FBI.

Hazardous Materials Incident Statistics

Hazmat incidents occur frequently. It is likely that all emergency first responders will have to deal with hazardous materials at some point in their careers. In fact, hazmat spills, releases, and incidents are so common that several different U.S. government agencies maintain databases to track them.

Table 1.3
Main Canadian Agencies Involved in the
Regulation of Hazardous Materials

Agency	Sphere of Responsibility	Important Legislation	Hazardous Material Terms/Definitions
Transport Canada (TC) Transport Dangerous Goods (TDG) Directorate	Transportation Safety	Transportation of Dangerous Goods Act	***Dangerous Goods:*** Any product, substance, or organism included by its nature, or by the regulation, in any of the classes listed in the schedule of the nine United Nations (UN) Classes of Hazardous Materials*.
Environment Canada	Public Health and the Environment	Canadian Environmental Protection Act 1999	***Toxic Substance:*** A substance that if it is entering or may enter the environment in a quantity or concentration or under conditions that it: (a) Has or may have an immediate or long-term harmful effect on the environment or its biological diversity (b) Constitutes or may constitute a danger to the environment on which life depends. (c) Constitutes or may constitute a danger in Canada to human life or health.
Transboundary Movement Division	Transportation of Hazardous Waste	Canadian Environmental Protection Act, 1999 Export and Import of Hazardous Wastes Regulations (EIHWR)	***Hazardous Waste:*** Any substance specified in Parts I, II, III, or IV of the List of Hazardous Wastes Requiring Export or Import Notification in Schedule III, *(déchets dangereux);* or any product, substance, or organism that is dangerous goods, as defined in Section 2 of the Transportation of Dangerous Goods Act, 1992, that is no longer used for its original purpose and that is recyclable material or intended for treatment or disposal, including storage prior to treatment or disposal, but does not include a product, substance or organism that is: (i) Household in origin (ii) Returned directly to a manufacturer or supplier of the product, substance, or organism for reprocessing, repackaging, or resale, including a product, substance or organism that is (A) Defective or otherwise not usable for its original purpose (B) In surplus quantities but still usable for its original purpose (iii) Included in Class 1 or 7 of the *Transportation of Dangerous Goods Regulations*

Continued

Table 1.3 (concluded)

Agency	Sphere of Responsibility	Important Legislation	Hazardous Material Terms/Definitions
Health Canada	Worker Safety	Hazardous Product Act	***Hazardous Product:*** Any product, material, or substance that is, or contains, a poisonous, toxic, flammable, explosive, corrosive, infectious, oxidizing, or reactive product, material, or substance (or other product, material or substance of a similar nature) that the Governor in Council is satisfied is (or is likely to be) a danger to the health or safety of the public.
Workplace Hazardous Materials Information System Division (WHMIS Division)	Worker Safety/ Chemicals Intended for the Workplace	Hazardous Product Act Workplace Hazardous Materials Information System (WHMIS)	***Controlled Product:*** Any product, material, or substance specified by the regulations to be included in any of the classes listed in Schedule II of the Hazardous Product Act.

*   Internationally, hazardous materials in transport are generally referred to as dangerous goods.

Table 1.4
Other Canadian Agencies Involved with Hazardous Materials

Canadian Authority	Program
Health Canada, Consumer Products Division	Chemicals on the retail market
Natural Resources Canada, Explosives Regulatory Division	Explosives
Environment Canada, Waste Management and Remediation	Management of hazardous wastes, assessment and remediation of contaminated sites, and the control of waste disposal at sea
Canadian Nuclear Safety Commission	Nuclear substances
Health Canada, Pest Management Regulatory Agency	Pesticides
Health Canada, Radiation Protection Bureau	Radioactive substances
National Energy Board	Transportation of chemical products (oil and natural gas) via pipeline

Agency	Sphere of Responsibility	Important Legislation	Hazardous Material Terms/Definitions
Secretaría de Communicaciones y Tranportes Ministry of Communications and Transportation	Transportation Safety	Mexican Hazardous Materials Land Transportation Regulation NOM-004-SCT-2000: System of Identification of Units Designated for the Transport of Hazardous Substances, Materials, and Wastes NOM-005-SCT 2000: Emergency Information for the Transport of Hazardous Substances, Materials, and Wastes	
Secretaría de Medio Ambiente y Recursos Naturales Ministry of the Environment and Natural Resources	Public Health and the Environment	La Ley General de Equilibrió Ecológico y Protección al Ambiente: Federal General Law of Ecological Equilibrium and the Protection of the Environment (LGEEPA) Regulation of LGEEPA in the area of hazardous wastes	
Secretaría del Trabajo y Previsión Social Ministry of Labor and Social Welfare	Worker Safety/ Labor	NOM-018-STPS-2000: System for the Identification and Communication of Hazards and Risks for Dangerous Chemical Substances in the Workplace Communications Wastes	***Sustancias qui'micas peligrosas (dangerous chemical substances):*** Those chemicals that through their physical and chemical properties upon being handled, transported, stored, or processed present the possibility of fire, explosion, toxicity, reactivity, radioactivity, corrosive action, or harmful biological action and can effect the health of the persons exposed or cause damage to installations and equipment.
		NOM-005-STPS-1998: Health and Safety Conditions in the Workplace for the Handling, Transport, and Storage of Hazardous Chemical Substances	***Sustancias tóxicas (toxic substances):*** Those chemicals in solid, liquid, or gaseous state that can cause death or damage to health if they are absorbed by the worker even in relatively small amounts.
		NOM-026-STPS-1998: Signs and Colors for Safety and Health, and Identification of Risk of Accidents by Fluids Conducted in Pipes	***Fluidos de bajo riesgo (dangerous fluids):*** Those liquids and gases that can cause injury or illness on the job because of their intrinsic hazards such as flammables, unstable combustibles that can cause explosion, irritants, corrosives, toxics, reactives, radioactives, biological agents, or those that are subjected to extreme pressure or temperature as part of a process.

Figure 1.19 Most hazmat incidents occur while trucks are transporting these materials on roads.

Because certain hazardous materials are more common than others, they are statistically more likely to be involved in incidents and accidents. Additionally, clandestine, illegal labs (particularly the ones associated with methamphetamine) with their hazardous products are problems in many jurisdictions. Records have shown that most hazmat incidents involve the following products (not necessarily in this order):

- Flammable/combustible liquids, such as petroleum products, paint products, resins, and adhesives

- Corrosives, such as sulfuric acid, hydrochloric acid, and sodium hydroxide

- Anhydrous ammonia

- Chlorine

Many incidents occur while hazardous materials are being transported. Statistics indicate that most transportation incidents occur while the materials are being transported via highway rather than by air, rail, or water (**Figure 1.19**).

Hazmat Incident Statistics

Several different U.S. governmental agencies maintain databases that track hazmat incidents. These agencies include the EPA, the DOT's Pipeline and Hazardous Materials Safety Administration (PHMSA), and OSHA. The Agency for Toxic Substances and Disease Registry (ATSDR) maintains the Hazardous Substances Emergency Events Surveillance (HSEES) database which gathers data from a few select states.

Chapter Review

Answer the following questions to review the information provided in this chapter.

1. How are hazardous materials incidents different from other types of emergency incidents?

2. What are the three levels of hazmat responders and what are their responsibilities?

3. What are the four main routes through which hazardous materials can enter the body and cause harm?

4. What are the three main primary mechanisms by which hazardous materials can cause bodily harm?

5. Why are there so many regulations regarding hazardous materials?

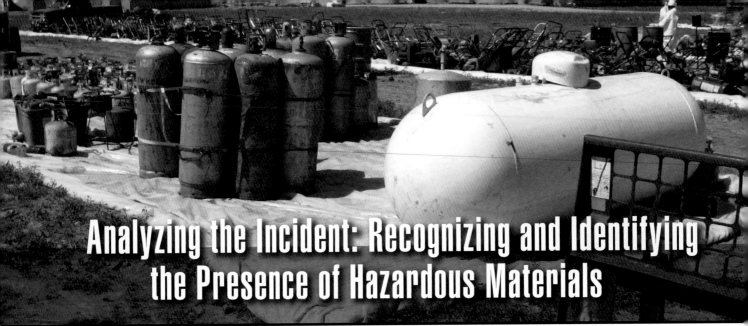

Analyzing the Incident: Recognizing and Identifying the Presence of Hazardous Materials

Photo courtesy of Rich Mahaney.

Chapter Contents

Key Terms

NFPA Job Performance Requirements

This chapter provides information that addresses the following job performance requirements of NFPA 1072, *Standard for Hazardous Materials/Weapons of Mass Destruction Emergency Response Personnel Professional Qualifications (2017).*

4.2

4.3

Analyzing the Incident: Recognizing and Identifying the Presence of Hazardous Materials

Learning Objectives

After reading this chapter, students will be able to:

1. Restate the seven clues to the presence of hazardous materials. (4.2)

2. Explain how preincident plans, occupancy types, and locations may indicate the presence of hazardous materials. (4.2)

3. Identify basic container shapes that indicate the presence and hazards of hazardous materials. (4.2)

4. Describe ways that U.S. transportation placards, labels, and markings indicate the presence and hazards of hazardous materials. (4.2)

5. Describe ways that Canadian transportation placards, labels, and markings indicate the presence and hazards of hazardous materials. (4.2)

6. Describe ways that Mexican transportation placards, labels, and markings indicate the presence and hazards of hazardous materials. (4.2)

7. Identify other markings and colors that indicate the presence of hazardous materials. (4.2)

8. Describe ways written resources are used to identify hazardous materials and their hazards. (4.2)

9. Explain the limited role of the five senses for identifying hazardous materials.

10. Explain the role of monitoring and detection devices for Awareness Level personnel.

Chapter 2
Analyzing the Incident: Recognizing and Identifying the Presence of Hazardous Materials

This chapter explains the following topics:

- Seven clues to the presence of hazardous materials
- Preincident plans, occupancy types, and locations
- Basic container shapes
- Transportation placards, labels, and markings
- Canadian placards, labels, and markings
- Mexican placards, labels, and markings
- Other markings and colors
- Written resources
- Senses
- Monitoring and detection devices

Seven Clues to the Presence of Hazardous Materials

Awareness level personnel and first responders must be able to analyze all incidents in order to detect and identify the presence of hazardous materials. Incidents involving hazmat can be controlled only when the personnel involved have sufficient information to make informed decisions. The time and effort devoted to identifying the contents of buildings, vehicles, and containers result in greater safety for first responders and the community. Historically, first responders' failure to recognize hazardous materials at accidents, fires, spills, and other emergencies has caused unnecessary injuries and deaths.

Once hazmat is detected, first responders can use a number of resources to identify it and its potential hazards. With that information, first responders can initiate appropriate response actions and perform them confidently.

Some hazmat identification clues are easily visible at a distance while others require responders to be much closer. The closer you need to be in order to identify the material, the greater your chances of being in an area where you could be exposed to its harmful effects. In general, distance often equates to safety when hazmat is involved.

The following are the seven clues to the presence of hazardous materials:

1. Occupancy types, locations, and preincident surveys
2. Container shapes
3. Transportation placards, labels, and markings
4. Other markings and colors (nontransportation)
5. Written resources
6. Senses
7. Monitoring and detection devices

This chapter will describe these seven clues in detail. The order of the clues also represents, in general, an increasing level of risk to responders **(Figure 2.1)**. For example, using monitoring and detection equipment to identify hazmat is more likely to place responders in dangerous areas than using occupancy types or container shapes to provide information. Discussion of terrorist attacks and weapons of mass destruction will be included in Chapter 8 of this manual.

Prepare for the Unexpected

While the seven clues provide many ways to recognize and identify hazardous materials, they do have limitations. You may be unable to clearly see placards, markings, labels, and signs from a safe distance. Identifying markings can be destroyed in the incident. Inventories may change from those identified during preincident surveys, or containers may be improperly labeled. Mixed loads in transportation incidents may not be marked at all **(Figure 2.2)**. Shipping papers may be inaccessible. You must always be prepared for the unexpected, including terrorist attacks.

Preincident Plans, Occupancy Types, and Locations

Simply stated, hazardous materials may be found anywhere. Not all locations or occupancies are as obvious as the local chemical manufacturing plant, and you may have little or no warning when hazmat is transported through your area by road, rail, or waterway. However, preincident surveys and a structure's occupancy type may provide the first clue that hazmat may be involved in an incident. The location and occupancy may also be an indicator that terrorism is involved.

In today's world, a new and emerging problem is clandestine laboratories and illegal or legal grow operations. These laboratories may be located in any occupancy or location, including vehicles, campgrounds, and hotel rooms. These labs may be haphazardly assembled and are often booby trapped.

The sections that follow will address:

- Preincident plans that identify where hazardous materials are used, stored, and transported
- Occupancy types which are likely to have hazardous materials
- Locations where hazmat incidents often occur

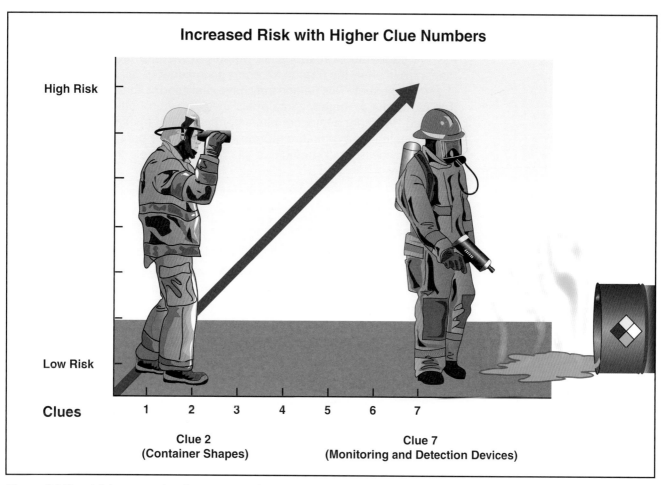

Figure 2.1 The risk to responders increases as they move closer to the hazardous material. It is much safer to identify a material from a distance based on a container shape than it is to physically sample the substance with a detection device.

Figure 2.2 First responders could have difficulty knowing what an unmarked truck without placards is transporting. *Courtesy of Rich Mahaney.*

Verbal Reports

A knowledgeable or responsible person at the site (such as a facility manager) is often the person to report an incident or tell responders that hazmat is present. This person may have vital information about the events that led to the emergency, materials involved, and humans or property exposed . Whether a telecommunicator/dispatcher questions this person over the telephone or first responders question the person at the scene, emergency personnel should use this resource to help them understand the incident.

Preincident Survey — Assessment of a facility or location made before an emergency occurs, in order to prepare for an appropriate emergency response. *Also known as* Preplan.

Preincident Plans

Because hazmat incidents can be quite volatile, first responders may need to make decisions quickly and accurately. Conducting **preincident surveys** (also called preplans) and being familiar with local emergency response plans (explained in Chapter 3) can simplify and reduce on-site decisions. With the groundwork laid, first responders can concentrate on the situation and operate more safely and efficiently. Preplanning reduces oversights, confusion, and effort duplication, and it results in a desirable outcome. Furthermore, preincident surveys identify the following items:

- Exposures such as people, property, and environment
- Hazmat types, quantities, dangers, and locations
- Building features such as the location of fixed fire suppression systems
- Site characteristics
- Possible access/egress difficulties
- Inherent limitations of response organizations when trying to control certain types of hazmat emergencies
- Twenty-four hour telephone numbers of responsible parties and site experts
- Site or occupancy response capability

Planning is an ongoing process that includes reviewing surveys and updating them regularly. Preincident surveys are not always accurate, however, because inventories, businesses, and other factors may change without notice. Compliance with existing reporting rules and regulations cannot be guaranteed. Always expect to find the unexpected.

Occupancy — (1) General fire and emergency services term for a building, structure, or residency. (2) Building code classification based on the use to which owners or tenants put buildings or portions of buildings. Regulated by the various building and fire codes. *Also known as* Occupancy Classification.

Occupancy Types

Certain **occupancies** are likely to have hazardous materials, including the following:

- Fuel storage facilities
- Gas/service stations and convenience stores
- Paint supply stores
- Plant nurseries, garden centers, and agricultural facilities
- Pest control and lawn care companies
- Medical facilities
- Photo processing laboratories

Figure 2.3 Feed/farm stores are likely to have hazardous materials in stock for customers. *Courtesy of Rich Mahaney.*

Figure 2.4 Large quantities of hazardous materials may pass through busy port facilities, making them a common location for hazmat incidents. *Courtesy of U.S. Customs and Border Protection, photo by Charles Csavossy.*

- Dry cleaners
- Plastics and high-technology factories
- Metal-plating businesses
- Mercantile concerns such as hardware stores, groceries stores, certain department stores
- Chemistry (and other) laboratories in educational facilities (including high schools)
- Lumberyards
- Feed/farm stores **(Figure 2.3)**
- Veterinary clinics
- Print shops
- Warehouses
- Industrial and utility plants
- Port shipping facilities (with changing cargo hazards) **(Figure 2.4)**
- Treatment storage disposal (TSD) facilities
- Abandoned facilities that may have contained or used hazardous materials
- Big box retail stores
- Shipping depots
- Military installations

Figure 2.5 Common household chemical products include gasoline, motor oil, paint, and insect repellant.

Figure 2.6 The presence of fume hood exhaust stacks on the roof or exterior of a building is a good indicator that hazardous materials are used inside.

Residential occupancies have hazardous chemicals such as drain cleaners, pesticides, fertilizers, paint products, flammable liquids (such as gasoline), swimming pool chemicals, propane tanks for gas grills, and other common household chemical products **(Figure 2.5)**. Propane tanks often provide heating fuel, and farms may have dangerous products such as pesticides and anhydrous ammonia. Any building with a fume hood exhaust stack (or stacks) on the roof (such as a research and development company or medical office building) probably has a functioning laboratory inside **(Figure 2.6)**.

Locations

Hazmat transportation accidents may be more likely to occur in certain areas. Places where hazmat is transferred or handled, such as trucking warehouses, are also likely locations for hazmat accidents. These locations include:

- Ports
- Docks or piers
- Railroad sidings
- Airplane hangars
- Truck terminals

Consult with local law enforcement officials to identify and determine potential problem spots or areas based on traffic studies. Each **transportation mode** has particular locations where accidents may occur more frequently:

- Roadways
  - Designated truck routes
  - Blind intersections
  - Poorly marked or poorly engineered interchanges
  - Areas frequently congested by traffic
  - Heavily traveled roads
  - Sharp turns
  - Steep grades
  - Highway interchanges and ramps
  - Bridges and tunnels

- Railways
  - Depots, terminals, and switch or classification yards
  - Sections of poorly laid or poorly maintained tracks
  - Steep grades and severe curves
  - Shunts and sidings
  - Uncontrolled crossings
  - Loading and unloading facilities
  - Bridges, trestles, and tunnels **(Figure 2.7)**

- Waterways
  - Difficult passages at bends or other threats to navigation
  - Bridges and other crossings
  - Piers and docks
  - Shallow areas
  - Locks
  - Loading/unloading stations

- Airways
  - Fueling ramps
  - Repair and maintenance hangars
  - Freight terminals
  - Crop duster planes and supplies

Transportation Mode — Technologies used to move people and/or goods in different environments; for example, rail, motor vehicles, aviation, vessels, and pipelines.

- Pipelines
  - Exposed crossings over waterways or roads
  - Pumping stations
  - Construction and demolition sites
  - Intermediate or final storage facilities

First responders should also pay attention to the water level in rivers and tidal areas. Be aware of the following facts:

- Many accidents occur because flow volume and tidal conditions were not considered. These flow and tidal variances affect clearance under bridges, many of which also have pipelines, water mains, gas lines, and the like attached to them.

- Occupancies in low-lying areas that may be affected by flood conditions must have a contingency plan to isolate and protect hazardous materials **(Figure 2.8)**.

Figure 2.7 Locations that experience more transportation accidents in general, such as railway bridges and trestles, are also more likely to be involved in hazmat incidents. *Courtesy of Phil Linder.*

Figure 2.8 Hazardous materials could be released when a flood hits occupancies in a low-lying area. A contingency plan must be in place to isolate and protect the hazardous materials. *Courtesy of Rich Mahaney.*

- Tidal and flow conditions are constantly changing. Areas that were once considered safe may become compromised by change of tide direction, flow rate, and back eddies.

- Once a hazardous material reaches an outside water source, it becomes a moving incident and is extremely difficult to contain, confine, and mitigate.

First responders should be familiar with the types of hazmat shipments that come through their jurisdictions. For example, farming communities may be more likely to see tanks of anhydrous ammonia passing through, whereas a port serving an industrial complex with many refineries might see more petroleum products.

Basic Container Shapes

Once you recognize that a location or occupancy may have hazardous materials, the presence of certain storage vessels, tanks, containers, packages, or vehicles can confirm their presence with certainty. These containers can provide useful information about the materials inside, so it is important for you to recognize the shapes of the different **containers** and **packaging** in which hazardous materials are stored and transported **(Figure 2.9)**.

The sections that follow will introduce hazmat container information:

- Container names by transport mode and capacity
- Pressure containers
- Cryogenic containers
- Liquid containers
- Solids containers
- Radioactive materials containers
- Pipelines
- Vessel cargo carriers
- Unit loading devices

Figure 2.9 A container's shape can tell first responders a great deal about the hazardous materials that might be inside. *Courtesy of Rich Mahaney.*

Container Names by Transport Mode and Capacity

Hazmat containers are sometimes classified according to their transport mode **(Table 2.1)**:

● Highway cargo trucks

● Rail cars

● Vessel cargo carriers

● Intermodal containers that transfer between modes

Containers can also be classified by their capacity. *Bulk packaging* refers to a packaging, other than that on a vessel (ship) or barge, in which materials are loaded with no intermediate form of containment. This packaging type includes a transport vehicle or freight container such as a cargo tank, railcar, or portable tank. Intermediate bulk containers (IBCs) and intermodal (IM) containers are also examples. To meet the criteria for bulk packaging, one of the following must be met:

● Maximum capacity is greater than 119 gallons (475 L) as a receptacle for a liquid

● Maximum net mass is greater than 882 pounds (440 kg) or maximum capacity is greater than 119 gallons (475 L) as a receptacle for a solid

● Water capacity is 1,001 pounds (500 kg) or greater as a receptacle for a gas

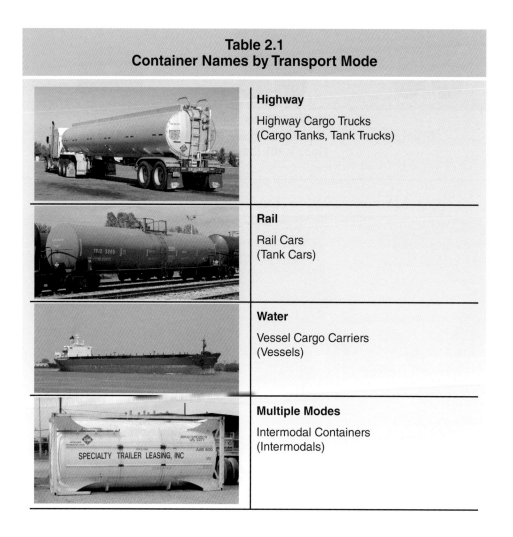

Table 2.1 Container Names by Transport Mode	
	Highway Highway Cargo Trucks (Cargo Tanks, Tank Trucks)
	Rail Rail Cars (Tank Cars)
	Water Vessel Cargo Carriers (Vessels)
	Multiple Modes Intermodal Containers (Intermodals)

Nonbulk packaging is smaller than the minimum criteria established for bulk packaging **(Figure 2.10)**. Drums, boxes, carboys, and bags are examples. Composite packages (packages with an outer packaging and an inner receptacle) and combination packages (multiple packages grouped together in a single outer container, such as bottles of acid packed inside a cardboard box) may also be classified as nonbulk packaging **(Figure 2.11)**.

Pressure Containers

Most people are familiar with compressed gas cylinders. Compressed gas cylinders are pressure containers, designed to hold product under pressure. The product may be a gas, liquefied gas, or a gas dissolved in a liquid **(Figure 2.12)**. Pressure containers have the potential to release a great deal of energy if involved in an incident **(Figure 2.13)**. For example, when stressed, pressure containers can rupture violently due to internal pressure. This can be accelerated if exposed to heat, flame, or mechanical damage. When released, the product will expand rapidly and will travel based on the product's physical

Figure 2.10 Bulk packaging allows large quantities of liquid, solid, or gas to be shipped.

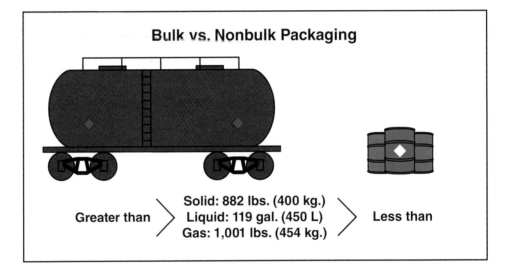

Figure 2.11 Composite packages are a form of nonbulk packaging.

and chemical properties and environmental conditions **(Figure 2.14)**. You may not be able to see if the contents of a pressure container are leaking, nor where they might be going.

WARNING!
STOP!!!! Products stored in pressure containers may kill you!
As a responder, your job is to stop and stop others! Isolate and deny entry.

Figure 2.12 Easily recognizable, compressed gas cylinders can hold gas, such as helium and nitrogen, under pressure.

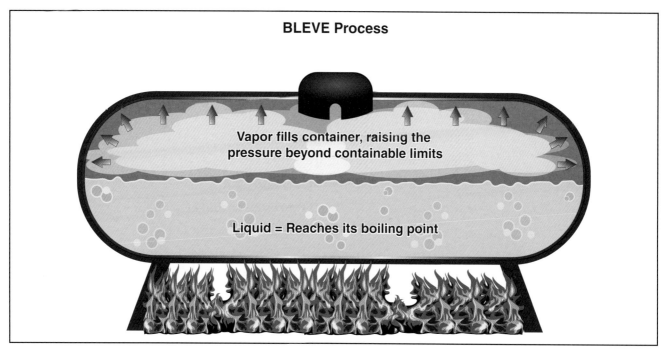

BLEVE Process

Vapor fills container, raising the pressure beyond containable limits

Liquid = Reaches its boiling point

Figure 2.13 There is a danger of BLEVE whenever pressurized containers are exposed to heat during an incident, even if the contents of the container are not flammable.

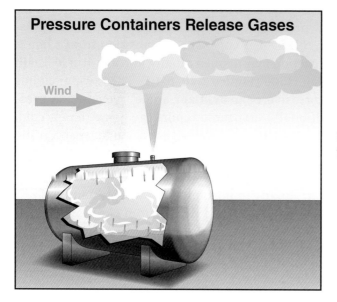

Pressure Containers Release Gases

Wind

Figure 2.14 A gas will expand rapidly when it is released into the environment because of a container rupture or failure.

WARNING!

When damaged or stressed by heat or flames, pressure containers may explode! Keep your distance!

WARNING!

Contents of pressure containers may ignite easily and will expand quickly if released! Keep your distance upwind, uphill, and upstream! Remove ignition sources if it can be done safely!

WARNING!

Contents of pressure containers may be extremely toxic and will expand quickly if released! Keep your distance upwind, uphill, and upstream!

WARNING!

Contents of pressure containers may be corrosive and will expand quickly if released! Keep your distance upwind, uphill, and upstream!

Manway — (1) Opening (hole) through which a person may go to gain access to an underground or enclosed structure. (2) Opening that is large enough to admit a person into a tank trailer or dry bulk trailer. This opening is usually equipped with a removable, lockable cover. *Also known as* Manhole.

Cryogen — Gas that is converted into liquid by being cooled below -130°F (-90°C). *Also known as* Refrigerated Liquid and Cryogenic Liquid.

Oxidizer — Any material that readily yields oxygen or other oxidizing gas, or that readily reacts to promote or initiate combustion of combustible materials. (Reproduced with permission from NFPA® 400-2010, *Hazardous Materials Code*, Copyright©2010, National Fire Protection Association®)

Regardless of size, transportation mode, or content, clues to pressure containers may include the following features:

- Rounded, almost spherical ends **(Figure 2.15)**
- Bolted **manways (Figure 2.16)**
- Bolted protective housings **(Figure 2.17)**
- Pressure relief devices **(Figures 2.18a and b, p. 58)**
- Pressure gauges **(Figure 2.19, p. 58)**

Pressure container examples are provided in **Table 2.2, p. 59**. Pressure containers will be explained in greater detail in Chapter 4.

Cryogenic Containers

Cryogenic containers are designed to store and transport cryogens. A **cryogen** (sometimes called *refrigerated liquefied gas*) is a gas that turns into a liquid at or below -130°F (-90°C) at 14.7 psi (101 kPa) {1.01 bar}.

These containers may be pressurized, though not to the degree that pressure containers are. When released, cryogens may transition from a liquid state to a vapor state. This reaction may happen rapidly, and a spill or leak will boil into a much larger vapor cloud **(Figure 2.20, p. 61)**. These vapor clouds may be flammable, toxic, corrosive, or an **oxidizer**. Some cryogens may present multiple hazards. Additionally, cryogenic vapors can be extremely cold, potentially causing freeze burns, which are treated as cold injuries according to their severity.

Any clothing saturated with a cryogenic material must be removed immediately. This action is particularly important if the vapor is flammable or an oxidizer. A first responder cannot escape flames from clothing-trapped vapors if they ignite.

Pressure Containers

Figure 2.15 Pressure containers can often be identified by their rounded ends. *Courtesy of Rich Mahaney.*

Figure 2.16 A bolted manway can be an identifying characteristic of a pressure container. *Courtesy of Rich Mahaney.*

Figure 2.17 First responders could look for bolted protective housings to help with their identification of pressure containers. *Courtesy of Rich Mahaney.*

WARNING!
Cryogens can displace oxygen and cause asphyxiation!

WARNING!
Cryogens are extremely cold and can severely injure you if contacted!

CAUTION
Immediately remove any clothing saturated with a cryogenic material.

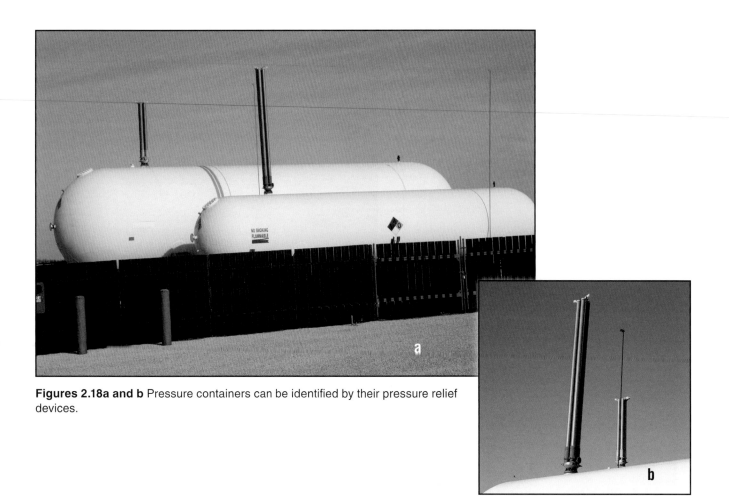

Figures 2.18a and b Pressure containers can be identified by their pressure relief devices.

Figure 2.19 Personnel can expect a pressure container to have a pressure gauge. *Courtesy of Rich Mahaney.*

Table 2.2
Pressure Containers

Fixed Facility (Bulk)	
Railway Tank Car	
Highway Cargo Tank	
Compressed Gas Tube-Trailer	
Intermodal	
Ton Container	

Table 2.2 (concluded)
Pressure Containers

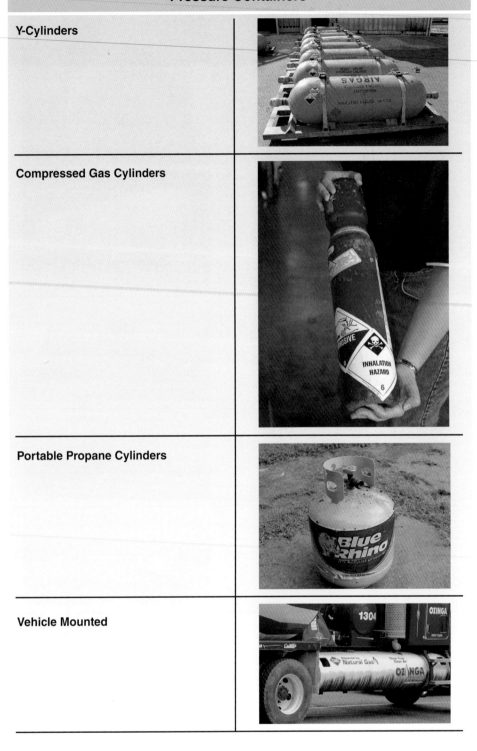

Y-Cylinders	
Compressed Gas Cylinders	
Portable Propane Cylinders	
Vehicle Mounted	

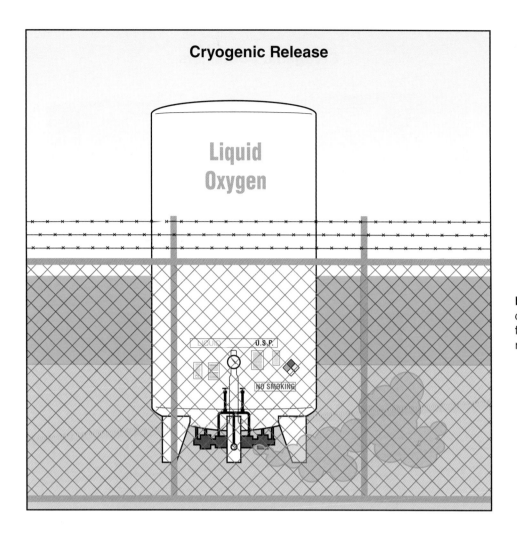

Figure 2.20 Cryogens may change, sometimes rapidly, from a liquid to a vapor when released.

Regardless of size, transportation mode, or content, clues indicating cryogenic containers may include the following features:

- Contents such as liquid oxygen (LOX), nitrogen, helium, hydrogen, argon, and liquefied natural gas (LNG) **(Figure 2.21, p. 62)**
- Box-like loading and unloading stations attached to transportation containers **(Figure 2.22, p. 63)**

Cryogenic container examples are provided in **Table 2.3, p. 64.** Additional cryogenic hazards will be discussed later in this chapter and in Chapter 4.

Liquid Containers

Common liquid containers include bottles, gasoline containers, paint pails, and drums. However, at fixed facilities, liquids can be stored in containers that hold many millions of gallons (liters). Large quantities are also transported in tanks via highway, rail, and other modes.

Many liquid containers will have some pressure due to the liquid's chemical and physical properties, but these pressures will be lower than a pressure container's. Although the pressures may be low, they can still be dangerous. Liquid containers may hold contents that have a variety of hazards including pressure, heat (often, their contents will burn), reactivity, corrosivity, and toxicity. Some liquid containers, when exposed to heat or flames, may rupture violently or explode.

Cryogenic Containers

Figure 2.21 Cryogenic containers are used to store liquefied gases, such as carbon dioxide, nitrogen, oxygen, and argon.

Figure 2.22 Cryogenic containers might be attached to box-like loading and unloading stations to make them easier to transport. *Courtesy of Rich Mahaney.*

WARNING!
STOP!!! Products in liquid containers may kill you!
Your job as a responder is to stop and stop others! Isolate and deny entry.

CAUTION
Many liquid containers will have low amounts of pressure. When released, this pressure may cause contents to splash or spray.

WARNING!
Liquid containers may explode when damaged or stressed by heat or flames. Keep your distance!

Regardless of size, transportation mode, or content, clues to liquid containers may include the following features: **(Figure 2.23, p. 66)**:

- Flat (or less rounded) ends on tanks
- Access hatches secured with easily removed latching devices
- Low pressure rail tank cars may have multiple fittings visible on top
- Intermodal, flexible **intermediate bulk containers**, and rigid intermediate bulk containers are designed to be stacked
- Flexible bladders filled with fluids
- Highway cargo tanks will have oval, upside-down horseshoe-shaped, or circular-shaped ends with less rounding than pressure tanks

Intermediate Bulk Container (IBC) — Rigid (RIBC) or flexible (FIBC) portable packaging, other than a cylinder or portable tank, that is designed for mechanical handling with a maximum capacity of not more than 3 cubic meters (3,000 L, 793 gal, or 106 ft³) and a minimum capacity of not less than 0.45 cubic meters (450 L, 119 gal, or 15.9 ft³) or a maximum net mass of not less than 400 kilograms (882 lbs).

Table 2.3
Cryogenic Containers

Fixed Facility (Bulk)	
Railway Tank Car	
Highway Cargo Tank	
Intermodal	
Cylinder	

Table 2.3 (concluded) Cryogenic Containers	
Dewar Flask	

Solids Containers

Many containers used to hold liquids may also be used for solids, for example, drums and bottles. Some transportation containers are specially designed for loading and unloading solids, and certain fixed facilities may store solids that are not typically deemed "hazardous," but may present a threat anyway, such as grain silos and storage facilities.

Hazardous solids may be dusts, powders, or small particles. Solids containers typically do not carry any pressure. A powder pesticide is an example of a potentially toxic solid. Boric acid and sodium hydroxide are corrosive solids. Dynamite is an energy-releasing solid. Calcium carbide is a reactive material that, when in contact with moisture, will release a flammable gas.

Small, airborne particles that burn (but may otherwise be harmless) can be dangerous if ignited in an enclosed location, causing a **dust explosion**. Grain, flour, sugar, coal, metal, and saw dust are examples of these particles. For that reason, you should be aware that fixed facilities where these materials are used, processed, or stored should be considered "containers" for purposes of this section.

Solid materials can also engulf you, causing suffocation and/or crushing injuries. These situations are typically associated with soil/dirt, sand, and gravel, but are also a concern at incidents involving large containers of grain, powdered substances, or any "flowing" solids.

Regardless of size, transportation mode, or content, clues to solids containers may include the following features:

- Transportation containers and systems designed for pneumatic loading and unloading **(Figures 2.24a and b, p. 66)**
- Open tops on hoppers, bins, or other containers, sometimes covered with tarps or plastic **(Figures 2.25, p. 66)**
- V-shaped sloping sides with bottom outlets

Dust Explosion — Rapid burning (deflagration), with explosive force, of any combustible dust. Dust explosions generally consist of two explosions: a small explosion or shock wave creates additional dust in an atmosphere, causing the second and larger explosion.

WARNING!
Dust explosions can kill you!

WARNING!
Solid materials can engulf and kill you!

Some Characteristics of Liquid Containers

Horseshoe Shape

Flat Ends

Stacked

Figure 2.23 First responders should be aware that liquid containers can have many identifying features, such as horseshoe shape, flat ends, and being stacked on top of each other, to list a few.

Figures 2.24a and b Systems are in place for the loading and unloading of dry bulk materials from railroad cars. *Courtesy of Rich Mahaney.*

Figure 2.25 Solids containers can be covered with tarps or left open to expose their content. *Courtesy of Rich Mahaney.*

Radioactive Materials Containers

All radioactive materials (RAM) shipments must be packaged and transported according to strict regulations designed to protect the public and emergency responders from radiation hazards **(Figure 2.26)**. The packaging used to transport radioactive materials is determined by the activity, type, and form of the material to be shipped. Depending upon these factors, radioactive material is shipped in one of five basic types, listed from least radioactive hazard to greatest:

1. **Excepted** — Excepted packaging is only used to transport materials with extremely low levels of radioactivity that present no risk to the public or environment.

2. **Industrial** — Container that retains and protects the contents during normal transportation activities such as laboratory samples and smoke detectors.

3. **Type A** — Packages that must demonstrate their ability to withstand a series of tests without releasing their contents.

4. **Type B** — Packages must demonstrate their ability to withstand tests simulating normal shipping conditions, and they must also withstand severe accident conditions without releasing their contents.

5. **Type C** — Very rare packages used for high-activity materials (including plutonium) transported by aircraft.

Figure 2.26 Type B packages are designed to withstand severe accident conditions. Type B packages contain materials with high levels of radioactivity that would present a radiation hazard to the public or the environment if there was a major release. *Courtesy of the National Nuclear Security Administration, Nevada Site Office.*

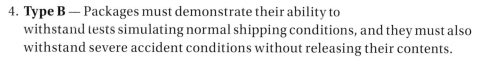
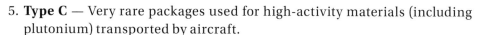

CAUTION
Radiation can travel in all directions for long distances and may pass through materials. It cannot be detected by the five human senses and requires a meter to detect.

Table 2.4, p. 68, provides examples of radioactive materials containers. Radioactive materials will be described in greater detail in Chapter 5.

Pipelines

Many hazardous materials, particularly petroleum varieties, are transported across both the U.S. and Canada in underground pipelines **(Figure 2.27, p. 69)**. Pipelines may transport liquids or gases.

Where pipelines cross under (or over) roads, railroads, and waterways, pipeline companies must provide markers. These markers are often the best way to identify that pipelines are present, as well as to identify their contents **(Figure 2.28, p. 69)**. Markers must be in sufficient numbers along the pipeline to identify the pipe's location. However, pipeline markers do not always mark the pipeline's exact location, and you should not assume that the pipeline runs in a perfectly straight line between markers.

Table 2.4
Radioactive Materials Containers

Excepted	
Industrial	
Type A	
Type B	
Type C	

Figure 2.27 Petroleum and other hazardous materials are shipped through underground pipelines that stretch throughout the United States and Canada. *Courtesy of Rich Mahaney.*

Figure 2.28 Pipeline markers in the U.S. and Canada include signal words, information describing the transported commodity, and the name and emergency telephone number of the carrier. *Courtesy of Rich Mahaney.*

WARNING!
Pipelines can transport high pressure materials and may explode!

WARNING!
Pipelines can transport a variety of very dangerous materials!

CAUTION
Pipelines may be buried in residential neighborhoods!

Vessel Cargo Carriers

Marine vessels transport over ninety percent of the world's cargo, and that amount is expected to increase in the future. Hazardous materials incidents involving vessels can be minor, such as a small spill at a port during loading or unloading, or major, such as a spill contaminating miles (kilometers) of river or coastline waters. Statistics on oil spills show that most spills are relatively small and result

Figure 2.29 Tankers, which can transport large quantities of liquid products, come in three types: Petroleum carriers, chemical carriers, and liquefied flammable gas carriers.

Tanker Types

Petroleum Carrier

Chemical Carrier

Liquefied Flammable Gas Carrier

from routine operations such as loading and unloading, which normally occur in ports or at oil/chemical terminals. Vessels that transport hazardous materials include the following:

- **Tankers (tank vessel)** — These vessels may transport very large quantities of liquid products. Tankers often carry different products in segregated tanks. There are three tanker types **(Figure 2.29)**:

  - *Petroleum carriers* transport crude or finished petroleum products.

  - *Chemical carriers* transport many different chemical products.

  - *Liquefied flammable gas carriers* transport liquefied natural gas (LNG) and liquefied petroleum gas (LPG).

- **Cargo Vessel** — Cargo is shipped in the following four vessel types:

  - *Bulk carriers* may transport liquids or solids.

  - *Break bulk carriers* may transport a variety of materials in many different containers such as pallets, drums, bags, boxes, and crates.

  - *Container vessels* transport cargo in standard intermodal containers with standard widths and varying heights and lengths **(Figure 2.30)**.

  - *Roll-on/roll-off vessels* have large stern and side ramp structures that are lowered to allow vehicles to be driven on and off the vessel **(Figure 2.31)**.

- **Barges** — Barges are typically box-shaped, flat-decked vessels used for transporting cargo **(Figures 2.32a and b)**. Towing or pushing vessels are usually used to move barges because they are not self-propelled. Virtually anything can be transported on a barge. Some barges are configured as floating barracks for military or construction crews; some are designed as bulk oil and chemical tankers. Other barges carry LNG in cylinders that may not be visible until a person is aboard. Barges may serve as floating warehouses with hazardous goods, vehicles, or rail cars inside.

WARNING!
Confined spaces in vessels can contain oxygen deficient atmospheres that cause asphyxiation!

Figure 2.30 Container vessels transport intermodal containers, including intermodal tanks.

Figure 2.31 Roll-on/roll-off vessels have large stern and side ramp structures that are lowered to allow vehicles to be driven on and off the vessel.

Figures 2.32a and b (a) Barges can travel waterways that large vessels cannot. They are versatile in their cargos, and some barges are designed to carry specific hazardous materials. **(b)** It may not be possible to tell from a distance if hazardous materials are being transported on a barge. *Courtesy of Rich Mahaney.*

Unit Loading Devices

Unit loading devices (ULDs) are containers and aircraft pallets used to consolidate air cargo into a single, transportable unit **(Figure 2.33)**. ULDs are designed and shaped to fit into airplane decks and compartments (particularly, commercial cargo planes), and in some cases they may be stacked. Hazardous materials may be shipped in ULDs provided they are in accordance with governmental regulations, including packaging and labeling requirements.

NOTE: Military aircraft or transport vehicles may transport Internal airlift and helicopter Slingable Units (ISUs) that transport everything including hazardous materials.

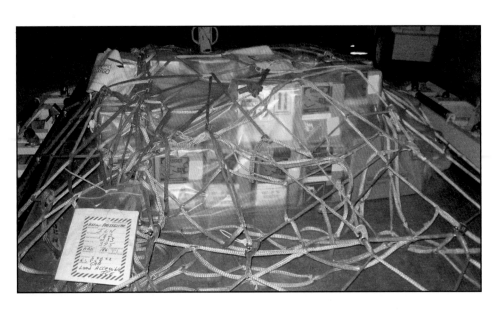

Figure 2.33 Unit loading devices (ULDs) are used to consolidate air cargo into a single, transportable unit. ULDs containing hazardous materials must be appropriately placarded and labeled. *Courtesy of John Deyman.*

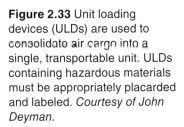

Label — Four-inch-square diamond-shaped marker required by federal regulations on individual shipping containers that contain hazardous materials, and are smaller than 640 cubic feet (18 m³).

Placard — Diamond-shaped sign that is affixed to each side of a structure or a vehicle transporting hazardous materials to inform responders of fire hazards, life hazards, special hazards, and reactivity potential. The placard indicates the primary class of the material and, in some cases, the exact material being transported; required on containers that are 640 cubic feet (18 m³) or larger.

Transportation Placards, Labels, and Markings

The U.S., Canada, and Mexico use a placarding, labeling, and marking system to identify hazmat during transportation. All three countries use the same source for their systems: the *Transport of Dangerous Goods – Model Regulations*, published by the United Nations (also known as the *U.N. Recommendations*). Therefore, with a few country-specific variations, the placards, labels, and markings used to identify hazardous materials during transport are very similar in all three countries.

Generally speaking, transportation **labels** are designed for nonbulk packages, whereas transportation **placards** are designed for bulk packages. They look similar and convey similar information. However, there are certain unique labels for which there are no equivalent placards.

Under the United Nations (U.N.) system, nine hazard classes are used to categorize hazardous materials:

- Class 1: Explosives
- Class 2: Gases
- Class 3: Flammable liquids
- Class 4: Flammable solids, substances liable to spontaneous combustion, substances that emit flammable gases on contact with water
- Class 5: Oxidizing substances and organic peroxides

- Class 6: Toxic and infectious substances
- Class 7: Radioactive materials
- Class 8: Corrosive substances
- Class 9: Miscellaneous dangerous substances and articles

Because most North American first responders primarily deal with DOT or Transport Canada (TC) placards, labels, and markings, the unique U.N. placards are not detailed in the sections that follow. Examples of the U.N. class placards and labels with brief explanations are found in **Appendix C**, U.N. Class Placards and Labels. Placards, labels, markings, and colors associated with other systems (such as NFPA 704, *Standard System for the Identification of the Hazards of Materials for Emergency Response*, and military markings) are explained in the section, Other Markings and Colors.

Four-Digit Identification Numbers

In addition to establishing hazard classes, the U.N. has assigned each individual hazardous material a unique four-digit number. This number is often displayed on placards, orange panels, and certain markings in association with materials being transported in cargo tanks, portable tanks, tank cars, or other containers and packages.

The four-digit identification (ID) number must be displayed on bulk containers in one of the three ways illustrated by **Figure 2.34**. In North America, the numbers must be displayed on the following containers/packages:

- Rail tank cars
- Cargo tank trucks
- Portable tanks
- Bulk packages
- Table 1 materials, regardless of quantity (see DOT Chart 15)
- Certain nonbulk packages (for example, poisonous gases in specified amounts)

Figure 2.34 Examples of how UN numbers will be displayed on bulk containers (such as cargo tank trucks and rail tank cars) and certain nonbulk containers.

Sample Displays of 4-Digit UN Identification Numbers

FLAMMABLE 3
1090

1090 3

1993 3

Emergency Response Guidebook (ERG) — Manual that aids emergency response and inspection personnel in identifying hazardous materials placards and labels; also gives guidelines for initial actions to be taken at hazardous materials incidents. Developed jointly by Transport Canada (TC), U.S. Department of Transportation (DOT), the Secretariat of Transport and Communications of Mexico (SCT), and with the collaboration of CIQUIME (Centro de Información Química para Emergencias).

The *Emergency Response Guidebook* (*ERG*) provides a key to the four-digit identification (ID) numbers in the yellow-bordered section (see Chapter 3, Awareness Level Actions at Hazardous Materials Incidents). Therefore, if the four-digit identification number is identified, first responders can use the *ERG* to determine appropriate initial response information based on the material involved. The four-digit identification number will also appear on shipping papers, and it should match the numbers displayed on tank or shipping container exteriors.

Common reference materials such as the *ERG* do not list all four-digit U.N. identification numbers. For example, the *ERG* does not list any numbers below 1000. In the U.S., the entire list is included in 49 *CFR* 172.101.

NOTE: *NA numbers* (**N**orth **A**merica), also known as *DOT numbers*, are issued by the United States Department of Transportation and are identical to U.N. numbers, except that some substances without a U.N. number may have an NA number. These additional NA numbers use the range NA8000 - NA9999.

Orange Panels

Do not be confused by an orange panel with two sets of numbers on intermodal tanks and containers. The four-digit ID number is on the bottom. The top number is a hazard identification number (or code) required under European and some South American regulations **(Figure 2.35)**. These numbers indicate the following hazards:

2 — Emission of a gas due to pressure or chemical reaction

3 — Flammability of liquids (vapors) and gases or self-heating liquid

4 — Flammability of solids or self-heating solid

5 — Oxidizing (fire intensifying) effect

6 — Toxicity or risk of infection

7 — Radioactivity

8 — Corrosivity

9 — Miscellaneous Dangerous Substance

Doubling a number (such as 33, 44, or 88) indicates an intensification of that particular hazard. When the hazard associated with a material is adequately indicated by a single number, it is followed by a zero (such as 30, 40, or 60). A hazard identification code prefixed by the letter X (such as X88) indicates that the material will react dangerously with water. When 9 appears as a second or third digit, this may present a risk of spontaneous violent reaction.

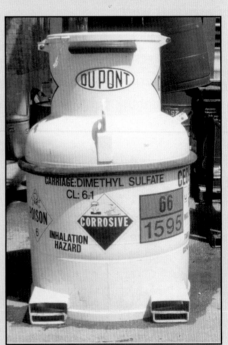

Figure 2.35 The UN ID number is the orange panel on the bottom. The top number is a hazard identification number (code) that some European and South American regulations require. *Courtesy of Rich Mahaney.*

Placards

Shippers provide diamond-shaped, color-coded signs (placards) on transportation containers to identify their contents. Each hazard class has a specific placard that identifies the material's hazard class. A material's hazard class is indicated either by its class (or division) number or name. **Figure 2.36** provides the required dimensions of transportation placards and summarizes the information conveyed by them. Placards may be found on the following types of containers:

- Bulk packages
- Rail tank cars
- Cargo tank vehicles
- Portable tanks
- Unit load devices containing hazardous materials over 640 cubic feet (18 m³) in capacity
- Certain nonbulk containers

You may see containers with more than one placard indicating that more than one hazard or product is present. **Figure 2.37, p. 76**, provides the U.S. DOT's Chart 15, Hazardous Materials Placarding Guide.

Unfortunately, improperly marked, unmarked, and otherwise illegal shipments are common. These shipments may include incompatible products, products that contravene local, state/provincial, and federal laws, and waste products shipped and disposed of without permit.

The following are important facts related to placards:

- A placard is not required for shipments of infectious substances, other regulated materials for domestic transport only (ORM-Ds),

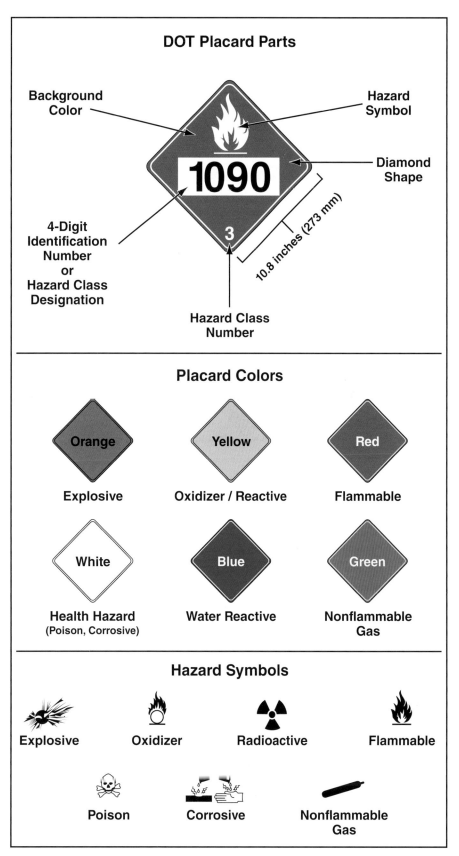

Figure 2.36 Placards provide many visual clues to the hazards that a material presents.

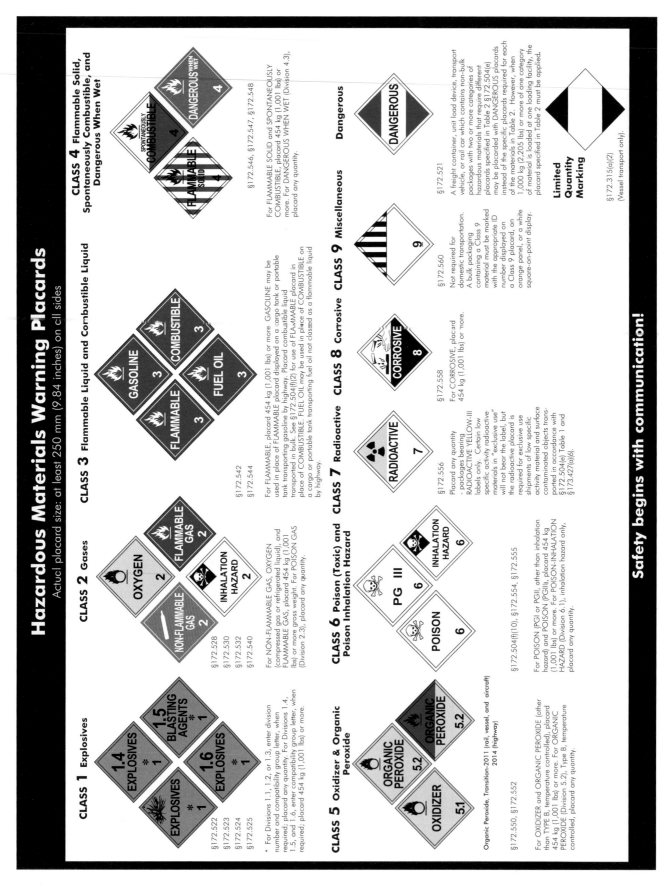

Figure 2.37 The U.S. Department of Transportation has a chart showing the different hazardous materials placards that might be posted on a container.

materials of trade (MOTs), limited quantities, small-quantity packages, radioactive materials (white label I or yellow label II; see Labels section), or combustible liquids in nonbulk packaging.

- Some private agriculture and military vehicles may not have placards, even though they are carrying significant quantities of hazardous materials. For example, farmers may carry fertilizer, pesticides, and fuel between fields of their farms or to and from their farms without any placarding.

- The hazard class or division number corresponding to the primary or subsidiary hazard class of a material must be displayed in the lower corner of a placard **(Figure 2.38)**.

- The DANGEROUS placard is for mixed loads where the transport vehicle contains non-bulk packages with two or more categories of hazardous materials that require different placards **(Figure 2.39)**.

- Other than Class 7 or the DANGEROUS placard, text indicating a hazard (for example, the word FLAMMABLE) is not required. Text may be omitted from the Oxygen placard only if the specific ID number is displayed.

- Drivers may have varying degrees of information about the hazardous materials in their vehicles.

- Containers may have placards even though they appear "empty" until they are certified as "clean."

Labels

Labels provide similar information as vehicle placards. Labels are 3.9-inch (100 mm), square-on-point diamonds, which may or may not have written text that identifies the hazardous material within the packaging. **Figure 2.40, p. 78**, shows the U.S. DOT's Chart 15, Hazardous Materials Labeling Guide.

Figure 2.38 The lower corner of a placard shows the hazard class or division number corresponding to the primary or subsidiary hazard class of a material.

Figure 2.39 The DANGEROUS placard indicates that the vehicle contains non-bulk packages with two or more categories of hazardous materials that require different placards. *Courtesy of Rich Mahaney.*

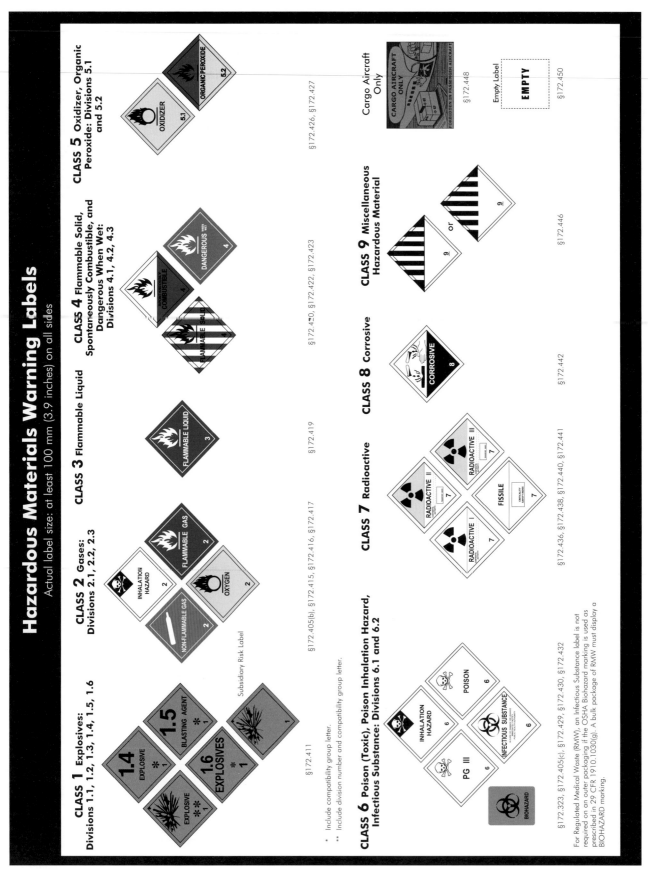

Figure 2.40 The U.S. Department of Transportation has a chart showing the different hazardous materials labels.

Class 7 Radioactive labels must always contain text. Most labels for the nine hazard classes and subdivisions are essentially the same as their placarding counterparts.

Packages with more than one label have more than one hazard or product. These packages contain a primary label and a subsidiary label for materials that meet the definition of more than one hazard class. In **Figure 2.41**, the toxic label is the primary label, while the flammable liquid label is the subsidiary.

The "Cargo Aircraft Only" label is not associated with a particular hazard class. This label is used to indicate materials that cannot be transported on passenger aircraft.

Markings

A marking is a descriptive name, an identification number, a weight, or a specification and includes instructions, cautions, or U.N. marks (or combinations thereof) required on the outer packaging of hazardous materials. This section, however, shows only those markings found on DOT Chart 15 **(Figure 2.42, p. 80)**. Markings on intermodal containers, tank cars, and other packaging are discussed in later sections.

Figure 2.41 The toxic label is the primary label (higher and to the left), while the flammable liquid label is the subsidiary. *Courtesy of Rich Mahaney.*

One marking you should note is the "Hot" marking for elevated-temperature materials. **Elevated temperature materials**, such as molten sulfur and molten aluminum, can present a thermal hazard in the form of heat **(Figure 2.43, p. 80)**. Molten aluminum, for example, is generally shipped at temperatures above 1,300°F (705°C). First responders must be extremely cautious around these materials to avoid being burned. Molten aluminum and other high-temperature materials can ignite flammable and combustible materials (including wood). Working around or near elevated-temperature materials can increase the effect of wearing personal protective equipment due to high ambient air temperatures (see Chapter 8, Personal Protective Equipment).

> **Elevated Temperature Material** — Material that when offered for transportation or transported in bulk packaging is (a) in a liquid phase and at temperatures at or above 212°F (100°C), (b) intentionally heated at or above its liquid phase flash points of 100°F (38°C), or (c) in a solid phase and at a temperature at or above 464°F (240°C).

The U.S. Department of Transportation (DOT) defines an elevated-temperature material as one that when offered for transportation or transported in bulk packaging has one of the following properties:

- Liquid phase at a temperature at or above 212°F (100°C)

- Liquid phase with a flash point at or above 100°F (38°C) that is intentionally heated and offered for transportation or transported at or above its flash point

- Solid phase at a temperature at or above 464°F (240°C)

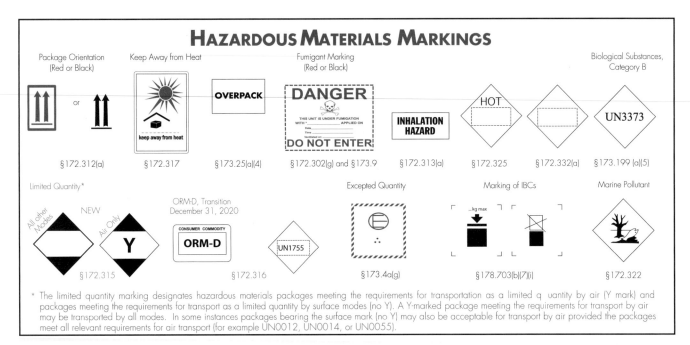

Figure 2.42 DOT haz mat markings.

Figure 2.43 Materials marked as "Hot" are being transported at an elevated temperature and should be treated as burn hazards. *Courtesy of Rich Mahaney.*

Canadian Placards, Labels, and Markings

Transport Canada (TC) and the Dangerous Goods Act govern transportation placards, labels, and markings in Canada. Like the U.S. HMR, the Dangerous Goods Act is based on the U.N. Recommendations and, therefore, is very similar. The nine hazard classes in both documents are identical. **Table 2.5, p. 82**. provides Canadian placards, labels, and markings divided by class. There are some differences, however, between Canadian and U.S. placards, labels, and markings such as the following:

- Most Canadian transport placards do not have any signal words written on them.

- Labels and markings may be in both English and French.

- Canada requires a unique placard for anhydrous ammonia and Inhalation Hazards.

- Radiation placard may have the four-digit U.N. number.

Mexican Placards, Labels, and Markings

Like Canada and the U.S., Mexican transportation placards, labels, and markings are based on the U.N. Recommendations and have the same hazard classes and subdivisions. In fact, Canadian and Mexican placards and labels are virtually the same, although Mexico does not recognize the Inhalation placard. However, because international regulations authorize the insertion of text (other than the class or division number) in the space below the symbol as long as the text relates to the nature of the hazard or precautions to be taken in handling, placards and labels in Mexico may have text that is in Spanish **(Figure 2.44)**. Likewise, information provided on markings is likely to be written in Spanish. English-speaking first responders in Mexico or along the U.S./Mexican border should familiarize themselves with the more common Spanish hazard warning terms such as *peligro* (danger).

Figure 2.44 Placards and labels in Mexico may have text that is written in Spanish. English-speaking responders should still be able to recognize symbols, shapes, and colors that provide them with information about the hazards associated with the contents of the package or container.

Other Markings and Colors

In addition to DOT placards, labels, and markings, a number of other markings, marking systems, labels, labeling systems, colors, color-codes, and signs may indicate the presence of hazardous materials at fixed facilities, on pipelines, on piping systems, and on other containers. These other markings may be as simple as the word *chlorine* stenciled on the outside of a fixed-facility tank or

Table 2.5
Canadian Transportation Placards, Labels, and Markings

Class 1: Explosives

Placard and Label	**Class 1.1** — Mass explosion hazard
Placard and Label	**Class 1.2** — Projection hazard but not a mass explosion hazard
	Class 1.3 — Fire hazard and either a minor blast hazard or a minor projection hazard or both but not a mass explosion hazard
Placard and Label	**Class 1.4** — No significant hazard beyond the package in the event of ignition or initiation during transport * = Compatibility group letter
Placard and Label	**Class 1.5** — Very insensitive substances with a mass explosion hazard
Placard and Label	**Class 1.6** — Extremely insensitive articles with no mass explosion hazard

Class 2: Gases

Placard and Label	**Class 2.1 — Flammable Gases**
Placard and Label	**Class 2.2 — Nonflammable and nontoxic Gases**

Continued

Table 2.5 (continued)

Class 2: Gases (continued)

	Class 2.3 — Toxic Gases
 Placard and Label	Anhydrous Ammonia
 Placard and Label	Oxidizing Gases

Class 3: Flammable Liquids

 Placard and Label	Class 3 — Flammable Liquids

Class 4: Flammable Solids, Substances Liable to Spontaneous Combustion, and Substances that on Contact with Water Emit Flammable Gases (Water-Reative Substances)

 Placard and Label	Class 4.1 — Flammable Solids
 Placard and Label	Class 4.2 — Substances Liable to Spontaneous Combustion
 Placard and Label	Class 4.3 — Water-Reactive Substances

Class 5: Oxidizing Substances and Organic Peroxides

 Placard and Label	Class 5.1 — Oxidizing Substances

Continued

Table 2.5 (continued)

Class 5: Oxidizing Substances and Organic Peroxides (continued)

Placard and Label

Class 5.2 — Organic Peroxides

Class 6: Toxic and Infectious Substances

Placard and Label

Class 6.1 — Toxic Substances

Label Only

Class 6.2 — Infectious Substances

Text:

INFECTIOUS

In case of damage or leakage, Immediately notify local authorities AND

INFECTIEUX

En cas de Dommage ou de fuite communiquer Immédiatement avec les autorités locales ET

CANUTEC

613-996-6666

Placard Only

Class 6.2 — Infectious Substances

Class 7: Radioactive Materials

**Label and
Optional Placard**

Class 7 — Radioactive Materials

Category I — White

RADIOACTIVE

CONTENTS.....................CONTENU

ACTIVITY......................... ACTIVITÉ

**Label and
Optional Placard**

Class 7 — Radioactive Materials

Category II — Yellow

RADIOACTIVE

CONTENTS.....................CONTENU

ACTIVITY........................ACTIVITÉ

INDICE DE TRANSPORT INDEX

Continued

Table 2.5 (concluded)

Class 7: Radioactive Materials (continued)

Label and Optional Placard

Class 7 — Radioactive Materials
Category III — Yellow
RADIOACTIVE
CONTENTS.....................CONTENU
ACTIVITY..........................ACTIVITÉ
INDICE DE TRANSPORT INDEX

Placard

Class 7 — Radioactive Materials
The word RADIOACTIVE is optional.

Class 8: Corrosives

Placard and Label

Class 8 — Corrosives

Class 9: Miscellaneous Products, Substances, or Organisms

Placard and Label

Class 9 — Miscellaneous Products, Substances, or Organisms

Other Placards, Labels, and Markings

Danger Placard

Elevated Temperature Sign

Fumigation Sign

Text is in both English and French

Marine Pollutant Mark

The text is MARINE POLLUTANT or POLLUANT MARIN.

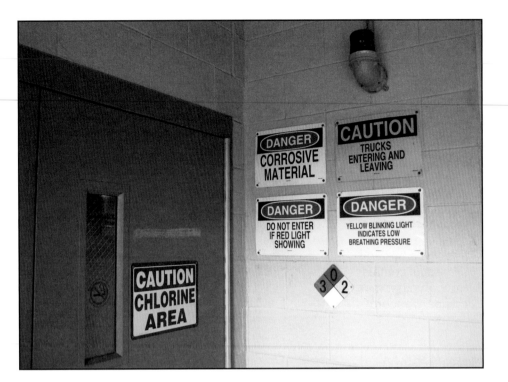

Figure 2.45 Fixed facilities may have a variety of other signs, markings, and color-codes that indicate the presence of hazardous materials. *Courtesy of Rich Mahaney.*

as complicated as a site-specific hazard communication system using a unique combination of labels, placards, emergency contact information, and color codes **(Figure 2.45)**. Some containers may be marked with special information, for example, *non-odorized*, meaning that the product will not have an intense smell by itself. Some fixed-facility containers may have identification numbers that correspond to site or emergency plans that provide details on the product, quantity, and other pertinent information.

The sections that follow highlight the most common specialized systems in North America, including the following:

- NFPA® 704 System
- Globally Harmonized System
- HMIS and other U.S. Hazard Communications labels and markings
- Canadian Workplace Hazardous Materials Information System
- Mexican Hazard Communication System
- CAS® numbers
- Military markings
- Pesticide labels
- Other symbols and signs
- ISO safety symbols
- Color codes

NFPA 704 System

The information in NFPA 704, *Standard System for the Identification of the Hazards of Materials for Emergency Response*, gives a widely recognized method for indicating the presence of hazardous materials at commercial, manufacturing, institutional, and other fixed-storage facilities. Use of this system is commonly required by local ordinances for all occupancies that contain hazardous materials. It is designed to alert emergency responders to health, flammability, instability, and related hazards (specifically, oxidizers and water-reactive materials) that may present as short-term, acute exposures resulting from a fire, spill, or similar emergency.

NFPA 704 Limitations

NFPA 704 markings provide very useful information, but the system does have its limitations. For example, an NFPA diamond does not tell exactly what chemical or chemicals may be present in specific quantities. Nor does it tell exactly where hazardous materials may be located when the sign is used for a building, structure, or area such as a storage yard rather than an individual container. Positive identification of the materials must be made through other means such as container markings, employee information, company records, and preincident surveys.

Specifically, the NFPA 704 system uses a rating system of numbers from 0 to 4. The number 0 indicates a minimal hazard, whereas the number 4 indicates a severe hazard. The rating is assigned to three categories: health, flammability, and instability. The rating numbers are arranged on a diamond-shaped marker or sign. The health rating is located on the blue background, the flammability hazard rating is positioned on the red background, and the instability hazard rating appears on a yellow background **(Figure 2.46)**. As an alternative, the

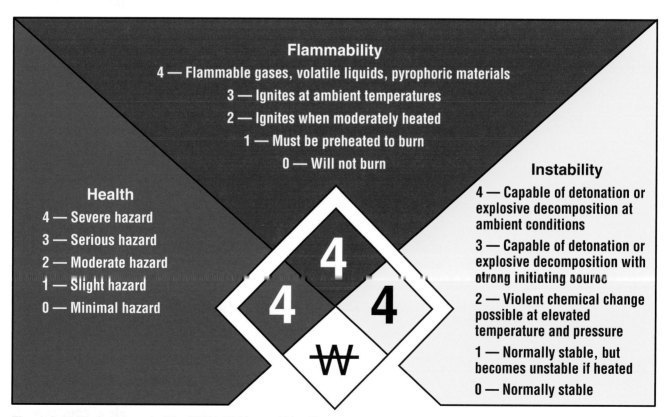

Flammability
4 — Flammable gases, volatile liquids, pyrophoric materials
3 — Ignites at ambient temperatures
2 — Ignites when moderately heated
1 — Must be preheated to burn
0 — Will not burn

Health
4 — Severe hazard
3 — Serious hazard
2 — Moderate hazard
1 — Slight hazard
0 — Minimal hazard

Instability
4 — Capable of detonation or explosive decomposition at ambient conditions
3 — Capable of detonation or explosive decomposition with strong initiating source
2 — Violent chemical change possible at elevated temperature and pressure
1 — Normally stable, but becomes unstable if heated
0 — Normally stable

Figure 2.46 Key and layout of the NFPA 704 hazard identification system.

backgrounds for each of these rating positions may be any contrasting color, and the numbers (0 to 4) may be represented by the appropriate color (blue, red, and yellow).

Special hazards are located in the six o'clock position and have no specified background color; however, white is most commonly used. Only two special hazard symbols are presently authorized for use in this position by the NFPA: W, indicating unusual reactivity with water, and OX, indicating that the material is an oxidizer. However, you may see other symbols in the white quadrant on old diamonds, including the trefoil radiation symbol. If more than one special hazard is present, multiple symbols may be seen.

NOTE: The NFPA 704 system may be used differently in countries outside of North America. For example, NFPA 704 symbols might be used on transportation containers.

Globally Harmonized System

Globally Harmonized System of Classification and Labeling of Chemicals (GHS) — International classification and labeling system for chemicals and other hazard communication information, such as safety data sheets.

Safety Data Sheet (SDS) — Form provided by chemical manufacturers, distributors, and importers; provides information about chemical composition, physical and chemical properties, health and safety hazards, emergency response procedures, and waste disposal procedures.

The U.S. and many other countries throughout the world have developed a **Globally Harmonized System of Classification and Labeling of Chemicals (GHS)**. The purpose of GHS is to promote common, consistent criteria for classifying chemicals according to their health, physical, and environmental hazards and encourage the use of compatible hazard labels, **safety data sheets** (formerly known as *material safety data sheets*) for employees, and other hazard communication information based on the resulting classifications. **Appendix C** provides an in-depth summary of the GHS system.

Several key harmonized information elements of GHS are as follows:

- Uniform classification of hazardous substances and mixtures
- Uniform labeling standards
    — Allocation of label elements
    — Symbols and pictograms **(Table 2.6)**
    — Signal words: *danger* (most severe hazard categories) and *warning* (less severe hazard categories)
    — Hazard statements
    — Precautionary statements and pictograms
    — Product and supplier identification
    — Multiple hazards and precedence of information
    — Arrangements for presenting GHS label elements
    — Special labeling arrangements
- Uniform safety data sheet (SDS) content and format

Table 2.6
Globally Harmonized System of Classification and Labeling of Chemicals (GHS)

Flammables/ Fire Hazard	Oxidizers	Explosives or Explosion Hazard	Corrosives	Compressed Gases
Warnings	Environmental Hazards	Poison/Toxic	Variety of Health Hazards	

HMIS and Other U.S. Hazard Communications Labels and Markings

OSHA's Hazard Communication Standard (HCS) requires employers to identify hazards in the workplace and train employees how to recognize those hazards. It also requires the employer to ensure that all hazardous material containers are labeled, tagged, or marked with the identity of the substances contained in them along with appropriate hazard warnings. The standard does not specify what system (or systems) of identification must be used, leaving that to be determined by individual employers. First responders, then, may encounter a variety of different (and sometimes unique) labeling and marking systems in their jurisdictions **(Figure 2.47)**. Conducting preincident surveys should assist responders in identifying and understanding these systems.

HMIS (Hazardous Materials Information System) is a commonly used proprietary system developed by the American Coatings Association in order to comply with HCS standards. It utilizes a numerical rating and color code system similar to NFPA 704 to convey the relative hazards of the product to employees.

Canadian Workplace Hazardous Materials Information System

Like the U.S. HCS, the Canadian Workplace Hazardous Materials Information System (WHMIS) requires that hazardous products be appropriately labeled and marked. It also spells out requirements for safety data sheets. As with the HCS, there are different ways for Canadian employers to meet the requirements of WHMIS; however, two types of labels will most commonly be used: the supplier label **(Figure 2.48, p. 90)** and the workplace label. These labels will include information such as the product name, a statement that an SDS is available, and other information that will vary depending on the type of label (supplier labels will include information about the supplier). **Table 2.7, p. 91**, shows the old WHMIS Symbols and Hazard Classes, which are being replaced by GHS.

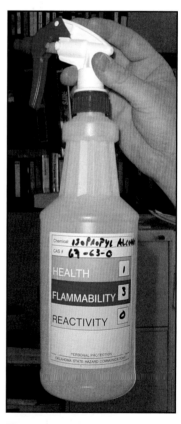

Figure 2.47 The OSHA Hazard Communications Standard requires employers to identify hazards in the workplace. First responders may encounter a variety of different identification systems that employers use in their area.

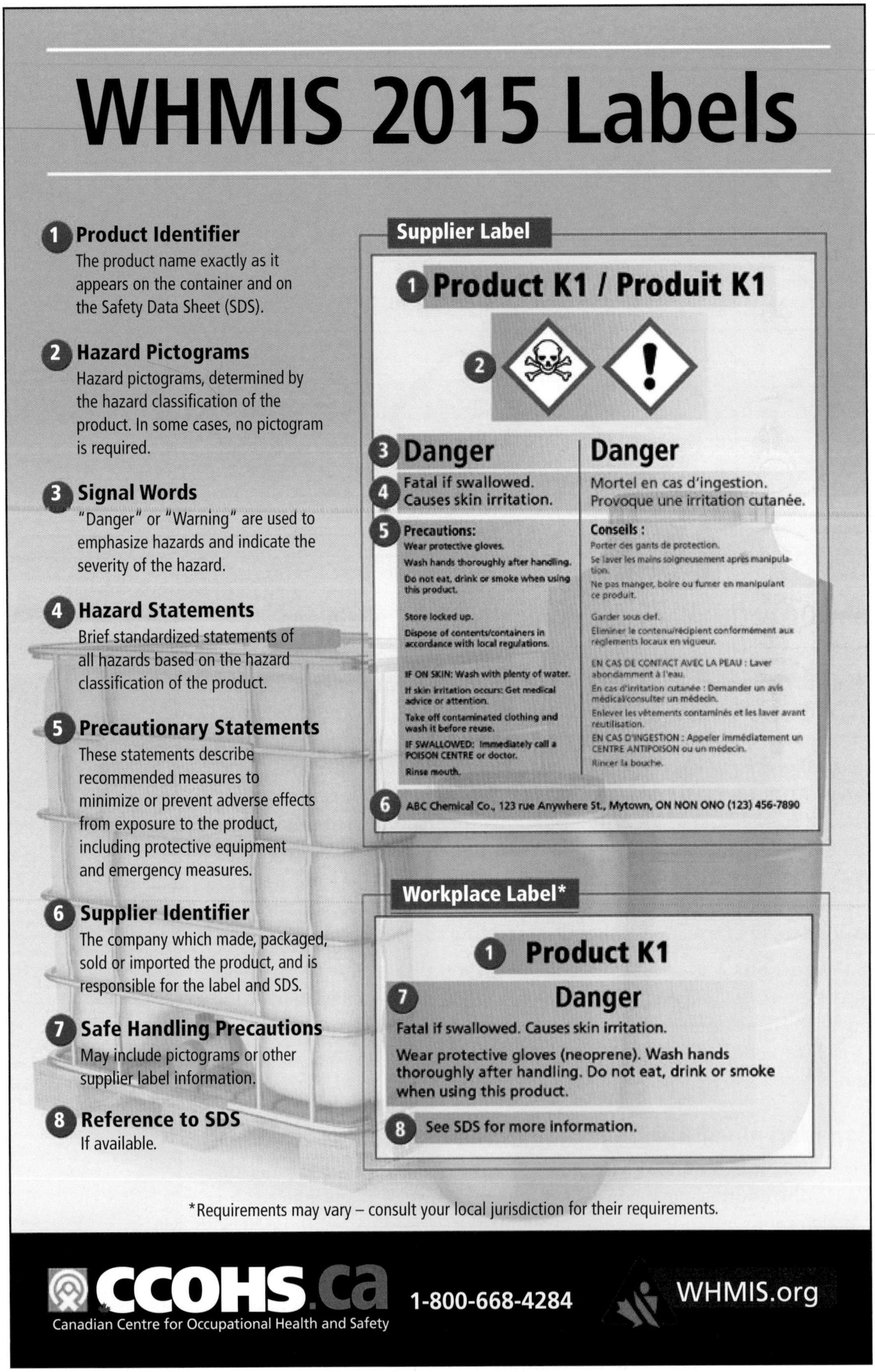

WHMIS 2015 Labels

1 **Product Identifier**
The product name exactly as it appears on the container and on the Safety Data Sheet (SDS).

2 **Hazard Pictograms**
Hazard pictograms, determined by the hazard classification of the product. In some cases, no pictogram is required.

3 **Signal Words**
"Danger" or "Warning" are used to emphasize hazards and indicate the severity of the hazard.

4 **Hazard Statements**
Brief standardized statements of all hazards based on the hazard classification of the product.

5 **Precautionary Statements**
These statements describe recommended measures to minimize or prevent adverse effects from exposure to the product, including protective equipment and emergency measures.

6 **Supplier Identifier**
The company which made, packaged, sold or imported the product, and is responsible for the label and SDS.

7 **Safe Handling Precautions**
May include pictograms or other supplier label information.

8 **Reference to SDS**
If available.

Supplier Label

1 **Product K1 / Produit K1**

2 (pictograms)

3 **Danger** | **Danger**

4 Fatal if swallowed. Causes skin irritation. | Mortel en cas d'ingestion. Provoque une irritation cutanée.

5 **Precautions:**
Wear protective gloves.
Wash hands thoroughly after handling.
Do not eat, drink or smoke when using this product.

Store locked up.
Dispose of contents/containers in accordance with local regulations.

IF ON SKIN: Wash with plenty of water.
If skin irritation occurs: Get medical advice or attention.
Take off contaminated clothing and wash it before reuse.
IF SWALLOWED: Immediately call a POISON CENTRE or doctor.
Rinse mouth.

Conseils :
Porter des gants de protection.
Se laver les mains soigneusement après manipulation.
Ne pas manger, boire ou fumer en manipulant ce produit.

Garder sous clef.
Éliminer le contenu/récipient conformément aux règlements locaux en vigueur.

EN CAS DE CONTACT AVEC LA PEAU : Laver abondamment à l'eau.
En cas d'irritation cutanée : Demander un avis médical/consulter un médecin.
Enlever les vêtements contaminés et les laver avant réutilisation.
EN CAS D'INGESTION : Appeler immédiatement un CENTRE ANTIPOISON ou un médecin.
Rincer la bouche.

6 ABC Chemical Co., 123 rue Anywhere St., Mytown, ON N0N 0N0 (123) 456-7890

Workplace Label*

1 **Product K1**

7 **Danger**

Fatal if swallowed. Causes skin irritation.

Wear protective gloves (neoprene). Wash hands thoroughly after handling. Do not eat, drink or smoke when using this product.

8 See SDS for more information.

*Requirements may vary – consult your local jurisdiction for their requirements.

CCOHS.ca 1-800-668-4284 WHMIS.org
Canadian Centre for Occupational Health and Safety

Figure 2.48 Canadian employers will most likely use either the supplier label or the workplace label to meet WHMIS requirements.

Table 2.7
Old WHMIS Symbols and Hazard Classes

Symbol	Hazard Class	Description
	Class A: **Compressed Gas**	Contents under high pressure; cylinder may explode or burst when heated, dropped, or damaged
	Class B: **Flammable and Combustible Material**	May catch fire when exposed to heat, spark, or flame; may burst into flames
	Class C: **Oxidizing Material**	May cause fire or explosion when in contact with wood, fuels, or other combustible material
	Class D, Division 1: **Poisonous and Infectious Material:** **Immediate and serious toxic effects**	Poisonous substance; a single exposure may be fatal or cause serious or permanent damage to health
	Class D, Division 2: **Poisonous and Infectious Material:** **Other toxic effects**	Poisonous substance; may cause irritation; repeated exposure may cause cancer, birth defects, or other permanent damage
	Class D, Division 3: **Poisonous and Infectious Material:** **Biohazardous infectious materials**	May cause disease or serious illness; drastic exposures may result in death
	Class E: **Corrosive Material**	Can cause burns to eyes, skin, or respiratory system
	Class F: **Dangerously Reactive Material**	May react violently, causing explosion, fire, or release of toxic gases when exposed to light, heat, vibration, or extreme temperatures

Source: WHMIS = Canadian Workplace Hazardous Materials Information System. Table adapted from Canadian Centre for Occupational Health and Safety (CCOHS) with pictograms from Health Canada.

Mexican Hazard Communication System

Mexico's equivalent to HCS is NOM-018-STPS-2000. It, too, requires employers to ensure that hazardous chemical substances in the workplace are appropriately and adequately labeled. Essentially, it adopts NFPA 704 and a related hazard communication label system as the official label and marking systems. However, employers can opt to use alternative systems so long as they comply with the objectives and purpose of the standard and are authorized by the Secretary of Labor and Social Welfare.

NOM-026-STPS-1998 ("Signs and Colors for Safety and Health") authorizes the use of some ISO safety symbols (ISO-3864, "Safety Colors and Safety Signs") on signs to communicate hazard information. General caution symbols in Mexico are triangular rather than round like those in Canada (WHMIS) or rectangular as typically found in the U.S **(Figure 2.49)**.

Figure 2.49 In Mexico, general caution symbols are triangular in shape. They are round in Canada (WHMIS) and typically rectangular in the United States.

CAS® Numbers

Chemical Abstract Service® (CAS®, a division of the American Chemical Society) registry numbers (often called **CAS® numbers**, CAS® #s, or CAS® RNs) are unique numerical identifiers assigned to individual chemicals and chemical compounds, polymers, mixtures, and alloys **(Figure 2.50)**. They may also be assigned to biological sequences. Over 100 million chemical substances and biological sequences have been registered. Most chemical databases are searchable by CAS® number, and they are typically included on safety data sheets (see Safety Data Sheets section) and other chemical reference materials such as the *NIOSH Pocket Guide*.

Military Markings

The U.S. and Canadian military services have their own marking systems for hazardous materials and chemicals in addition to DOT and TC transportation markings **(Figure 2.51)**. These markings are used on fixed facilities, and they may be seen on military vehicles, although they are not required. Exercise caution, however, because the military placard system is not necessarily uniform. For security reasons, some buildings and areas that store hazardous materials may not be marked. **Table 2.8, p. 94**, provides the U.S. and Canadian military markings for explosive ordnance and fire hazards, chemical hazards, and PPE requirements.

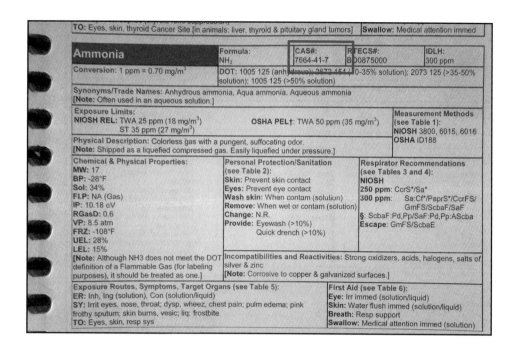

TO: Eyes, skin, thyroid Cancer Site [in animals: liver, thyroid & pituitary gland tumors]			Swallow: Medical attention immed	

Ammonia	Formula: NH₃	CAS#: 7664-41-7	RTECS#: BO0875000	IDLH: 300 ppm
Conversion: 1 ppm = 0.70 mg/m³	DOT: 1005 125 (anhydrous); 2672 154 (10-35% solution); 2073 125 (>35-50% solution); 1005 125 (>50% solution)			

Synonyms/Trade Names: Anhydrous ammonia, Aqua ammonia, Aqueous ammonia
[**Note:** Often used in an aqueous solution.]

Exposure Limits: **NIOSH REL:** TWA 25 ppm (18 mg/m³) ST 35 ppm (27 mg/m³)	**OSHA PEL†:** TWA 50 ppm (35 mg/m³)	Measurement Methods (see Table 1): NIOSH 3800, 6015, 6016 OSHA ID188

Physical Description: Colorless gas with a pungent, suffocating odor.
[**Note:** Shipped as a liquefied compressed gas. Easily liquefied under pressure.]

Chemical & Physical Properties: MW: 17 BP: -28°F Sol: 34% Fl.P: NA (Gas) IP: 10.18 eV RGasD: 0.6 VP: 8.5 atm FRZ: -108°F UEL: 28% LEL: 15% [**Note:** Although NH3 does not meet the DOT definition of a Flammable Gas (for labeling purposes), it should be treated as one.]	Personal Protection/Sanitation (see Table 2): **Skin:** Prevent skin contact **Eyes:** Prevent eye contact **Wash skin:** When contam (solution) **Remove:** When wet or contam (solution) **Change:** N.R. **Provide:** Eyewash (>10%) Quick drench (>10%) **Incompatibilities and Reactivities:** Strong oxidizers, acids, halogens, salts of silver & zinc [**Note:** Corrosive to copper & galvanized surfaces.]	Respirator Recommendations (see Tables 3 and 4): **NIOSH** **250 ppm:** CcrS*/Sa* **300 ppm:** Sa:Cf*/PaprS*/CcrFS/ GmFS/ScbaF/SaF **§:** ScbaF:Pd,Pp/SaF:Pd,Pp:AScba **Escape:** GmFS/ScbaE

Exposure Routes, Symptoms, Target Organs (see Table 5): **ER:** Inh, Ing (solution), Con (solution/liquid) **SY:** Irrit eyes, nose, throat; dysp, wheez, chest pain; pulm edema; pink frothy sputum; skin burns, vesic; liq: frostbite **TO:** Eyes, skin, resp sys	First Aid (see Table 6): **Eye:** Irr immed (solution/liquid) **Skin:** Water flush immed (solution/liquid) **Breath:** Resp support **Swallow:** Medical attention immed (solution)

Figure 2.50 The Chemical Abstract Service® (CAS®) registry number for ammonia is listed next to the chemical formula for ammonia. *Courtesy of Rich Mahaney.*

Figure 2.51 Unique military markings can be found on fixed facilities and vehicles, but the military placard system is not uniform. Secure locations that contain hazardous materials may not be marked. *Courtesy of Rich Mahaney.*

Pesticide Labels

The EPA regulates the manufacture and labeling of pesticides. In accordance with GHS, pesticide labels in the U.S. and Canada now include the following (Figure 2.52, p. 94).

- **EPA number** or **Canadian PCP number**
- **Hazard statement(s)** — Phrase assigned to each hazard category that describes the nature of the hazard. Examples of hazard statements are: "Harmful if swallowed," "Highly flammable liquid and vapor," and "Harmful to aquatic life." GHS hazard statements are based in part on current EPA requirements and are generally similar, but there are some differences.

Table 2.8
U.S. and Canadian Military Symbols

Symbol	Fire (Ordnance) Divisions
1	**Division 1: Mass Explosion** Fire Division 1 indicates the greatest hazard. This division is equivalent to DOT/UN Class 1.1 Explosives Division **Also, this exact symbol may be used for:** **Division 5: Mass Explosion — very insensitive explosives (blasting agents)** This division is equivalent to DOT/UN Class 1.5 Explosives Division
2	**Division 2: Explosion with Fragment Hazard** This division is equivalent to DOT/UN Class 1.2 Explosives Division **Also, this exact symbol may be used for:** **Division 6: Nonmass Explosion — extremely insensitive ammunition** This division is equivalent to DOT/UN Class 1.6 Explosives Division
3	**Division 3: Mass Fire** This division is equivalent to DOT/UN Class 1.3 Explosives Division
4	**Division 4: Moderate Fire — no blast** This division is equivalent to DOT/UN Class 1.4 Explosives Division

Symbol	Chemical Hazards
"Red You're Dead"	**Wear Full Protective Clothing (Set One)** Indicates the presence of highly toxic chemical agents that may cause death or serious damage to body functions.
"Yellow You're Mellow"	**Wear Full Protective Clothing (Set Two)** Indicates the presence of harassing agents (riot control agents and smokes).
"White is Bright"	**Wear Full Protective Clothing (Set Three)** Indicates the presence of white phosphorus and other spontaneously combustible material.

Continued

Table 2.8 (concluded)

Symbol	Chemical Hazards
	Wear Breathing Apparatus Indicates the presence of incendiary and readily flammable chemical agents that present an intense heat hazard. This hazard and sign may be present with any of the other fire or chemical hazards/symbols.
	Apply No Water Indicates a dangerous reaction will occur if water is used in an attempt to extinguish the fire. This symbol may be posted together with any of the other hazard symbols.

Symbol	Supplemental Chemical Hazards
	G-Type Nerve Agents — persistent and nonpersistent nerve agents *Examples: sarin (GB), tabun (GA), soman (GD)*
	VX Nerve Agents — persistent and nonpersistent V-nerve agents *Example: V-agents (VE, VG, VS)*
	Incapacitating Nerve Agent *Examples: lacrymatory agent (BBC), vomiting agent (DM)*
	H-Type Mustard Agent/Blister Agent *Example: persistent mustard/lewisite mixture (HL)*
	Lewisite Blister Agent *Examples: nonpersistent choking agent (PFIB), nonpersistent blood agent (SA)*

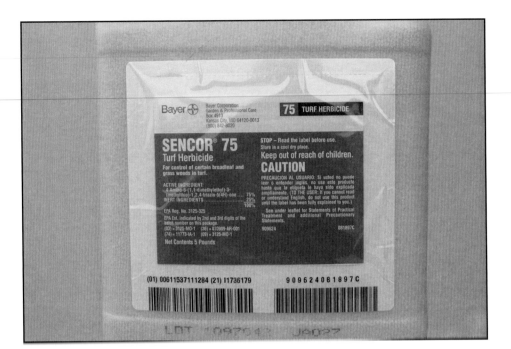

Figure 2.52 A container of turf herbicide is labelled according to EPA regulations.

- **Pictogram(s)** — A symbol inside a diamond with a red border, denoting a particular hazard class such as acute toxicity/lethality, and skin irritation/corrosion.

- **Precautionary statement(s)** — Phrases that describe recommended measures that should be taken to minimize or prevent adverse effects resulting from exposure to a hazardous product, or improper storage or handling of a hazardous product. These phrases cover prevention, response, storage, and disposal of products. GHS provides guidance on precautionary statements and includes a list of statements that may be used. These statements are similar to the precautionary statements that EPA currently uses. Work to increase standardization of precautionary statements may be undertaken in the future.

- **Product identifiers** — Names or numbers used on a hazardous product label or in a safety data sheet. They provide a unique means by which the product user can identify the chemical substance or mixture. Under the GHS, labels for substances should include the chemical identity of the substance. Labels for mixtures should include the identities of the ingredients that are responsible for certain hazards on the label, except that regulatory authorities may establish rules for protection of Confidential Business Information that preclude ingredient disclosure. (The hazard information would still appear on the label even if the ingredients are not named.) Current EPA requirements for product identifiers are consistent with GHS.

- **Signal word** — One word used to indicate the relative severity of hazard and alert the reader to a potential hazard on the label and safety data sheet. The GHS includes two signal words:
  — "Warning" for less severe hazard categories and;
  — "Danger" for more severe hazard categories.

- **Supplier identification** — Under the GHS supplier identification, it would include the name, address and telephone number of the manufacturer or supplier of the substance. Current EPA requirements for product identifiers are generally consistent with GHS. EPA encourages, but does not require, telephone contact numbers on pesticide labels.

NOTE: Lower categories of classification and unclassified products would not require pictograms or signal words under GHS. The current EPA system includes a third signal word "Caution" which is used in addition to "Warning" and "Danger."

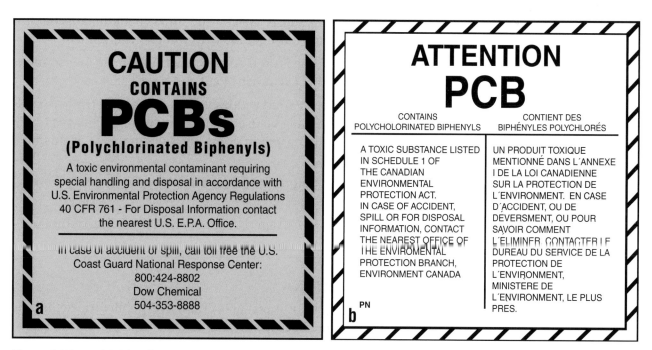

Other Symbols and Signs

Every facility may have its own system and its own symbols, signs, and markings. Responders should familiarize themselves with the signs and symbols used in their areas.

The EPA requires a warning label on any containers, transformers, or capacitors that contain polychlorinated biphenyl (PCB), which is considered hazardous because it may cause cancer. **Figure 2.53a** shows a typical U.S. PCB warning label, whereas **Figure 2.53b** shows a Canadian PCB warning label.

Figure 2.53a and b (a) U.S. PCB warning labels are found on containers, transformers, or capacitors that contain polychlorinated biphenyls. **(b)** This example is one of several different styles of Canadian PCB warning labels.

Table 2.9
Sample ISO-3864 Type Symbols*

Corrosive	Explosive	Flammable	Toxic/ Poisonous
Biological Hazard	Radiation	Oxidizer	Irritant

* ISO = International Organization for Standardization. This table is not comprehensive.

ISO Safety Symbols

The International Organization for Standardization (ISO) defines the design criteria for international safety signs in their standard, ISO-3864. These symbols are being used more frequently in the U.S. in conjunction with OSHA-required hazard signs (designed per ANSI Standard Z535.4, "Product Safety Signs and Labels") as well as in Mexico, so first responders should be able to recognize the more common symbols that are used to indicate hazardous materials **(Table 2.9)**.

Color Codes

Colors can sometimes provide clues to the nature of hazardous materials in North America. For example, even if a DOT placard is too far away to clearly read the number, a first responder can deduce that the material inside is some kind of oxidizer if the placard background color is yellow. If the placard color is red, the material is flammable. Preincident surveys can assist in identifying color systems used by local industries, for example, to identify materials used in piping systems.

ANSI Z535.1 sets forth the following safety color code that is recommended for use in the U.S. and Canada:

- **Red** — Means Danger or Stop; is used on containers of flammable liquids, emergency stop bars, stop buttons, and fire protection equipment
- **Orange** — Means Warning; is used on energized equipment or hazardous machinery with parts that can crush or cut
- **Yellow** — Means Caution; solid yellow, yellow and black stripes, or yellow and black checkers may be used to indicate physical hazards such as tripping hazards; also used on containers of corrosive or unstable materials
- **Green** — Marks safety equipment such as first-aid stations, safety showers, and exit routes
- **Blue** — Marks safety information signage such as labels or markings indicating the type of required personal protective equipment (PPE)

Written Resources

A variety of written resources are available to assist responders in identifying hazardous materials at both fixed facilities and transportation incidents. Fixed facilities should have safety data sheets, inventory records, and other facility documents in addition to signs, markings, container shapes, and other labels. At transportation incidents, first responders should be able to use the current *ERG* as well as shipping papers.

Shipping Papers

Shipments of hazardous materials must be accompanied by shipping papers that describe them. The information can be provided on a **bill of lading**, waybill, or similar document. The general location and type of paperwork change according to the mode of transport **(Table 2.10)**. However, the exact location of the documents varies. Hazardous waste shipments must be accompanied by a Uniform Hazardous Waste Manifest document that is typically attached to the shipping papers.

The Basic Description provided in shipping papers will follow a sequence best remembered by the acronym, ISHP:

- I = Identification Number
- S = Proper Shipping Name
- H = Hazard Class or Division
- P = Packing Group

When you know that a close approach to an incident is safe, you can then examine the cargo shipping papers. The information provided, such as proper shipping name and the hazard class, can then be used to identify hazards such as potential fire, explosion, and health hazards. Precautions to protect emergency responders and the public can be identified using the *Emergency Response Guidebook* (see Emergency Response Guidebook section) or other reference source.

You may need to check with the responsible party in order to locate shipping papers. If the responsible party is not carrying them, you will need to check the appropriate locations. In trucks and airplanes, these papers are placed near the driver or pilot. On ships and barges, the papers are placed on the bridge or in the pilothouse of a controlling tugboat.

The train crew should have train consists (entire train's cargo lists), train list, and/or wheel reports. Look for the train crew first, as they should have the paperwork (train list), however, if they cannot be located, contact the railroad through their emergency phone number for a copy of the train list. It is possible there may also be a copy of the current train list in the engine. On the train list, most railroad companies will count and list their train cars from the front of

Bill of Lading — Shipping paper used by the trucking industry (and others) indicating origin, destination, route, and product; placed in the cab of every truck tractor. This document establishes the terms of a contract between a shipper and a carrier. It serves as a document of title, contract of carriage, and receipt for goods.

Table 2.10 Shipping Paper Identification			
Transportation Mode	**Shipping Paper Name**	**Location of Papers**	**Party Responsible**
All	Air Bill	Cockpit	Pilot
Highway	Bill of Lading	Vehicle Cab	Driver
Rail	Trainlist/Consist	Engine (or Caboose)	Conductor
Water	Dangerous Cargo Manifest	Bridge or Pilot House	Captain or Master

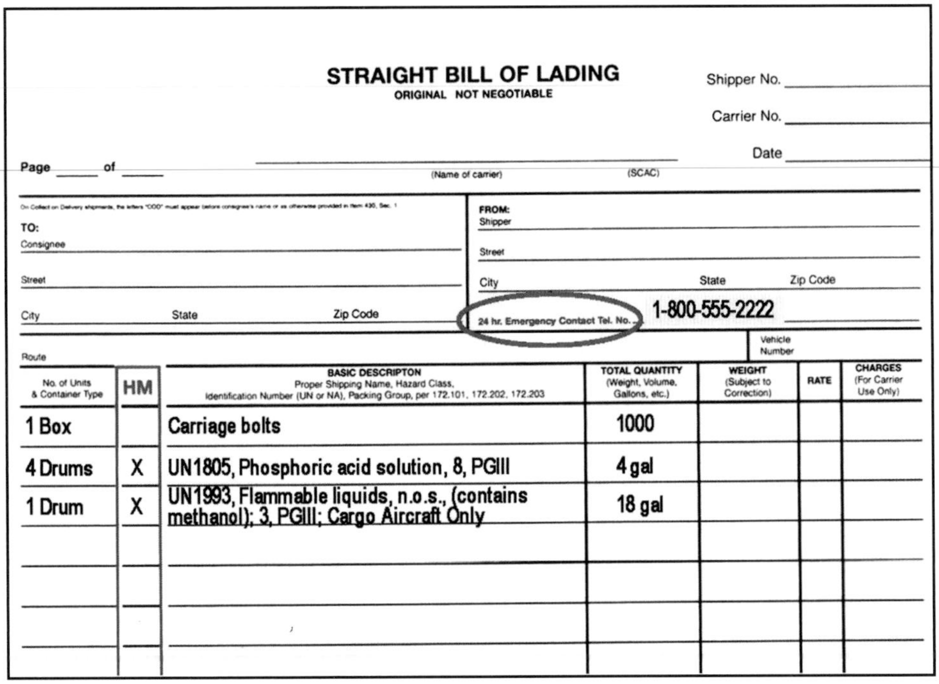

STRAIGHT BILL OF LADING
ORIGINAL NOT NEGOTIABLE

Shipper No. _____

Carrier No. _____

Date _____

Page _____ of _____

_____
(Name of carrier)   (SCAC)

On Collect on Delivery shipments, the letters "COD" must appear before consignee's name or as otherwise provided in Item 430, Sec. 1

TO:
Consignee _____

Street _____

City _____ State _____ Zip Code _____

FROM:
Shipper _____

Street _____

City _____ State _____ Zip Code _____

24 hr. Emergency Contact Tel. No. **1-800-555-2222**

Route _____

Vehicle Number _____

No. of Units & Container Type	HM	BASIC DESCRIPTON Proper Shipping Name, Hazard Class, Identification Number (UN or NA), Packing Group, per 172.101, 172.202, 172.203	TOTAL QUANTITY (Weight, Volume, Gallons, etc.)	WEIGHT (Subject to Correction)	RATE	CHARGES (For Carrier Use Only)
1 Box		Carriage bolts	1000			
4 Drums	X	UN1805, Phosphoric acid solution, 8, PGIII	4 gal			
1 Drum	X	UN1993, Flammable liquids, n.o.s., (contains methanol); 3, PGIII; Cargo Aircraft Only	18 gal			

Figure 2.54 Shipping paper requirements. Note the "X" placed in the column captioned "HM" for hazardous material. *Courtesy of PHMSA.*

the train to the back. During preincident surveys, the location of the papers (and how to read them) for a specific rail line can be determined. **Figure 2.54** provides a summary of shipping paper requirements.

NOTE: Shipping paper information may be provided in a variety of formats such as FAX and email.

Standard Transportation Commodity Code Numbers

Every railroad car is marked with an identifying mark that works as a serial number to identify the car independently of every other. This number on a railcar is known as the **reporting mark**. Identification indicates the car itself, the owner, and whether or not it is owned by a railroad.

Railroads and railroad paperwork also use Standard Transportation Commodity Code numbers (STCC numbers) to identify chemicals. These are a seven-digit number. If the seven-digit number starts with 48, it is a hazardous waste. If it starts with 49, it's a hazardous material. These numbers may also be found in some hazardous materials reference sources. Phone applications for smart phones are increasingly available. These apps may be helpful in finding the information for a specific car.

Transborder shipments between the U.S. and Mexico may be accompanied by shipping documents in both English and Spanish. To satisfy the emergency response information requirements in the U.S. or Mexico, a shipper may attach a copy of the appropriate guide page from the current *ERG* to the shipping papers. The information must be provided in Spanish when the material is shipped to Mexico and in English when shipped to the U.S. so that emergency responders in each country will be able to understand the appropriate initial response procedures in the event of a hazardous material release.

Safety Data Sheets

A safety data sheet (SDS) is a detailed information bulletin prepared by a chemical's manufacturer or importer that provides specific information about the product. SDSs are formatted according to Globally Harmonized System (GHS) specifications.

SDSs are often the best sources of detailed information about a particular material to which emergency responders have access. The sheets can be acquired from the manufacturer of the material, the supplier, the shipper, an emergency response center such as CHEMTREC®, or the facility hazard communication plan **(Figure 2.55)**. SDSs are sometimes attached to shipping papers and containers. SDS sheets are used worldwide.

Relevant sections of SDSs can be used to identify potential fire, explosion, and health hazards as well as precautions to be taken to protect responders and the public. Per OSHA, the list below describes SDS sections:

Section 1: Identification — This section identifies the chemical on the SDS as well as the recommended uses. It also provides the essential contact information of the supplier.

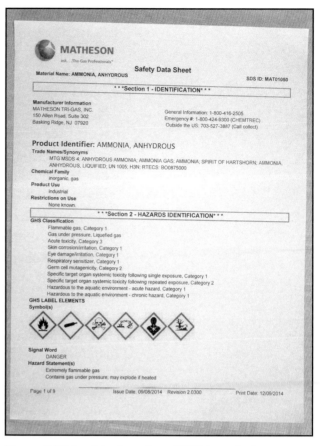

Figure 2.55 A safety data sheet (SDS) for anhydrous ammonia provides detailed information about the product, including that it is an extremely flammable gas that may explode if heated. *Courtesy of Rich Mahaney.*

Section 2: Hazard(s) identification — This section identifies the hazards of the chemical presented on the SDS and the appropriate warning information associated with those hazards.

Section 3: Composition / information on ingredients — This section identifies the ingredient(s) contained in the product indicated on the SDS, including impurities and stabilizing additives. This section includes information on substances, mixtures, and all chemicals where a trade secret is claimed.

Section 4: First aid measures — This section describes the initial care that should be given by untrained responders to an individual who has been exposed to the chemical.

Section 5: Fire fighting measures — This section provides recommendations for fighting a fire caused by the chemical.

Section 6: Accidental release measures — This section provides recommendations on the appropriate response to spills, leaks, or releases, including containment and cleanup practices to prevent or minimize exposure to people, properties, or the environment. It may also include recommendations distinguishing between responses for large and small spills where the spill volume has a significant impact on the hazard.

Section 7: Handling and storage — This section provides guidance on the safe handling practices and conditions for safe storage of chemicals.

Section 8: Exposure controls / personal protection — This section indicates the exposure limits, engineering controls, and personal protective measures that can be used to minimize worker exposure.

Section 9: Physical and chemical properties — This section identifies physical and chemical properties associated with the substance or mixture.

Section 10: Stability and reactivity — This section describes the reactivity hazards of the chemical and the chemical stability information. This section is broken into three parts: reactivity, chemical stability, and other.

Section 11: Toxicological information — This section identifies toxicological and health effects information or indicates that such data are not available.

Section 12: Ecological information — This section provides information to evaluate the environmental impact of the chemical(s) if it were released to the environment.

Section 13: Disposal considerations — This section provides guidance on proper disposal practices, recycling or reclamation of the chemical(s) or its container, and safe handling practices. To minimize exposure, refer to Section 8 (Exposure Controls/Personal Protection) of the SDS.

Section 14: Transport information — This section provides guidance on classification information for shipping and transporting of hazardous chemical(s) by road, air, rail, or sea.

Section 15: Regulatory information — This section identifies the safety, health, and environmental regulations specific for the product that is not indicated anywhere else on the SDS.

Section 16: Other information — This section indicates when the SDS was prepared or when the last known revision was made. The SDS may also state where the changes have been made to the previous version. Other useful information also may be included here.

Emergency Response Guidebook

The *Emergency Response Guidebook* (*ERG*) was developed to provide guidance to firefighters, law enforcement, and other emergency services personnel who may be the first to arrive at the scene of a transportation incident involving hazardous materials. The *ERG* will help you identify the material's specific or generic hazards, and it will also provide you with basic guidance on how to protect yourself and the general public during the incident's initial response phase. For information on how to use the *ERG*, see Chapter 3.

The *ERG* does not address all possible circumstances that may be associated with a hazardous materials incident. It is primarily designed for use at incidents occurring on a highway or railroad. There may be limited value in its application at fixed-facility locations.

NOTE: Operations level responders at the scene of a hazmat incident should seek additional, specific information about any material in question as soon as possible. The information received by contacting the appropriate emergency response agency, calling the emergency response number on the shipping document, or consulting the information on or accompanying the shipping document may be more specific and accurate than the guidebook in providing direction for managing the materials involved.

Facility Documents

The Hazard Communication Standard (HCS) requires U.S. employers to maintain Chemical Inventory Lists (CILs) of all their hazardous substances **(Figure 2.56)**. Because CILs usually contain information about the locations of materials within a facility, they can be useful tools in identifying containers that may have damaged or missing labels or markings (such as a label or marking made illegible because of fire damage). Several other documents and records may provide information about hazardous materials at a facility such as the following:

- Shipping and receiving documents
- Inventory records
- Risk management and hazardous communication plans
- Chemical inventory reports (known as Tier II reports)
- Facility Preplans

The **Local Emergency Planning Committee (LEPC)** is another potential source of information. LEPCs were designed to provide a forum for emergency management agencies, responders, industry, and the public to work together to evaluate, understand, and communicate chemical hazards in the community and develop appropriate emergency plans in case these chemicals are accidentally released. These plans are called **Local Emergency Response Plans (LERPs)**.

Local Emergency Planning Committee (LEPC) — Community organization responsible for local emergency response planning. Required by SARA Title III, LEPC's are composed of local officials, citizens, and industry representatives with the task of designing, reviewing, and updating a comprehensive emergency plan for an emergency planning district; plans may address hazardous materials inventories, hazardous material response training, and assessment of local response capabilities.

Local Emergency Response Plan (LERP) — Plan detailing how local emergency response agencies will respond to community emergencies; required by U.S. Environmental Protection Agency (EPA) and prepared by the Local Emergency Planning Committee (LEPC).

<table>
<tr><td colspan="16" align="right">Page 1 of 3</td></tr>
<tr>
<td colspan="4">Dept:
Invntry Supv:
Campus Addr:</td>
<td colspan="4" align="center">**OSU ENVIRONMENTAL HEALTH & SAFETY**
Hazard Communications
Chemical Inventory</td>
<td></td>
<td colspan="2">Phone #</td>
<td colspan="3">Building Name:
Building Number:
Date of Inventory:</td>
</tr>
<tr>
<td>Act
Count</td>
<td>Max
Amt</td>
<td>Chemical Name</td>
<td>Common Name</td>
<td colspan="3" align="center">Container</td>
<td>CAS Number</td>
<td>Manufacturer</td>
<td colspan="4" align="center">N.F.P.A. Rating</td>
<td>Location</td>
<td colspan="2">MSDS?</td>
</tr>
<tr>
<td></td><td></td><td></td><td></td><td>Size</td><td>Type</td><td>PS</td><td></td><td></td><td>H</td><td>F</td><td>R</td><td>S</td><td>Room #</td><td>Yes</td><td>No</td>
</tr>
</table>

Figure 2.56 If container markings have been damaged in a fixed-facility incident, the chemical inventory list (CIL) may help technicians identify the product.

Electronic Technical Resources

Technical resources and references have advanced with technology. Many common written resources and references are now available in an electronic format, for example, the *ERG*. Electronic resources have the added benefit of search features that allow information to be accessed in a more efficient manner than print resources. If the product has been identified, some of these resources can be used to determine precautions to be taken to protect responders and the public.

Many references can also be accessed on smartphones or mobile devices. While electronic resource use is increasing, it may still be necessary to have print resources available should there be issues in accessing the electronic data.

There are a variety of mobile weather applications (apps) that can be used. These provide up-to-date weather information on mobile devices.

Computer-Aided Management of Emergency Operations (CAMEO)

Computer-Aided Management of Emergency Operations (CAMEO) is a resource designed by the National Oceanic and Atmospheric Administration (NOAA). CAMEO is a system of software applications that helps emergency responders develop safe response plans. It can be used to access, store, and evaluate information critical in emergency response.

Wireless Information System for Emergency Responders (WISER)

The **Wireless Information System for Emergency Responders (WISER)** is an electronic resource that brings a wide range of information to the hazmat responder including:

- Chemical identification support
- Characteristics of chemicals and compounds
- Health hazard information
- Containment advice

WISER is available in different formats depending on the operating system. It may be downloaded free of charge.

911 Toolkit

Available only for Apple devices, the 911 Toolkit app provides a variety of information that may be useful for first responders. It provides information on hydraulics, water delivery, EMS, hazmat, NIMS/ICS. It also provides checklists and quizzes.

Hazmat IQ eCharts

Based on the popular HazMat IQ™ training course, the Hazmat IQ eCharts app provides the charts used to determine the appropriate response to chemical incidents. The app is available for Android and Apple devices.

Senses

Vision is definitely the safest of the five senses to use to detect a hazardous material. Hearing can also be used to detect information from a distance. While it may be perfectly safe to observe an overturned cargo tank from a distance through binoculars, emergency responders have to come into close or actual physical contact with a hazardous material (or its mists, vapors, dusts, or fumes, and the like) in order to smell, taste, or feel a release. While many products release odors well below dangerous levels, there is a good chance that if you are this close to a hazardous material, you are too close for safety's sake.

It should also be noted that many hazardous materials are invisible, have no odor, and cannot readily be detected by the senses. Hydrogen sulfide and certain other chemicals may cause **olfactory fatigue** (in other words, you may cease to smell it even though it is still present). However, any smells, tastes, or symptoms reported by victims and witnesses may prove to be helpful. Warning properties of chemicals include visible gas clouds, pungent odors, and irritating fumes.

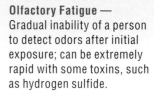

Olfactory Fatigue — Gradual inability of a person to detect odors after initial exposure; can be extremely rapid with some toxins, such as hydrogen sulfide.

You should be aware of visual/physical chemical indicators that provide tangible evidence that hazardous materials are present. Unusual noises (such as the hiss of a gas escaping a valve at high pressure) may also alert you to the presence of hazards. Some hazardous materials have odorants added to them to aid in detection; for example, the distinct odor normally associated with natural gas (an odorless gas) is actually caused by mercaptan, an additive.

Direct, visible evidence that physical and/or chemical actions and reactions are taking place include the following:

- Spreading vapor cloud or smoke **(Figure 2.57)**
- Unusual colored smoke
- Flames
- PPE fails
- Dying or discolored vegetation
- Container deterioration
- Containers bulging
- Sick humans
- Dead or dying birds, animals, insects, or fish
- Discoloration of valves or piping

Physical actions are processes that do not change the elemental composition of the materials involved. Several indications of a physical action are as follows:

- Rainbow sheen on water surfaces **(Figure 2.58, p. 106)**
- Wavy vapors over a volatile liquid
- Frost or ice buildup near a leak
- Containers deformed by the force of an accident
- Activated pressure-relief devices
- Pinging or popping of heat- or cold-exposed vessels

Chemical reactions convert one substance to another. Visual and sensory evidence of chemical reactions include the following:

- Heat
- Unusual or unexpected temperature drop (cold)

Figure 2.57 Spreading smoke or a vapor cloud is a visual indicator that a chemical reaction is taking place.

Figure 2.58 A rainbow sheen on water surfaces is a good indication that a hazardous material is present. *Courtesy of FEMA News Photos, photo by Liz Roll.*

- Extraordinary fire conditions
- Peeling or discoloration of a container's finish
- Spattering or boiling of unheated materials
- Distinctively colored vapor clouds
- Smoking or self-igniting materials
- Unexpected deterioration of equipment
- Peculiar smells
- Unexplained changes in ordinary materials
- Symptoms of chemical exposure

Physiological signs and symptoms of chemical exposure may also indicate the presence of hazardous materials. Symptoms can occur separately or in clusters, depending on the chemical. Symptoms of chemical exposure include the following **(Figure 2.59)**:

- **Changes in respiration** — Difficult breathing, increase or decrease in respiration rate, tightness of the chest, irritation of the nose and throat, and/or respiratory arrest

- **Changes in level of consciousness** — Dizziness, lightheadedness, drowsiness, confusion, fainting, and/or unconsciousness

- **Abdominal distress** — Nausea, vomiting, and/or cramping

- **Change in activity level** — Fatigue, weakness, stupor, hyperactivity, restlessness, anxiety, giddiness, and/or faulty judgment

- **Visual disturbances** — Double vision, blurred vision, cloudy vision, burning of the eyes, and/or dilated or constricted pupils

- **Skin changes** — Burning sensations, reddening, paleness, fever, and/or chills, itchiness, blisters

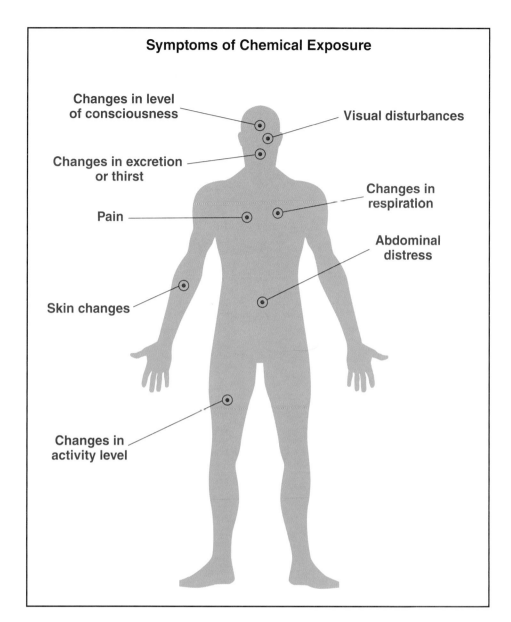

Symptoms of Chemical Exposure

Changes in level of consciousness

Visual disturbances

Changes in excretion or thirst

Changes in respiration

Pain

Abdominal distress

Skin changes

Changes in activity level

Figure 2.59 There are many symptons of chemical exposure.

- **Changes in excretion or thirst** — Uncontrolled tears, profuse sweating, mucus flowing from the nose, diarrhea, frequent urination, bloody stool, and/or intense thirst

- **Pain** — Headache, muscle ache, stomachache, chest pain, and/or localized pain at sites of substance contact

Monitoring and Detection Devices

Monitoring and detection devices can be useful in determining the presence of hazardous materials as well as the concentration(s) present. They can also be used to determine the scope of the incident. As with the senses, effectively using the monitoring and detection devices requires actual contact with the hazardous material (or its mists, dusts, vapors, or fumes) in order to measure it, and it is therefore outside the scope of action for Awareness level personnel. Chapter 11, Mission Specific Detection, Monitoring, and Sampling, addresses monitoring and detection devices for Operations level responders.

Chapter Review

Answer the following questions to review the information provided in this chapter.

1. List the seven clues to the presence of hazardous materials.

2. What types of occupancies are most likely to have hazardous materials?

3. How can you differentiate pressure, cryogenic, liquid, and solids containers from a distance, and what types of hazardous materials are each likely to contain?

4. List the five basic types of containers used to transport radioactive materials and briefly explain what these packages are designed to withstand.

5. What types of hazardous materials are transported in pipelines?

6. What are unit loading devices and can hazardous materials be shipped in them?

7. What are the nine hazard classes used by the U.N. to categorize hazardous materials?

8. How does the U.S. transportation system of placards, labels, and markings indicate the hazards posed by the hazardous materials carried?

9. How do Canadian placards, markings, and labels differ from the U.S. system?

10. How do Mexican placards, markings, and labels differ from the U.S. system?

11. Where might you find other types of markings, marking systems, labels, labeling systems, colors, color-codes, and signs that indicate the presence of hazardous materials?

12. Describe the colors and number system used in NFPA 704, *Standard System for the Identification of the Hazards of Materials for Emergency Response*.

13. What are they key elements of the Globally Harmonized System?

14. How do pesticide labels differ from other hazardous materials labeling systems?

15. Where might you find other symbols and signs for hazardous materials not covered above?

16. Describe common symbols and colors that indicate the presence of hazardous materials as set forth by ANSI Standard Z535.4.

17. What written resources are first responders most likely to utilize, and why?

18. How can using the senses for hazardous materials identification be dangerous?

19. Why will Awareness Level personnel not use monitoring and detection devices?

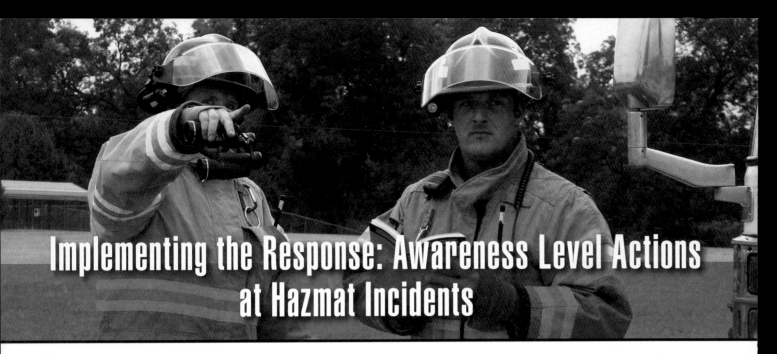

Implementing the Response: Awareness Level Actions at Hazmat Incidents

Chapter Contents

Key Terms

NFPA Job Performance Requirements

This chapter provides information that addresses the following job performance requirements of NFPA 1072, *Standard for Hazardous Materials/Weapons of Mass Destruction Emergency Response Personnel Professional Qualifications (2017).*

4.2.1

4.3.1

4.4.1

Implementing the Response: Awareness Level Actions at Hazmat Incidents

Learning Objectives

After reading this chapter, students will be able to:

1. Recognize notification procedures. (4.4.1)

2. Describe ways first responders use the Emergency Response Guidebook at hazardous materials incidents. (4.2.1, 4.3.1)

3. Explain the role of first responders in initiating protective actions. (4.3.1)

4. Identify actions that Awareness level personnel should take when responding to terrorist incidents. (4.2.1, 4.3.1)

5. Skill Sheet 3-1: Make appropriate notifications of a hazardous materials incident. (4.4.1)

6. Skill Sheet 3-2: Identify indicators and hazards present at a hazardous materials incident using approved reference sources. (4.2.1)

7. Skill Sheet 3-3: Implement protective actions at a hazardous materials incident. (4.2.1, 4.3.1)

Chapter 3
Implementing the Response: Awareness Level Actions at Hazmat Incidents

This chapter explains the roles and responsibilities of Awareness level personnel at emergency incidents. Awareness level personnel must:

- Understand the roles in the notification process

- Have the skills to use the *Emergency Response Guidebook*

- Provide basic procedures to isolate the incident

- Know the steps to take in case an accident involves terrorism or criminal activity

Notification

Predetermined procedures, such as standard operating procedures (SOPs) and the emergency response plan, should define roles in the notification process (see Chapter 6, Identifying Action Options) and methods for communication. For Awareness level personnel, notification may be as simple as dialing 9-1-1 to report an incident and request emergency assistance **(Figure 3.1)**. Fixed-facility responders may have their own internal procedures to follow such as calling on the radio for an internal fire brigade

Figure 3.1 The actions of Awareness level personnel at hazmat incidents can be beneficial. Calling for help quickly and isolating the area can save lives. *Courtesy of Rich Mahaney.*

or hazmat response team. If criminal or terrorist activity is suspected, personnel should notify law enforcement immediately. Skills associated with making appropriate notifications of a hazardous materials incident are shown in **Skill Sheet 3-1**.

NOTE: Additional information about notification requirements for Operations level responders is covered in Chapter 7, Incident Management.

Departmental SOPs usually cover communication methods (both externally and internally) at incidents, whether by radio, cell phone, hand signals, or other methods. Personnel and responders must be able to communicate the need for assistance through their department/organization's communications equipment. Some of these communications might be requests for additional personnel or special equipment, or to notify others at the incident of any apparent hazards. Personnel and responders must be trained to use the communication equipment assigned to them in accordance with policies and procedures.

NOTE: All of the actions described in this chapter are also applicable to Operations level responders.

Using the Emergency Response Guidebook

The *Emergency Response Guidebook* (*ERG*) is a guide to aid emergency responders in quickly identifying the initial hazmat hazards involved in a transportation emergency incident. This allows you to help protect yourself and others during the incident's initial response phase by avoiding and minimizing hazards.

The *ERG* is primarily designed to be used at hazardous materials incidents occurring on a highway or railroad. Isolation and protective distances in the *ERG* are based on conditions commonly associated with transportation incidents in open areas and may not be useful when applied to fixed-facility locations or in urban settings.

NOTE: The *ERG* does **NOT** address all possible circumstances that may be associated with a dangerous goods/hazardous materials incident.

You can locate the appropriate initial action guide page in the *ERG* in several different ways:

- Identify the four-digit U.N. identification number on a placard or shipping papers and then look up the appropriate guide in the yellow-bordered pages.

- Reference the name of the material involved (if known) in the blue-bordered pages. Many chemical names differ only by a few letters, so exact spelling

- Identify the material's transportation placard and then reference the three-digit guide code associated with the placard in the *Table of Placards and Initial Response Guide to Use On-Scene* located in the front of the *ERG*.

- Reference the container profiles provided in the white pages in the front of the book. First responders can identify container shapes, and then cross-reference the guide number to the orange-bordered page provided in the nearest circle **(Figure 3.3)**.

Using the four-digit ID number or the chemical name allows responders to locate the most specific initial action guide. Skills associated with identifying indicators and hazards present at a hazardous materials incident using approved reference sources are shown in **Skill Sheet 3-2**. The sections that follow describe the *ERG*'s design and layout.

Figure 3.2 The exact spelling of a chemical name is important when using the *Emergency Response Guidebook*. Misspellings can cause inappropriate and potentially dangerous actions to be taken. *Courtesy of Rich Mahaney.*

Figure 3.3 As a last resort, if placards or 4-digit ID numbers are not visible, first responders can use container profiles to identify the proper *ERG* page. *Courtesy of Rich Mahaney.*

Multiple Information Sources

At an incident, first responders should always seek additional specific information about any material in question as soon as possible. Do not rely on the *ERG* alone. Instead, performing any of the following actions may provide information about the materials involved which may be more specific or more accurate than the guidebook:

- Contacting the appropriate emergency response agency
- Calling the emergency response number on the shipping document
- Consulting the accompanying shipping document

When consulting chemical reference sources for information about a particular substance, consult more than one reference source to ensure information is complete and accurate. Reference books may leave out many chemicals if they were written for a specific purpose, such as compiling information about the most dangerous workplace chemicals. Absence from one reference book does not mean that the substance is safe. Check multiple sources.

ERG Instructions (White Pages)

The white pages provide instructions for using the *ERG*. There are two white-page sections, one in front and one in back.

The front section provides information on the following:

- Shipping documents (papers)
- How to use the guidebook
- Local emergency telephone numbers
- Safety precautions (see Safety Box)
- Notification and request for technical information
- Hazard classification system
- Introduction to the Table of Markings, Labels, and Placards
- Table of Markings, Labels, and Placards and Initial Response Guide to use on-scene
- Rail Car Identification Chart
- Road Trailer Identification Chart
- Globally Harmonized System of Classification and Labeling of Chemicals (GHS)
- Hazard information numbers displayed on some intermodal containers
- Pipeline transportation

The back section provides information about:

- *ERG* User's Guide
- Protective clothing
- Fire and spill control
- BLEVE safety precautions
- Criminal/terrorist use of chemical/biological/radiological agents
- Improvised Explosive Device (IED) safe standoff distances
- Glossary

- Canada and United States National Response Centers
- Emergency Response Assistance Plans (ERAP)
- Emergency response telephone numbers

ERG Safety Precautions

RESIST RUSHING IN!

APPROACH CAUTIOUSLY FROM *UPWIND, UPHILL* OR *UPSTREAM*:

- Stay clear of *Vapor, Fumes, Smoke* and *Spills*
- Keep vehicle at a safe distance from the scene

SECURE THE SCENE:

- Isolate the area and protect yourself and others

IDENTIFY THE HAZARDS USING ANY OF THE FOLLOWING:

- Placards
- Container labels
- Shipping documents
- Rail Car and Road Trailer Identification Chart
- Safety Data Sheets (SDS)
- Knowledge of persons on scene
- Consult applicable guide page

ASSESS THE SITUATION:

- Is there a fire, a spill or a leak?
- What are the weather conditions?
- What is the terrain like?
- Who/what is at risk: people, property or the environment?
- What actions should be taken – evacuation, shelter in-place or dike?
- What resources (human and equipment) are required?
- What can be done immediately?

OBTAIN HELP:

- Advise your headquarters to notify responsible agencies and call for assistance from qualified personnel

RESPOND:

- Enter only when wearing appropriate protective gear
- Rescue attempts and protecting property must be weighed against you becoming part of the problem
- Establish a command post and lines of communication
- Continually reassess the situation and modify response accordingly
- Consider safety of people in the immediate area first, including your own safety

ABOVE ALL: Do not assume that gases or vapors are harmless because of lack of a smell—odorless gases or vapors may be harmful. Some products cannot be detected using the five human senses. Use **CAUTION** when handling empty containers because they may still present hazards until they are cleaned and purged of all residues.

ERG ID Number Index (Yellow-Bordered Pages)

The *ERG*'s yellow-bordered pages provide a four-digit UN/NA ID number index list in numerical order. The four-digit ID number is followed by its assigned three-digit Emergency Response Guide number (referred to hereafter as the "Guide," orange-bordered pages) and the material's name **(Figure 3.4)**.

The yellow-bordered section in the *ERG* enables first responders to identify the Guide number to consult for the substance involved. Green highlighting in the yellow-bordered index indicates that the substance releases gases that are **toxic inhalation hazard (TIH)**. These materials require enhanced emergency response distances. A "P" following the Guide number indicates that a material polymerizes. **Polymerization** is a violent reaction that releases great amounts of heat and energy.

Toxic Inhalation Hazard (TIH) — Volatile liquid or gas known to be a severe hazard to human health during transportation.

Polymerization — Chemical reactions in which two or more molecules chemically combine to form larger molecules; this reaction can often be violent.

Figure 3.4 The yellow-bordered guide pages provide a four-digit UN/NA ID number for hazardous materials in numerical order.

ERG Material Name Index (Blue-Bordered Pages)

The blue-bordered pages of the *ERG* provide an index of dangerous goods in alphabetical order by material name so that the first responder can quickly identify the Guide to consult for the name of the material involved. This list displays the material's name followed by its assigned three-digit emergency response Guide and four-digit UN/NA ID number **(Figure 3.5)**. The user needs to be very careful when looking up a product's name because a subtle mistake in the spelling can lead a responder toward a substantial misunderstanding of a product's behavior. As in the yellow-bordered pages, green highlighting on substances listed in the blue-bordered pages indicates the release of TIH gases, and a "P" following the Guide number indicates that the material polymerizes.

NOTE: Many agencies use a phonetic alphabet to spell chemical names.

ERG Initial Action Guides (Orange-Bordered Pages)

The book's orange-bordered section is the most useful because it provides safety recommendations and general hazards information. The orange-bordered pages comprised of individual Guides presented in a two-page format, featuring three sections **(Figure 3.6)**. The left-hand page lists potential hazards and public safety information. The right-hand page provides emergency response information. Each Guide is designed to cover a group of materials that possess similar chemical and toxicological characteristics. The Guide title identifies the general hazards of the materials or dangerous goods addressed.

Potential Hazards Section

The Potential Hazards section addresses two hazard types under separate headers: Health hazards and Fire or Explosion hazards **(Figure 3.7, p. 120)**. The highest potential hazard is listed first.

This section should be consulted first because it will assist in making decisions regarding the protection of individuals at the incident. Types of warnings found in this section include: TOXIC, HIGHLY FLAMMABLE, and CORROSIVE.

Figure 3.5 The blue-bordered pages of the *ERG* provide an index of hazardous materials in alphabetical order.

Figure 3.6 The orange-bordered guide pages provide safety recommendations, general hazard information, and basic emergency response actions.

Initial Isolation Distance — Distance within which all persons are considered for evacuation in all directions from a hazardous materials incident.

Initial Isolation Zone — Circular zone, with a radius equivalent to the initial isolation distance, within which persons may be exposed to dangerous concentrations upwind of the source and may be exposed to life-threatening concentrations downwind of the source.

Street Clothes — Clothing that is anything other than chemical protective clothing or structural firefighters' protective clothing, including work uniforms and ordinary civilian clothing.

Structural Firefighters' Protective Clothing — General term for the equipment worn by fire and emergency services responders; includes helmets, coats, pants, boots, eye protection, gloves, protective hoods, self-contained breathing apparatus (SCBA), and personal alert safety system (PASS) devices.

Self-Contained Breathing Apparatus (SCBA) — Respirator worn by the user that supplies a breathable atmosphere that is either carried in or generated by the apparatus and is independent of the ambient atmosphere. Respiratory protection is worn in all atmospheres that are considered to be Immediately Dangerous to Life and Health (IDLH). *Also known as* Air Mask or Air Pack.

Chemical Protective Clothing (CPC) — Clothing designed to shield or isolate individuals from the chemical, physical, and biological hazards that may be encountered during operations involving hazardous materials.

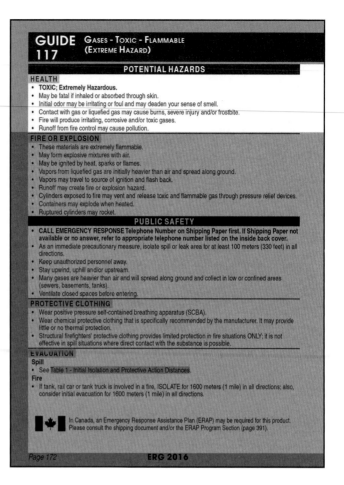

Figure 3.7 The *Potential Hazards* section describes health hazards, with the highest potential hazard listed first.

Public Safety Section

The Public Safety section provides general information regarding immediate isolation of the incident site and protective clothing and respiratory protection recommendations. This section also lists suggested evacuation distances for small and large spills and for fire situations such as distances for fragmentation hazards for tanks that might explode.

Isolation distances are provided in the bullet points immediately below the Public Safety section heading **(Figure 3.8)**. The **initial isolation distance** is a distance within which all persons should be considered for evacuation in all directions from the hazmat spill or leak source **(Figure 3.9)**. This distance can be used to identify and establish an **initial isolation zone**. If safe to do so, you should evacuate people from the initial isolation zone to outside the safe distance (at a minimum) **(Figure 3.10)**. You should then secure the scene and deny entry/access to the isolation zone from anyone on the outside.

Protective Clothing Section. This section describes the type of personal protective clothing and equipment that should be worn at incidents involving these products **(Figure 3.11)**. Examples include the following:

- **Street clothing** and work uniforms
- **Structural firefighters' protective clothing** (also called bunker gear or turnouts)
- Positive pressure **self-contained breathing apparatus** (SCBA)
- **Chemical protective clothing** (CPC)

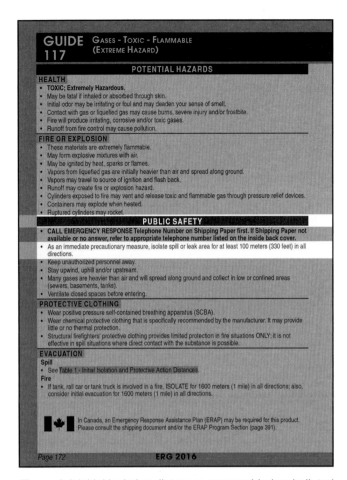

Figure 3.8 Initial isolation distances are provided as bulleted points in the *Public Safety* section.

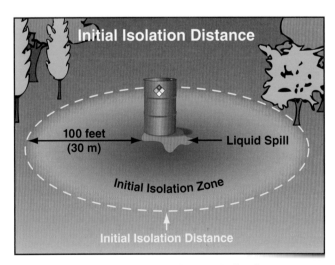

Figure 3.9 The initial isolation distance is the distance within which all persons should be considered for evacuation in all directions.

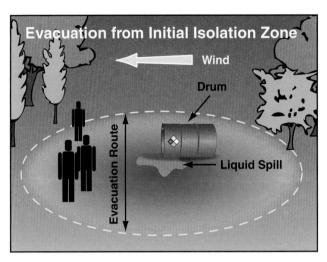

Figure 3.10 When downwind from the incident, evacuation from the initial isolation zone should be conducted at right angles to the prevailing wind direction, if possible.

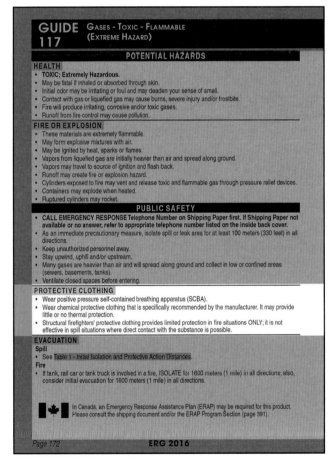

Figure 3.11 The *Protective Clothing* section recommends the type of personal protective clothing and equipment that should be worn.

NOTE: More information about protective clothing is provided in Chapter 9, Mission Specific Personal Protective Equipment.

Evacuation Section. This section provides **evacuation** recommendations for spills and fires **(Figure 3.12)**. When the material is a green-highlighted chemical in the yellow-bordered and blue-bordered pages, this section also directs the reader to consult the tables on the green-bordered pages listing TIH materials and water-reactive materials (see *ERG Table of Initial Isolation and Protective Action Distances* [Green Pages] section). Awareness level personnel will probably not be involved in evacuations beyond the initial isolation phase.

NOTE: Evacuation, sheltering in place, and protecting/defending in place are explained in greater detail in Chapter 7, Incident Management.

Figure 3.12 Evacuation distances for known spills or fires are provided in the Evacuation section. These distances may differ from the ones provided in the *Public Safety* section. In this case, when a spill is involved, the user is referred to the *Table of Initial Isolation and Protective Action Distances* (green-bordered pages) for more information.

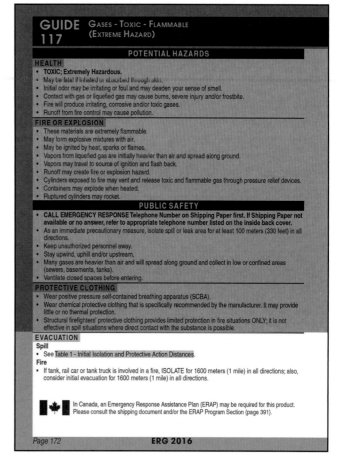

Emergency Response Section

The third section, Emergency Response, describes emergency response topics, including precautions for incidents involving fire, spills or leaks, and first aid. Several recommendations are listed under each section to further assist in the decision-making process. The first aid information provides general guidance before seeking medical care. Awareness level responders need to be aware of the information in this section, but may not be qualified to mitigate the incident. This section is described in more detail for Ops level responders in Chapter 6, Identifying Response Options.

Fire Section. This section recommends the extinguishing agent to use on large fires, small fires, and fires involving bulk containers **(Figure 3.13)**. Examples might include foam or water, or a specific type of fire extinguisher for small fires. If foam is recommended, it will specify the type of foam to be used. Recommendations vary by Guide, but may include such things as cooling containers with flooding quantities of water or using unmanned hose holders.

Spill or Leak Section. This section provides actions to take in regards to spills and leaks **(Figure 3.14)**. If a flammable liquid is involved, for example, it would recommend eliminating all ignition sources. It will also provide basic information needed to mitigate a spill, such as what materials to use to absorb the spill.

CAUTION

Before attempting to conduct *ERG*-recommended actions, you must be properly trained and have the correct equipment.

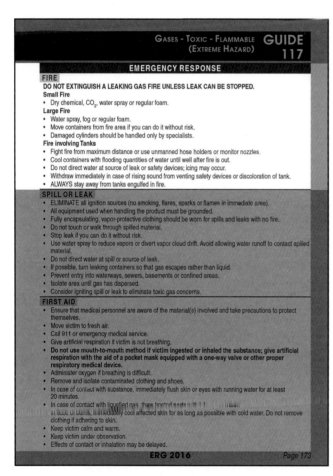

Figure 3.13 The *Fire* section provides information for firefighters, including appropriate extinguishing agents, types of foam to use, and actions to take or avoid.

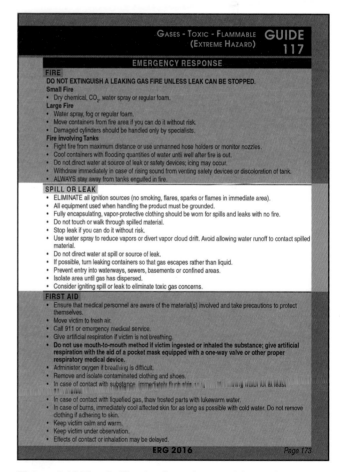

Figure 3.14 The *Spill or Leak* section provides actions to take to mitigate a spill, such as what materials to use to absorb the material.

First Aid Section. This section provides basic steps to help victims affected by the hazardous material involved **(Figure 3.15)**. Common recommendations include calling for emergency medical service assistance, moving victims to fresh air, and flushing contaminated skin and eyes with running water **(decontamination)**. Avoiding direct contact with the hazardous material is also emphasized.

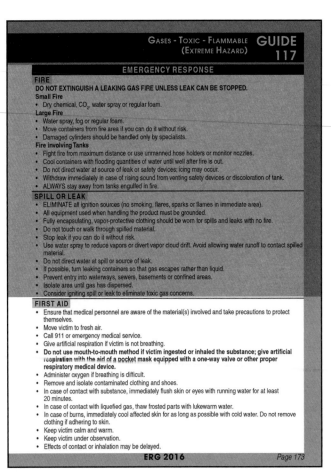

Decontamination — Process of removing a hazardous foreign substance from a person, clothing, or area. *Also known as* Decon.

Cross Contamination — Contamination of people, equipment, or the environment outside the hot zone without contacting the primary source of contamination. *Also known as* Secondary Contamination.

Many recommendations provided in this section will be beyond the scope of Awareness level personnel due to the need for specialized training and personal protective equipment, the dangers of **cross contamination**, and the necessity of decontaminating victims before first aid is provided (see Chapter 10, Decontamination, and Chapter 12, Mission Specific Victim Rescue and Recovery). For example, Awareness level personnel should never enter hazardous atmospheres or potentially contaminated areas.

Figure 3.15 The *First Aid* section provides basic steps to help victims, such as calling for medical assistance, moving them to fresh air, and flushing contaminated skin and eyes with running water.

Victims at hazmat incidents may present serious hazards to rescuers because they may be contaminated with the hazardous material **(Figure 3.16)**. Only first responders with appropriate training and wearing the appropriate personal protective clothing and equipment should touch or handle these victims. Awareness level personnel should not handle or touch contaminated or potentially contaminated victims at hazmat incidents, even to provide basic first aid.

WARNING!
Awareness level personnel should not handle or touch contaminated or potentially contaminated victims!

Figure 3.16 Because individuals may be contaminated at a hazmat incident, only responders with appropriate training and wearing the appropriate personal protective clothing and equipment should handle these victims.

ERG Tables of Initial Isolation and Protective Action Distances (Green-Bordered Pages)

The green-bordered section consists of three tables:

- Table 1, Initial Isolation and Protective Action Distances
- Table 2, Water Reactive Materials which Produce Toxic Gases
- Table 3, Initial Isolation and Protective Action Distances for Different Quantities of Six Common TIH (PIH in the US) Gases

Table 1, Initial Isolation and Protective Action Distances

Table 1 lists TIH materials by their 4-digit UN/NA ID number. The table provides two different types of recommended safe distances: initial isolation distances and **protective action distances (Figure 3.17)**. These materials are highlighted for easy identification in both numeric (yellow-bordered) and alphabetic (blue-bordered) *ERG* indexes **(Figure 3.18, p. 126)**.

Protective Action Distance — Downwind distance from a hazardous materials incident within which protective actions should be implemented.

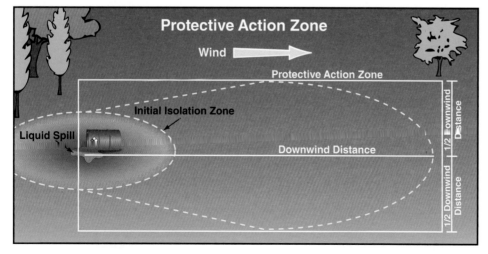

Figure 3.17 The *protective action distance* is the downwind distance from a hazardous materials incident within which protective actions should be implemented.

The table provides isolation and protective action distances for both small and large spills **(Figure 3.19)**. A small spill (approximately 55 gallons [220 L] or less) involves a single, small package, small cylinder, or small leak from a large package. A large spill (more than 55 gallons [220 L]) is one that involves a spill from a large package or multiple spills from many small packages. The list is further subdivided into daytime and nighttime situations because atmospheric conditions are often different depending on the time of day. Atmospheric conditions can significantly affect the size of a chemically hazardous area.

The warmer, more active atmosphere common during the day disperses chemical contaminants more readily than the cooler, calmer conditions common at night. Therefore, during the day, lower toxic concentrations may be spread over a larger area than at night when higher concentrations may exist in a smaller area. The quantity of material spilled or released and the area affected are both important, but the single most critical factor is the concentration of the contaminant in the air.

As with the isolation distances provided in the orange-bordered pages, the initial isolation distances provided in the green-bordered pages are the distance within which all persons should be considered for evacuation in all directions from an actual hazardous materials spill/leak source. This distance will always be at least 100 feet (30 m).

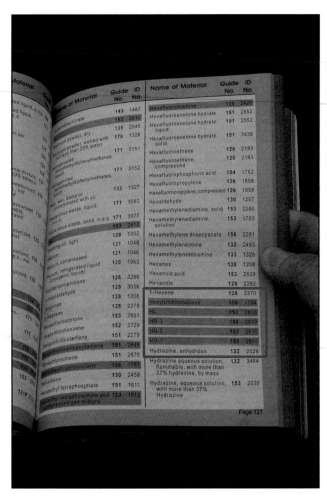

Figure 3.18 Both the numeric (yellow-bordered) and alphabetic (blue-bordered) *ERG* indexes can help make identification of Toxic Inhalation Hazard materials easier.

Figure 3.19 The *Table of Initial Isolation and Protective Action Distances* provides several variables, including small and large spills at night and during the day.

			SMALL SPILLS (From a small package or small leak from a large package)			LARGE SPILLS (From a large package or from many small packages)		
			First ISOLATE in all Directions	Then PROTECT persons Downwind during		First ISOLATE in all Directions	Then PROTECT persons Downwind during	
ID No.	Guide	NAME OF MATERIAL	Meters (Feet)	DAY Kilometers (Miles)	NIGHT Kilometers (Miles)	Meters (Feet)	DAY Kilometers (Miles)	NIGHT Kilometers (Miles)
1005 1005	125 125	Ammonia, anhydrous Anhydrous ammonia	30 m (100 ft)	0.1 km (0.1 mi)	0.2 km (0.1 mi)	Refer to table 3		
1008 1008	125 125	Boron trifluoride Boron trifluoride, compressed	30 m (100 ft)	0.1 km (0.1 mi)	0.7 km (0.4 mi)	400 m (1250 ft)	2.2 km (1.4 mi)	4.8 km (3.0 mi)
1016 1016	119 119	Carbon monoxide Carbon monoxide, compressed	30 m (100 ft)	0.1 km (0.1 mi)	0.2 km (0.1 mi)	200 m (600 ft)	1.2 km (0.7 mi)	4.4 km (2.8 mi)
1017	124	Chlorine	60 m (200 ft)	0.3 km (0.2 mi)	1.1 km (0.7 mi)	Refer to table 3		
1026	119	Cyanogen	30 m (100 ft)	0.1 km (0.1 mi)	0.4 km (0.3 mi)	60 m (200 ft)	0.3 km (0.2 mi)	1.1 km (0.7 mi)
1040 1040	119P 119P	Ethylene oxide Ethylene oxide with Nitrogen	30 m (100 ft)	0.1 km (0.1 mi)	0.2 km (0.1 mi)	Refer to table 3		
1045 1045	124 124	Fluorine Fluorine, compressed	30 m (100 ft)	0.1 km (0.1 mi)	0.2 km (0.1 mi)	100 m (300 ft)	0.5 km (0.3 mi)	2.2 km (1.4 mi)
1048	125	Hydrogen bromide, anhydrous	30 m (100 ft)	0.1 km (0.1 mi)	0.2 km (0.2 mi)	150 m (500 ft)	0.9 km (0.6 mi)	2.6 km (1.6 mi)
1050	125	Hydrogen chloride, anhydrous	30 m (100 ft)	0.1 km (0.1 mi)	0.3 km (0.2 mi)	Refer to table 3		
1051	117	AC (when used as a weapon)	60 m (200 ft)	0.3 km (0.2 mi)	1.0 km (0.6 mi)	1000 m (3000 ft)	3.7 km (2.3 mi)	8.4 km (5.3 mi)
1051 1051 1051	117 117 117	Hydrocyanic acid, aqueous solutions, with more than 20% Hydrogen cyanide Hydrogen cyanide, anhydrous, stabilized Hydrogen cyanide, stabilized	60 m (200 ft)	0.2 km (0.2 mi)	0.9 km (0.6 mi)	300 m (1000 ft)	1.1 km (0.7 mi)	2.4 km (1.5 mi)

TABLE 1 - INITIAL ISOLATION AND PROTECTIVE ACTION DISTANCES

Page 296

Protective actions are those steps taken to preserve the health and safety of emergency responders and the public. People in this area could be evacuated and/or sheltered in-place.

If hazardous materials are on fire or have been leaking for longer than 30 minutes, this *ERG* table does not apply. Seek more detailed information on the involved material on the appropriate orange-bordered page in the *ERG*. Also, the orange-bordered pages in the *ERG* provide recommended isolation and evacuation distances for nonhighlighted chemicals with poisonous vapors and situations where the containers are exposed to fire.

Materials with the text, "when spilled in water," following their names in Table 1 are considered water reactive and are covered in more detail in Table 2. Table 1 lists the chemicals produced when the material is spilled into water at the bottom of the page. Some Water Reactive materials are also TIH materials themselves (Bromine trifluoride, Thionyl chloride). In these instances, entries are provided in Table 1 to differentiate water-based spills from land-based spills ("when spilled on land").

Table 2, Water Reactive Materials which Produce Toxic Gases

Table 2 lists water reactive materials which produce large amounts of TIH gases when the material is spilled in water. Table 2 also identifies the TIH gases produced as a result of the spill. The materials are listed in ID number order. If the Water Reactive material is **NOT** a TIH and this material is **NOT** spilled in water, Table 1 and Table 2 do not apply and safety distances will be found within the appropriate orange guide.

Table 3, Initial Isolation and Protective Action Distances for Different Quantities of Six Common TIH (PIH in the US) Gases

Table 3 lists Toxic Inhalation Hazard materials that may be more commonly encountered. The selected materials are:

- Ammonia (UN1005)
- Chlorine (UN1017)
- Ethylene oxide (UN1040)
- Hydrogen chloride (UN1050) and Hydrogen chloride, refrigerated liquid (UN2186)
- Hydrogen fluoride (UN1052)
- Sulfur dioxide/Sulphur dioxide (UN1079)

The materials are presented in alphabetical order and provide Initial Isolation and Protective Action Distances for large spills involving different container types (therefore different volume capacities) for daytime and nighttime situations and different wind speeds.

Initiating Protective Actions

Protective actions, including isolation and scene control, can ensure your own safety and the safety of others. These techniques protect people by separating them from the potential source of harm and preventing the spread of hazardous materials through cross contamination.

Isolation involves physically securing and maintaining the emergency scene (scene control) by establishing isolation perimeters or cordons and denying entry to unauthorized persons **(Figure 3.20)**. It also includes preventing contaminated or potentially contaminated individuals (or animals) from leaving the scene in order to stop the spread of hazardous materials. The **isolation perimeter** (*outer perimeter* or *outer cordon*) is the boundary established to prevent unauthorized access to and egress from the scene. If an incident is inside a building, personnel posted at entrances can deny entry and exit from the building in order to set the isolation perimeter. If the incident is outdoors, the perimeter might be set at the surrounding intersections with response vehicles or law enforcement officers diverting vehicular traffic and pedestrians **(Figure 3.21)**. Ropes, cones, and barrier tape can also be used. In some cases, a traffic cordon may be established beyond the outer cordon to prevent unauthorized vehicle access while still allowing for pedestrian traffic. The isolation process may continue with evacuation, **defending in place**, or **sheltering in place** of people located within protective-action zones (see Chapter 7, Implementing and Evaluating the Action Plan: Incident Management and Response Objectives and Options).

Figure 3.20 Keeping other individuals from entering the hazardous area is one of the more important tasks that Awareness-Level personnel can do. Using barrier tape is an effective way to prevent entrance. *Courtesy of Ron Moore/McKinney (TX) Fire Department.*

Figure 3.21 Vehicles and barricades can block streets and intersections.

The isolation perimeter can be expanded or reduced as needed. For example, when additional resources arrive, the initial isolation perimeter may be expanded to accommodate new apparatus, equipment, and personnel. Law enforcement officers are often tasked to establish and maintain isolation perimeters.

Once the scene is secure and an isolation perimeter has been established and controlled, Awareness level responders are likely not trained to the necessary levels to continue the mitigation of the incident. There is one exception to this general rule. Awareness level responders may be trained to mitigate **incidental releases** without calling for additional assistance. For more information about response to incidental releases, see 29 *CFR* 1910.120. Skills associated with implementing protective actions at a hazardous materials incident are shown in **Skill Sheet 3-3**.

> **Incidental Release** — Spill or release of a hazardous material where the substance can be absorbed, neutralized, or otherwise controlled at the time of release by employees in the immediate release area, or by maintenance personnel who are not considered to be emergency responders.

Terrorist Incidents

Because terrorist and criminal incidents may differ from ordinary hazmat incidents, there are some specific, unique actions that need to be taken, such as immediately notifying law enforcement **(Figure 3.22)**. You should also be alert for secondary devices and booby traps since terrorists and criminals may deliberately target first responders or crowds. Terrorism is covered in more depth in Chapter 8 of this manual.

As true at all hazmat incidents, Awareness level personnel should do all of the following:

- Protect yourself and others by isolating the incident and denying entry.

- Prevent contaminated persons and animals from leaving the scene, if possible, and direct them to a safe area to wait for help.

- Avoid contacting contaminants or contaminated surfaces.

- Remember that WMD agents may be deadly in very small amounts, and biological agents may not cause symptoms for several days.

Figure 3.22 If personnel suspect a crime or terrorist attack at a hazmat incident, they should immediately notify law enforcement. *Courtesy of August Vernon.*

Finally, Awareness level personnel are likely to be on or near the scene when an incident or attack occurs, and therefore they make important witnesses. Law enforcement will want to know what you saw and when. In addition, if it can be done safely, you should do the following:

- Document your observations.

- Take pictures, if possible.

- Make note of other witnesses and observers at the scene.

- Protect evidence at the crime scene as best you can.

WMD/Terrorism Incident Resources

The first responder community should be encouraged to get to know its WMD Coordinator. There is at least one in every field office (56 field offices across the US). You can always reach one by calling the local FBI office and asking for the WMD Coordinator, who is the conduit for all things WMD / terrorism on the federal side of the house. The WMD Coordinator can leverage the resources of the FBI and other government agencies as well as bring training to the community. These coordinators are a great resource for information and support.

Chapter Review

Answer the following questions to review the information provided in this chapter.

1. Where should the notification procedures for Awareness level personnel be defined?

2. What are the sections of the ERG and what information do the different sections contain?

3. Explain the difference between isolation and scene control.

4. What are the responsibilities of Awareness level personnel at terrorist incidents?

3-1
Make appropriate notifications of a hazardous materials incident.

Step 1: Operate approved communication equipment.

Step 2: Communicate all necessary information about incident following policies and procedures.

Step 3: Ensure notification process has been properly initiated.

Step 3: Use approved reference sources to identify:
- Hazardous material(s) by name(s)
- Emergency response information
- Potential fire, explosion, and health hazards

Step 1: Recognize indicators of the presence of hazardous materials/WMD.

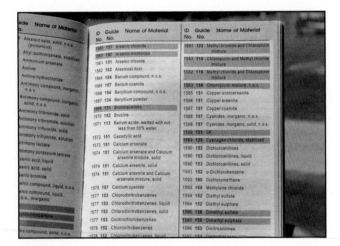

Step 2: Use approved reference sources to identify one or more of the following:
- The name of hazardous materials/WMD
- The UN/NA ID number
- The applied placard
- The container shape

Step 1: Identify an isolation area.

Step 2: Secure and evacuate isolation area.

Step 3: Follow precautions for protecting responders and the public.

Step 4: Follow personal safety procedures.

Step 5: Minimize/avoid hazards.

Step 6: Ensure additional people are not harmed or injured.

Step 7: Ensure unauthorized personnel and public do not enter isolation area.

Analyzing the Incident: Identifying Potential Hazards

Chapter Contents

Key Terms

Continued on page 136

NFPA Job Performance Requirements

This chapter provides information that addresses the following job performance requirements of NFPA 1072, *Standard for Hazardous Materials/Weapons of Mass Destruction Emergency Response Personnel Professional Qualifications (2017)*.

5.2.1

5.3.1

Learning Objectives

After reading this chapter, students will be able to:

1. Identify states of matter as they relate to hazardous materials. (5.2.1)

2. Explain physical properties that aid in identifying potential hazards and predicting behavior of hazardous materials. (5.2.1)

3. Explain chemical properties that aid in identifying potential hazards and predicting behavior of hazardous materials. (5.2.1)

4. Define the hazard classes. (5.2.1)

5. Describe actions taken to gather sufficient information to identify the hazardous material(s)/substance(s) involved in a hazmat incident. (5.2.1)

Chapter 4
Analyzing the Incident: Identifying Potential Hazards

This chapter explains the following topics:

- States of matter
- Physical properties
- Chemical properties
- Hazard classes
- Additional information

States of Matter

An uncontrolled hazmat release from a container can create many problems. The material's physical and chemical properties affect how it behaves, determine the harm it can cause, and influence the effect it may have on all it contacts, including people, other living organisms, other chemicals, and the environment. A material's physical and chemical properties also influence how a container will behave if it is damaged or ruptured.

First responders need to know how to collect hazard and response data that provide information about the substance's physical and chemical properties. The proper resources can greatly assist the responder in determining the present hazards, estimating the potential harm, and predicting how the incident may progress.

Matter is found in three states **(Figure 4.1, p. 138)**:

- **Gas**
- **Liquid**
- **Solid**

At a hazmat incident, try to identify the material's physical state as early as possible, because gaseous, liquid, and solid hazardous materials behave differently. This behavior influences the material's potential hazards. Once you understand how matter behaves in each state, you can better predict where the hazardous material is going, what exposures it may affect, and what those effects may be **(Figure 4.2, p. 138)**. The material's state of matter will indicate how mobile that material may become and can help determine if there will be far-reaching hazardous properties. Awareness of hazardous material mobility helps rescuers determine control zones and evacuation distances. The *Emergency Response Guidebook* establishes separate initial isolation distances based solely on the involved product's state of matter:

- Solids – 75 feet (25m)
- Liquids – 150 feet (50m)
- Gases – 330 feet (100m)

Gas — Compressible substance, with no specific volume, that tends to assume the shape of a container. Molecules move about most rapidly in this state.

Liquid — Incompressible substance with a constant volume that assumes the shape of its container; molecules flow freely, but substantial cohesion prevents them from expanding as a gas would.

Solid — Substance that has a definite shape and size; the molecules of a solid generally have very little mobility.

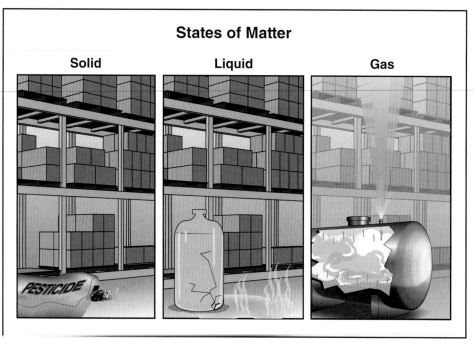

States of Matter

Solid   Liquid   Gas

Figure 4.1 Gases, liquids, and solids behave very differently. Knowing the state of matter will provide clues on how the incident may progress.

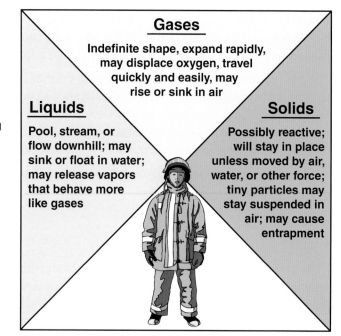

Gases
Indefinite shape, expand rapidly, may displace oxygen, travel quickly and easily, may rise or sink in air

Liquids
Pool, stream, or flow downhill; may sink or float in water; may release vapors that behave more like gases

Solids
Possibly reactive; will stay in place unless moved by air, water, or other force; tiny particles may stay suspended in air; may cause entrapment

Figure 4.2 Understanding a material's behavior enables responders to identify potential hazards as well as protective actions that should be taken.

In general, solids are the least mobile and gases have the greatest mobility. Liquids may be mobile depending on the properties of the substance. A substance's state may change if the temperature changes. A solid may change to a liquid if the temperature increases. You should consider the temperature's effect on a substance if the incident is located outside because air temperature and weather factors can strongly influence a substance's state of matter and subsequently its behavior **(Figure 4.3)**.

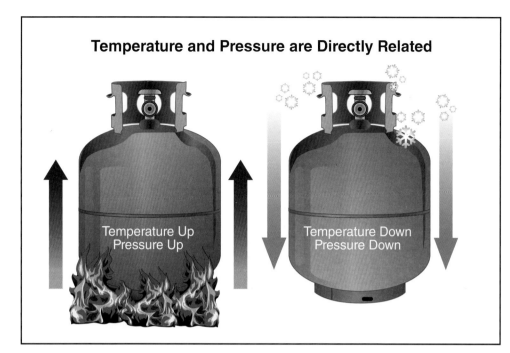

Temperature and Pressure are Directly Related

Temperature Up
Pressure Up

Temperature Down
Pressure Down

Figure 4.3 Temperature may strongly affect how a material, and its container, behaves. For example, rising temperatures may increase the pressure in liquid, cryogenic, and pressure containers.

Is It a Solid, Liquid, or Gas? Industry Terms to Describe Air Contaminants.

A first responder may be unable to distinguish between these types of contaminants at an incident scene. Though air contaminants are commonly classified as either particulate or gas and vapor contaminants, visible releases of particulates may also include:

- **Dust** — Solid particle that is formed or generated from solid organic or inorganic materials by reducing its size through mechanical processes, such as crushing, grinding, drilling, abrading, or blasting. Example: Grain elevators with airborne grain dust.

- **Fume** — Suspension of particles that form when material from a volatilized (vapor state) solid condenses in cool air. In most cases, the solid, smokelike particles resulting from the condensation react with air to form an oxide. Examples: Paint, smoke.

- **Mist** — Finely divided liquid suspended in the atmosphere. Mists are generated by liquids condensing from a vapor back to a liquid or by breaking up a liquid into a dispersed state by splashing, foaming, or atomizing. Mists may also be generated during temperature differentials, such as temperature inversions. Mists are not usually pressurized. Example: Acids such as sulfuric acid.

- **Aerosol** — Form of pressurized mist characterized by highly respirable, minute liquid or solid particles. Usually identifiable by its high speed of travel. Example: Released anhydrous ammonia. Leaking thermanol is an example of a high-temperature aerosol.

- **Fiber** — Solid particle whose length is several times greater than its diameter is formed by a disruption of the natural state. Usually not visibly identifiable in the air. Example: Asbestos.

- **Vapor** — Gaseous form of a material that is normally in a solid or liquid state at room temperature and pressure. Vapors are formed by evaporation from a liquid or sublimation from a solid, and are visible as atmospheric disturbances (wavy lines) over a surface. Vapors are volatile. Examples: Gasoline, solvents.

- **Fog** — Visible aerosol of a liquid formed by condensation. Liquefied gases that auto-refrigerate at low pressure will form fogs. Fog particulates have a smaller droplet size than mists. Usually identifiable and discernable from an aerosol by its relatively low speed of travel that is dependent on wind speed. Examples: Chlorine, anhydrous ammonia.

Gases

Incidents involving gases are potentially the most dangerous for emergency responders. Many hazmat-related injuries are due to the inhalation of vapors or gases. Gaseous materials could have many variables and hazards, such as:

- May have an odor (such as chlorine)

- May be colorless, odorless, and/or tasteless (such as carbon monoxide)

- May be separately, or any combination of: toxic (such as phosgene), corrosive (such as ammonia), or flammable (such as methane, natural gas)

- May have high pressure in excess of 15,000 psi (103 421 kPa) (such as liquid helium)

- May be extremely cold upon release and/or may have a large expansion ratio if liquefied

Gases have an undefined shape and volume and keep expanding if uncontained. As a result, it is difficult to detect where they are, where they are not, and where they may be going **(Figure 4.4)**. A gas leak in a building has many potential directions to spread. Depending on ventilation and other factors, the gas may spread:

- Throughout the building.

- To other buildings.

- Through access shafts.

- Into the soil.

- Into the street, where it will drift wherever the wind may take it.

Gases are difficult (if not impossible) to contain for mitigation purposes **(Figure 4.5)**. **Compressed gases** and **liquefied gases** expand rapidly when released, potentially threatening large areas. If a gas is invisible and/or has little or no odor, it may be impossible to detect without specialized detection equipment, such as a **combustible gas detector** or another instrument **(Figure 4.6)**. **NOTE:** Research has shown that 1- and 2-ton quantities of chlorine and anhydrous ammonia released in an open area both initially spread in a 360° radius before being dispersed downwind in some situations.

Compressed Gas — Gas that, at normal temperature, exists solely as a gas when pressurized in a container, as opposed to a gas that becomes a liquid when stored under pressure.

Liquefied Gas — Confined gas that at normal temperatures exists in both liquid and gaseous states.

Combustible Gas Detector — Device that detects the presence and/or concentration of predefined combustible gases in a defined area. May require additional features to indicate the results to an operator.

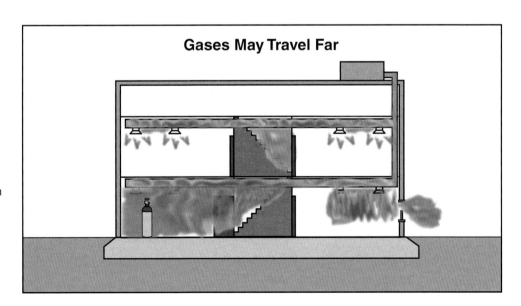

Figure 4.4 Gases may travel in unexpected ways, depending on many factors including air movement, topography, and barriers such as walls or buildings.

Gases May Travel Far

Figure 4.5 Incidents involving gases are often difficult to mitigate because they can't be easily contained. Large perimeters may be necessary, even for incidents involving small containers.

Figure 4.6 Gases may be impossible to detect without appropriate monitoring and detection devices.

Figure 4.7 Some gases are corrosive. These include Bromine, Chlorine, and Anhydrous Ammonia. *Courtesy of Rich Mahaney.*

Materials kept under pressures and/or temperatures higher or lower than ambient conditions may change state upon release. The ratio that a gas will expand (expansion ratio) from its liquid state is a significant factor in mitigating a hazmat incident involving materials under specific conditions (especially cryogenic liquids and liquefied gases).

NOTE: For additional information on the behavior of gases, refer to Chapter 5.

WARNING!
Expanding gases can displace oxygen, creating an asphyxiating atmosphere.

If a hazardous material is a gas, it will be present in the air and will potentially present a breathing/inhalation hazard. Some gases may also present a contact hazard **(Figure 4.7)**. In general, if an incident involves a gas, it has the potential to be much harder to mitigate and affect larger areas than incidents involving other states of matter. Incidents involving gas require complex and difficult actions to protect responders and the public.

Liquids

Liquids are usually visible, even if their vapors are not, so it may be easier to detect their presence and determine the hazard area **(Figure 4.8)**. Liquids typically do not travel as far as gases unless they spill into a path or channel, such as a storm drain, stream, river, or other waterway, that transports liquids quickly and efficiently **(Figure 4.9)**. Responders may be able to predict the paths that spilled liquids will most likely follow.

Liquids will flow or pool according to surface contours and topography, permitting opportunities for containment or confinement **(Figure 4.10)**. Liquids present a splash or contact hazard **(Figure 4.11)**.

Liquids may pose unique challenges to responders because they may take on the additional characteristics of a gas by emitting vapors **(Figure 4.12)**. The conversion of liquid to vapor increases both the hazardous material's mobility and the challenges responders face when dealing with the material.

Vapors from liquids may travel much like gases, although typically not as far from their source, and they may be much more difficult to detect than the liquid itself **(Figure 4.13)**. Be cautious and alert to vapors from liquids as they may be:

- Contact hazards
- Inhalation hazards
- Flammable
- Corrosive
- Toxic

CAUTION

Vapors from liquids behave like gases, and may be flammable, corrosive, or toxic.

Figure 4.8 Liquids are usually visible even if their vapors are not. This may assist in determining the hazard area. *Courtesy of Rich Mahaney.*

Figure 4.9 Liquids tend not to travel as far as gases unless topography assists them; for example, they reach a storm drain or stream.

Figure 4.10 Because liquids follow topography, they can be contained. *Courtesy of Phil Linder.*

Figure 4.11 Liquids present a splash hazard.

Figure 4.12 Liquid vapors behave more like a gas.

Figure 4.13 Liquid vapors tend not to travel as far from their source as a gas, but monitoring and detection devices should be used to determine where they are located.

Solids

Solids are the least mobile of the three states of matter. They typically will remain in place unless acted upon by exterior forces, such as wind, water, and gravity **(Figure 4.14, p. 144)**. The particle size of solids, such as dusts, fumes, or powders, may influence their behavior. Larger par-

Micron — Unit of length equal to one-millionth of a meter.

ticles will probably settle out of the air fairly quickly. Smaller particles may remain suspended longer and travel further than larger particles **(Figure 4.15)**. **Micron** is the unit of measure typically used to express particle size.

Solids may have the following dangerous properties:

- Inhalation or contact hazards
- Small, combustible particles that, if ignited, may explode
- Entrapment hazard in the form of loose solids confined to large containers **(Figure 4.16)**
- Flammable, reactive, radioactive, corrosive, toxic

You can usually detect a solid visually, unless it has microscopic particles. This visibility makes detecting the presence of solids easier than detecting gases or vapors from liquids. Solids such as dry ice, elemental iodine and naphthalene may sublimate/sublime (transition directly from a solid to a gas). Sublimating materials present the same hazards and concerns as liquids that emit vapors.

With some exceptions, incidents involving solid materials are confined to limited areas, with less likelihood of undetected travel. Solid

Figure 4.14 Solids tend to stay in place unless something else moves them. *Courtesy of David Alexander with the Texas Commission on Fire Protection.*

Figure 4.15 Tiny particles may stay suspended in air and travel with air currents whereas larger particles tend to settle out.

Figure 4.16 Confined solids can cause entrapment.

incidents may require less complicated mitigation and protective actions than gas and liquid incidents. This response depends on the chemical and physical properties of the material involved.

Physical Properties

Physical properties are the characteristics of a material that do not involve the chemistry or chemical nature of the material. Physical properties describe how a material behaves in relation to physical influences, such as temperature and pressure, or how a material behaves when mixed with, or compared to, another material. Materials can be characterized by the following physical properties:

- Vapor pressure
- Boiling point
- Melting point/freezing point/sublimation
- Vapor density
- Solubility/miscibility
- Specific gravity
- Persistence
- Appearance and odor

Vapor Pressure

Vapor pressure is the pressure exerted by a saturated vapor above its own liquid in a closed container. More simply, it is the pressure produced or exerted by the vapors released by a liquid. Vapor pressure can be viewed as the measure of the tendency of a substance to evaporate.

Vapor pressures reported in reference materials may use any of the following units:

- Pounds per square inch (psi)
- Kilopascals (kPa)
- Bars
- Millimeters of mercury (mmHg) (used in older-style material safety data sheets [MSDS])
- Atmospheres (atm)
- Hectopascals (hPa) (used on new GHS safety data sheets [SDS])

Be aware of the following facts regarding vapor pressure:

- Materials with a vapor pressure over 760 mmHg will be gases under normal conditions.
- The higher the temperature of a substance, the higher its vapor pressure will be **(Figure 4.17)**. The vapor pressure of a material at 100°F (38°C) will always be higher than the vapor pressure of the same material at 68°F (20°C). Higher temperatures provide more

Physical Properties — Properties that do not involve a change in the chemical identity of the substance, but affect the physical behavior of the material inside and outside the container, which involves the change of the state of the material. Examples include boiling point, specific gravity, vapor density, and water solubility.

Vapor Pressure — The pressure at which a vapor is in equilibrium with its liquid phase for a given temperature; liquids that have a greater tendency to evaporate have higher vapor pressures for a given temperature.

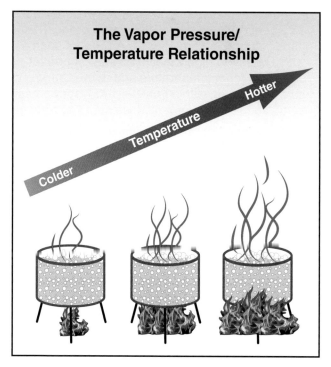

The Vapor Pressure/ Temperature Relationship

Colder · Temperature · Hotter

Figure 4.17 A liquid's vapor pressure increases as the temperature rises.

energy to a liquid, which allows more liquid to escape into a gaseous form. The gas rises above the liquid and exerts a downward pressure.

- Atmospheric pressure is the baseline measurement for pressure. Other measurements are listed in the Information Box in this section.

- The lower the boiling point (the temperature at which a liquid turns to a gas) of a material, the higher its vapor pressure will be. If a material has a low boiling point, it requires less heat to change from liquid into a gas.

NOTE: Water requires a lot of heat to boil (212°F [100°C]), but some substances boil at room temperature (68°F [20°C]).

Units of Pressure Measurement

- 1 atmosphere
  - —760 mmHg = 760 torr (1mmHg = 1 torr)
  - —29.9 inHg (inches of mercury)
  - —407 inches of water
  - —14.7 psi
  - —1.01 bar
  - —101.3 kPa
  - —1013.25 hPa (hPa = millibar)

- 1 bar = 14.5 psi
- 1 bar = 100,000 pascals
- 1 foot water = 0.43 psi

What This Means To You

Vapor Pressure as an Evaporation Indicator

If you know the vapor pressure of a material, you can use it as a general gauge to tell how fast a product will evaporate under normal circumstances. A product such as acetone, with a relatively high vapor pressure, evaporates much more quickly at room temperature and normal atmospheric pressure than water, with a relatively low vapor pressure. In the same conditions, motor oil does not evaporate easily.

Under most normal conditions, a spill of a liquid with a high vapor pressure (such as isopropylamine) will produce a greater concentration of vapors than a substance with a low vapor pressure (such as sarin). These fumes or vapors could then be carried by the wind or travel distances on air currents. Due to the increased mobility of vapors over liquids, the hazardous vapors could cause problems far from the spill itself (such as toxic or flammable vapors being blown into a residential neighborhood).

Figure 4.18 Chlorine has a very high vapor pressure. If it escapes its container, it will do so mostly as a gas. *Courtesy of Rich Mahaney.*

Vapor pressure may indicate the state of matter of a product. For example, chlorine, with its extremely high vapor pressure, is likely to be released as a gas since it instantly evaporates at normal atmospheric pressure and temperatures **(Figure 4.18)**.

Vapor pressure may also indicate the degree to which a material will present an inhalation hazard. Materials with lower vapor pressure are less likely to produce fumes or vapors. Substances with higher vapor pressures are much more likely to produce fumes or vapors.

Boiling Point

Boiling point is the temperature at which a liquid changes to a gas at a given pressure. Boiling point is usually expressed in degrees Fahrenheit (Celsius) at sea level air pressure **(Figure 4.19)**. For mixtures, the initial boiling point or boiling-point range may be given. Flammable materials with low boiling points generally present special fire hazards.

A **boiling liquid expanding vapor explosion (BLEVE)** (also called *violent rupture*) can occur when a liquid within a container is heated, causing the material inside to boil or vaporize (such as in the case of a liquefied petroleum gas tank exposed to a fire). If the resulting increase in internal vapor pressure exceeds the vessel's ability to relieve (or retain) the excess pressure, the container can fail catastrophically. As the vapor is released, it expands rapidly and ignites, sending flames and pieces of tank flying. BLEVEs most commonly occur when flames contact a tank shell above the liquid level or when insufficient water is applied to keep a tank shell cool.

Melting Point/Freezing Point/Sublimation

Melting point is the temperature at which a solid substance changes to a liquid state at normal atmospheric pressure. An ice cube melts at just above 32°F (0°C) — its melting point.

Freezing point is the temperature at which a liquid becomes a solid at normal atmospheric pressure. Water freezes at 32°F (0°C) — its freezing point. Some substances will actually *sublimate* or change directly from a solid into a gas without going into a liquid state in between **(Figure 4.20)**. Dry ice (solid-state carbon dioxide) and mothballs sublimate rather than melt.

Temperatures change throughout the day due to weather patterns and exposure to the sun. A material that begins the day as a solid may change to a liquid if heated sufficiently. Since materials are typically easier to control as a solid than a liquid, this may affect mitigation strategies.

Boiling Point — Temperature of a substance when the vapor pressure equals atmospheric pressure. At this temperature, the rate of evaporation exceeds the rate of condensation. At this point, more liquid is turning into gas than gas is turning back into a liquid.

Boiling Liquid Expanding Vapor Explosion (BLEVE) — Rapid vaporization of a liquid stored under pressure upon release to the atmosphere following major failure of its containing vessel. Failure is the result of overpressurization caused by an external heat source, which causes the vessel to explode into two or more pieces when the temperature of the liquid is well above its boiling point at normal atmospheric pressure.

Boiling Point Example

212° (100° C) at Sea Level

Figure 4.19 Boiling point is the temperature at which a liquid will boil at sea level. Flammable liquids with low boiling points are especially hazardous because they turn into a gas at normal temperatures. Liquids with high boiling points have to be heated before they start to boil.

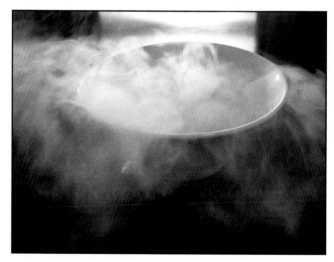

Figure 4.20 Dry ice sublimates from a solid into a gas without transitioning into a liquid.

Vapor Density

Vapor density is the weight of a given volume of pure vapor or gas compared to the weight of an equal volume of dry air at the same temperature and pressure. A vapor density less than one indicates a vapor lighter than air; while a vapor density greater than one indicates a vapor heavier than air. Lighter than air gases and vapors will rise, while heavier than air gases and vapors will sink **(Figure 4.21)**. Examples of materials with a vapor density less than one include helium, neon, acetylene, and hydrogen (see information box). Gases with a vapor density less than one will rise quickly and spread to a wide geographical area.

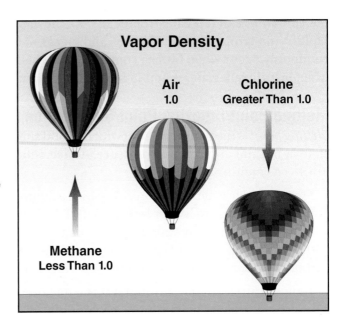

Figure 4.21 Materials with vapor densities less than one will rise in air while materials with vapor densities greater than one will sink.

Lighter-than-Air Gases

The list of gases that are lighter than air is short compared to the list of gases heavier than air. Thirteen common chemicals have a vapor density lighter than or equal to air:

- Acetylene (.9)
- Ammonia (.59)
- Carbon monoxide (.96)
- Diborane (.96)
- Ethylene (.96)
- Helium (.14)
- Hydrogen (.07)
- Hydrogen cyanide (.95)
- Hydrogen fluoride (.34)
- Illuminating gases (.6)
- Methane (.55)
- Neon (.34)
- Nitrogen (.96)

Most gases have a vapor density greater than one; they will sink in relation to ambient air and will displace oxygen at low elevations. Heavier vapors and gases are likely to concentrate in low places along or under floors; in sumps, sewers, and manholes; and in trenches and ditches where they may create fire or health hazards **(Figure 4.22)**. Examples of common materials with a vapor density greater than one include:

- Propane
- Hydrogen sulfide
- Ethane
- Butane
- Chlorine
- Sulfur dioxide

Vapor density varies with the temperature of the vapor or gas. Hot vapors will rise, but unless totally dispersed, they will sink once they have cooled. Cold vapors are dense and will stay low but will rise when they warm.

Personnel cannot precisely predict the spread of vapors from the vapor density because topography, weather conditions, and the vapor mixture with air affect vapors. However, knowing the vapor density gives a general idea of what to expect from a specific gas or vapor.

Figure 4.22 Vapors and gases that are heavier than air may concentrate in low-lying areas such as trenches, sewers, manholes, and other confined spaces.

NOTE: All vapors and gases will mix with air, but the lighter materials (unless confined) tend to rise and dissipate.

Solubility/Miscibility

Solubility in water expresses the percentage of a material (by weight) that will dissolve in water at ambient temperature. A substance's solubility affects whether it mixes in water. Solubility information can be useful in determining spill cleanup methods and extinguishing agents. When a non-water-soluble liquid such as a hydrocarbon (gasoline, diesel fuel, pentane) combines with water, the two liquids remain separate **(Figure 4.23, p. 150)**. When a water-soluble liquid such as a **polar solvent** (alcohol, methanol, methyl ethyl ketone [MEK]) combines with water, the two liquids mix.

NOTE: Emergency responders may find that some materials, such as hydrocarbons (gasoline, oil), will initially float, but will sink over time. Degradation, chemical reactions, exposure, and time will all affect materials and change their characteristics.

Water solubility is also an important contributor for symptom development. Irritant agents that are water-soluble usually cause early upper respiratory tract irritation, resulting in coughing and throat irritation. Partially water-soluble chemicals will penetrate into the lower respiratory system and cause delayed (12 to 24 hours) symptoms that include breathing difficulties, pulmonary edema, and coughing up blood.

Solubility — Degree to which a solid, liquid, or gas dissolves in a solvent (usually water).

Polar Solvent — 1) A material in which the positive and negative charges are permanently separated, resulting in their ability to ionize in solution and create electrical conductivity. Examples include water, alcohol, esters, ketones, amines, and sulfuric acid. 2) Flammable liquids with an attraction for water.

Figure 4.23 Non-water-soluble liquids, including most hydrocarbons, don't dissolve in water.

Degrees of Solubility

Materials with higher degrees of solubility are easier to control using water. The following terms describe degrees of solubility:

- Negligible (insoluble) — Less than 0.1 percent
- Slight (slightly soluble) — From 0.1 to 1 percent
- Moderate (moderately soluble) — From 1 to 10 percent
- Appreciable (partly soluble) — More than 10 to 25 percent
- Complete — Soluble at all protortions

NOTE: Some materials may be soluble at percentages higher than 100 percent. For example, NIOSH lists materials that are over 300 percent soluble.

Miscibility — Two or more liquids' capability to mix together.

Immiscible — Incapable of being mixed or blended with another substance.

Miscibility describes the ability of two or more gases or liquids to mix with or to dissolve into each other. Two liquids or gases are miscible if they mix or dissolve into each other in any proportion. Typically, two materials that do not readily dissolve into each other are considered **immiscible**. For example, water and fuel oil are immiscible. Immiscible materials can create a hazard because oil (which weighs less than water) will float on top of water and could ignite and burn **(Figure 4.24)**.

Figure 4.24 Most hydrocarbons are also immiscible. Oil weighs less than water, so it will float on top, where it can ignite. *Courtesy of U.S. Coast Guard.*

Specific Gravity

Specific gravity is the ratio of the density (mass per volume) of a material to the density of a standard material, usually an equal volume of water, at standard conditions of pressure and temperature. If a volume of a material weighs 8 pounds (3.6 kg), and an equal volume of water weighs 10 pounds (4.5 kg), the material is said to have a specific gravity of 0.8. Materials with specific gravities less than one will float in (or on) water. Materials with specific gravities greater than one will sink in water.

Solubility plays an important role in specific gravity. Highly soluble materials will mix or dissolve more completely in water (distributing themselves more evenly throughout), rather than sinking or floating (without dissolving) according to their specific gravities. Most (but not all) flammable liquids have specific gravities less than one and will float on water **(Figure 4.25)**. An important consideration for fire-suppression activities is that flammable liquids will float on water.

Specific Gravity — Mass (weight) of a substance compared to the weight of an equal volume of water at a given temperature. A specific gravity less than one indicates a substance lighter than water; a specific gravity greater than one indicates a substance heavier than water.

Most Flammable Liquids
Float on Water

FLAMMABLE LIQUID

Figure 4.25 Water alone is not very effective at putting out flammable liquid fires because flammable liquids have specific gravities less than water and will float to the surface where they will continue burning.

Heptane

Heptane is a major component of gasoline and has the following physical and chemical properties:

- Vapor Pressure: 45 mmHg
- Flash Point: 25°F (-4°C)
- Boiling Point: 210°F (98°C)
- Vapor Density: 3.5
- Solubility in Water: Negligible
- Specific Gravity: 0.7

By understanding how to interpret this information, you can predict how the material is likely to behave. If a significant amount of heptane spills into a pond or waterway, you might follow this chain of thought:

- First, note that the heptane was spilled in water. Consider what the material will do relative to the water. Will it mix with the water? Will it sink or float? Since heptane's solubility in water is negligible, you can gather that it will not dissolve or mix in water. Because heptane's specific gravity is less than one, it will float on the surface of water.

- You know that heptane will not mix with water, will float on top of the water, and it will burn. You now will want to know whether it will emit vapors or fumes that could accidentally ignite. Its vapor pressure of 45 mmHg (higher than that of water) tells you that it will likely evaporate under most normal conditions. Its flashpoint of 25°F (-4°C) tells you that those vapors will burn if exposed to most ignition sources. Therefore, prioritize and keep ignition sources away from the vapors.

- What are the vapors doing and where might they be going? Are they rising in the air or staying close to the surface of the water? Heptane's vapor density of 3.5 tells you that they will tend to stay low or close to the surface of the water (assuming there is no wind or other disturbances).

Persistence — Length of time a chemical agent remains effective without dispersing.

Dispersion — Act or process of being spread widely.

Viscosity — Measure of a liquid's internal friction at a given temperature. This concept is informally expressed as thickness, stickiness, and ability to flow.

Persistence and Viscosity

The **persistence** of a chemical is its ability to remain in the environment. Chemicals that remain in the environment for a long time are more persistent than chemicals that quickly dissipate or break down **(Figure 4.26)**. Persistent nerve agents will remain effective at their point of **dispersion** (release) for a much longer time than nonpersistent nerve agents.

NOTE: Persistence is not often referenced on an SDS.

Viscosity is the measure of the thickness or flowability of a liquid at a given temperature **(Figure 4.27)**. Numerical values sometimes describe viscosity, with higher numbers indicating higher viscosity. Viscosity determines the ease with which a product will flow; it is greatly affected by temperature. Usually, the hotter a liquid, the thinner or more fluid it becomes. Likewise, the cooler a liquid, the thicker or less fluid it becomes. Liquids with high viscosities, such as heavy oils, have to be heated to increase their fluidity. Viscous materials tend to be more persistent and may have a lower vapor pressure. Examples of materials with differences in viscosity are acetone, water, oil, and honey. First

Persistence

Figure 4.26 Persistent chemicals stick around in the environment before dispersing.

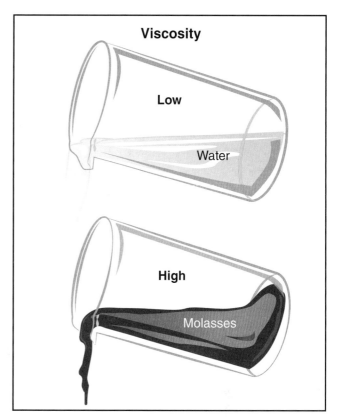

Viscosity

Low

Water

High

Molasses

Figure 4.27 Molasses are very viscous at room temperature, unlike water. Viscosity changes with temperature.

responders use viscous materials to reference the way the viscosity of these materials will affect decontamination or collection.

Appearance and Odor

A safety data sheet (SDS) typically contains a description of the material's appearance (such as physical state or color) and odor. Referring to the SDS could allow first responders to quickly gain important information about the state of matter and potential behavior of the substance or material.

Appearance may help you detect a substance/material. You should evaluate a change in appearance, which may indicate a change in the behavior of the substance or material. For many industrial products, the color listed on the SDS may represent an "average," and the product shipped may vary significantly in color and still be the same product. In other instances, a significant difference in color may also show contamination or high levels of impurities that may have their own hazards.

Responders detecting a chemical via odor could indicate that they are too close. Some chemicals have little or no odor, while others have a strong characteristic odor **(Figure 4.28)**. Some characteristic odors can help identify a material. The smell of natural gas, based on an additive, **mercaptan**, is similar to that of rotten eggs or sewage. An unexpected odor may be a warning that a substance has escaped from its container.

NON-ODORIZED
LIQUEFIED
PETROLEUM GAS
DOT 112 J 340 W

1075
2

Figure 4.28 Odorants may be added to some hazardous materials to make them easier to detect, but many products have no odor. *Courtesy of Rich Mahaney.*

Mercaptan — A sulfur-containing organic compound often added to natural gas as an odorant. Natural gas is odorless; natural gas treated with mercaptan has a strong odor. *Also known as a* Thiol.

WARNING!
If you can smell a chemical, you are exposed.
Move out of the area and reassess the situation.

The ability to smell or sense an odor is highly dependent on the individual. *Odor threshold* is the concentration (in air) at which the "average person" can smell a particular compound. Some people can smell a given compound at an extremely low level. Other people may not be able to smell a particular compound even at very high concentrations in the air.

Never use odors to determine safe or unsafe areas; some highly toxic products may cause significant damage at a concentration below the odor threshold. Responders spending too much time exposed to some compounds may become desensitized to the smell of a chemical and may no longer be able to determine its presence.

NOTE: Visual indicators and chemical odors were described in greater detail in Chapter 2, *Senses* section.

Chemical Properties

Chemical properties describe the chemical nature of a material and the behaviors and interactions that occur at a molecular level. While not always grouped with other chemical properties, toxicity and biological hazards will also be addressed in this section and in greater detail than in Chapter 1. This section also explains the following important chemical properties in order of commonality at incidents:

- Flammability
- Corrosivity
- Reactivity
- Radioactivity

Flammability

Most hazardous materials incidents involve flammable materials. Flammable materials can damage life and property when they ignite, burn, or explode. Use a hazard's flammability to help determine incident strategies and tactics. A flammable hazard depends on properties, including its:

- Flash point
- Autoignition temperature (sometimes called the *autoignition point*)
- Flammable (explosive or combustible) range

Flash Point

Flash point is the minimum temperature at which a liquid or volatile solid gives off sufficient vapors at its lower explosive limit (LEL) to form an ignitable mixture with air near its surface **(Figure 4.29)**. At its flash point, a material's vapors will flash in the presence of an ignition source but will not continue to burn. Do not confuse flash point with fire point. **Fire point** is the temperature at which a liquid or volatile substance gives off enough vapors to support continuous burning. A material's fire point is usually only a few degrees (10-30 degrees Celsius) higher than its flash point.

Only the vapors burn. The liquid or volatile solid that produces the vapors does not burn. As the liquid's temperature increases, it emits more vapors. Vapors are emitted below the flash point but not in sufficient quantities to ignite. A substance will not burn if it is not at its flash point temperature. Flammable gases have extremely low flash points so they are flammable all the time.

Chemical Properties — Relating to the way a substance is able to change into other substances. Chemical properties reflect the ability to burn, react, explode, or produce toxic substances hazardous to people or the environment.

Flash Point — Minimum temperature at which a liquid gives off enough vapors to form an ignitable mixture with air near the surface of the liquid.

Fire Point — Temperature at which a liquid fuel produces sufficient vapors to support combustion once the fuel is ignited. Fire point must exceed five seconds of burning duration during the test. The fire point is usually a few degrees above the flash point.

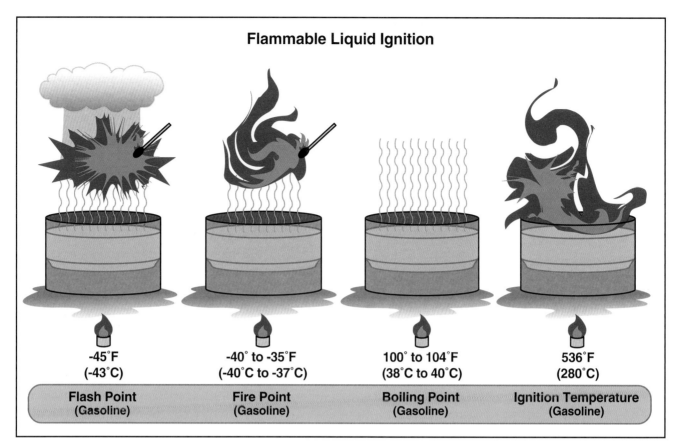

Flammable Liquid Ignition

-45°F (-43°C)	-40° to -35°F (-40°C to -37°C)	100° to 104°F (38°C to 40°C)	536°F (280°C)
Flash Point (Gasoline)	**Fire Point** (Gasoline)	**Boiling Point** (Gasoline)	**Ignition Temperature** (Gasoline)

Figure 4.29 The minimum temperature at which a liquid gives off enough vapors to form an ignitable mixture with air is called its *flash point*.

Is it Flammable, Inflammable, or Combustible?

In everyday language, the terms *flammable* and *combustible* can be used interchangeably to denote a substance that will burn. These terms have more technical meanings when referring to hazardous materials, particularly liquids. The flash point is commonly used to determine the flammability of a liquid. Liquids that have low flash points and burn easily are designated as **flammable liquids**, whereas liquids with higher flash points that do not burn as easily are called **combustible liquids**. However, different U.S. agencies use different flash points as the threshold to designate flammable and combustible substances.

The term *inflammable* means the same thing as *flammable* in many parts of the world. In contrast, a **nonflammable** material does not ignite easily. In Mexico, for example, a tank truck carrying flammable liquids may read either *flammable* or *inflammable* **(Figure 4.30, p. 156)**. Transport Canada (TC) allows only the term *flammable*, but inflammable is the French word for flammable.

Flammable Liquid — Any liquid having a flash point below 100°F (37.8°C) and a vapor pressure not exceeding 40 psi absolute (276 kPa) {2.76 bar}, per NFPA.

Combustible Liquid — Liquid having a flash point at or above 100°F (37.8°C) and below 200°F (93.3°C), per NFPA.

Nonflammable — Incapable of combustion under normal circumstances; normally used when referring to liquids or gases.

Autoignition Temperature — The lowest temperature at which a combustible material ignites in air without a spark or flame. (NFPA 921)

Ignition Temperature — Minimum temperature to which a fuel (other than a liquid) in air must be heated in order to start self-sustained combustion independent of the heating source.

Autoignition Temperature

The **autoignition temperature** of a substance is the minimum temperature to which the fuel in air must be heated to initiate self-sustained combustion without initiation from an independent ignition source. This temperature, also known as the **ignition temperature**, is the point at which a fuel spontaneously ignites. All flammable materials have autoignition temperatures,

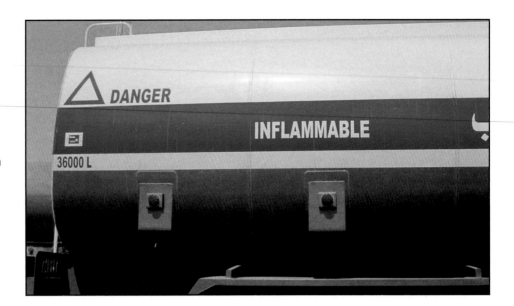

Figure 4.30 *Inflammable* is another word for *flammable* in many countries. *Courtesy of Rich Mahaney.*

and these are considerably higher than the flash and fire points. For example, the autoignition temperature of gasoline is about 536°F (280°C), but the flash point of gasoline is -45°F (-43°C). This difference means that at -45°F (-43°C), gasoline will temporarily ignite if a match is waved through its vapors, whereas at 536°F (280°C) it ignites all by itself. The terms *autoignition temperature* and *ignition temperature* are often used synonymously; they are always the same temperature. However, the NFPA defines these terms separately.

Flammable, Explosive, or Combustible Range

The flammable, explosive, or combustible range is the percentage of the gas or vapor concentration in air that will burn or explode if ignited. The LEL or **lower flammable (explosive) limit (LFL)** of a vapor or gas is the lowest concentration (or lowest percentage of the substance in air) that will produce a flash of fire when an ignition source is present. At concentrations lower than the LEL, the mixture is too lean to burn.

The upper explosive limit (UEL) or **upper flammable limit (UFL)** of a vapor or gas is the highest concentration (or highest percentage of the substance in air) that will produce a flash of fire when an ignition source is present. At higher concentrations, the mixture is too rich to burn **(Figure 4.31)**. Within the upper and lower limits, the gas or vapor concentration will burn rapidly if ignited. Atmospheres within the flammable range are particularly dangerous. **Table 4.1** provides the flammable ranges for some selected materials.

> **Lower Flammable (Explosive) Limit (LFL)** — Lower limit at which a flammable gas or vapor will ignite and support combustion; below this limit the gas or vapor is too *lean* or *thin* to burn (too much oxygen and not enough gas, so lacks the proper quantity of fuel). *Also known as* Lower Explosive Limit (LEL).

> **Upper Flammable Limit (UFL)** — Upper limit at which a flammable gas or vapor will ignite. Above this limit, the gas or vapor is too rich to burn (lacks the proper quantity of oxygen). *Also known as* Upper Explosive Limit (UEL).

What This Means To You

Lower and Upper Explosive Limits

Products with a low LEL and products with a wide range between the LEL and UEL are especially dangerous. Concentrations above the UEL do not guarantee safety. If the concentration drops for any reason, you could still be in an explosive atmosphere. The addition of fresh air may dilute the concentration, or the concentration may be lower than the UEL in places where you did not measure.

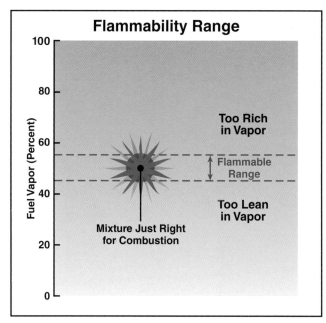

Flammability Range

Figure 4.31 Flammable vapors and gases can burn or explode when they are mixed with the right concentration of air. If there is too little air, the mixture is too lean to ignite; with too much flammable vapor or gas, the mixture is too rich to ignite.

Table 4.1
Flammable Ranges for Selected Materials

Material	Lower Flammable Limit (LFL) (percent by volume)	Upper Flammable Limit (UFL) (percent by volume)
Acetylene	2.5	100.0
Carbon Monoxide	12.5	74.0
Ethyl Alcohol	3.3	19.0
Fuel Oil No. 1	0.7	5.0
Gasoline	1.4	7.6
Methane	5.0	15.0
Propane	2.1	9.5

Source: *NIOSH Pocket Guide to Chemical Hazards*

Corrosivity

Chapter 1 introduced corrosives as materials that destroy living tissue and damage or destroy metal. Corrosives are commonly divided into two broad categories: acids and bases (bases are sometimes called *alkalis* or *caustics*). However, some corrosives (such as hydrogen peroxide) are neither acids nor bases. The corrosivity of acids and bases is often measured or expressed in terms of **pH (Figure 4.32, p. 158)**. Acids and bases have the following characteristics:

- **Acid** — Any chemical that ionizes (dissociates) to yield hydrogen **ions** (hydronium) in water. Acids have pH values of 0 to 6.9. An acid may cause severe chemical burns to flesh and permanent eye damage. Contact with an acid typically causes immediate pain. Hydrochloric acid, nitric acid, and sulfuric acid are examples of common acids.

- **Base** (alkalis) — A water-soluble compound that chemically **dissociates** in water to form a negatively charged hydroxide ion. Bases react with an acid to form a salt by releasing an unshared pair of electrons to the acid or by receiving a proton (hydrogen ion) from the acid. Bases have pH values of 7.1 to 14. A base breaks down fatty skin tissues and can penetrate deeply into the body. Bases tend to adhere to the tissues in the eye, which makes them difficult to remove. Bases often cause more eye damage than acids due to the longer duration of exposure. Contact with a base does not normally cause immediate pain. A common sign of exposure to a base is a greasy or slick feeling of the skin, which is caused by **saponification**, the breakdown of fatty tissues. Examples of bases include caustic soda, potassium hydroxide, and other alkaline materials commonly used in drain cleaners.

pH — Measure of the acidity or alkalinity of a solution.

Acid — Compound containing hydrogen that reacts with water to produce hydrogen ions; a proton donor; a liquid compound with a pH less than 7. Acidic chemicals are corrosive.

Ion — Atom that has lost or gained an electron, thus giving it a positive or negative charge.

Base — Any alkaline or caustic substance; corrosive water-soluble compound or substance containing group-forming hydroxide ions in water solution that reacts with an acid to form a salt.

Dissociation (Chemical) — Process of splitting a molecule or ionic compounds into smaller particles, especially if the process is reversible. *Opposite of* Recombination.

Saponification — Reaction between an alkaline and a fatty acid that produces soap.

pH Scale

Concentration of Hydrogen Ions Compared to Distilled Water	pH Scale	Examples of Solutions at this pH
Acids	0	Hydrochloric Acid
	1	Battery Acid
	2	Vinegar
	3	Orange Juice
	4	Acid rain, Wine
	5	Black Coffee
	6	Milk
Neutral	7	Distilled Water
Bases	8	Seawater
	9	Baking Soda
	10	Milk of Magnesia
	11	Ammonia
	12	Lime
	13	Lye
	14	Sodium Hydroxide

Figure 4.32 pH measures acidity and alkalinity.

Is it Corrosive, Caustic, an Acid, a Base, or an Alkali?

Some people make this differentiation: acids are corrosive, while bases are caustic. In the world of emergency response, however, both acids and bases are called *corrosives*. The U.S. Department of Transportation (DOT) and Transport Canada (TC), for example, do not differentiate between the two. These agencies consider ANY materials that destroy skin tissue or metal as corrosives.

The terms *base* and *alkali* are often used interchangeably, but some chemical dictionaries define alkalis as *strong bases*. **Basic solutions** are usually referred to as *alkaline* rather than *basic*, but, again, the two terms are often used synonymously. Just be aware that if you hear the terms *caustic*, *alkali*, or *alkaline*, they are referring to bases or basic solutions.

NOTE: For the purposes of this manual, the phrase *basic solutions* will refer to both basic and alkaline solutions.

Basic Solution — Solution that has a pH between 7 and 14.

Reactivity — Ability of a substance to chemically react with other materials, and the speed with which that reaction takes place.

Reactive Material — Substance capable of chemically reacting with other substances; for example, material that reacts violently when combined with air or water.

Reactivity

The chemical **reactivity** of a substance describes its relative ability to undergo a chemical reaction with itself or other materials. As a result, pressure buildup, temperature increase, and/or formation of noxious, toxic, or corrosive by-products may occur. **Reactive materials** commonly react vigorously or violently with air, water, heat, light, each other, or other materials.

Many first responders are familiar with the fire tetrahedron or the four elements necessary to produce combustion: oxygen, fuel, heat, and a chemical chain reaction. Fire is just one type of chemical reaction. A reactivity triangle can be used to explain the basic components of many (though not all) chemical reactions: an oxidizing agent (oxygen), a reducing agent (fuel), and an activation energy source (often, but not always, heat) **(Figure 4.33)**.

All reactions require some energy to get them started **(Figure 4.34)** (commonly referred to as **activation energy**). How much energy is needed depends on the particular reaction. In some cases, heat from an external source provides the energy-added heat from an external source (such as when starting a fire with a match). In some instances, radio waves, radiation, or another waveform of energy may provide the activation energy to the molecules (such as when food is heated in a microwave oven). In other reactions, the energy could come from a shock or pressure change (such as might occur when nitroglycerin is jostled).

Activation Energy — Minimum energy that starts a chemical reaction when added to an atomic or molecular system.

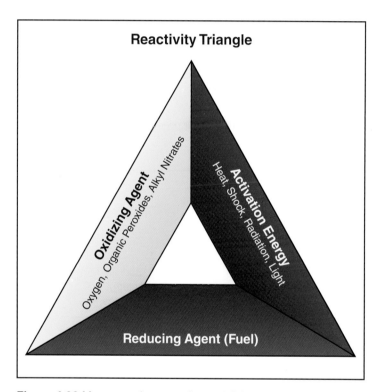

Figure 4.33 Many reactions need an oxidizing agent, a reducing agent, and some kind of activation energy to get them started.

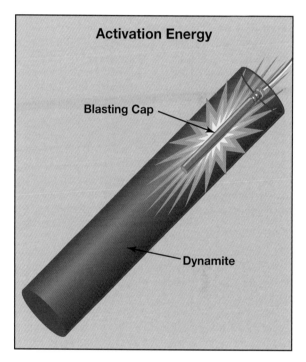

Figure 4.34 Activation energy is the energy needed to start the reaction, much like a blasting cap activating dynamite.

Reactions that have low-activation energies need little help to begin reacting. Materials that are generally classified as water-reactive typically react with water easily at room temperature because the ambient temperature is sufficient to start the reaction. First responders may see terms such as light-sensitive, heat-sensitive, or shock-sensitive on SDSs and/or manufacturers' labels, indicating that those products have an increased susceptibility to those sources of activation energy. See **Table 4.2, p. 160**, for a summary of the different ways in which chemicals can be reactive. This table supplies the definition and chemical examples of nine reactive hazard classes.

Table 4.2
Nine Reactive Hazard Classes

Reactive Hazard Class	Definition	Chemical Examples
Highly Flammable	Substances having flash points less than 100°F (38°C) and mixtures that include substances with flash points less than 100°F (38°C).	Gasoline, Acetone, Pentane, Ethyl Ether, Toluene, Methyl Ethyl Ketone (MEK), Turpentine
Explosive	A material synthesized or mixed deliberately to allow the very rapid release of chemical energy; also, a chemical substance that is intrinsically unstable and liable to detonate under conditions that might reasonably be encountered.	Dynamite, Nitroglycerin, Perchloric Acid, Picric Acid, Fulminates, Azide
Polymerizable	Capable of undergoing self-reactions that release energy; some polymerization reactions generate a great deal of heat. (The products of polymerization reactions are generally less reactive than the starting materials.)	Acrylic Acid, Butadiene, Ethylene, Styrene, Vinyl Chloride, Epoxies
Strong Oxidizing Agent	Oxidizing agents gain electrons from other substances and are themselves thereby chemically reduced, but strong oxidizing agents accept electrons particularly well from a large range of other substances. The ensuing oxidation-reduction reactions may be vigorous or violent and may release new substances that may take part in further additional reactions. Keep strong oxidizing agents well separated from strong reducing agents. In some cases, the presence of a strong oxidizing agent can greatly enhance the progress of a fire.	Hydrogen Peroxide, Fluorine, Bromine, Calcium Chlorate, Chromic Acid, Ammonium Perchlorate
Strong Reducing Agent	Reducing agents give up electrons to other substances and are thereby oxidized, but strong reducing agents donate electrons particularly well to a large range of other substances. The ensuing oxidation-reduction reactions may be vigorous or violent and may generate new substances that take part in further additional reactions.	Alkali metals (Sodium, Magnesium, Lithium, Potassium), Beryllium, Calcium, Barium, Phosphorus, Radium, Lithium Aluminum Hydride
Water-Reactive	Substances that may react rapidly or violently with liquid water and steam, producing heat (or fire) and often toxic reaction products.	Alkali metals (Sodium, Magnesium, Lithium, Potassium), Sodium Peroxide, Anhydrides, Carbides
Air-Reactive	Likely to react rapidly or violently with dry air or moist air; may generate toxic and corrosive fumes upon exposure to air or catch fire.	Finely divided metal dusts (Nickel, Zinc, Titanium), Alkali metals (Sodium, Magnesium, Lithium, Potassium), Hydrides (Diborane, Barium Hydrides, Diisobutyl Aluminum Hydride)
Peroxidizable Compound	Apt to undergo spontaneous reaction with oxygen at room temperature, to form peroxides and other products. Most such auto-oxidations are accelerated by light or trace impurities. Many peroxides are explosive, which makes peroxidizable compounds a particular hazard. Ethers and aldehydes are particularly subject to peroxide formation (the peroxides generally form slowly after evaporation of the solvent in which a peroxidizable material had been stored).	Isopropyl Ether, Furan, Acrylic Acid, Styrene, Vinyl Chloride, Methyl Isobutyl Ketone, Ethers
Radioactive Material	Spontaneously and continuously emitting ions or ionizing radiation. Radioactivity is not a chemical property, but an additional hazard that exists in addition to the chemical properties of a material.	Radon, Uranium

Source: U.S. Environmental Protection Agency's CEPPO (Chemical Emergency Preparedness and Prevention Office) Computer-Aided Management of Emergency Operations (CAMEO) software was used to identify this information.

The oxidizing agent in the reactivity triangle provides the oxygen necessary for the chemical reaction. **Strong oxidizers** are materials that encourage a strong reaction (by readily accepting electrons) from reducing agents (fuels). The greater the concentrations of oxygen present in the atmosphere, the hotter, faster, and brighter a fire will burn. The same principle applies to oxidation reactions — in general, the stronger the oxidizer, the stronger the reaction. Many organic materials ignite spontaneously when they come into contact with a strong oxidizer. An asphalt roadway could explode if liquid oxygen (a cryogenic liquid) spills on it and is accompanied by sufficient activation energy (from shock or friction such as someone stepping on it) **(Figure 4.35)**.

The **reducing agent** in the fire tetrahedron acts as the fuel source for the reaction. It combines with the oxygen (or losing electrons to the oxidizer) in such a way that energy is being released. Oxidation-reduction (redox) reactions can be extremely violent and dangerous because they release a tremendous amount of energy. Some reducing agents (fuels) are more volatile than others.

NOTE: Wood is not as prone to undergo rapid oxidation (it will not burn as easily) as a highly flammable liquid such as MEK.

Polymerization is a chemical reaction in which simple molecules combine to form long chain molecules. Catalysts will increase the rate of polymerization and decrease the activation energy necessary for further polymerization. Examples of catalysts include light, heat, water, acids, or other chemicals. Uncontrolled polymerization often results in a tremendous release of energy. Materials that may undergo violent polymerization if subjected to heat or contamination are designated with a *P* in the blue and yellow sections of the *ERG* **(Figure 4.36, p. 162)**.

NOTE: Potential for polymerization may not be included on any type of reference material other than the *ERG*, and the *ERG* may not be fully inclusive of all polymerizing materials.

Inhibitors are materials that are added to products that easily polymerize in order to control or prevent an undesired reaction. Inhibitors increase the needed activation energy. They may be exhausted over a period of time or when exposed to circumstances or unexpected contamination that causes them to be consumed more rapidly, such as exposure to heat or other reaction triggers. Shipments of polymerizing materials may become unstable if

Strong Oxidizer — Substance that readily gives off large quantities of oxygen, thereby stimulating combustion; produces a strong reaction by readily accepting electrons from a reducing agent (fuel).

Reducing Agent — Fuel that is being oxidized or burned during combustion. *Also known as* Reducer.

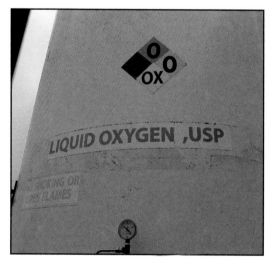

Figure 4.35 Liquid oxygen is a powerful oxidizer. If spilled on asphalt, very little activation energy is needed to cause an explosion.

Inhibitor — Material that is added to products that easily polymerize in order to control or prevent an undesired reaction. *Also known as* Stabilizer.

ID No.	Guide No.	Name of Material
1086	**116P**	Vinyl chloride, stabilized
1087	**116P**	Vinyl methyl ether, stabilized
1088	**127**	Acetal
1089	**129P**	Acetaldehyde
1090	**127**	Acetone
1091	**127**	Acetone oils
1092	**131P**	Acrolein, stabilized
1093	**131P**	Acrylonitrile, stabilized
1098	**131**	Allyl alcohol

Figure 4.36 Materials designated with a *P* in the *ERG* may undergo violent polymerization.

Figure 4.37 Time-sensitive inhibitors are added to liquid styrene before it is shipped. If involved in an emergency incident, it is very important to find out how long the inhibitor will be effective. *Courtesy of Rich Mahaney.*

delayed during transport or involved in accidents. For example, time-sensitive inhibitors are added to liquid styrene before it is shipped in order to prevent the styrene from polymerizing during transport **(Figure 4.37)**. If containers holding the styrene rupture or emergency responders add water, the inhibitor becomes exhausted (often 20-30 days), and the polymerization reaction begins. The sudden loss of containment due to polymerization is a chemical process that may not require an external heat source.

Under emergency conditions, reactive materials can be extremely destructive and dangerous. Keep people and equipment upwind, uphill, and back a safe distance or in protected locations until pertinent facts are established and definite plans can be formulated. With advances in modern technology, more and more reactive and unstable materials are being used for various processes, and you must be prepared to deal with them.

Radioactivity

In addition to recognizing radioactive material packaging, as explained in Chapter 2, first responders need to understand basic protection strategies if radioactive materials or radiation is present at an incident. Radiation comes in different forms, some more energetic than others **(Figure 4.38)**. The least energetic form of radiation is **nonionizing radiation** such as visible light and radio waves. The most energetic (and hazardous) form of radiation is **ionizing radiation**.

The following sections will address:

- Types of ionizing radiation
- Radioactive material exposure and contamination

Nonionizing Radiation — Series of energy waves composed of oscillating electric and magnetic fields traveling at the speed of light. Examples include ultraviolet radiation, visible light, infrared radiation, microwaves, radio waves, and extremely low frequency radiation.

Ionizing Radiation — Radiation that causes a chemical change in atoms by removing their electrons.

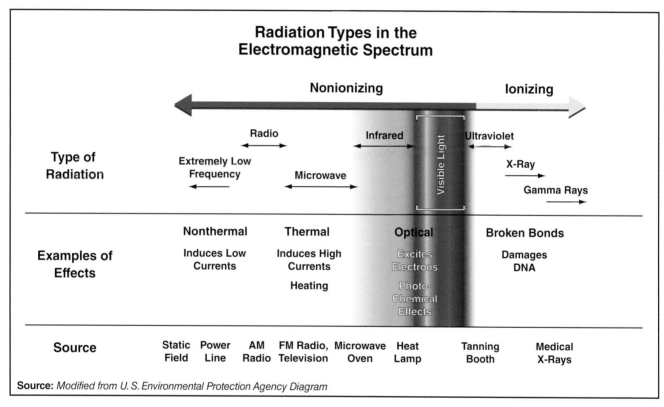

Radiation Types in the Electromagnetic Spectrum

	Nonionizing			Ionizing
Type of Radiation	Extremely Low Frequency	Radio / Microwave	Infrared / Visible Light	Ultraviolet / X-Ray / Gamma Rays
Examples of Effects	**Nonthermal** Induces Low Currents	**Thermal** Induces High Currents / Heating	**Optical** Excites Electrons / Photo-Chemical Effects	**Broken Bonds** Damages DNA
Source	Static Field / Power Line	AM Radio / FM Radio, Television / Microwave Oven	Heat Lamp	Tanning Booth / Medical X-Rays

Source: *Modified from U. S. Environmental Protection Agency Diagram*

Figure 4.38 Nonionizing radiation and ionizing radiation are on opposite sides of the spectrum.

Types of Ionizing Radiation

Ionizing radiation can be divided into four types: alpha, beta, gamma, and neutron. Each type will be explained in the following **(Figure 4.39, p.164)**:

- **Alpha** — Energetic, positively charged alpha particles (helium nuclei) emitted from the nucleus during radioactive decay that rapidly lose energy when passing through matter **(Figure 4.40, p. 164)**. They are commonly emitted in the radioactive decay of some manmade elements and the heaviest radioactive elements such as uranium and radium. Alpha particles do not travel far in open air; you may have to get very close to the source for the equipment to detect particles. Details:

  — Alpha particles lose energy rapidly when travelling through matter and do not penetrate deeply. They can cause damage over a short path through human tissue. They are usually completely blocked by the outer, dead

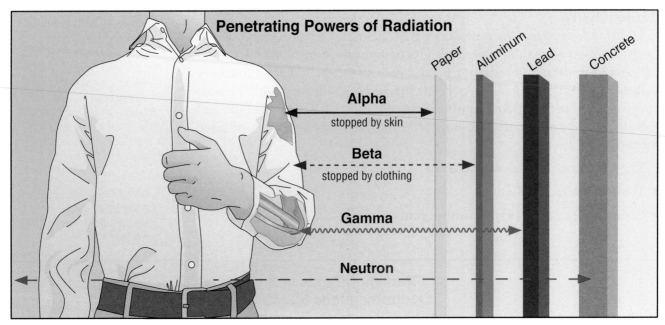

Source: *Modified from U.S. Environmental Protection Agency*

Figure 4.39 Neutron radiation is the hardest to protect against because it is highly penetrating.

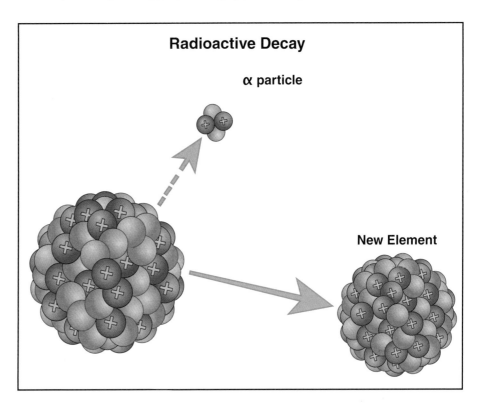

Figure 4.40 During radioactive decay, alpha particles are emitted from the nucleus of an atom, forming a new element.

layer of the human skin, so alpha-emitting radioisotopes are not a hazard outside the body. However, they can be very harmful if the material emitting the alpha particles is ingested or inhaled.

— Alpha particles can be stopped completely by a sheet of paper.

- **Beta** — Fast-moving, positively charged protons or negatively charged **electrons** emitted from the atom's nucleus during radioactive decay. Humans are exposed to beta particles from manufactured and natural sources such as tritium, carbon-14, and strontium-90.

Electron — Subatomic particle with a physical mass and a negative electric charge.

Details:

- — Beta particles penetrate further than alpha particles but cause less damage over equally traveled distances. Beta particles are capable of penetrating the skin and causing radiation damage; however, as with alpha emitters, beta emitters are generally more hazardous when inhaled or ingested.

- — Beta particles travel appreciable distances in air but can be reduced or stopped by a layer of clothing, a thin sheet of metal, or thick Plexiglass. Detection distances for beta particles vary based on the activity of the source. Compared to alpha radiation, beta radiation will travel farther. Shielding beta emitters with dense metals can result in the release of X-rays (Bremsstrahlung radiation).

- **Gamma** — High-energy **photons** (weightless packets of energy like visible light and X-rays). Gamma rays often accompany the emission of alpha or beta particles from a nucleus. They have neither a charge nor a mass but are penetrating. One source of gamma radiation in the environment is naturally occurring potassium-40. Common industrial gamma emitting sources include cobalt-60, iridium-192, and cesium-137. Details:

  - — Gamma radiation can easily pass completely through the human body or be absorbed by tissue. It constitutes a radiation hazard for the entire body.

  - — Gamma radiation levels vary depending on the isotope and activity **(Figure 4.41)**. Materials such as concrete, earth, and lead may be useful as a shield against radiation. Standard fire fighting protective clothing provides no protection against gamma radiation.

- **Neutron** — Particles that have a physical mass but have no electrical charge. Neutrons are highly penetrating. Fission reactions produce neutrons along with gamma radiation. Neutron radiation can be measured in the field using specialized equipment. Details:

  - — Soil moisture density gauges, often used at construction sites, are a common source of neutron radiation. Neutrons may also be encountered in research laboratories or operating nuclear power plants.

  - — Shielding from neutron radiation requires materials with high amounts of hydrogen, such as oil, water, and concrete.

X-rays and gamma rays are high energy electromagnetic radiation commonly referred to as *photons*. The hazards of these types of radiation are directly correlated to their activity. For the purposes of this manual, they are identical and should be treated the same. Machines such as those found in medical facilities and airports are almost exclusively the sole source of terrestrial X-ray radiation. Since machines can only produce X-rays when powered on, the chances of encountering X-rays at a hazardous materials incident are remote.

Radioactive Material Exposure and Contamination

Radioactive materials (RAM) emit ionizing radiation. Incidents involving radioactive materials are uncommon because they are strictly governed in use, packaging, and transportation. However, there is some concern that radioactive materials could be used in a terrorist attack **(Figure 4.42, p. 166)**.

Photon — Weightless packet of electromagnetic energy, such as X-rays or visible light.

Figure 4.41 Activity refers to the number of atoms in a radioactive material that will decay and emit radiation in a second. The higher the number is, the more radiation that is emitted.

Radioactive Material (RAM) — Material with an atomic nucleus that spontaneously decays or disintegrates, emitting radiation as particles or electromagnetic waves at a rate of greater than 0.002 microcuries per gram (Ci/g).

Exposure — (1) Contact with a hazardous material, causing biological damage, typically by swallowing, breathing, or touching (skin or eyes). Exposure may be short-term (acute exposure), of intermediate duration, or long-term (chronic exposure). (2) People, property, systems, or natural features that are or may be exposed to the harmful effects of a hazardous materials emergency.

Dose — Quantity of a chemical material ingested or absorbed through skin contact for purposes of measuring toxicity

Radiation **exposure** occurs when a person is near a radiation source and is exposed to the energy from that source. Exposure and damage are not necessarily related. A first responder will need to know the types of radiation that will cause damage and what proximity or level of exposure will cause what kinds of harm **(Figure 4.43)**.

A person may receive a **dose** of radiation based upon the length of exposure, energy, and type of source (alpha, beta, gamma, or neutron). Exposure to radioactive material does not make a person or object radioactive. Damage is often described in terms of dosage, indicating the amount of energy absorbed by matter.

Figure 4.42 Incidents involving radiation are rare, but there are concerns that radioactive materials could be used in a terrorist attack.

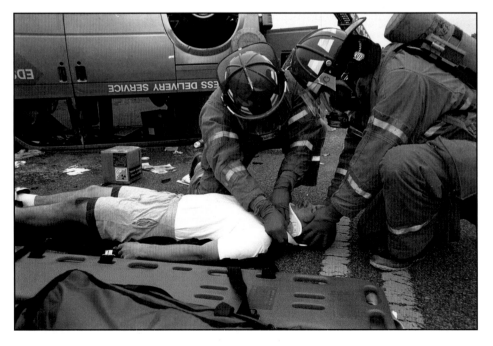

Figure 4.43 First responders must understand how to protect themselves from exposure when radioactive materials are involved in an incident. *Courtesy of Tom Clawson.*

Radioactive **contamination** occurs when radioactive material is deposited on surfaces, skin, clothing, or any place where it is not desired. Radiation does not spread; radioactive material and contamination spread.

Exposure to radiation alone does not contaminate a person. Contamination only occurs when the radioactive material remains on a person or the person's clothing after coming into contact with a **contaminant**. A person can become contaminated externally, internally, or both. Radioactive material can enter the body via one or more routes of entry. An unprotected person contaminated with radioactive material receives radiation exposure until the source of radiation (radioactive material) is removed. Radiation detectors capable of detecting alpha and beta contamination can detect radioactive contamination. Note the following examples:

- A person is externally contaminated (and receives external exposure) when radioactive material is on the skin or clothing.

- A person is internally contaminated (and receives internal exposure) when radioactive material is breathed, swallowed, or absorbed through wounds.

- The environment is contaminated when radioactive material is spread about or is unconfined. Environmental contamination is another potential source of external exposure.

NOTE: Some contamination, such as alpha contamination, often requires the detector to be almost touching the source.

Contamination —Impurity resulting from mixture or contact with a foreign substance.

Contaminant — Foreign substance that compromises the purity of a given substance.

Radiation Health Hazards

The effects of ionizing radiation occur at the cellular level. Ionizing radiation can negatively affect the normal operation of the cells that compose human organs.

Radiation may cause damage to any material by ionizing the atoms in that material. When atoms are ionized, the chemical properties of those atoms are altered. This change in chemical properties can result in a change in the chemical behavior of the atoms and/or molecules in the cell. If a person receives a sufficiently high dose of radiation and many cells are damaged, this may cause observable health effects, including genetic mutations and cancer.

The biological effects of ionizing radiation depend on how much and how fast a radiation dose is received. The two categories of radiation doses are acute and chronic.

Acute doses. Exposure to radiation received in a short period of time is an acute dose. Acute exposures are usually associated with large doses. Some acute doses of radiation are permissible and have no long-term health effects. However, high levels of radiation received over a short time can produce serious health effects, including reduced blood count, hair loss, nausea, vomiting, diarrhea, and fatigue. Extremely high levels of acute radiation exposure (such as those received by victims of a nuclear bomb) can result in death within a few hours, days, or weeks.

Chronic doses. Small amounts of radiation received over a long period of time. The body is better equipped to handle a chronic dose of radiation than an acute radiation dose. After a chronic dose, the body has enough time to replace dead or nonfunctioning cells with healthy ones. Chronic doses do not result

in the same detectable health effects seen with acute doses. However, chronic exposure to radiation causes cancer. Examples of chronic radiation doses include the everyday doses received from natural background radiation, and those received by workers in nuclear and medical facilities.

First responders at most hazmat incidents are unlikely to encounter exposures that cause any health effects, especially if proper precautions are taken. Even at terrorist incidents, it is unlikely that first responders will encounter dangerous or lethal doses of radiation **(Figure 4.44)**.

Figure 4.44 It is unlikely that responders will encounter lethal doses of radiation, especially if monitoring and detection is used appropriately. *Courtesy of the U.S. Department of Energy.*

Units Used for Measuring Radioactivity

- Radioactivity — Quantifiable measurement of activity in a sample of material over time. Measured in curie (Ci) and becquerel (Bq).

- Exposure — Amount of radiation in the ambient air of a specific place. Measured in roentgen (R) and coulomb/kilogram (C/kg).

- Absorbed dose — Amount of radiation energy deposited in a material. Measured in radiation absorbed dose (rad) and gray (Gy).

- Dose equivalent — Absorbed dose plus medical effects. Measured in roentgen equivalent man (rem) and sievert (Sv). Biological dose equivalents are measured in rem or sieverts (Sv).

Protection from Radiation

Because radiation is invisible, it may be difficult to determine if it is involved in an incident.

Class 7 radioactive materials packages should have the appropriate placard or label in transport **(Figure 4.45)**. If responders note their presence at an incident, they should initiate radiation detection and monitoring. Responders should conduct radiation monitoring if an incident is a suspected terrorist attack or explosion. **NOTE:** Chapter 11 provides more information about detecting and monitoring radiation.

While most incidents involving radioactive materials present minimal risks to emergency responders, it is still necessary to take appropriate precautions to prevent unnecessary exposures. One basic protection strategy uses time, distance, and shielding **(Figure 4.46)**:

- **Time** — Decrease the amount of time spent in areas where there is radiation. At a minimum, the time required includes:
  - Entering the zone
  - Staying within the zone
  - Exiting the zone

Figure 4.45 Radioactive materials are placarded/labeled *Class 7* in transport. *Courtesy of Rich Mahaney.*

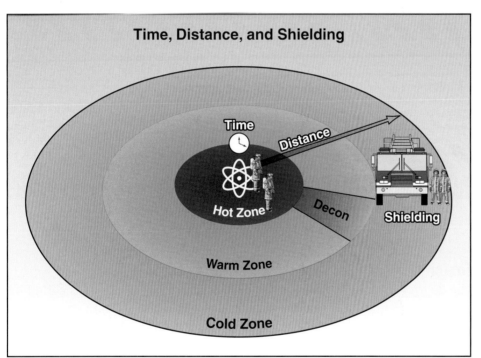

Figure 4.46 For protection, responders should limit the time they are exposed to radiation, increase their distance from the radiation source, and use shielding.

• **Distance** — Know your dose rate to know the safe distances from the radioactive material. Increase the distance from a radiation source. Doubling the distance from a point source divides the dose by a factor of four. This calculation is sometimes referred to as the **inverse square law**. When the radius doubles, the radiation spreads over four times as much area, so the dose is only one-fourth as much **(Figure 4.47, p. 170)**. If sheltered in a contaminated area, keep a distance from exterior walls and roofs. This calculation is only a rule of thumb, and the information must be supplemented with information from your meter.

Inverse Square Law — Physical law that states that the amount of radiation present is inversely proportional to the square of the distance from the source of radiation.

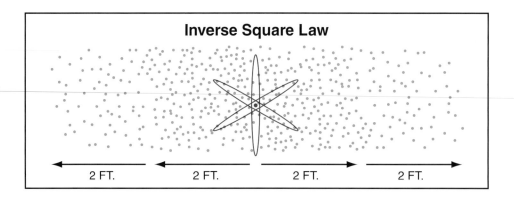

Figure 4.47 Doubling the distance from a radiation source divides the dose by a factor of four.

Inverse Square Law

2 FT.   2 FT.   2 FT.   2 FT.

- **Shielding** — Create a barrier between responders and the radiation source with a building, earthen mound, or vehicle. Buildings, especially those made of brick or concrete, provide considerable shielding from radiation. For example, exposure from fallout is reduced by about 50 percent inside a one-story building and by about 90 percent at a level belowground.

NOTE: Using time, distance, and shielding to limit exposure to radiation is sometimes referred to as the ALARA (As Low As Reasonably Achievable) method or principle.

CAUTION
Limit your time to limit the dose!

CAUTION
Maximize your distance to limit the dose!

CAUTION
Use shielding to limit your dose!

Toxicity — Degree to which a substance (toxin or poison) can harm humans or animals. Ability of a substance to do harm within the body.

Systemic Effect — Damage spread through an entire system; opposite of a local effect, which is limited to a single location.

Toxicity

The degree to which a substance causes harm within the body is called its **toxicity**. A chemical injury at the site of contact (typically the skin and mucous membranes of the eyes, nose, mouth, or respiratory tract) is termed a *local toxic effect*. Irritant gases such as chlorine and ammonia can, for example, produce a localized toxic effect in the respiratory tract. In addition, toxic materials may be absorbed into the bloodstream and distributed to other parts of the body, producing **systemic effects**. Many pesticides absorb through the skin, distribute to other sites in the body, and produce adverse effects such as seizures or cardiac, pulmonary, or other problems.

Exposure to toxic materials can result not only in the development of a single systemic effect but also in the development of multiple systemic effects or a combination of systemic and local effects. Some of these effects may be delayed in a range anywhere between seconds and decades. **Table 4.3** gives types of chemical toxins, their target organs, and chemical examples.

Table 4.3
Types of Toxicants and Their Target Organs

Toxin	Targets	Chemical Examples
Nephrotoxicant	Kidney	Halogenated Hydrocarbons, Mercury, Carbon Tetrachloride
Hemotoxicant	Blood	Carbon Monoxide, Cyanides, Benzene, Nitrates, Arsine, Naphthalene, Cocaine
Neurotoxicant	Nervous System	Organophosphates, Mercury, Carbon Disulphide, Carbon Monoxide, Sarin
Hepartoxicant	Liver	Alcohol, Carbon Tetrachloride, Trichloroethylene, Vinyl Chloride, Chlorinated HC
Immunotoxicant	Immune System	Benzene, Polybrominated Biphenyls (PBBs), Polychlorinated Biphenyls (PCBs), Dioxins, Dieldrin
Endocrine Toxicant	Endocrine System (including the pituitary, hypothalamus, thyroid adrenals, pancreas, thymus, ovaries, and testes)	Benzene, Cadmium, Chlordane, Chloroform, Ethanol, Kerosene, Iodine, Parathion
Musculoskeletal Toxicant	Muscles/Bones	Fluorides, Sulfuric Acid, Phosphine
Respiratory Toxicant	Lungs	Hydrogen Sulfide, Xylene, Ammonia, Boric Acid, Chlorine
Cutaneous Hazards	Skin	Gasoline, Xylene, Ketones, Chlorinated Compounds
Eye Hazards	Eyes	Organic Solvents, Corrosives, Acids
Mutagens	DNA	Aluminum Chloride, Beryllium, Dioxins
Teratogens	Embryo/Fetus	Lead, Lead Compounds, Benzene
Carcinogens	All	Tobacco Smoke, Benzene, Arsenic, Radon, Vinyl Chloride

Following exposure to a toxic substance, the likelihood of an adverse health effect occurring and the severity of the effect depend on the following:

- Toxicity of the chemical or biological substance
- Exposure pathway or route
- Nature and extent of the exposure
- Person's susceptibility to illness or injury, affected by such factors as their age or other health concerns (including chronic diseases)

CAUTION
All personnel working at hazardous materials incidents must use appropriate personal protective equipment, including appropriate respiratory protection equipment.

Eating and Drinking Can be Dangerous . . .

. . . especially on the scene of a hazardous material incident. If hazardous materials at an incident site contaminate food or water, the chemicals can be ingested into the body where they can cause harm. Therefore, never eat or drink in areas where hazardous materials may be present. Make sure that water comes from a clean source and is dispensed in disposable cups. Always place rehabilitation areas far away from any sources of contamination. Finally, wash your hands and be certain that you are completely decontaminated before eating or drinking.

Asphyxiant — Any substance that prevents oxygen from combining in sufficient quantities with the blood or from being used by body tissues.

Irritant — Liquid or solid that, upon contact with fire or exposure to air, gives off dangerous or intensely irritating fumes. *Also known as* Irritating Material.

Convulsant — Poison that causes convulsions.

The following are some specific toxic chemical hazard categories:

- **Asphyxiants** — Asphyxiants prevent access to sufficient volumes of oxygen. They can be divided into two classes: simple and chemical. Simple asphyxiants are gases that displace oxygen **(Figure 4.48)**. These gases may dilute or displace the oxygen concentration below the level required to sustain life. Chemical asphyxiants are materials that prohibit the body's cells from using oxygen. Some chemical asphyxiants may be used in terrorist attacks.

Figure 4.48 Nitrogen is a simple asphyxiant because it can dilute and displace oxygen.

Figure 4.49 Irritants often attack the skin, eyes, nose, mouth, throat, and lungs.

- **Irritants** — Irritants cause temporary, sometimes severe, inflammation to the eyes, skin, or respiratory system **(Figure 4.49)**. Irritants often attack the body's mucous membranes, such as the surfaces of the eyes or the linings of the nose, mouth, throat, and lungs.

- **Convulsants** — Convulsants cause convulsions (involuntary muscle contractions). Convulsants can kill if the victim asphyxiates or succumbs to exhaustion while convulsing. Examples of convulsants include strychnine, organophosphates, carbamates, and infrequently used drugs such as picrotoxin **(Figure 4.50)**.

Figure 4.50 Exposure to some organophosphate pesticides may cause convulsions. *Courtesy of Rich Mahaney.*

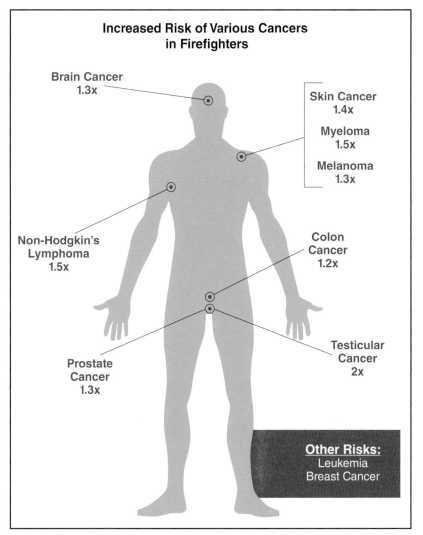

Increased Risk of Various Cancers in Firefighters

Brain Cancer
1.3x

Skin Cancer
1.4x

Myeloma
1.5x

Melanoma
1.3x

Non-Hodgkin's Lymphoma
1.5x

Colon Cancer
1.2x

Testicular Cancer
2x

Prostate Cancer
1.3x

Other Risks:
Leukemia
Breast Cancer

Figure 4.51 Firefighters have a higher risk for certain cancers than the general population.

- **Carcinogens** — Carcinogens are known, or suspected, to cause cancer. While exact exposure data is unknown for most carcinogens, exposures to only small amounts of some substances may have long-term consequences. Disease and complications can occur as many as 10 to 40 years after exposure. Statistics reveal 1 in 3 firefighters will be diagnosed with cancer during their career **(Figure 4.51)**. Another 45% of firefighters will have cancer diagnosed within seven years of retirement. Meta studies in the U.S. and Australia show firefighters have an elevated risk of suffering from leukemia and testicular cancer (114%-202% respectively) to name just two. Known or suspected carcinogens include:

— Arsenic

— Asbestos

— Benzene

— Many plastics

— Nickel

— Polyvinyl chloride

— Some chlorinated hydrocarbons

— Some pesticides

- **Allergens** and sensitizers — Allergens cause allergic reactions in people or animals. Sensitizers are chemicals that cause a substantial proportion of exposed people or animals to develop an allergic reaction after one or more exposures to the chemical. Some individuals exposed to a material

Carcinogen — Cancer-producing substance.

Allergen — Material that can cause an allergic reaction of the skin or respiratory system.

may not be abnormally affected at first but may experience significant and dangerous effects when exposed to the material again. Common examples of sensitizers and allergens include latex, bleach, and urushiol (the chemical found in the sap of poison ivy, poison oak, and poison sumac) **(Figure 4.52)**.

Figure 4.52 After repeated exposure to bleach, some people become sensitized to it.

Toxic Products of Combustion

While the heat energy from a fire is a danger to anyone directly exposed to it, exposure to toxic smoke can cause both acute and chronic health effects. Smoke is an aerosol comprised of gases, vapor, and solid particulates. Fire gases, such as carbon monoxide, are generally colorless, while vapor and particulates give smoke its varied colors. Most components of smoke are toxic and many are carcinogenic. Almost all present a significant threat to human life. The materials that compose smoke vary from fuel to fuel. Generally, consider all smoke toxic and carcinogenic. In addition to the gases listed below, fires and smoke may expose responders to a wide range of potentially carcinogenic substances such as asbestos, soot, and creosote. **Table 4.4** lists some of the more common products of combustion and their toxic effects. Three of the more common products of combustion are:

- **Carbon monoxide (CO)** is a chemical asphyxiant that is a byproduct of the incomplete combustion of organic (carbon-containing) materials. This gas is probably the most common product of combustion encountered in structure fires. Exposure to it is frequently identified as the cause of death for civilian fire fatalities and for firefighters who have run out of air in their SCBAs.

- **Hydrogen cyanide (HCN)**, produced in the combustion of materials containing nitrogen, is also commonly encountered in smoke, although at lower concentrations than CO. HCN also acts as a chemical asphyxiant. HCN is a significant byproduct of the combustion of polyurethane foam, which is commonly used in furniture and bedding.

- **Carbon dioxide (CO_2)** is a product of complete combustion of organic materials. It acts as a simple asphyxiant by displacing oxygen. Carbon dioxide also increases respiratory rate.

Carbon Monoxide (CO) — Colorless, odorless, dangerous gas (both toxic and flammable) formed by the incomplete combustion of carbon. It combines with hemoglobin more than 200 times faster than oxygen does, decreasing the blood's ability to carry oxygen.

Hydrogen Cyanide (HCN) — Colorless, toxic, and flammable liquid until it reaches 79° F (26° C). Above that temperature, it becomes a gas with a faint odor similar to bitter almonds; produced by the combustion of nitrogen-bearing substances.

Carbon Dioxide (CO_2) — Colorless, odorless, heavier than air gas that neither supports combustion nor burns; used in portable fire extinguishers as an extinguishing agent to extinguish Class B or C fires by smothering or displacing the oxygen. CO_2 is a waste product of aerobic metabolism.

Biological Hazards

Biological (or *etiological*) hazards are microorganisms, such as viruses or bacteria (or their toxins), that may cause severe, disabling disease or illness. Many of these hazards can be transferred from the blood or other bodily fluids of an infected individual. Additionally, some biological hazards cause illness through their toxicity. Always wear appropriate PPE to prevent potential transmission.

Types of biological hazards include:

- **Viruses** — Viruses are the simplest types of microorganisms that can only replicate themselves in the living cells of their hosts **(Figure 4.53, p. 176)**. Viruses do not respond to antibiotics.

- **Bacteria** — Bacteria are microscopic, single-celled organisms **(Figure 4.54, p. 176)**. Bacteria may cause disease in people either by invading the tissues or by producing toxins (poisons).

Table 4.4
Common Products of Combustion and Their Toxic Effects

Acetaldehyde	Colorless liquid with a pungent choking odor, which is irritating to the mucous membranes and especially the eyes. Breathing vapors will cause nausea, vomiting, headache and unconsciousness.
Acrolein	Colorless to yellow volatile liquid with a disagreeable choking odor, this material is irritating to the eyes and mucous membranes. This substance is extremely toxic; inhalation of concentrations as little as 10 ppm may be fatal within a few minutes.
Asbestos	A magnesium silicate mineral that occurs as slender, strong flexible fibers. Breathing of asbestos dust causes asbestosis and lung cancer.
Benzene	Colorless liquid with a petroleum-like odor. Acute exposure to benzene can result in dizziness, excitation, headache, difficulty breathing, nausea and vomiting. Benzene is also a carcinogen.
Benzaldehyde	Colorless to clear yellow liquid with a bitter almond odor. Inhalation of concentrated vapor is irritating to the eyes, nose, and throat.
Carbon Monoxide	Colorless, odorless gas. Inhalation of carbon monoxide causes headache, dizziness, weakness, confusion, nausea, unconsciousness, and death. Exposure to as little as 0.2% carbon monoxide can result in unconsciousness within 30 minutes. Inhalation of high concentration can result in immediate collapse and unconsciousness.
Formaldehyde	Colorless gas with a pungent irritating odor that is highly irritating to the nose. 50-100 ppm can cause severe irritation to the respiratory track and serious injury. Exposure to high concentrations can cause injury to the skin. Formaldehyde is a suspected carcinogen.
Glutaraldehyde	Light yellow liquid that causes severe irritation of the eyes and irritation of the skin.
Hydrogen Chloride	Colorless gas with a sharp, pungent odor. Mixes with water to form hydrochloric acid. Hydrogen chloride is corrosive to human tissue. Exposure to hydrogen chloride can result in irritation of skin and respiratory distress.
Isovaleraldehyde	Colorless liquid with a weak, suffocating odor. Inhalation causes respiratory distress, nausea, vomiting and headache.
Nitrogen Dioxide	Reddish brown gas or yellowish-brown liquid, which is highly toxic and corrosive.
Particulates	Small particles that can be inhaled and be deposited in the mouth, trachea, or the lungs. Exposure to particulates can cause eye irritation, respiratory distress (in addition to health hazards specifically related to the particular substances involved).
Polycyclic Aromatic Hydrocarbons (PAH)	PAH are a group of over 100 different chemicals that generally occur as complex mixtures as part of the combustion process. These materials are generally colorless, white, or pale yellow-green solids with pleasant odor. Some of these materials are human carcinogens.
Sulfur Dioxide	Colorless gas with a choking or suffocating odor. Sulfur dioxide is toxic and corrosive and can irritate the eyes and mucous membranes.

Source: *Computer Aided Management of Emergency Operations (CAMEO) and Toxicological Profile for Polycyclic Aromatic Hydrocarbons.*

- **Rickettsias** — Rickettsias are specialized bacteria that live and multiply in the gastrointestinal tract of arthropod carriers (such as ticks and fleas) **(Figure 4.55)**. They are smaller than most bacteria, but larger than viruses. Like bacteria, they are single-celled organisms with their own metabolisms, and they are susceptible to broad-spectrum antibiotics. However, like viruses, they only reproduce in living cells. Most rickettsias spread only through the bite of infected arthropods (such as ticks) and not through human contact. Two types of rickettsia have been weaponized as bioterrorism agents.

- **Biological toxins** — Biological toxins are produced by living organisms; however, the biological organism itself is usually not harmful to people **(Figure 4.56)**.

Infectious diseases are caused by the reproduction and spread of microorganisms (**pathogens**) in the body. They may be **contagious**.

Infectious — Transmittable; able to infect people.

Pathogen — Biological agent that causes disease or illness.

Contagious — Capable of transmission from one person to another through contact or close proximity.

Figure 4.53
Viruses like the deadly Ebola virus picturod here are unaffected by antibiotics. *Courtesy of the CDC Public Health Image Library.*

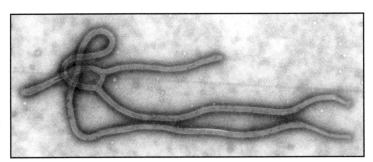

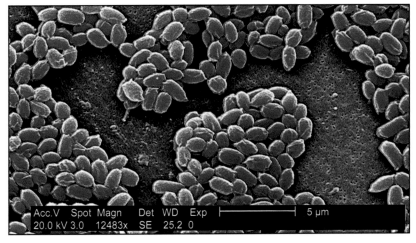

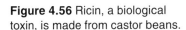

Figure 4.54 Anthrax is a type of bacteria. Bacterial infections can be treated with antibiotics. *Courtesy of the CDC Public Health Image Library.*

Figure 4.55 Ticks, fleas, and other arthropods may carry rickettsias. *Courtesy of the U.S. Department of Agriculture.*

Figure 4.56 Ricin, a biological toxin, is made from castor beans.

Exposure to biological hazards may occur in biological and medical laboratories, agricultural facilities, or when dealing with people or animals who are carriers of such diseases. Some of these diseases are carried in body fluids and are transmitted by contact with the fluids. For instance, in 2014, health care providers in Dallas, TX, were infected with a naturally occurring strain of Ebola after contact with one patient exposed in the outbreak in Africa. Examples of diseases associated with biological hazards or threats are:

- Malaria

- Tuberculosis

- Hepatitis B

- Measles

- Ebola

- Influenza

- Typhoid

First responders may also be exposed to biological agents used as weapons in terrorist attacks and criminal activities. These biological attacks could produce death and disease in people, animals, and plants. The 2001 anthrax attacks in the United States were an example of a biological attack. Biological attacks use weaponized forms of disease-causing organisms and/or their toxins. Examples of potential biological weapons include:

- Smallpox (virus) **(Figure 4.57)**

- Anthrax (bacteria)

- Botulism (toxin from the bacteria Clostridium botulinum)

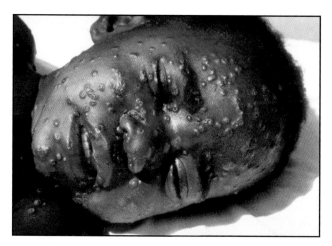

Figure 4.57 Smallpox virus could be used as a biological weapon. *Courtesy of the CDC Public Health Image Library.*

Hazard Classes

Transportation hazard classes, introduced in Chapter 2, are easier to understand once you have a better understanding of the basic physical and chemical properties explained in the previous sections. In general, a product's hazard class is assigned based on its most dangerous chemical and/or physical properties. Flammable gases combine the physical hazard of a gaseous state, which expands rapidly, spreads easily, and is difficult to confine, with flammability.

This section describes:

- Class 1—Explosives

- Class 2—Gases

- Class 3—Flammable liquids (and combustible liquids in the U.S.)

- Class 4—Flammable solids, spontaneously combustible, and dangerous when wet

- Class 5—Oxidizers and organic peroxides

- Class 6—Poisons, poison inhalation hazards, and **infectious substances**
- Class 7—Radioactive
- Class 8—Corrosives
- Class 9—Miscellaneous hazardous materials

NOTE: Chapter 2 describes the difference between placards and labels.

Class 1: Explosives

Explosives are reactive. An **explosive** is any substance or article with a great deal of potential energy that may rapidly expand and release upon activation (undergo an explosion) **(Figure 4.58)**. Explosives may release energy in the form of light, gas, and/or heat. Some explosives may not be specifically designed to explode. Chapter 8 describes the explosives as a hazard that responders face.

Explosive placards list both a **division number** and a **compatibility group letter** on them. First responders should pay particular attention to the division number, which assigns the level of explosion hazard to the product. Compatibility group letters categorize different types of explosive substances and articles for purposes of stowage and segregation **(Figure 4.59)**.

Explosives will typically be packaged as solids in individual packages or boxes. However, some explosives are liquids such as certain **binary explosives**. Some transportation vehicles and certain storage areas are specifically designed for explosives **(Figures 4.60)**.

Figure 4.58 Explosives will undergo an extremely fast self-propagation reaction when subjected to the necessary activation energy.

Figure 4.59 Explosive placards include a division number and compatibility group letter. This vehicle is used to transport fireworks. *Courtesy of Rich Mahaney.*

Figure 4.60 Certain containers and storage areas are specifically designed for explosives. *Courtesy of David Alexander with the Texas Commission on Fire Protection.*

The primary hazards of explosives are thermal and mechanical. These hazards may manifest in the following conditions:

- **Blast-pressure wave (shock wave)** — Rapidly released gases can create a shock wave that travels outward from the center. As the wave increases in distance, the strength decreases. This blast-pressure wave is the primary reason for injuries and damage. The blast-pressure wave has a positive and negative phase, both of which can cause damage (**Figures 4.61a, b, and c**).

Positive Pressure Phase

Figure 4.61a An explosion's blast pressure will compress the surrounding atmosphere into a rapidly expanding shock front. Depending on its force, this positive pressure wave can be extremely destructive.

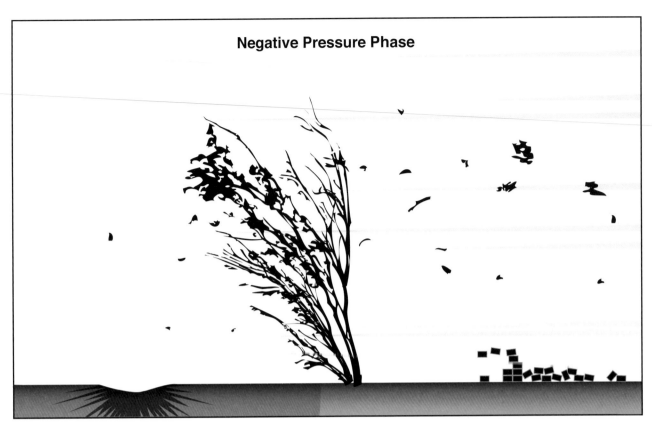

Figure 4.61b Typically less destructive than the positive pressure phase, additional damage can be done during the negative pressure phase, particularly to buildings and structures damaged in the initial blast.

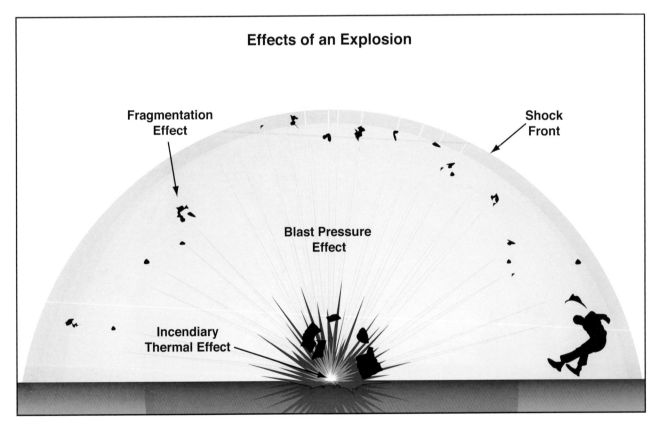

Figure 4.61c Effects of an explosion include the blast pressure effect, incendiary/thermal effects, the shock front, and the fragmentation effect.

- **Shrapnel and fragmentation** — Small pieces of debris thrown from a container or structure that ruptures during an explosion from containment or restricted blast pressure. Shrapnel and fragmentation may be thrown over a wide area and great distances, causing personal injury and other types of damage to surrounding structures or objects. Shrapnel and fragmentation can result in bruises, punctures, or even avulsions (part of the body being torn away) when they strike a person.

- **Seismic effect** — Vibration is similar to an earthquake. Explosions can cause a seismic effect. When a blast occurs at or near ground level, the air blast creates a ground shock or crater. As the shock waves move across or underground, they form a seismic disturbance. The distance the shock wave travels depends on the type and size of the explosion and type of soil.

- **Incendiary thermal effect** — Occurs during an explosion when thermal heat energy forms a fireball. Fireballs result from the interactions among burning combustible gases or flammable vapors and ambient air at high temperatures. The thermal heat fireball is present for a limited time after the explosive event.

Additional hazards unrelated to the explosion include:

- Chemical hazards will probably result from the production of toxic gases and vapors.

- Explosives may self-contaminate as they age, which increases their sensitivity and instability.

- Explosives may have high sensitivity to shock and friction.

Table 4.5 provides the U.S. DOT's explosive divisions' definitions, with examples.

	Table 4.5 Class 1 Divisions	
Division Number	**Definition**	**Examples**
Division 1.1	Explosives that have a mass explosion hazard. A mass explosion is one that affects almost the entire load instantaneously.	Dynamite, mines, wetted mercury fulminate
Division 1.2	Explosives that have a projection hazard but not a mass explosion hazard.	Detonation cord, rockets (with bursting charge), flares, fireworks
Division 1.3	Explosives that have a fire hazard and either a minor blast hazard or a minor projection hazard or both. Not a mass explosion hazard.	Liquid-fueled rocket motors, smokeless powder, practice grenades, aerial flares
Division 1.4	Explosives that present a minor explosion hazard. The explosive effects are largely confined to the package and no projection of fragments of appreciable size or range is expected. An external fire must not cause virtually instantaneous explosion of almost the entire contents of the package.	Signal cartridges, cap type primers, igniter fuses, fireworks
Division 1.5	Substances that have a mass explosion hazard but are so insensitive that there is little probability of initiation or of transition from burning to detonation under normal transportation conditions.	Prilled ammonium nitrate fertilizer or fuel oil (ANFO) mixtures and blasting agents
Division 1.6	Extremely insensitive articles that do not have a mass explosive hazard. This division is comprised of articles that contain only extremely insensitive detonating substances and that demonstrate a negligible probability of accidental initiation or propagation.	Low vulnerability military weapons

Source: *49 CFR 173.50*

Class 2: Gases

Gases are materials that are in a gaseous state at normal temperatures and pressures **(Figure 4.62)**. Gases are transported or stored in pressure containers or cryogenic containers **(Figures 4.63 a and b)**. Gas division numbers are assigned according to the type of potential hazard gases pose, such as flammability. The potential hazards of gas include energy, toxicity (including asphyxiation), and corrosivity **(Figure 4.64).** Other potential hazards include:

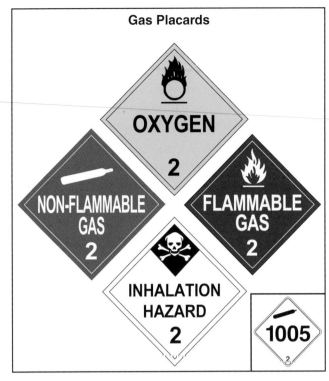

Figure 4.62 Class 2 materials are gases at normal temperatures and pressures.

- **Heat hazards** — Fires, particularly associated with Division 2.1 and oxygen; gases can travel great distances to an ignition source

- **Asphyxiation hazards** — Leaking or released gases displacing oxygen in a confined space

- **Cold hazards** — Exposure to Division 2.2 cryogens

- **Mechanical hazards** — A BLEVE (boiling liquid expanding vapor explosion) for containers exposed to heat or flame; a ruptured cylinder rocketing after exposure to heat or flame

- **Chemical hazards** — Toxic and/or corrosive gases and vapors, particularly associated with Division 2.3

Figure 4.63a and b Gases are transported in **(a)** pressure containers and **(b)** cryogenic containers.

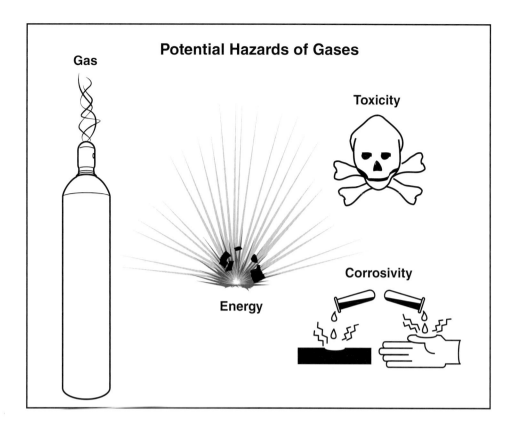

Potential Hazards of Gases

Gas

Toxicity

Energy

Corrosivity

Figure 4.64 Gases can cause harm via energy release, toxicity, and corrosivity.

Table 4.6, p. 184, provides the U.S. DOT's Class 2 divisions' placards, definitions, and examples.

Class 3: Flammable Liquids (and Combustible Liquids [U.S.])

Flammable and combustible liquids ignite and burn with relative ease (**Figure 4.65, p. 185**). Most hazmat incidents, such as gasoline and diesel fuel spills, involve this hazard class. In addition to burning, all flammable and combustible liquids exhibit varying degrees of toxicity. Some flammable liquids are also corrosive.

While these materials are transported in liquid containers, they may give off hazardous vapors, much like gases, that may travel from their source (**Figure 4.66, p. 185**). These vapors will burn if ignited.

The primary hazards of flammable and combustible liquids are energy, corrosivity, and toxicity. They often manifest in the following conditions:

- **Thermal hazards (heat)** — Fires and **vapor explosions (Figure 4.67, p. 185)**

- **Asphyxiation** — Heavier-than-air vapors displacing oxygen in low-lying and/or confined spaces

- **Chemical hazards** — Toxic and/or corrosive gases and vapors; these may be produced by fires

- **Mechanical hazards** — A BLEVE, for containers exposed to heat or flame; caused by a vapor explosion

- **Vapors** — Can mix with air and travel great distances to an ignition source

- **Environmental hazards (pollution)** — Caused by runoff from fire control

Vapor Explosion — Occurrence when a hot liquid fuel transfers heat energy to a colder, more volatile liquid fuel. As the colder fuel vaporizes, pressure builds in a container and can create shockwaves of kinetic energy.

Table 4.6
Class 2 Divisions, Placards, Definitions, and Examples

Division Number and Placard	Definition
Division 2.1 FLAMMABLE GAS 2	**Flammable Gas** — Consists of any material that is a gas at 68°F (20°C) or less at normal atmospheric pressure or a material that has a boiling point of 68°F (20°C) or less at normal atmospheric pressure and that (1) Is ignitable at normal atmospheric pressure when in a mixture of 13 percent or less by volume with air, or (2) Has a flammable range at normal atmospheric pressure with air of at least 12 percent, regardless of the lower limit. *Examples:* compressed hydrogen, isobutene, methane, and propane
Division 2.2 NON-FLAMMABLE GAS 2	**Nonflammable, Nonpoisonous Gas** — Nonflammable, nonpoisonous compressed gas, including compressed gas, liquefied gas, pressurized cryogenic gas, and compressed gas in solution, asphyxiant gas and oxidizing gas; means any material (or mixture) which exerts in the packaging an absolute pressure of 40.6 psi (280 kPa) or greater at 68°F (20°C) and does not meet the definition of Divisions 2.1 or 2.3. *Examples:* carbon dioxide, helium, compressed neon, refrigerated liquid nitrogen, cryogenic argon
Division 2.3 INHALATION HAZARD 2	**Gas Poisonous by Inhalation** — Material that is a gas at 68°F (20°C) or less and a pressure of 14.7 psi (101.3 kPa) (a material that has a boiling point of 68°F [20°C] or less at 14.7 psi [101.3 kPa]), and that is known to be so toxic to humans as to pose a hazard to health during transportation; or (in the absence of adequate data on human toxicity) is presumed to be toxic to humans because of specific test criteria on laboratory animals. Division 2.3 has *ERG*-designated hazard zones associated with it, determined by the concentration of gas in the air: • Hazard Zone A — LC50 less than or equal to 200 ppm • Hazard Zone B — LC50 greater than 200 ppm and less than or equal to 1,000 ppm • Hazard Zone C — LC50 greater than 1,000 ppm and less than or equal to 3,000 ppm • Hazard Zone D — LC50 greater than 3,000 ppm and less than or equal to 5,000 ppm *Examples:* cyanide, diphosgene, germane, phosphine, selenium hexafluoride, and hydrocyanic acid
OXYGEN 2	**Oxygen Placard** — Oxygen is not a separate division under Class 2, but first responders may see this oxygen placard on containers with 1,001 lbs (454 kg) or more gross weight of either compressed gas or refrigerated liquid.

Source: *49* CFR *173.115*

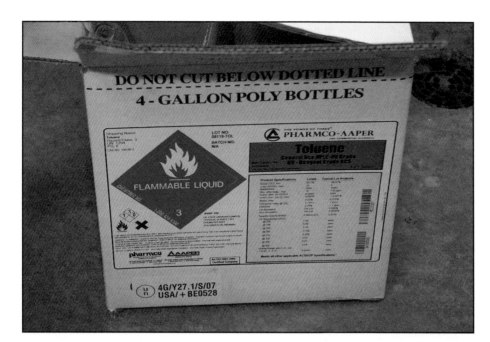

Figure 4.65 Class 3 materials ignite and burn easily.

Figure 4.66 Although liquids, in most conditions, Class 3 materials will give off flammable vapors that behave much like gases.

Figure 4.67 While flammable liquids may be corrosive and/or toxic, their primary hazard is flammability. *Courtesy Williams Fire & Hazard Control Inc., Brent Gaspard.*

Table 4.7 provides the U.S. DOT's Class 3 divisions' placards, definitions, and examples.

Table 4.7
Class 3 Divisions, Placards, Definitions, and Examples

Placard	Definition
FLAMMABLE 3	**Flammable** A *flammable liquid* is generally a liquid having a flash point of not more than 140°F (60°C), or any material in a liquid state with a flash point at or above 100°F (37.8°C) that is intentionally heated and offered for transportation or transported at or above its flash point in a bulk packaging. *Examples:* gasoline, methyl ethyl ketone
GASOLINE 3	**Gasoline Placard** — May be used in the place of a flammable placard on a cargo tank or a portable tank being used to transport gasoline by highway
COMBUSTIBLE 3	**Combustible** A *combustible liquid* is any liquid that does not meet the definition of any other hazard class and has a flash point above 140°F (60°C) and below 200 °F (93 °C). A flammable liquid with a flash point at or above 100°F (37.8°C) that does not meet the definition of any other hazard class may be reclassified as a combustible liquid. This provision does not apply to transportation by vessel or aircraft, except where other means of transportation is impracticable. An elevated temperature material that meets the definition of a Class 3 material because it is intentionally heated and offered for transportation or transported at or above its flash point may not be reclassified as a combustible liquid. *Examples:* diesel, fuel oils, pine oil
FUEL OIL 3	**Fuel Oil Placard** — May be used in place of a combustible placard on a cargo tank or portable tank being used to transport fuel oil by highway. *Examples:* Bunker fuel, heating fuel

Source: *49 CFR 173.120*

Class 4: Flammable Solids, Spontaneously Combustible, and Dangerous When Wet

Class 4 materials are divided into three different divisions (**Figure 4.68**):

- 4.1 Flammable Solids
- 4.2 Spontaneously Combustible Materials
- 4.3 Dangerous When Wet

It may be difficult for responders to extinguish fires involving Class 4 materials. Class 4 materials are often solids (metals) that react violently in unexpected ways. For example:

- Some flammable solids will react to friction.

- Spontaneously combustible materials may ignite after contact with air.

- Dangerous when wet materials, if involved in a fire, may burn more intensely if firefighters attempt to extinguish the fire with water (**Figure 4.69**).

- Fires involving Class 4 materials may be difficult to extinguish.

- Incidents involving these materials can be difficult to manage. Even more experienced responders may not fully understand the hazards, and the typical response may make the situation worse.

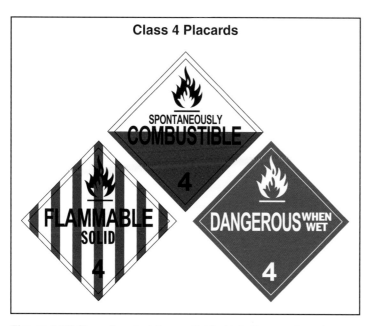

Figure 4.68 Class 4 materials are divided into three categories based on type of reactivity.

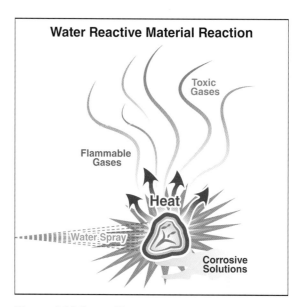

Figure 4.69 Some Class 4 materials react violently when contacted by water.

The primary hazards of Class 4 materials are: chemical energy, mechanical energy, corrosivity, and toxicity. Some examples include:

- Thermal hazards (heat)

- Fires that may start or reignite spontaneously or upon contact with air or water

- Fires and vapor explosions

- Molten substances

- Chemical hazards from irritating, corrosive, and/or highly toxic gases and vapors produced by fire or decomposition

- Severe chemical burns

- Mechanical effects from a BLEVE (if containers exposed to heat, or flame or Division 4.3 contaminated with water) or other unexpected, violent chemical reactions and explosions

- Chemical hazards from:
  — Production of hydrogen gas from contact with metal
  — Production of corrosive solutions on contact with water, for Division 4.3
  — Production of flammable gas on contact with water, for Division 4.3 (such as calcium carbide)
- Environmental hazards (pollution) caused by runoff from fire control

Table 4.8 provides the U.S. DOT's Class 4 divisions' placards, definitions, and examples.

Table 4.8
Class 4 Divisions, Placards, Definitions, and Examples

Division Number and Placard	Definition
Division 4.1	**Flammable Solid Material** — Includes (1) wetted explosives, (2) self-reactive materials that can undergo a strongly exothermal decomposition, and (3) readily combustible solids that may cause a fire through friction, certain metal powders that can be ignited and react over the whole length of a sample in 10 minutes or less, or readily combustible solids that burn faster than 2.2 mm/second: • Wetted explosives: Explosives with their explosive properties suppressed by wetting with sufficient alcohol, plasticizers, or water • Self-reactive materials: Materials liable to undergo a strong exothermic decomposition at normal or elevated temperatures due to excessively high transport temperatures or to contamination • Readily combustible solids: Solids that may ignite through friction or any metal powders that can be ignited *Examples:* phosphorus heptasulfide, paraformaldehyde, magnesium alloys
Division 4.2	**Spontaneously Combustible Material** — Includes (1) a pyrophoric material (liquid or solid) that, without an external ignition source, can ignite within 5 minutes after coming in contact with air and (2) a self-heating material that, when in contact with air and without an energy supply, is liable to self-heat *Examples:* sodium sulfide, potassium sulfide, phosphorus (white or yellow, dry), aluminum and magnesium alkyls, charcoal briquettes when shipped in bulk
Division 4.3	**Dangerous-When-Wet Material** — Material that, by contact with water, is liable to become spontaneously flammable or to release flammable or toxic gas at a rate greater than 1 liter per kilogram of the material per hour *Examples:* magnesium powder, lithium, ethyldichlorosilane, calcium carbide, potassium

Source: *49 CFR 173.124*

Class 5: Oxidizers and Organic Peroxides

Class 5 is divided into two divisions **(Figure 4.70)**:

- 5.1 Oxidizers — Typically solids or aqueous solutions

- 5.2 Organic Peroxides — Liquids or solids.

Oxidizers vigorously support combustion, may be explosive, and when combined with fuel, may burn continuously **(Figure 4.71)**. Some oxidizers, in conjunction with a fuel, have the ability to burn continuously without air being present. Oxidizers may also be explosive. Oxygen is an example of an oxidizer.

Organic peroxides are oxidizers with a specific chemical composition that make them prone to reactivity. When these materials are involved in an incident, you will need a small amount of heat to start a fire or explosion. Organic peroxides are both a fuel and an oxidizer. Because of this, they are reactive. Store organic peroxides below the **maximum safe storage temperature (MSST)**.

Organic Peroxide — Any of several organic derivatives of the inorganic compound hydrogen peroxide.

Maximum Safe Storage Temperature (MSST) — Temperature below which the product can be stored safely. This is usually 20-30 degrees cooler than the SADT temperature, but may be much cooler depending on the material.

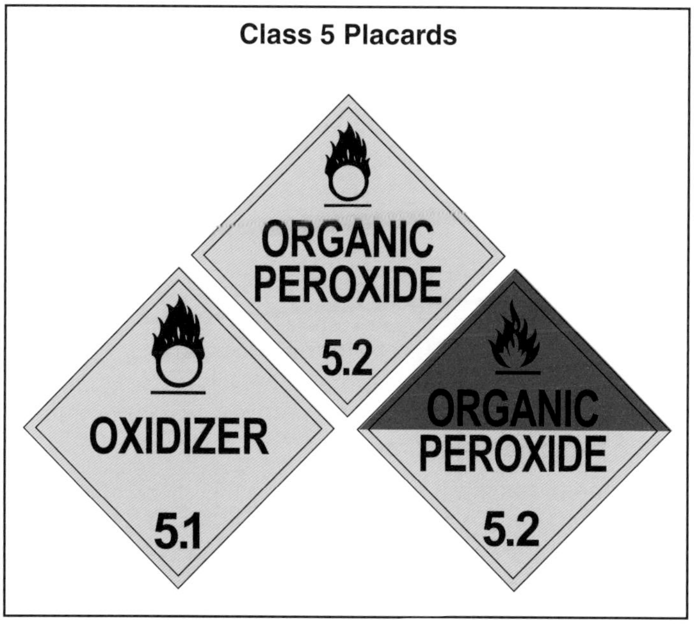

Class 5 Placards

ORGANIC PEROXIDE 5.2

OXIDIZER 5.1

ORGANIC PEROXIDE 5.2

Figure 4.70 Class 5 materials are oxidizers and organic peroxides.

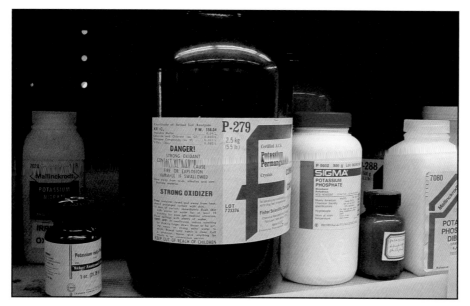

Figure 4.71 Oxidizers support combustion.

If organic peroxides reach the **self-accelerating decomposition temperature (SADT)**, they undergo a chemical change and may violently release from their packaging. The time before reaction depends upon how much the SADT is exceeded, which can greatly accelerate the decomposition.

WARNING!

Immediately evacuate the area if the SADT is reached. If decomposition occurs, observe it from a safe distance and take only those measures necessary to preserve life and nearby property.

Organic peroxides are not the only materials to have SADTs. Many polymerization initiators or reactive chemicals have an SADT. The responder should recognize these materials by using an SDS or other reference source to obtain this data. Many times the SADT is written into the SDS as *decomposition temperature*.

Relevance of SADT and MSST

Benzoyl peroxide is used in several different formulations and chemical reactions in industry and college chemistry laboratories. It has an SADT at the melting point of 71°C (160°F). Its MSST is 30°C (86°F). The laboratory stores the material in an explosion-proof refrigerator.

At a university, a graduate assistant was using benzoyl peroxide in a research experiment. The material was stored in a normal small laboratory refrigerator set at 40°F (4.5°C). The assistant left a stainless steel scoopula in the container with the benzoyl peroxide. One evening, the fire alarm went off in the chemical lab.

The university and the local fire department responded rapidly. Responders entered the building wearing turnouts and breathing apparatus and made their way to the laboratory on the third floor. The third floor was full of smoke, but without any active fire. There was no loss of life, but damage to the laboratory was extensive. The refrigerator door had been violently thrown to the back wall of the laboratory, through several sets of oak benches. The far wall of the laboratory was black from the flash fire that had occurred some 40 feet (12 m) away.

An investigation showed that the refrigerator had lost power and the temperature in the refrigerator had reached room temperature at 70°F (21°C). Even though the temperature was below the MSST of 86°F (30°C), the metal scoopula acted as a catalyst. It dramatically lowered the SADT of the benzoyl peroxide. Responders should take steps to prevent contamination when responding to hazardous materials incidents involving peroxides.

The primary hazards of Class 5 materials are thermal, mechanical, chemical. Some examples include:

- Thermal hazards (heat) from fires that may explode or burn hot and fast or materials'/substances' sensitivity to heat, friction, shock, and contamination.
- Explosive reactions to contact with hydrocarbons (fuels)
- Mechanical hazards
  - Violent reactions and explosions
  - Sensitivity to heat, friction, shock, and/or contamination with other materials
- Chemical hazards
  - From toxic gases, vapors, dust
  - From products of combustion
  - Resulting in burns
- Thermal hazards from ignition of combustibles (including paper, cloth, wood, etc.)
- Asphyxiation hazards from accumulation of toxic fumes and dusts in confined spaces

Table 4.9 provides the U.S. DOT's Class 5 divisions' placards, definitions, and examples.

Table 4.9 Class 5 Divisions, Placards, Definitions, and Examples	
Division Number and Placard	**Definition**
Division 5.1	**Oxidizer** — Material that may, generally by yielding oxygen, cause or enhance the combustion of other material *Examples:* chromium nitrate; copper chlorate; calcium permanganate, ammonium nitrate fertilizer
Division 5.2	**Organic Peroxide** — Any organic compound containing oxygen (O) in the bivalent -O-O- structure and which may be considered a derivative of hydrogen peroxide, where one or more of the hydrogen atoms has been replaced by organic radicals *Examples:* liquid organic peroxide type B

Source: *49 CFR 173.127 and 128*

Class 6: Poisons, Poison Inhalation Hazards, and Infectious Substances

Class 6 materials and substances include **poisons**, poison **inhalation hazards**, and infectious substances **(Figure 4.72)**. Poisonous materials are known to be toxic to humans. Avoid contact with these materials **(Figure 4.73)**.

Inhalation hazards are toxic vapors that can be lethal if inhaled. These materials can be extremely dangerous at hazmat incidents because they can travel great distances and harm or kill anyone who breathes them **(Figure 4.74)**.

Infectious substances and biohazards are materials that have the potential to cause disease in humans or animals. Infectious materials are typically shipped in small containers, so there is no placard for them, only a label. A biohazard label is used for large and small quantities of regulated medical waste.

The secondary hazards of Class 6 materials are:

- Toxic hazards
- Chemical hazards from toxic and/or corrosive products of combustion
- Thermal hazards (heat) from substances transported in molten form
- Thermal hazards (heat) from flammability and fires

WARNING!
Do not inhale or come into contact with Class 6 materials.

Table 4.10 provides the U.S. DOT's Class 6 divisions' placards, definitions, and examples. **Table 4.11, p. 194**, provides the U.S. DOT's Class 6 unique labels, definitions, and examples.

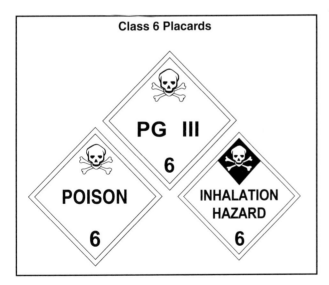

Figure 4.72 Class 6 materials are toxic to humans.

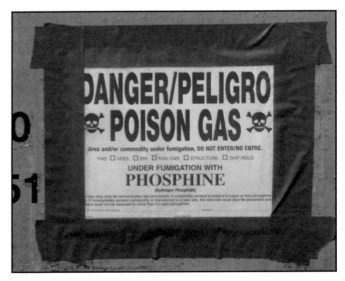

Figure 4.73 Class 6 materials may be harmful to touch, breathe, ingest or contact via any of the routes of entry. *Courtesy of Rich Mahaney.*

Figure 4.74 Inhalation hazards may be lethal if inhaled. *Courtesy of Rich Mahaney.*

Table 4.10
Class 6 Divisions, Placards, Definitions, and Examples

Division Number and Placard	Definition
Division 6.1 POISON 6	**Poisonous Material** — Material, other than a gas, that is known to be so toxic to humans as to afford a hazard to health during transportation or that is presumed to be toxic to humans based on toxicity tests on laboratory animals *Examples:* aniline, arsenic, liquid tetraethyl lead
PG III 6	**PG III** — For Division 6.1, packing group III* (PG III) materials, a POISON placard may be modified to display the text "PG III" below the mid line of the placard rather than the word "POISON." *A packing group is a DOT packaging category based on the degree of danger presented by the hazardous material. Packing Group I indicates great danger; Packing Group II, medium danger; and Packing Group III, minor danger. The PG III placard, then, might be used for materials that are not as dangerous as those that would be placarded with the "POISON" placard.* *Examples:* chloroform, alkaloid solids
INHALATION HAZARD 6	**Inhalation Hazard Placard** — Used for any quantity of Division 6.1, Zones A or B inhalation hazard only (see Division 2.3 for hazard zones) *Examples:* nerve agents, cyanide

Source: *49 CFR 173.132 and 134*

Table 4.11
Class 6 Unique Labels, Definitions, and Examples

Division Number and Label	Definition
Division 6.2 *(INFECTIOUS SUBSTANCE label, 6)*	**Infectious Substance** — Material known to contain or suspected of containing a pathogen. A pathogen is a virus or microorganism (including its viruses, plasmids, or other genetic elements, if any) or a proteinaceous infectious particle (prion) that has the potential to cause disease in humans or animals. *Examples:* anthrax, hepatitis B virus, *escherichia coli* (e coli)
(BIOHAZARD label)	**Biohazard Label** — Marks bulk packaging containing a regulated medical waste as defined in 49 CFR 173.134(a)(5). *Examples:* used needles/syringes, human blood or blood products, human tissue or anatomical waste, carcasses of animals intentionally infected with human pathogens for medical research

Class 7: Radioactive Materials

Radioactive materials cannot be detected with the senses **(Figure 4.75)**. While Class 7 placards and labels can indicate that radioactive materials are present, without specialized monitoring and detection equipment, it is not possible to determine if a container is actually emitting radiation. It is impossible to tell if radiation is involved in an incident, such as a terrorist attack, where no placards or labels are evident.

Small packages of radioactive materials must be labeled on two opposite sides, with a distinctive warning label. Each of the three label categories — RADIOACTIVE WHITE-I, RADIOACTIVE YELLOW-II, or RADIOACTIVE YELLOW-III — bears the unique trefoil symbol for radiation.

Class 7 Radioactive I, II, and III labels must always contain the following additional information **(Figure 4.76)**:

- **Isotope** name **(Figure 4.77)**
- Radiation activity

Radioactive II and III labels will also provide the Transport Index (TI), which indicates the carrier's degree of control during transportation. The number in the transport index box indicates the maximum radiation level measured in (mrem/hr) at one meter from the surface of the package. Packages with the Radioactive I label have a Transport Index of 0.

NOTE: Items placarded as Radioactive II and III have a maximum allowed TI rating of 50 mrem/hr at 1 meter.

Table 4.12 provides the U.S. DOT's Class 7 divisions' placards, definitions, and examples. **Table 4.13, p. 196**, provides the U.S. DOT's Class 7 unique labels, definitions, and examples.

Isotope — Atoms of a chemical element with the usual number of protons in the nucleus, but an unusual number of neutrons; has the same atomic number but a different atomic mass from normal chemical elements.

Class 7 Placard

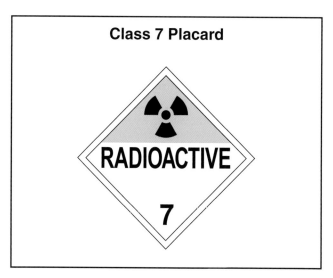

Figure 4.75 Class 7 materials are radioactive and cannot be detected with the senses.

Class 7 Radioactive Placard Information

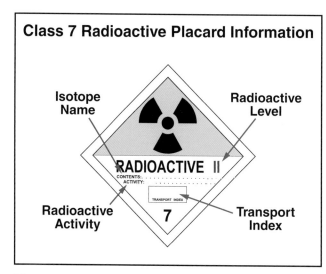

Isotope Name

Radioactive Level

RADIOACTIVE II
CONTENTS:
ACTIVITY:
TRANSPORT INDEX

Radioactive Activity

Transport Index

Figure 4.76 Class 7 labels will always provide the isotope name, activity level, transport index, and radioactive level.

Activity

The energy of radiation gives it the ability to penetrate matter. Higher energy radiation will be able to penetrate a higher volume and denser matter than lower energy radiation. The strength of a radioactive source is called its *activity*. The activity of a radioactive source can be defined as the rate at which a number of atoms will decay and emit radiation in one second.

The International System (SI) unit for activity is the Becquerel (Bq), which is the quantity of radioactive material in which one atom transforms per second. The Becquerel tends to be a small unit. The curie (Ci) is also used as the unit for activity of a particular source material. The curie is a quantity of radioactive material in which 1 Ci = 3.7 x 10^{10} atoms disintegrate per second.

Common Isotopes

Industrial	Medical
Cs-137	TI-201
Co-60	Tc-99m
Ir-192	I-131
Am-241	I-125
	Pd-103
	Ru-106

Figure 4.77 Common isotope names that might be seen on Class 7 labels.

Table 4.12
Class 7 Divisions, Placards, Definitions, and Examples

Division Number and Label	Definition
Division 7 RADIOACTIVE 7	**Radioactive Placard** — In required on certain shipments of radioactive materials; vehicles with this placard are carrying "highway route controlled quantities" of radioactive materials and must follow prescribed, predetermined transportation routes *Examples:* solid thorium nitrate, uranium hexafluoride

Source: *49 CFR 173.403*

Table 4.13
Class 7 Unique Labels, Definitions, and Examples

Class Number and Label	Definition
Class 7 **RADIOACTIVE I** 7	**Radioactive I Label** — Label with an all-white background color that indicates that the external radiation level is low and no special stowage controls or handling are required.
Class 7 **RADIOACTIVE II** 7	**Radioactive II Label** — Upper half of the label is yellow, which indicates that the package has an external radiation level or fissile (nuclear safety criticality) characteristic that requires consideration during stowage in transportation.
Class 7 **RADIOACTIVE III** 7	**Radioactive III Label** — Yellow label with three red stripes indicates the transport vehicle must be placarded RADIOACTIVE
Class 7 **FISSILE** CRITICALITY SAFETY INDEX 7	**Fissile Label** — Used on containers of fissile materials (materials capable of undergoing fission such as uranium-233, uranium-235, and plutonium-239). The Criticality Safety Index (CSI) must be listed on this label. The CSI is used to provide control over the accumulation of packages, overpacks, or freight containers containing fissile material.
Class 7 **EMPTY**	**Empty Label** — Used on containers that have been emptied of their radioactive materials, but still contain residual radioactivity

Class 8: Corrosives

Corrosives are either a liquid or solid that cause full thickness destruction of human skin at the site of contact within a specific period of time, or a liquid that has a severe corrosion rate on steel or aluminum **(Figure 4.78)**. Corrosives can also cause a fire or an explosion if they come in contact with other materials because their corrosive actions can generate enough heat to start a fire. Some can react with metal to form (explosive) hydrogen gas. Different types of corrosives (acids and bases) can react violently when mixed together or when combined with water.

Corrosives can be toxic, flammable, reactive, and/or explosive and some are oxidizers **(Figure 4.79)**. Because of the wide variety of hazards presented by corrosives, do not focus solely on the corrosive properties when considering appropriate actions at incidents involving these materials.

The primary hazards of Class 8 materials are chemical, toxic, thermal and mechanical. Some examples include:

- Chemical hazards such as chemical burns
- Toxic hazards due to exposure via all routes of entry into a body
- Thermal hazards (heat), including fire, caused by chemical reactions generating heat
- Mechanical hazards caused by BLEVEs and violent chemical reactions

Table 4.14 provides the U.S. DOT's Class 8 divisions' placards, definitions, and examples.

Figure 4.78 Corrosives cause damage to metal and skin.

Figure 4.79 Class 8 corrosives are liquids or solids.

Table 4.14
Class 8 Division, Placard, Definition, and Examples

Corrosive Placard

A corrosive material means a liquid or solid that causes full thickness destruction of human skin at the site of contact within a specific period of time or a liquid that has a severe corrosion rate on steel or aluminum.

Examples: battery fluid, chromic acid solution, soda lime, sulfuric acid, hydrochloric acid (muriatic acid), sodium hydroxide, potassium hydroxide

Class 9 Miscellaneous Hazardous Materials

A miscellaneous dangerous good is a material that **(Figure 4.80)**:

- Has an anesthetic, noxious, or other similar property that could cause distraction or discomfort to crew members during transportation
- Is a hazardous substance or a hazardous waste
- Is an elevated temperature material
- Is a marine pollutant

Miscellaneous dangerous goods will primarily have thermal and chemical hazards. For example, elevated temperature materials may present some thermal hazards, and polychlorinated biphenyls (PCBs) are carcinogenic. However, hazardous wastes may present any of the hazards associated with the materials in normal use.

Table 4.15 provides the U.S. DOT's Class 9 divisions' placard and examples.

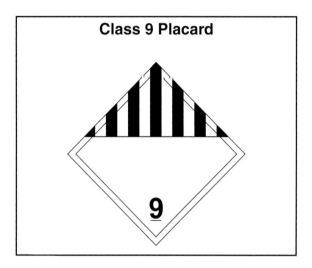

Class 9 Placard

Figure 4.80 Class 9 materials mostly present thermal and chemical hazards.

Table 4.15
Class 9 Divisions, Placards, Definitions, and Examples

A miscellaneous dangerous good is a material that (1) has an anesthetic, noxious, or other similar property that could cause extreme annoyance or discomfort to flight crew members and would prevent their correct performance of assigned duties; (2) is a hazardous substance or a hazardous waste; or (3) is an elevated temperature material; or (4) is a marine pollutant.

Miscellaneous dangerous goods will primarily have thermal and chemical hazards. For example, polychlorinated biphenyls (PCBs) are carcinogenic, while elevated temperature materials may present some thermal hazards. However, hazardous wastes may present any of the hazards associated with the materials in normal use.

Miscellaneous Placard

Examples: blue asbestos, polychlorinated biphenyls (PCBs), solid carbon dioxide (dry ice)

Dangerous Placard — A freight container, unit load device, transport vehicle, or railcar that contains nonbulk packaging with two or more DOT Chart 12, Table 2 categories of hazardous materials may be placarded *DANGEROUS*. However, when 2,205 lbs (1,000 kg) or more of one category of material is loaded at one loading facility, the placard specified in DOT Chart 12, Table 2 must be applied.

Other Regulated Materials (ORM-Ds) and Materials of Trade (MOTs)

ORM-Ds are consumer commodities that present a limited hazard during transportation due to their form, quantity, and packaging. No placards are required for ORM-Ds, but they are otherwise subject to the requirements of the Hazardous Materials Regulations (HMR). Examples of ORM-Ds include consumer commodities and small arms cartridges.

A MOT is a hazardous material, other than a hazardous waste, that is carried on a motor vehicle for the purposes listed below. MOTs do not require placards, shipping papers, emergency response information, formal record keeping, or formal training. MOT purposes include:

- To protect the health and safety of motor vehicle operators or passengers. Examples: insect repellant, fire extinguishers, and self-contained breathing apparatus (SCBA)

- To support the operation or maintenance of motor vehicles, including its auxiliary equipment. Examples: spare batteries, gasoline, and engine starting fluid

- To directly support principal nontransportation businesses (by private motor carriers). Examples: lawn care, pest control, plumbing, welding, painting, and door-to-door sales

Many ORM-Ds (such as hairspray) may qualify as MOTs. However, self-reactive materials, poison inhalation hazard materials, and hazardous wastes never qualify as MOTs.

Additional Information

As soon as first responders identify an incident as a hazmat incident, they should take steps to gather additional information to identify the hazardous material(s)/substance(s) involved, take into account additional hazards or complications caused by surrounding conditions, and contact any additional resources that may provide additional technical information **(Figure 4.81)**. The following sections address each of these concerns:

- Collecting hazard and response information

- Surrounding conditions

- Emergency response centers

Collecting Hazard and Response Information

After identifying a hazardous material/substance, use the following sources (described in Chapter 2) to gather information about its physical and chemical properties:

Figure 4.81 Responders should contact additional resources that can provide technical information.

- *Emergency Response Guidebook (ERG)*

- Shippers and shipping papers (see Table 2.11 for typical locations)

- Safety Data Sheets (SDS) (available at fixed facilities where the products are stored or used)

- Pipeline operators

- Computer apps such as CAMEO and Wiser
- Placards and labels
- Manufacturers

NOTE: Local, state, and governmental authorities may also provide assistance and will be explained in Chapter 6, Notification section.

Responders can use the previous sources listed to determine a product's hazards and the way it is likely to behave based on its chemical and physical properties. Additional information gathered from these sources may include:

- Potential health hazards
- Signs and symptoms of exposure
- Responsible party contact information
- Precautions for safe handling and control measures including PPE and spill cleanup procedures
- Emergency and first aid procedures

Surrounding Conditions

In addition to identifying hazmat containers and their contents, first responders need to survey surrounding conditions. While conducting this survey, first responders should identify relevant information, including **(Figure 4.82)**:

- Potential site hazards, such as overhead power lines, highway traffic, and rail lines
- Potential ignition sources
- Potential victims and exposures
- Weather and time of day
- Topography
- Information about the building and building components, if indoors

Figure 4.82 Always survey surrounding conditions for hazards such as overhead power lines, oncoming traffic, rail lines, weather, and topography. *Courtesy of Rich Mahaney.*

Site Hazards

Hazmat incidents can occur anywhere and, often, the location itself will present its own hazards **(Figure 4.83)**. For example, if the incident occurs on a road or highway, responders will need to take protective actions against traffic and other highway hazards such as falls from overpasses, bridges, and other heights. If the incident occurs on or near rail lines, responders should protect themselves, victims, and property from passing trains and other rail hazards. Overhead power lines may have been knocked down during the incident or present a hazard to elevating equipment such as aerial apparatus or cranes. Other site-specific hazards could present potential contamination, environmental, or thermal hazards specific to the hazardous materials involved in the incident.

Figure 4.83 The locations of hazmat incidents often have their own hazards. *Courtesy of South Wales Fire Brigades.*

Potential Ignition Sources

If the incident involves a flammable or combustible material — and the majority of hazmat incidents do — responders must avoid igniting these materials. Even if the material involved at the incident has not been identified, remove as many ignition sources as possible. Flammable gases and vapors can travel to unexpected places and tend to settle in low-lying areas.

Many potential ignition sources may exist at the scene of a hazardous materials incident including **(Figure 4.84)**:

- Open flames
- Static electricity
- Pilot lights
- Electrical sources including non-explosion-proof electrical equipment
- Internal combustion engines in vehicles and generators

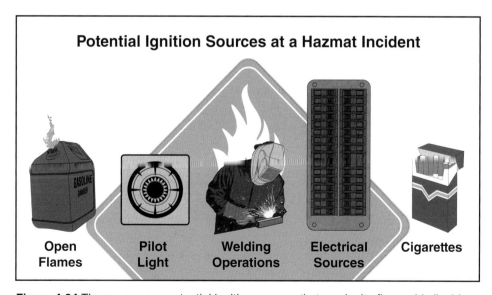

Potential Ignition Sources at a Hazmat Incident

Open Flames Pilot Light Welding Operations Electrical Sources Cigarettes

Figure 4.84 There are many potential ignition sources that can ignite flammable liquids, vapors, and gases. They can also provide activation energy to initiate other reactions.

- Heated surfaces
- Cutting and welding operations
- Radiant heat
- Heat caused by friction or chemical reactions
- Cigarettes and other smoking materials
- Cameras/cellular phones
- Road flares

The following actions can ignite flammable/explosive atmospheres (**Figure 4.85**):

- Opening or closing a switch or electrical circuit such as a light-switch
- Turning on a flashlight
- Operating a radio
- Activating a cell phone

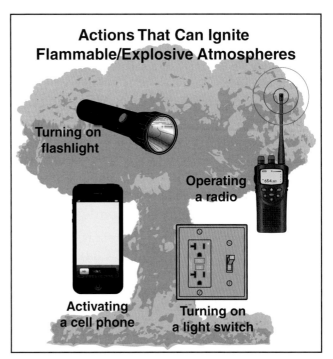

Actions That Can Ignite Flammable/Explosive Atmospheres

Turning on flashlight

Operating a radio

Activating a cell phone

Turning on a light switch

Figure 4.85 Common actions can ignite flammable/explosive atmospheres.

Potential Victims and Exposures

Responders must quickly identify potential victims and exposures. Potential exposures include people, property, and the environment. The potential exposures will determine the need for rescue and protective actions. The nature and extent of injuries may give clues to the product(s) involved and the hazards present as well as determine the need for decontamination and medical care. Exposures are addressed in greater detail in the Exposure/Contact section of this chapter; protective actions are detailed in Chapter 7.

Weather

If an incident is outdoors, the weather can dramatically affect how an incident progresses and is mitigated. For instance, if temperatures are below freezing, it may be impractical or impossible to use water for decontamination or dilution processes. Hot temperatures may cause liquids to evaporate more rapidly, producing more vapors or potentially raising a flammable material's temperature to its ignition point. Wind direction may determine where and how far gases, vapors, or solid particulates travel. Rain or high humidity may cause water-reactive materials to burn or explode.

As explained in Chapter 3, Table 1 section, the time of day can also influence chemical behavior due to the conditions typically present. At night, winds tend to be lighter, so gases and vapors will not typically travel as far. Nights also tend to be cooler, so liquids tend not to evaporate as rapidly. In addition, temperature gradients may be significantly different in an area due to topography and bodies of water.

Topography

Topography makes a significant difference in the considerations needed to determine the appropriate isolation distance. The *ERG* (green-bordered pages) defines isolation distances. Topography is a factor in rural environments, such as flat plains or passes through mountains, as well as in developed environments such as wind tunnels between tall buildings and chemical processing areas. Topography may play an important role in where liquid and gaseous hazardous materials travel. If an outdoor incident involves a liquid, topography and gravity determine where the liquid might go, such as into culverts and ditches. These drainage areas may lead to the following environmentally sensitive areas that require protection:

- Streams and rivers
- Ponds, lakes, or wetlands
- Storm and sewer drains

Topography may also affect the travel of gases and vapors, with heavier-than-air vapors and gases following the contours of the land. When determining potential movement of hazardous vapors and gases, consider:

- Local thermal winds
- Upslope winds
- Downslope winds
- Breezes
- Aspect (for instance, if the incident aspect is facing the sun, a rise in temperature may affect the material and/or the container)
- Mountain or valley elevation, which can become an issue in relation to vapor density

Building Information

For incidents occurring indoors, the following information may be relevant:

- Location of floor drains
- Air handling ducts, returns, and units
- Location and components of fire protection and detection equipment
- Location of gas, electric, and water shut-off locations
- Presence of potential backup generators

Emergency Response Centers

Emergency response centers can provide useful information and guidance to first responders. The *ERG* provides contact information for emergency response centers in the U.S., Canada, Mexico, Argentina, Brazil, and Colombia. Contact

Figure 4.86 Many manufacturers and shippers use CHEMTREC and CANUTEC as their emergency response contact numbers. *Courtesy of Rich Mahaney.*

numbers are provided in the white pages in both the front and the back of the *ERG*.

In the U.S., several emergency response centers, such as the Chemical Transportation Emergency Center (CHEMTREC®), are not government-operated. CHEMTREC® was established by the chemical industry as a public service hotline for firefighters, law enforcement responders, and other emergency service responders to obtain information and assistance for emergency incidents involving chemicals and hazardous materials **(Figure 4.86)**. The experts staffing these centers can provide 24-hour assistance to personnel responding to hazmat incidents.

Transport Canada operates the Canadian Transport Emergency Centre (CANUTEC). This national, bilingual (English and French) advisory center is part of the Transportation of Dangerous Goods Directorate. CANUTEC has a scientific data bank on chemicals manufactured, stored, and transported in Canada and is staffed by professional scientists who specialize in emergency response.

Mexico has two emergency response centers: (1) National Center for Communications of the Civil Protection Agency (CENACOM) and (2) Emergency Transportation System for the Chemical Industry (SETIQ), which is operated by the National Association of Chemical Industries.

NOTE: CENACOM has phone numbers dedicated to calls originating in Mexico City and its metropolitan area. Do not call these numbers if you are not in that area.

Before you contact the emergency response center, collect as much of the following information as safely possible:

- Caller's name, callback telephone number, and fax number
- Location and nature of problem (such as spill or fire)
- Name and identification number of material(s) involved
- Shipper/consignee/point of origin
- Carrier name, railcar reporting marks (letters and numbers), or truck number
- Container type and size
- Quantity of material transported/released
- Local conditions (such as weather, terrain, proximity to schools, hospitals, or waterways)
- Injuries, exposures, current conditions involving spills, leaks, fires, explosions, and vapor clouds
- Local emergency services that have been notified

The emergency response center will:

- Confirm that a chemical emergency exists.
- Record details electronically and in written form.

- Provide immediate technical assistance to the caller.

- Contact the shipper of the material or other experts.

- Provide the shipper/manufacturer with the caller's name and callback number so that the shipper/manufacturer can deal directly with the party involved.

Chapter Review

Answer the following questions to review the information provided in this chapter.

1. What are the different hazards for gases, liquids, and solids?

2. List the physical properties of materials and explain how they help to determine hazards.

3. List the chemical properties of materials and explain how they help to determine hazards.

4. List the hazard classes and give examples of each class that a first responder might commonly encounter.

5. What types of information do you need to collect at a hazmat incident?

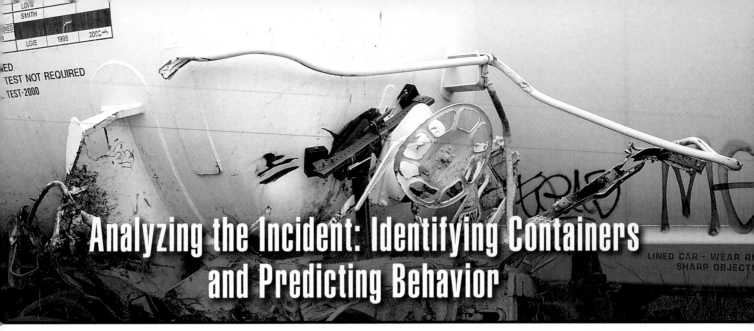

Chapter Contents

Key Terms

NFPA Job Performance Requirements

This chapter provides information that addresses the following job performance requirements of NFPA 1072, *Standard for Hazardous Materials/Weapons of Mass Destruction Emergency Response Personnel Professional Qualifications (2017).*

5.2.1

Analyzing the Incident: Identifying Containers and Predicting Behavior

Learning Objectives

After reading this chapter, students will be able to:

1. Describe methods of identifying potential outcomes. (5.2.1)

2 Explain the role of the General Hazardous Materials Behavior Model in predicting the behavior of containers. (5.2.1)

3. Recognize general container types and their associated behaviors. (5.2.1)

4. Describe the types of bulk facility storage tanks and their associated hazards. (5.2.1)

5. Describe the types of cargo tank trucks and their associated hazards. (5.2.1)

6. Describe the types of tank cars and their associated hazards. (5.2.1)

7. Describe the types of intermodal tanks and their associated hazards. (5.2.1)

8. Describe the types of bulk transportation containers and their associated hazards. (5.2.1)

9. Describe the other types of bulk and nonbulk containers and their associated hazards. (5.2.1)

10. Analyze a hazardous materials scenario to identify potential hazards (5.2.1, Skill Sheet 5-1)

Chapter 5
Analyzing the Incident: Identifying Containers and Predicting Behavior

This chapter explains the following topics:

- Identifying potential outcomes
- General hazardous materials behavior model
- Containers and hazmat behavior
- General container types and behavior
- Bulk facility storage tanks
- Bulk transportation containers: Cargo tanks
- Bulk transportation containers: Tank cars
- Bulk transportation containers: Intermodal tanks
- Bulk transportation containers: Ton containers
- Other bulk and nonbulk containers

Identifying Potential Outcomes

As explained in Chapter 4, an uncontrolled release from a container can expose people, animals, and the environment to many hazards. When first responders collect information about the physical and chemical properties of released hazardous materials, they can:

- Determine the present hazards
- Estimate potential harm
- Predict how the incident may progress

The same material physical and chemical properties that create hazards upon release will influence how a container will behave when damaged or ruptured. First responders need to know how to account for these factors when attempting to understand the problem posed by a hazmat incident.

The first step in mitigating or solving any hazmat incident is understanding the problem within the framework of incident priorities, IMS, and predetermined procedures. First responders can form an overall plan of action by understanding the problem and its entire component parts.

Figure 5.1 Skillful Incident Commanders synthesize information quickly in order to form a clear picture of the incident.

The initial survey should answer the following questions:

- Where is the incident scene in relation to population and environmental and property exposures?

- What are the hazardous materials involved?

  — What are their classes?

  — What are their quantities?

  — What are their concentrations?

  — How could they react?

- How is the material likely to behave?

  — Is it a liquid, solid, or a gas?

  — Is something burning?

- What kind of container holds the material?

- What is the condition of the container?

- How much time has elapsed since the incident began?

- What resources are available?
  - What personnel, equipment, and extinguishing agents are available?
  - Is there private fire protection or other help available?
- What hazards does the site present?
  - What effect can the weather have?
  - Are there nearby lakes, ponds, streams or other bodies of water?
  - Are there overhead wires, underground pipelines, or other utilities?
  - Where are the nearest storm and sewer drains?
  - Is the incident scene inside or outside a building?
- What has already been done?

All of the previously described factors and others can affect the incident. The first responder must make reasonable determinations as to the amount or level of hazard present and the risks associated with dealing with the incident. Gather this information during the incident scene analysis or **size-up** and then analyze it through a hazard and risk assessment model.

Size-Up — Ongoing evaluation of influential factors at the scene of an incident.

To identify endangered area(s), personnel must gather and correctly interpret information, including the following:

- **Size** — Is the endangered area changing or moving (expanding, being blown by wind, or flowing)? How wide should the initial isolation zone be?
- **Shape** — Is the endangered area a square room or floor of a building? Is it a cone-shaped area downwind? Is it a loading dock? A small stretch of highway and surrounding drainage ditches?
- **Exposures** — Are people, animals, or property in the endangered area? Is the environment in danger of being contaminated or damaged? Are rescues needed?
- **Physical, health, and safety hazards** — What potential hazards do the material and its container present? What other hazards are present? What are the surrounding conditions?

If possible, verify all information through contact with emergency response agencies, manufacturers, shippers, and/or other resources that can confirm handling procedures, product identity, and appropriate response information. These resources can often assist the first responder in estimating the size of the endangered area (for example, the *ERG* lists recommended isolation distances) and the potential harm posed by the material and its container.

General Hazardous Materials Behavior Model

In order for first responders to protect themselves and others, they must understand how a hazardous material and its container are likely to behave in any given situation. This behavior typically follows a general pattern. The General Hazardous Materials Behavior Model, often referred to as the **General Emergency Behavior Model (GEBMO)**, describes this general pattern. This model is based on Ludwig Benner Jr.'s definition of hazardous materials as "things that can escape from their containers and hurt or harm the things that they touch."

General Emergency Behavior Model (GEBMO) — Model used to describe how hazardous materials are accidentally released from their containers and how they behave after the release.

The model assumes that hazardous materials incidents have the following common elements:

- Material or materials presenting hazards to people, the environment, or property
- Container or containers that have failed or have the potential to fail
- Exposure or potential exposure to people, the environment, and/or property

Given these three elements (material (s), container, and exposure), a common sequence generally occurs **(Figure 5.2)**:

- **Stress** — The container undergoes physical, thermal, or other types of damage that reduces its ability to function and leads to breach or failure.

- **Breach** — The container becomes open to the environment. This opening depends on its construction material, type of stress that it undergoes, and pressure inside the container at the time that it fails. A breach or failure of the container may be partial (as in a puncture) or total (as in disintegration).

- **Release** — When a container breaches or fails, contents, stored energy, and pieces of the container may be expelled into the environment (release). A release always involves the hazardous material product and may (depending on the product, container, and incident conditions) involve the release of energy and container parts.

- **Dispersion/engulf** — This occurs as the hazardous material inside the container and any stored energy release and move away from the container. Patterns of dispersion are influenced by chemistry, physics, environmental factors, and the chemical and physical characteristics of the product.

- **Exposure/contact** — Anything (such as persons, the environment, or property) that is in the area of the release is exposed to the hazardous material.

- **Harm** — Depending on the container, hazardous material, and energy involved, exposures may result in harm or damage.

The previous sequence is expanded in the following paragraphs. The behavior of explosives, chemical and biological agents, and radiological materials used for purposes of terror attacks will be explained in Chapter 8, Implementing the Response: Terrorist Attacks, Criminal Activities, and Disasters.

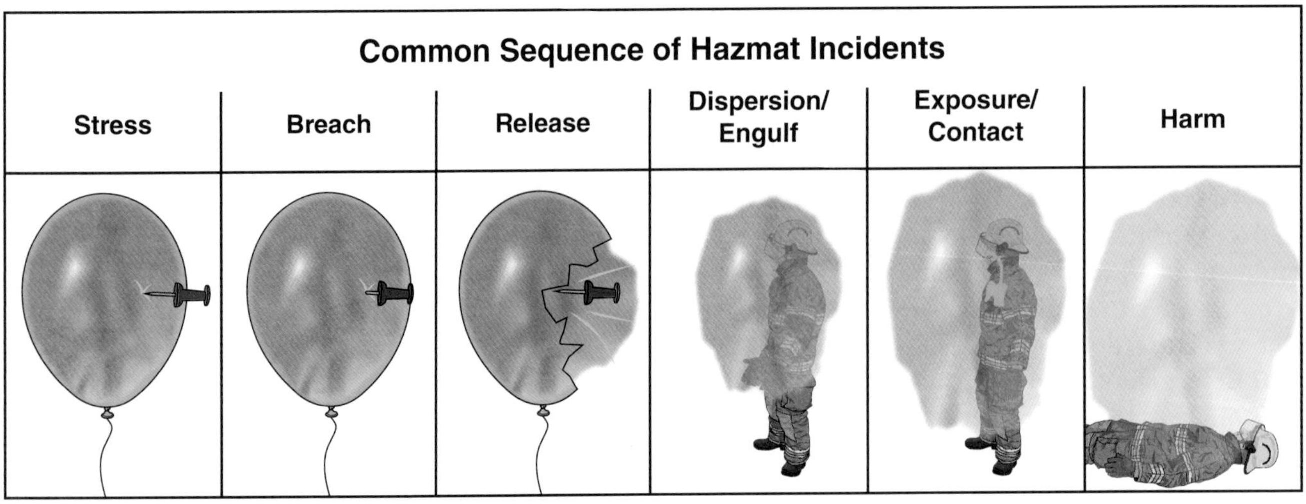

Common Sequence of Hazmat Incidents

Stress	Breach	Release	Dispersion/ Engulf	Exposure/ Contact	Harm

Figure 5.2 Most hazmat incidents follow a common pattern.

Stress

Container stress is caused by thermal energy, chemical energy, and mechanical energy:

- **Thermal energy** — Excessive heat or cold could cause intolerable expansion, contraction, weakening (loss of temper), or consumption of the container and its parts. Thermal stress may increase internal pressure and reduce container shell integrity, resulting in sudden failure. Thermal stress may result from the heating or cooling of the container.

  — A container undergoing excessive heat may be:

    o Extremely close to flames

    o Undergoing the operation of a relief device

    o Making noises of expansion or contraction

    o Subject to changing environmental conditions (such as increased temperature)

  — A container succumbing to cold may exhibit:

    o Excessive frosting **(Figure 5.3)**

    o Visible cold vapors (white clouds)

    o Changes in steel structure (smooth to grainy)

    o Pools of cold liquid

Figure 5.3 The appearance of frost is an indicator that a container is under thermal stress. *Courtesy of Barry Lindley.*

- **Chemical energy** — Uncontrolled reactions/interactions of the container and its contents. Chemical reactions/interactions could result in the following:

  — Sudden or long-term deterioration of the container.

  — Excess heat and/or pressure, causing deterioration of the container.

  — Corrosive or other incompatible interactions between the hazardous materials and the container material.

  — Visible corrosion or other degradation of container surfaces, including bulging, cracking, and/or popping noises **(Figure 5.4)**.

  — The interior of a container may experience chemical stress with no visible indication from the exterior.

- **Mechanical energy** — Physical application of energy could result in container/attachment damage. Mechanical stress may:

  — Change the shape of the container (crushing).

  — Reduce the thickness of the container surface (abrading or scoring).

  — Crack or produce gouges.

  — Unfasten (sheer) or disengage valves and piping, or penetrate the container wall **(Figure 5.5)**.

Common causes of mechanical stress include collision, impact, or internal overpressure. Clues of mechanical stress include physical damage, the mechanism of injury (forces placed on the container), or operation of relief devices.

According to U.S. Department of Transportation (DOT) records, from 2006-2014, nearly 41 percent of all reported hazmat incidents were attributed to container failure. Responders may encounter one or all three

Figure 5.4 Chemical reactions may cause a container to bulge, which is a sign of significant stress. *Courtesy of Barry Lindley.*

Figure 5.5 Mechanical energy can crush or damage a container. *Courtesy of Phil Linder.*

of the stressors at any hazmat incident. For instance, heat (thermal stress) can initiate or speed a chemical reaction while weakening a container and increasing its internal pressure. Similarly, a mechanical blow can initiate a violent chemical reaction in an unstable chemical while simultaneously damaging the container.

When evaluating container stress, consider the following:

- Type of container
- Product in the container
- Type and amount of stress
- Potential duration of the stress

Container stress may involve a single factor or several stressors acting on the container simultaneously. Preventing container failure may require reducing or eliminating the factors placing stress on a container. Those factors may be readily visible, such as a collision or a fire impinging on a container surface, or they may not be directly observable and must be predicted based on conditions or other indirect indicators. If the container has already failed, think about other containers that may be exposed and evaluate the impact of product contact with hazardous materials.

The material's state of matter will affect the stress experienced by containers. For example, containers holding gases are inherently subject to stress. Heating or cooling may increase or reduce this stress. These containers may fail catastrophically and/or BLEVE if they are damaged or subjected to additional stress (such as heat from a fire or even hot daytime temperatures). Liquid containers, especially those holding liquids with high vapor pressures, may also fail when subjected to fires.

Liquid containers may also transport materials that polymerize. The stress created by an uncontrolled polymerization (chemical stress) may cause container failure. This failure may be explosive. Most solids containers will be damaged via mechanical stressors rather than the physical properties of the materials contained in them. Exceptions would include reactive materials from hazard classes, such as explosives, oxidizers, peroxides, and water reactive materials.

WARNING!
Use extreme caution when working with containers that have been involved in an accident.

Waverly Overpressurization Incident (1978)

At about 10:30 p.m. on February 22, 1978, twenty-four cars of a Louisville and Nashville (L&N) Railroad freight train derailed in downtown Waverly, Tennessee. Initially, local emergency services handled the accident, including inspecting the wreck for signs of any hazardous material leaks.

At 5:10 a.m. on February 23, the Tennessee Office of Civil Defense (now the Tennessee Emergency Management Agency) sent out a hazmat team to assess the situation. The team concurred with the local officials' decision to keep the tank cars cool by spraying them with streams of water. The decision was made to evacuate a ¼-mile (0.4 km) area around the derailment zone and shut off gas and electric service to the area. By this time, L&N wreck crews were beginning to clear debris. Water spray was discontinued while crews removed the wrecked cars. UTLX 83013, the car that would eventually explode, was moved to clear the tracks. The rail line partially reopened at about 8 p.m. on February 23.

Temperatures during the first two days of the incident had remained below freezing, and light snow was on the ground. However, by midday on February 24, the temperature had risen to around 55°F (around 12.5°C) with clear skies.

A tanker truck and a crew specializing in liquefied petroleum gas (LPG) cleanup arrived on scene about 1 p.m. on February 24. Responders tested the area and found no leaks about 20 minutes before the LPG removal was to begin. Then at 2:58 p.m., with Waverly police and fire chiefs on the scene and the hazmat crew moving equipment for the transfer, testers discovered vapor leaking from the tank car. A BLEVE occurred before any action could be taken.

The blast was felt for hundreds of feet and seen for miles. Sixteen people died as a result of the blast and the aftermath; six were killed instantly and 43 others were injured to various degrees. The blast destroyed most of the fire fighting equipment at the site. One piece of the tank car launched over 330 feet (100 m) and landed in front of a house. The explosion started numerous fires in nearby buildings and torched a number of road vehicles and other rail cars. The blast destroyed sixteen structures in Waverly and seriously damaged another twenty.

Lessons Learned:

The National Transportation Safety Board (NTSB) eventually blamed the blast on the car itself. The car developed a crack during the derailment. It is believed that this crack expanded when the car was moved off the tracks, and the rising temperature caused overpressurization in the tank. In response to this incident, the Tennessee Office of Civil Defense (TOCD) developed standards and training for hazmat responders that have been actively utilized in a variety of capacities in the years since.

Limits of Recovery —
A container's design strength or ability to hold contents at pressure.

Breach — To make an opening in a structural obstacle (such as a masonry wall) without compromising the overall integrity of the wall to allow access into or out of a structure for rescue, hoseline operations, ventilation, or to perform other functions.

Breach

When a container is stressed beyond its **limits of recovery**, it opens or **breaches** and releases its contents **(Figure 5.6)**. Different container types breach in different ways based on a variety of factors (including internal pressure). The type and extent of breach depends upon the type of container and the stress applied. First responders should try to predict the type of damage that may result from the stress that is being or has been applied. The nature of a breach is a major factor in planning offensive product control operations. Types of breaches include:

- **Disintegration** — Occurs in containers that are made of a brittle material (or that have been made more brittle by some form of stress). The container suffers a general loss of integrity. Examples of disintegration include a glass bottle shattering or an exploding grenade **(Figure 5.7)**.

- **Runaway cracking** — Breaks the container into two or more relatively large pieces (fragmentation) or large tears **(Figure 5.8)**. A crack develops in a container and continues to grow rapidly. Runaway cracking often occurs in closed containers, such as drums, tank cars, or cylinders. Runaway linear cracking is commonly associated with BLEVEs.

- **Attachments (closures) open or break** — May fail, open, or break off when subjected to stress, leading to a total failure of a container **(Figure 5.9)**. When evaluating an attachment (such as a pressure-relief device, discharge valve, or other related equipment) that failed, first responders should consider the entire system and the effect of failure at a given point.

- **Puncture** — Occurs when foreign objects penetrate through a container, such as forklifts puncturing drums and couplers puncturing a rail tank car **(Figure 5.10, p. 218)**.

- **Split or tear** — Containers may also breach through a split, such as a welded seam on a tank or when a drum fails. Mechanical or thermal stressors may cause splits or tears, such as when a seam on a bag of fertilizer rips **(Figure 5.11, p. 218)**.

Figure 5.6 A container breaches when it is stressed beyond its limits of recovery. *Courtesy of Phil Linder.*

Figure 5.7 This chlorine cylinder disintegrated due to corrosion. *Courtesy of Barry Lindley.*

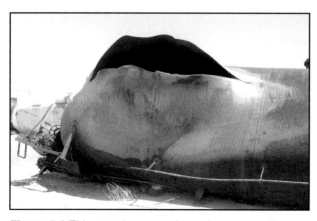

Figure 5.8 This container experienced a runaway linear crack. *Courtesy of Barry Lindley.*

Figure 5.9 Damage to attachments is a common breach. *Courtesy of Barry Lindley.*

Figure 5.10 This tank car has been punctured. *Courtesy of Barry Lindley.*

Figure 5.11 The welded seam has split on this intermodal container. *Courtesy of Rich Mahaney.*

Release

When a container fails, its contents, energy, and the container itself (whole or in pieces) may release. If a cylinder of pressurized, flammable gas suffers an attachment failure at the valve due to mechanical stress, the product releases along with a substantial amount of energy (because of stored pressure), which rapidly accelerates the valve and/or cylinder in the opposite direction from the release **(Figure 5.12)**. Depending on the situation, this release can occur quickly or over an extended time period. Generally, large amounts of stored chemical/mechanical energy result in a more rapid release, presenting a greater risk to first responders. Releases are classified according to how fast they occur:

- **Detonation** — Instantaneous and explosive release of stored chemical energy of a hazardous material. The duration of a detonation can be measured in hundredths or thousandths of a second. An explosion is an example of a detonation. This release could result in fragmentation, disintegration, or shattering of the container; extreme overpressure; and considerable heat release.

- **Violent rupture** — Immediate release of chemical or mechanical energy caused by runaway cracks. Violent ruptures occur within a timeframe of one second or less. These releases result in ballistic behavior of the container and its contents and/or localized projection of container pieces/parts and hazardous material. A BLEVE is an example of a violent rupture.

- **Rapid relief** — Fast release of a pressurized hazardous material through properly operating safety devices. This action may occur in a period of several seconds to several minutes. Damaged valves, damaged piping, damaged attachments, or holes in the container can result in rapid relief **(Figure 5.13)**.

- **Spill/leak** — Slow release of a hazardous material under atmospheric or **head pressure** through holes, rips, tears, or usual openings/attachments. Spills and leaks can occur in a period lasting from several minutes to several days.

Head Pressure — Pressure exerted by a stationary column of water, directly proportional to the height of the column.

Release Potential

When evaluating release potential, remember the total amount of product in the container. A valve blowout in a pressurized container causes a rapid release. If this size breach occurs in a 150-pound (68 kg) cylinder, the contents quickly release. If this same type of release occurs in a ton container, cargo tank, or tank car, the release occurs over a longer period of time and may have a substantially greater effect.

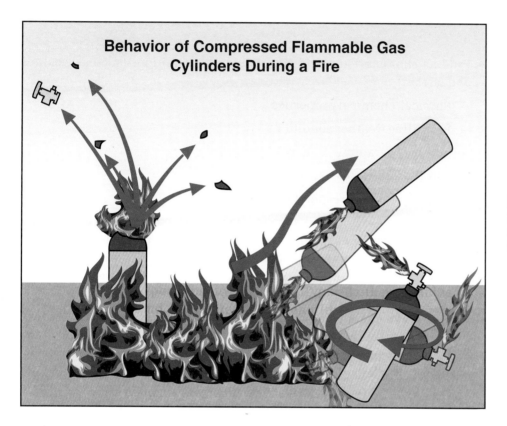

Figure 5.12 Cylinders of pressurized flammable gases can behave like erratic rockets if they suffer a release.

Figure 5.13 Rapid relief occurs when pressurized hazmat is released through properly operating safety devices. *Courtesy of Rich Mahaney.*

Dispersion and Engulfment

The dispersion of material is sometimes referred to as *engulfment* (**Figure 5.14, p. 220**). Dispersion of the hazardous material, energy, and container components depends on the type of release, which include:

- A solid, liquid, or gas/vapor

- Mechanical, thermal, or chemical energy and ionizing radiation

Engulfment — Dispersion of material as defined in the General Emergency Behavior Model (GEBMO); an engulfing event occurs when matter and/or energy disperses and forms a danger zone.

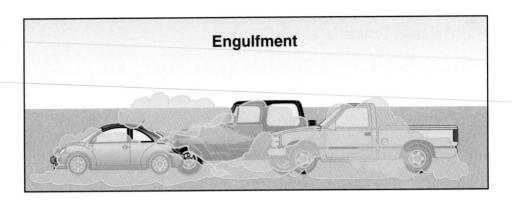

Engulfment

Figure 5.14 Engulfment occurs when a product disperses, forming a danger zone.

- Product characteristics and environmental conditions (such as weather and terrain)
    — Physical/chemical properties
    — Prevailing weather conditions
    — Local topography
    — Duration of the release
    — Control efforts of responders

The shape and size of the dispersing hazardous material also depends on how the material emerges from its container — whether as an instantaneous "puff," a continuous plume, or a sporadic fluctuation. The outline of the dispersing hazardous material, sometimes called its *dispersion pattern*, can be described in a number of ways. Common dispersion patterns include:

- **Hemispheric** — Semicircular or dome-shaped pattern of airborne hazardous material that is still partially in contact with the ground or water **(Figure 5.15)**. A **hemispheric release** generally results from a rapid release of energy (such as detonation, deflagration, and violent rupture). The following elements are common to hemispheric releases:
    — **Energy** — Generally travels outward in all directions from the point of release.
    — **Dispersion of energy** — Affected by terrain and cloud cover. Solid cloud cover can reflect the detonation shock wave, increasing the explosion impact.
    — **Energy release** — May propel the hazardous material and container parts; however, this dispersion may not be hemispherical. Large container parts generally (but not always) travel in line with the long axis of the container.

- **Cloud** — Ball-shaped pattern of the airborne hazardous material that collectively rises above the ground or water **(Figure 5.16)**. Gases, vapors, and finely divided solids that release quickly (puff release) can disperse in cloud form under minimal wind conditions. Terrain and/or wind effects can transform a **cloud** into a plume.

- **Plume** — Irregularly shaped pattern of an airborne hazardous material where wind and/or topography influence the downrange course from the point of release **(Figure 5.17)**. Dispersion of a **plume** (generally composed of gases and vapors) is affected by vapor density and terrain (particularly

Hemispheric Release — Semicircular or dome-shaped pattern of airborne hazardous material that is still partially in contact with the ground or water.

Cloud — Ball-shaped pattern of an airborne hazardous material where the material has collectively risen above the ground or water at a hazardous materials incident.

Plume — Irregularly shaped pattern of an airborne hazardous material where wind and/or topography influence the downrange course from the point of release.

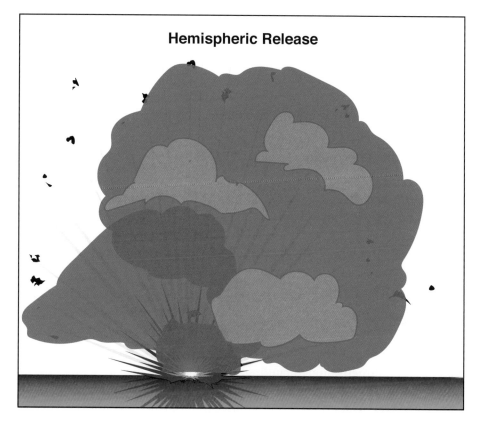

Hemispheric Release

Figure 5.15 A *hemispheric release* is a semicircular or dome-shaped pattern of an airborne hazardous material that is still partially in contact with the ground or water.

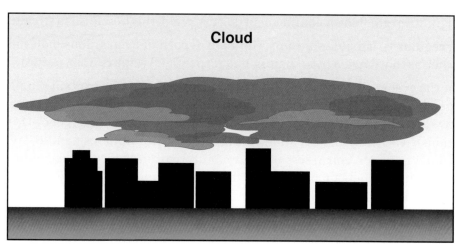

Cloud

Figure 5.16 A *cloud* is a pattern of an airborne hazardous material where the material has collectively risen above the ground or water.

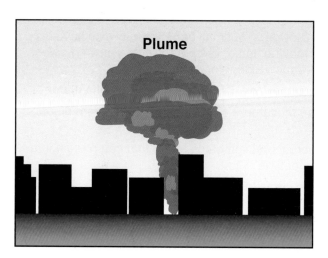

Plume

Figure 5.17 A *plume* is an irregularly shaped pattern of an airborne hazardous material influenced by wind and/or topography in its downrange course.

if vapor density is greater than 1) as well as wind speed and direction. **Figure 5.18** provides several general guidelines in regards to plume modeling behavior in urban environments. Other plume dispersion elements include the following:

— **Puff release** — When all of the material is released at one time, the concentration of gas or vapor in the cloud or plume decreases over time.

— **Ongoing release** — Concentration increases over time until the leak stops or all of the product releases; then it decreases.

- **Cone** — Triangular-shaped pattern of a hazardous material with a point source at the breach and a wide base downrange **(Figure 5.19)**. An energy release may be directed (based on the nature of the breach) and may project solid, liquid, or gaseous material in a three-dimensional **cone**-shaped dispersion. Examples of cone-shaped dispersions include container failures in a BLEVE or a pressurized liquid or gas release.

- **Stream** — Surface-following pattern of liquid hazardous material that is affected by gravity and topographical contours **(Figure 5.20, p. 224)**. Liquid releases flow downslope whenever there is a gradient away from the point of release.

- **Pool** — Three-dimensional (including depth), slow-flowing liquid dispersion. Liquids assume the shape of their container and pool in low areas **(Figure 5.21, p. 224)**. As the liquid level rises above the confinement provided by the terrain, the substance flows outward from the point of release. If there is a significant gradient or confinement due to terrain, this flow forms a stream.

- **Irregular** — Irregular or indiscriminate deposit of a hazardous material (such as that carried by contaminated responders) **(Figure 5.22, p. 224)**.

In the event of a release, facility preincident surveys may contain plume dispersion models to help estimate the size of an endangered area. Computer software such as CAMEO (Computer-Aided Management of Emergency Operations), ALOHA (Area Locations of Hazardous Atmospheres), and HPAC (Hazard Prediction and Assessment Capability) can also assist in the prediction of plume dispersion patterns. First responders may consult the *ERG* for isolation and evacuation distances.

Dispersion of Solids

Solids in the form of dusts, powders, or small particles may also have dispersion patterns. Wind, a moving liquid, or contact with a moving object can disperse the spilled product. Examples include:

- Release of a pesticide powder and dispersed in the form of a plume
- Disperse the solid by the movement of the liquid in the stream dispersion pattern.
- Suspended airborne microscopic asbestos fibers that can remain in a *cloud* or for long periods of time

Hazardous materials also come in solid form, and there is the potential for dispersion of solids. First responders should give this top priority when planning an appropriate response.

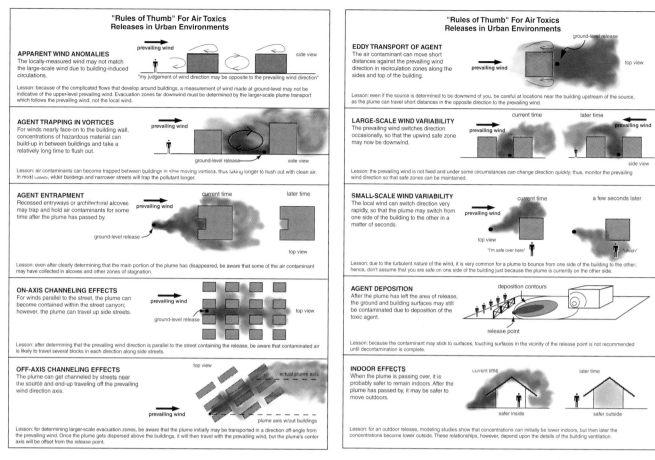

Figure 5.18. Responders should be aware of these rules of thumb regarding plume modeling behavior in urban environments. *Courtesy of Los Alamos National Laboratory.*

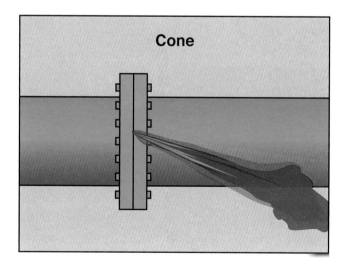

Figure 5.19 A *cone* is a triangular-shaped pattern of a hazardous material with a point source at the breach and a wide base.

Exposure/Contact

As a container releases, it risks dispersing its contents and the container itself on different types of exposures: people, the environment, and property. In some cases, you can use the *ERG* to estimate the size of an endangered area. Some hazardous materials may present a threat to one specific exposure type (such as marine pollutants that threaten fish and other marine plants and animals) and others present a threat to all types. When you evaluate

Figure 5.20 A *stream* is pulled by gravity, following the topographical contours of the surface.

Figure 5.21 In a *pool*, liquids assume the shape of their container, typically accumulating in low areas.

Figure 5.22 *Irregular dispersion* results from indiscriminate deposit of a hazardous material such as that caused by contaminated vehicles or responders.

the severity of exposures, consider the hazards present, concentration of the material, and duration of contact. Consider the following exposures in hazard and risk assessment **(Figure 5.23)**:

- **People** — Includes responders and others in the path of a hazardous material.

- **Environment** — Includes the air, water, ground, and life forms other than humans. The potential effect on the environment varies with the location in which the product is released as well as its characteristics.

- **Property** — Includes things threatened directly by the hazardous material or the energy liberated at the time of release.

Exposures in Hazard and Risk Assessment

People　　Environment　　Property

Figure 5.23 Potential exposures include people, the environment, and property.

Contacts (impingements) are associated with the following general timeframes:

- **Immediate** — Milliseconds, seconds (deflagration, explosion, or detonation)
- **Short-term** — Minutes, hours (gas or vapor cloud)
- **Medium-term** — Days, weeks, months (lingering pesticide)
- **Long-term** — Years, generations (permanent radioactive source)

Harm

Harm is defined as injury or damage caused by exposure to a hazardous material. The three mechanisms of harm in a hazardous materials incident were explained in Chapter 1:

- Energy release (thermal, mechanical, chemical, pressure, electrical, radiological)
- Corrosivity
- Toxicity

General Container Types and Behavior

Chapter 2 introduced container shapes and provided basic information regarding basic container types and contents for Awareness Level personnel. The following sections will provide additional container recognition information based upon transportation mode or fixed facility with a brief description of common stressors leading to breaches and releases. Some basic generalizations may be made about a hazmat incident based on the type(s) of container involved. For example, if the incident involves a pressure container, any product released is likely to be a gas or a liquid that rapidly evaporates and expands into a gas or vapor. Once released, the hazardous material will behave like a gas depending on its properties and the environmental conditions at the scene. **Table 5.1** provides a basic overview of the four major types of containers and their relationship to aspects of the General Behavior Model. **Skill Sheet 5-1** provides steps to analyze a hazardous materials incident to identify potential hazards.

NOTE: Table 5.1 is not intended to be comprehensive.

CAUTION
Approach each hazmat incident as a unique situation, regardless of commonalities between incidents.

In addition to the general container shapes, first responders should be familiar with the basic concepts of pressure and their measurements:

- **Pounds per square inch (psi), kilopascal (kPa), bar** — Common measurements for pressure in the English or Customary System, the International System of Units (SI), and the metric (non-SI) unit, respectively (**Table 5.2, p. 228**). This manual uses *psi (kPa)* to describe pressure.

- **Atmospheric pressure** — Force exerted by the weight of the atmosphere at the surface of the earth. Atmospheric pressure is greatest at low altitudes; consequently, its pressure at sea level is used as a standard. At sea level, the atmosphere exerts a pressure of 14.7 psi (101 kPa). A common method of measuring atmospheric pressure is to compare the weight of the atmosphere with the height of a column of mercury: the greater the atmospheric pressure, the taller the column of mercury (**Figure 5.24, p. 229**).

- **Pressure at gauge** — This describes a unit of pressure relative to the surrounding atmosphere. The Customary System unit is pounds per square inch gauge (psig). The International System of Units (SI) unit is kPaG; the metric (non-SI) unit is bar. For example, at sea level, a reading of 30 psig (207 kPaG){2.07 bar} on a tire gauge represents an absolute pressure of 44.7 psi (308 kPa){3.08 bar} because the gauge was calibrated to zero in atmospheric pressure of approximately 14.7 psi (101 kPa){1.01 bar}.

Pressure Containers

Pressure containers carry gases or liquids. In general, all pressure containers are constructed to prevent accidental releases. Pressure containers are subject to stress when holding materials. Heating or cooling may increase or reduce this stress. If pressure containers are damaged or subjected to additional stress, they may fail catastrophically and/or BLEVE. Hazmat incidents involving pressure containers are dangerous to emergency responders and the public.

Table 5.1
General Hazardous Materials Behavior Model by Container

State of Matter of Release	Pressure Containers	Cryogenic Containers	Liquid-Holding Containers	Solids-Holding Containers
Gas Release	Yes	Cold vapors that expand as they warm	Vapors from liquids depending on vapor pressure and temperature	Reactive solids may release vapors/gases
Liquid Release	Cold liquid that rapidly expands into gas / vapor	Cold liquid that rapidly expands into gas	Yes	No
Solid Release	No	No	No	Yes

Common Stressors	Pressure Containers	Cryogenic Containers	Liquid-Holding Containers	Solids-Holding Containers
Thermal	High temperatures cause extreme stress	Contents are extremely cold and leaks may cause cold stress to container or container supports	- High temperatures may cause extreme stress - Polymerization may cause heat build-up	High temperatures may cause extreme stress
Chemical	Corrosive materials may damage container components if released	May be highly oxidizing or flammable	- Corrosive materials may damage container components if released - Polymerization may occur	- Corrosive materials may damage container components if released - Decomposition may occur
Mechanical	-Contents under high pressure; - Accidents may cause mechanical damage	Accidents may cause mechanical damage	- Accidents may cause mechanical damage - Polymerization may cause pressure build-up	Accidents may cause mechanical damage

Common Breaches	Pressure Containers	Cryogenic Containers	Liquid-Holding Containers	Solids-Holding Containers
Disintegration*	Yes	Uncommon	Yes	Uncommon
Runaway Cracking*	Yes	Uncommon	Yes	Uncommon
Attachments	Yes	Yes	Yes	Yes
Punctures	Uncommon	No	Yes	Yes
Splits or tears	Yes	Yes	Yes	Yes

* The higher the pressure of the container, the more likely a catastrophic failure will occur if the container is damaged.

Continued

Table 5.1 (concluded)

Common Releases	Pressure Containers	Cryogenic Containers	Liquid-Holding Containers	Solids-Holding Containers
Detonation	No	No	Liquid explosives	Explosive solids
Violent Rupture	Yes	Yes	Yes	Yes
Rapid Relief	Yes	Yes	Yes	Uncommon
Spill/Leak	Yes	Yes	Yes	Yes

Common Dispersion Patterns	Pressure Containers	Cryogenic Containers	Liquid-Holding Containers	Solids-Holding Containers
Hemispheric	Yes	Yes, if they Rupture	Yes	Yes (with explosives)
Cloud	Yes	Yes	Yes	Yes
Plume	Yes	Yes	Yes	No
Cone	Yes	Yes	Yes	Yes
Stream	Yes	No	Yes	No
Pool	Yes	Yes	Yes	No
Irregular	No	No	Yes	Yes

Table 5.2
Common Bulk Storage Tank Pressures

Type of Tank	Pressure in psi, kPa, and bar
Pressure Tanks	Above 15 psi, 103 kPa, 1.03 bar
Cryogenic Tanks	Pressures may be very low or very high
Low Pressure Tanks	Between 0.5 psi, 3.45 kPa, 0.03 bar and 15 psi, 103 kPa, 1.03 bar
Nonpressure/Atmospheric Tanks	Up to 0.5 psi, 3.45 kPa, 0.03 bar

First responders at hazmat incidents involving pressure containers need to consider the following common stressors:

- **Thermal** — Exposure to heat or flame can cause pressure containers to BLEVE.

- **Chemical** — Released corrosive gases can cause additional damage to the container; pressure on the container can increase by the reactions of the contents.

- **Mechanical** — Accidents may cause mechanical stress, particularly to the container fittings. Severe accidents may cause damage to container walls.

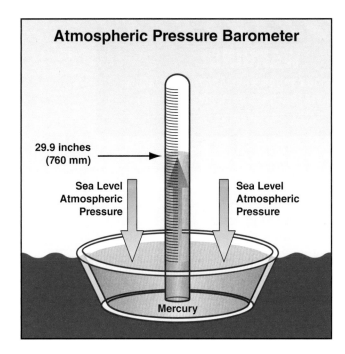

Atmospheric Pressure Barometer

29.9 inches
(760 mm)

Sea Level
Atmospheric
Pressure

Sea Level
Atmospheric
Pressure

Mercury

Figure 5.24 Atmospheric pressure measures the force exerted by the weight of the atmosphere. In this example, the greater the atmospheric pressure, the taller the column of mercury.

NFPA Pressure Tank Definitions

The NFPA uses the term *pressure tank* to cover both low pressure storage tanks and pressure vessels (with higher pressures). Per NFPA definitions, low pressure storage tanks have operating pressures from 0.5 to 15 psi (3.45 kPa to 103 kPa). Pressure vessels (including many large cryogenic liquid storage tanks) have pressures of 15 psi (103 kPa) or greater.

Pressure Relief Device (PRD) — An engineered valve or other device used to control or limit the pressure in a system or vessel, often by venting excess pressure.

Pressure containers can experience any type of breach, although releases from **pressure relief devices** and/or damaged fittings are most common. Runaway cracking is associated with BLEVEs. Punctures, splits, and tears are rare, although violent accidents such as railway derailments or highway accidents may involve enough force to damage welded seams or cause punctures through pressure container walls.

Pressure containers release rapidly expanding gases or liquids that quickly evaporate and expand into gases or vapors. Releases can occur through pressure relief devices (rapid relief) or as leaks through damaged attachments and fittings. Common dispersion patterns from pressure containers include:

- **Hemispheric pattern** — BLEVE
- **Cloud** — A cloud above the container if there is little wind and the release is intermittent or short in duration
- **Plume** — Depending on vapor density, terrain (particularly if vapor density is greater than 1), and wind speed
- **Cone** — A steady release of product from a pressure container as it expands from the point of release outward. The cone will be directed downwind in accordance with the prevailing wind direction.

Exploding Pressure Container

On January 19, 2003, a 32-year-old volunteer firefighter in Texas died while fighting a structure fire at a specialized vehicle restoration shop. A four-member crew responded to the scene and began interior attack operations. Soon, the fire intensified and rolled over their heads. Within minutes, the nozzleman had to exit the building due to burning hands and another firefighter took the nozzle. As the nozzleman was exiting, an air horn was sounded warning the crew to exit the building. Two of the three remaining crew members made it to safety. Less than a minute after they exited, a nitrous oxide cylinder that was attached to a race car in the building exploded **(Figure 5.25)**.

A Rapid Intervention Team (RIT) assembled to rescue the missing firefighter (the victim). The RIT made two attempts to rescue the victim but had to exit because of the intensity of the fire. After approximately forty minutes of master stream application, three teams entered the structure and found the victim lying near the office door. The alarm for his Personal Alert Safety System (PASS) device was functioning but was not audible due to his prone position.

Lessons Learned:

The preliminary autopsy findings indicated that the victim had received significant blast injuries. Both eardrums were ruptured and there was concussive damage to his lungs. A subsequent NIOSH investigation determined that the nitrous oxide cylinder ruptured and exploded with a force equivalent to as much as 4 pounds (2 kg) of TNT. At a distance of 10 feet (3 m) from the exploding cylinder, the firefighter could have been exposed to a shock wave of up to 30 psi (207 kPa), well above the threshold level for eardrum rupture and internal lung damage. The effects of a blast overpressure shock wave increase when explosions occur in closed or confined spaces, such as inside a building or a vehicle. In addition, blast waves are reflected by solid surfaces. As a result, a person standing next to a wall or vehicle may suffer increased primary blast injury. *Source: NIOSH*

Figure 5.25 A Texas firefighter was killed by this nitrous oxide cylinder that exploded. *Courtesy of NIOSH.*

Cryogenic Containers

Like pressure containers, cryogenic containers are solidly constructed to prevent accidental releases. Cryogenic containers are durable containers that insulate their contents, keeping them cold. They maintain cold temperatures through the use of insulating materials and a vacuum space between the two container walls **(Figure 5.26)**. Insulation between the exterior and the interior walls of the container keeps the product cool but does not provide protection against breaches. The container's external support structure is not designed to handle the temperatures inside the container.

Common stressors to consider at hazmat incidents involving cryogenic containers include:

- **Thermal** — If a breach occurs, the extreme cold of the released product may cause damage to the container or fittings. When exposed to heat or flame, cryogenic containers can BLEVE, although this is unusual. If vacuum is lost, the product will heat rapidly and may blow the rupture disc/relief valve. If the system cannot relieve pressure quickly enough, the container may rupture. Damage to the outer jacket and insulation may result in loss of vacuum.

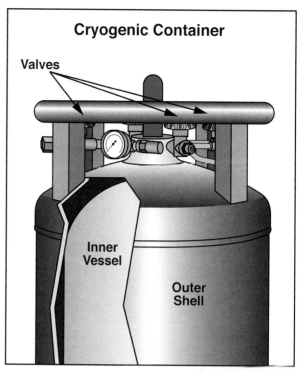

Figure 5.26 Cryogenic cylinders are insulated by a vacuum between an inner vessel and an outer shell.

- **Chemical** — Normal stresses are due to exposure from the outside contact on the container. An incorrect chemical added to a cryogenic tank may cause corrosion.

- **Mechanical** — Accidents may cause mechanical stress, particularly to the container fittings. Severe accidents may cause damage to container walls, creating loss of vacuum within the vacuum space of the container.

NOTE: Venting is a function of some cryogenic containers and may not indicate a system failure.

Cryogenic containers can experience any type of breach, although releases from pressure relief devices and/or damaged fittings are most common. Punctures, splits, and tears are rare, although violent accidents, such as railway derailments or highway accidents, may involve enough force to damage both exterior and interior container walls.

CAUTION
Cold stress and heat stress can result in similar failures.

Table 5.3, p. 232, provides the expansion ratios of several common cryogenic materials.

Table 5.3
Expansion Ratios of Common Cryogenic Materials

Gas	Nitrogen	Oxygen	Argon	Hydrogen	Helium
Boiling Point, °F	-320	-297	-303	-423	-452
Boiling Point, °C	-196	-183	-186	-253	-268
Volume Expansion	696	860	696	850	745

Cryogenic containers are designed to handle extremely cold temperatures; however, the external support structure of the container is not designed to do so. If cryogenic materials leak, the support structure can fail without warning due to stress failures. If other containers are involved in a wreck or derailment with cryogenic containers, leaks from the cryogenic containers may cause failure of other containers due to cold stress.

Cryogenic containers release cold liquids or vapors that rapidly boil into expanding gases **(Figure 5.27)**. Common releases occur through pressure relief devices (rapid relief) or as leaks through damaged attachments and fittings.

Typical dispersion patterns from cryogenic containers include:

- Hemispheric
- Cloud
- Plume
- Pool

Figure 5.27 Released cryogens quickly boil into gas. *Courtesy of Steve Irby, Owasso (OK) Fire Department.*

Liquid-Holding Containers
Liquid-holding containers come in a variety of designs and construction types depending on:

- Size
- Mode of transport
- Material contained

- Use
- Other factors

Liquid-holding containers may have the following characteristics:

- Extremely durable, such as a tank car
- Fragile, such as a glass bottle
- May fail when subjected to fires
- Less likely to fragment as they BLEVE
- Transport materials that polymerize
- Uncontrolled polymerization (chemical stress) may create enough stress to cause container failure
- Explosive in nature

While streams and pools are common dispersion patterns for liquids, many liquids also release vapors that act like gases. All dispersion patterns can be associated with liquids depending on the product and container.

Solids-Holding Containers

Solids-holding containers also come in a variety of designs and construction types depending on size, mode of transport, material contained, use, and other factors. Many containers used for liquids, such as drums, are also used for solids; but some solid hazardous materials have unique containers.

Most solids-holding containers will be damaged via mechanical stressors rather than the physical properties of the materials contained in them. Exceptions include reactive materials such as oxidizers, peroxides, explosives, and water reactive materials.

Common breaches of solids-holding containers include punctures, splits, and tears. Pneumatic loading and unloading attachments may be damaged in accidents. Common releases result in the following:

- Spills and leaks (**Figure 5.28**)
- Detonation — Occurs when oxidizers, peroxides, explosives, and water-reactive materials are involved
- Violent ruptures — Reactive solids release
- Clouds, cones, or via irregular dispersion — Solids may disperse
- Explosions — Detonated explosives disperse in a hemispheric dispersion

Figure 5.28 Solids typically release as a spill. *Courtesy of Barry Lindley.*

Bulk Facility Storage Tanks

Bulk facility storage tanks come in many varieties, depending on the type(s) of materials that will be stored and the pressures at which they need to be maintained. Common bulk storage tanks are described in the following sections.

Pressure Tanks

Pressure tanks (also called *pressure containers* or *pressure vessels*) are designed to hold contents under pressure. Pressure tanks have pressures of 15 psi (103 kPa) or greater. Examples of pressure containers include horizontal pressure vessels and spherical pressure vessels **(Table 5.4)**.

These tanks typically release their products as gases and vapors during rapid relief events, slow leaks from valves and fittings, or violent ruptures **(Figure 5.29)**. Pressure tanks may contain a variety of flammable, toxic, and/or corrosive gases. Pressure tanks are especially dangerous when subjected to heating or fire because they can BLEVE.

Contents leaking from these containers may expand rapidly and displace oxygen, especially in confined spaces. Flammable gases may travel far distances where they can ignite if exposed to an ignition source. Toxic gases may travel far distances and affect people and animals well away from the incident scene.

Table 5.4 Pressure Vessels	
Vessel Type	**Descriptions**
	**Horizontal Pressure Vessel*** Have high pressures and capacities from 500 to over 40,000 gallons (1 893 L to over 151 416 L). They have rounded ends and are not usually insulated. They usually are painted white or some other highly reflective color. **Contents:** LPG, anhydrous ammonia, vinyl chloride, butane, ethane, compressed natural gas (CNG), chlorine, hydrogen chloride, and other similar products
	**Spherical Pressure Vessel** Have high pressures and capacities up to 600,000 gallons (2 271 240 L). They are often supported off the ground by a series of concrete or steel legs. They usually are painted white or some other highly reflective color. **Contents:** Liquefied petroleum gases and vinyl chloride

* It is becoming more common for horizontal propane tanks to be buried underground. Underground residential tanks usually have capacities of 500 or 1,000 gallons (1 893 L or 3 785 L). Once buried, the tank may be noticeable only because of a small access dome protruding a few inches (millimeters) above the ground.

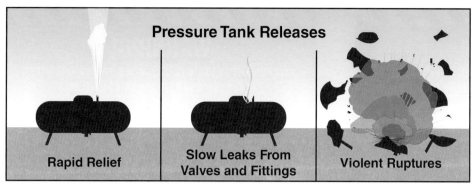

Figure 5.29 Pressure containers release their products during rapid relief events, slow leaks from valves and fittings, or violent ruptures.

Cryogenic Liquid Tanks

Cryogenic liquid storage tanks may have pressures of 15 psi (103 kPa) {1.03 bar} or greater, with capacities from 300 to 400,000 gallons (1 136 L to 1 514 165 L). They have insulated, vacuum-jacketed tanks with safety relief valves and rupture disks. These tanks may contain cryogenic carbon dioxide, liquid oxygen, liquid nitrogen, or other materials **(Figures 5.30a, b, and c)**.

NOTE: Some new Liquefied Natural Gas (LNG) tanks have a capacity of 30 to 60 million gallons (114 to 227 million L).

Materials released from bulk cryogenic liquid tanks will be very cold, and therefore will tend to pool close to the ground. Typically, they will initially be visible as a fog or cloud. Most types of cryogenic leaks will displace oxygen. Some will create an explosive environment.

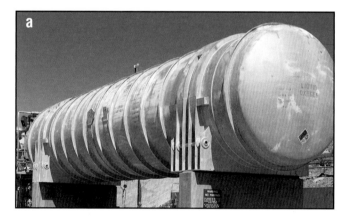

Figures 5.30a, b, and c Cryogenic tanks may contain cryogenic carbon dioxide, liquid oxygen, liquid nitrogen or other products. **a)** and **c)** *Courtesy of Rich Mahaney*; **b)** *Courtesy of Barry Lindley.*

Low Pressure Storage Tanks

Low pressure storage tanks have operating pressures from 0.5 to 15 psi (3.45 kPa to 103 kPa). Types of low pressure storage tanks include **(Table 5.5)**:

- Dome roof tanks
- Spheroid tanks
- Noded spheroid tanks

Low pressure storage tanks typically contain flammable and combustible liquids with low vapor pressures, such as solvents. Flammability and/or toxicity are often hazards associated with the products stored in these containers. Low pressure tanks can release their products as liquids or as gases and vapors depending on the product contained. The priority is eliminating ignition sources at incidents involving these containers.

Table 5.5
Low Pressure Storage Tanks

Tank Type	Descriptions
	**Dome Roof Tank** Generally classified as low pressure tanks with operating pressures as high as 15 psi (103 kPa). They have domes on their tops. **Contents:** Flammable liquids, combustible liquids, fertilizers, solvents, etc.
	**Spheroid Tank** Low pressure storage tanks. They can store 3,000,000 gallons (11 356 200 L) or more of liquid. **Contents:** Liquefied petroleum gas (LPG), methane, and some flammable liquids such as gasoline and crude oil
	**Noded Spheroid Tank** Low pressure storage tanks. They are similar in use to spheroid tanks, but they can be substantially larger and flatter in shape. These tanks are held together by a series of internal ties and supports that reduce stresses on the external shells. **Contents:** LPG, methane, and some flammable liquids such as gasoline and crude oil

Nonpressure/Atmospheric Storage Tanks

Nonpressure/atmospheric storage tanks are designed to hold contents under little or no pressure **(Table 5.6, p. 238)**. The maximum pressure under which an atmospheric tank is capable of holding its contents is 0.5 psi (3.45 kPa). Common types of atmospheric tanks include:

- Horizontal tanks
- Ordinary cone roof tanks
- Open and closed roof floating-roof tanks
- Lifter roof tanks
- Vapordome roof tanks

Nonpressure/atmospheric storage tanks typically hold liquids, most often hydrocarbons. This liquid may be flammable/combustible products, such as fuel oil or other petroleum products, or corrosive and/or toxic, such as sulfuric acid and aniline.

Damaged nonpressure/atmospheric storage tanks release their contents via spilling or leaking through container walls, valves, fittings, and attachments. Depending on the product, vapors, which are often heavier than air, may travel some distance from their liquid source.

Treat interiors of bulk nonpressure/atmospheric storage tanks as **confined spaces**. Even when empty, these containers are likely to have dangerous atmospheres **(Figure 5.31)**. Follow special procedures when operating in and around them.

> **Confined Space** — Space or enclosed area not intended for continuous occupation, having limited (restricted access) openings for entry or exit, providing unfavorable natural ventilation and the potential to have a toxic, explosive, or oxygen-deficient atmosphere.

Figure 5.31 The interiors of bulk containers are confined spaces.

Atmospheric Storage Tank Failures

Catastrophic failures of aboveground atmospheric storage tanks can occur when flammable vapors in a tank explode and break either the shell-to-bottom or side seam. These failures have caused tanks to rip open and (in rare cases) hurtle through the air. Shell-to-bottom seam failures are more common among old storage tanks. Steel storage tanks built before 1950 generally do not conform to current industry standards for explosion and fire-venting situations. A properly designed and maintained storage tank will break along the shell-to-top seam, which is more likely to limit the fire to the damaged tank and prevent the contents from spilling.

Atmospheric tanks used to store flammable and combustible liquids should be designed to fail along the shell-to-roof seam when an explosion occurs in the tank. This feature prevents the tank from propelling upward or splitting along the side.

Many safety issues arise once atmospheric tanks become involved in or are exposed to fire. Emergency response planning is essential to prevent injuries or deaths caused by the special problems presented by tank fires and emergencies.

Table 5.6
Atmospheric/Nonpressure Storage Tanks

Tank Type	Descriptions

Horizontal Tank

Cylindrical tanks sitting on legs, blocks, cement pads, or something similar; typically constructed of steel with flat ends. Horizontal tanks are commonly used for bulk storage in conjunction with fuel-dispensing operations. Old tanks (pre-1950s) have bolted seams, whereas new tanks are generally welded. A horizontal tank supported by unprotected steel supports or stilts (prohibited by most current fire codes) may fail quickly during fire conditions.

Contents: Flammable and combustible liquids, corrosives, poisons, etc.

Cone Roof Tank

Have cone-shaped, pointed roofs with weak roof-to-shell seams that break when or if the container becomes overpressurized. When it is partially full, the remaining portion of the tank contains a potentially dangerous vapor space.

Contents: Flammable, combustible, and corrosive liquids

Open Top Floating Roof Tank

Large-capacity, aboveground holding tanks. They are usually much wider than they are tall. As with all floating roof tanks, the roof actually floats on the surface of the liquid and moves up and down depending on the liquid's level. This roof eliminates the potentially dangerous vapor space found in cone roof tanks. A fabric or rubber seal around the circumference of the roof provides a weather-tight seal.

Contents: Flammable and combustible liquids

Covered Top Floating Roof Tank

Have fixed cone roofs with either a pan or deck-type float inside that rides directly on the product surface. This tank is a combination of the open top floating roof tank and the ordinary cone roof tank.

Contents: Flammable and combustible liquids

Vents around rim provide differentiation from Cone Roof Tanks

Continued

Table 5.6 (concluded)

Tank Type	Descriptions
	**Covered Top Floating Roof Tank with Geodesic Dome** Floating roof tanks covered by geodesic domes are used to store flammable liquids.
	**Lifter Roof Tank** Have roofs that float within a series of vertical guides that allow only a few feet (meters) of travel. The roof is designed so that when the vapor pressure exceeds a designated limit, the roof lifts slightly and relieves the excess pressure. **Contents:** Flammable and combustible liquids
	**Vapordome Roof Tank** Vertical storage tanks that have lightweight aluminum geodesic domes on their tops. Attached to the underside of the dome is a flexible diaphragm that moves in conjunction with changes in vapor pressure. **Contents:** Combustible liquids of medium volatility and other nonhazardous materials
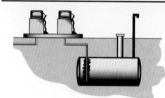 **Fill Connections Cover** 	**Atmospheric Underground Storage Tank** Constructed of steel, fiberglass, or steel with a fiberglass coating. Underground tanks will have more than 10 percent of their surface areas underground. They can be buried under a building or driveway or adjacent to the occupancy. This tank has fill and vent connections located near the tank. Vents, fill points, and occupancy type (gas/service stations, private garages, and fleet maintenance stations) provide visual clues. Many commercial and private tanks have been abandoned, some with product still in them. These tanks are presenting major problems to many communities. **Contents:** Petroleum products Rare and technically are not "tanks." First responders should be aware that some natural and manmade caverns are used to store natural gas. The locations of such caverns should be noted in local emergency response plans.

Underground Storage Tanks

Underground storage tanks are typically constructed of steel, fiberglass, or steel with a fiberglass coating **(Figure 5.32)**. Other features of underground storage tanks include:

- Usually contain liquids (typically, gasoline).
- Some horizontal propane pressure tanks have been buried underground.
- Classified as low pressure or nonpressure/atmospheric.
- More than ten percent of their surface areas are underground.
- Can be buried under or adjacent to a building or driveway.

Underground tanks have fill and vent connections located nearby **(Figure 5.33)**. Vents, fill points, and occupancy type (gas/service stations, private garages, and fleet maintenance stations) provide visual clues to the presence of underground tanks.

Many commercial and private underground tanks have been abandoned, some with product still in them. Leaking underground storage tanks may go undetected until they leak into an undesirable location in liquid or vapor form. Fires and/or explosions can occur if flammable materials contact an ignition source.

Figure 5.32 An underground storage tank. Typically, these contain liquids such as gasoline.

Figure 5.33 Underground tanks will have vents nearby.

Bulk Transportation Containers: Cargo Tank Trucks

Highway vehicles that transport hazardous materials include **(Figure 5.34)**:

- Cargo tank trucks (also called *tank motor vehicles, cargo tanks,* and *tank trucks*)
- Dry bulk containers
- Compressed gas tube trailers
- Mixed load containers (also called *box trucks* or *dry van trucks*)

Figure 5.34 Cargo tanks transport hazardous materials via roads and highways. *Courtesy of Rich Mahaney.*

These vehicles transport all types of hazardous materials in a wide range of quantities. Depending on the type and quantity of hazardous material, these vehicles may have DOT/TC placards. Even unplacarded highway vehicles may be carrying quantities of hazardous materials, such as fuel in saddle tanks or other materials in levels below placarding requirements.

Cargo tank trucks are recognizable because they have construction features, fittings, attachments, or shapes characteristic of their uses. Even if first responders recognize one of the cargo tank trucks described in this section, the process of positive identification must proceed from placards to shipping papers or other formal sources of information.

Cargo tank trucks commonly transport bulk amounts of hazardous materials via roadway. Most cargo tank trucks that haul hazardous materials are designed to meet government tank safety specifications. These specifications set minimum tank construction material thicknesses, required safety features, and maximum allowable working pressures.

Cargo tank specification and name plates provide information about the standards to which the container/tank was built **(Figure 5.35, p. 242)**. These plates are usually found on the roadside/driver's side of the vehicle near the dolly leg (landing gear).

The two specifications in use are the motor carrier (MC) standards and DOT/TC standards. Cargo tank trucks built to a given specification are designated using the MC or DOT/TC initials followed by a three-digit number identifying the specification (such as MC 306 and DOT/TC 406). Emergency responders can recognize these cargo tank trucks by their required construction features, fittings, valves, attachments, and shapes, such as **ring stiffeners** on corrosive liquid tanks and bolted manways or manholes on high pressure tanks.

Tanks not constructed to meet one of the common MC or DOT/TC specifications are commonly referred to as *nonspec tanks*. Nonspec tanks may haul nonregulated hazardous materials if the tank was designed for a specific purpose and exempted from the DOT/TC requirements. Examples of nonregulated hazards include: molten sulphur, asphalt, and milk. Nonhazardous materials may be hauled in either nonspec cargo tank trucks or cargo tank trucks that meet a designated specification **(Figure 5.36)**.

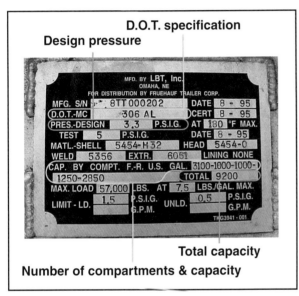

Figure 5.35 Manufacturers' specification plates provide information about the tank's pressure, capacity, and number of compartments. *Courtesy of Rich Mahaney.*

Figure 5.36 Nonhazardous materials may be transported in cargo tanks.

Milk Spill, 2002

While seeming harmless, milk can actually be an environmental and biological hazard. If a large amount of milk enters a waterway and is not immediately removed, bacteria will feed on the milk and deplete the water of oxygen needed for fish to survive.

Firefighters faced an environmental hazard in July of 2002 when a dairy tanker truck crashed into a bridge over a stream in England, spilling around 5,000 gallons of milk (around 19 000 liters). The stream flows into Rudyard Lake, and the local fish population was at risk if the milk reached the lake and remained there long enough for bacteria to feed on it.

Firefighters worked for hours, using pumps to remove the milk from the stream and building dams to prevent the milk from flowing into the lake. The cleanup allowed the popular lake, which was home to around 50,000 fish, to avoid an environmental hazard.

High Pressure Tank Trucks

High pressure tank trucks are also known as MC-331 (or equivalent) cargo tanks. These tank trucks have pressures typically between 100 to 500 psi (690 to 3 448 kPa) with typical capacities between 3,000 to 11,000 gallons (11 356 to 41 640 L). High pressure cargo tanks have a single, steel compartment.

High pressure tank trucks transport liquefied gases, such as propane, anhydrous ammonia, and butane, or high vapor pressure liquids and highly hazardous materials such as parathion. High-pressure "Bobtail Tanks" are used for local delivery of liquefied petroleum gas and anhydrous ammonia **(Figure 5.37)**.

MC-331 cargo tank trucks have the following features:

- Bolted manway
- Inlet and outlet valves
- White or other reflective paint scheme (typically)
- Large hemispherical heads on both ends
- Guard cage around the bottom loading/unloading piping **(Figure 5.38)**
- Uninsulated tanks, single-shell vessels
- Emergency shut-offs (typically located in the left front and right rear)
- Permanent markings such as *FLAMMABLE GAS*, *COMPRESSED GAS*, shipping name, or identifiable manufacturer or distributor names

NOTE: See Chapters 2 and 4 for markings, labels, placards.

High pressure tank trucks may experience disintegration, runaway cracking, damage to attachments, punctures, splits, or tears. They may release via violent rupture, rapid relief, or leaks. When exposed to heat or flames, they may BLEVE **(Figure 5.39, p. 244)**. Flammable gases/vapors may explode/ignite when coming into contact with an ignition source.

Figure 5.37 Bobtail tanks are used to deliver propane and other products.

Figure 5.38 High pressure tank trucks have a cage guarding the bottom fittings. *Courtesy of Barry Lindley.*

Cryogenic Tank Trucks

Cryogenic tank trucks are also known as MC-338, TC-338, or SCT-338 (or equivalent) cargo tanks. These tank trucks have pressures that can be less than 25 and up to 500 psi (172 to 3 447 kPa), and capacities of 8,000 to 10,000 gallons (30 283 to 37 854 L). These trucks have well-insulated aluminum or steel tanks with vacuum-sealed shells.

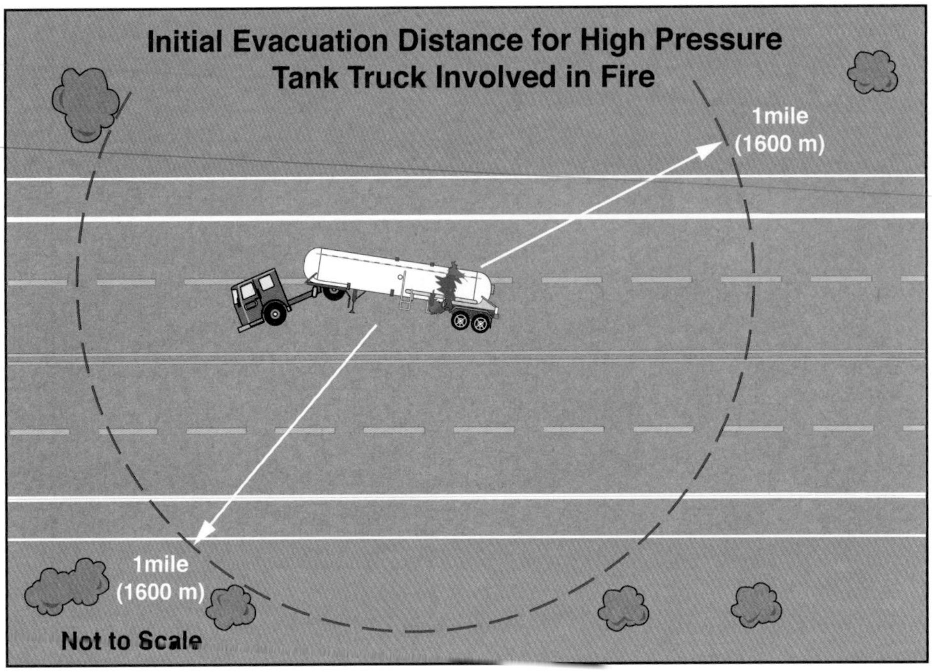

Figure 5.39 The initial evacuation distance for high pressure tank trucks involved in fire is one mile (1 600 meters) because they may BLEVE.

These tank trucks transport the following gases that have been liquefied by lowering their temperatures, including:

- Liquefied oxygen
- Liquefied nitrogen **(Figure 5.40)**
- Liquefied carbon dioxide
- Liquefied hydrogen

When released, the product will be extremely cold and therefore will tend to pool close to the ground. Typically, the product will initially be visible as a fog or cloud because of the condensation of humidity and ice formation. As they warm, these liquids will change into gas state and expand. Gases will expand as they warm.

Figure 5.40 Cryogenic tank trucks have a wide range of pressures, from very low to very high.

WARNING!
Gases will displace oxygen when released.

WARNING!
Many released liquefied gases can kill a person with one breath due to oxygen deprivation leading to asphyxiation.

MC-338 cryogenic tank trucks have the following features:

- Relief valves that may be discharging nonhazardous vapor such as nitrogen or oxygen **(Figure 5.41)**
- Round tank with flat ends
- Large and bulky double shelling and heavy insulation
- Loading/unloading station attached either at the rear or in front of the rear dual wheels
- Permanent markings such as *REFRIGERATED LIQUID* or an identifiable manufacturer name, proper shipping name
- Emergency shutoffs on the left-front and right-rear

Figure 5.41 Cryogenic tank trucks may discharge nonhazardous vapors through their relief valves. *Courtesy of Rich Mahaney.*

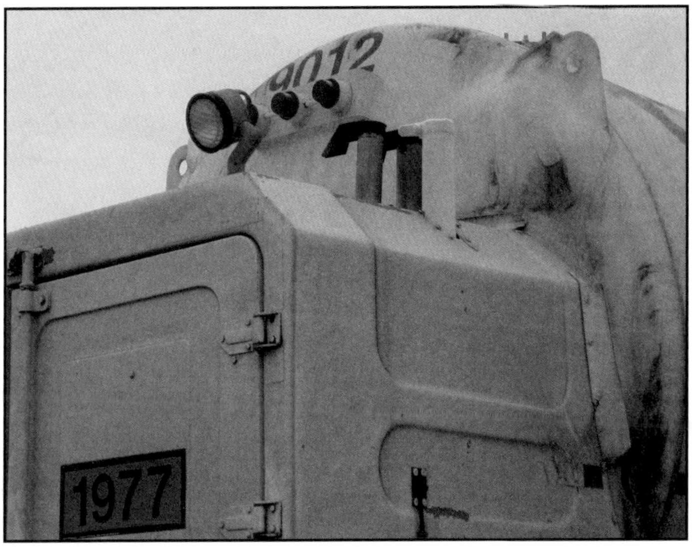

Cryogenic tank trucks may experience disintegration, runaway cracking, damage to attachments, punctures, splits or tears. They may release via violent rupture, rapid relief, or leaks. When exposed to heat or flames, they may BLEVE **(Figure 5.42)**.

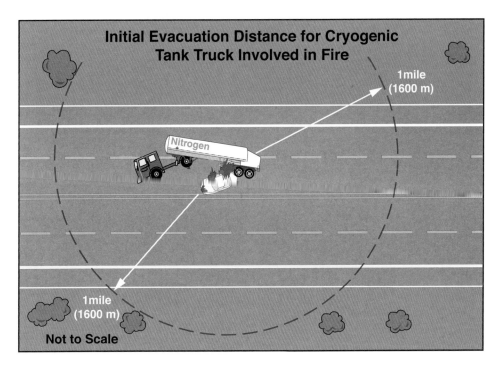

Figure 5.42 Because some cryogenic tank trucks have high pressures, if they are involved in fire, their initial evacuation distance is one mile (1 600 meters). Like pressure tank trucks, they may BLEVE.

Low Pressure Chemical Tank Trucks

Low pressure chemical tank trucks are also known as MC-307 or DOT/TC-407 (or equivalent) cargo tanks depending on the specifications to which they were built **(Figure 5.43)**. These tank trucks typically have a pressure of 25 to 35 psi (172 to 241 kPa), with typical capacities of 5,500 to 7,000 gallons (20 820 L to 26 500 L). Most have a stainless steel, mild steel, or aluminum tank and some may have a rubber lining, rubber coating, or polymer coating.

Figure 5.43 Low pressure chemical tank trucks may carry flammable liquids, mild corrosives, and some toxics/poisons *Courtesy of Rich Mahaney.*

Low pressure chemical tank trucks carry liquids, including flammable/combustible liquids, mild corrosives, and some toxics/poisons. Liquids and vapors may have a variety of hazards depending on the product(s) being transported.

MC-307 or DOT/TC-407 cargo tanks have the following features:

Figure 5.44 Stiffening rings are visible on this low pressure chemical tank truck. *Courtesy of Rich Mahaney.*

- Rounded or horseshoe-shaped ends

- Stiffening rings that may be visible or covered **(Figure 5.44)**

- Rollover/turnover protection

- Single or multiple compartments

- Manway assembly protected by the rollover/turnover protection (crash box)

- Discharge piping at midship or rear

- Rear or middle unloading

- Pressure and vacuum protection

- Drain hose from the rollover/turnover protection down the side of the tank

- Emergency shutoffs (hydraulic or pneumatic) located on the left front of the tank

Low pressure chemical tanks can be breached by damage to attachments, punctures, splits or tears. When involved in fire or unusual chemical reactions, they are unlikely to rupture violently. More commonly, these cargo tanks are involved in liquid spills or leaks **(Figure 5.45)**.

Nonpressure Cargo Tank Trucks

Nonpressure cargo tank trucks are also known as MC-306 or DOT/TC-406 (or equivalent) cargo tanks depending on the specifications to which they were built. New tanks are constructed of aluminum or steel, but older tanks are made of steel.

Typical pressure in these tanks is less than 4 psi (28 kPaG). The maximum capacity of these tanks is 14,000 gallons (53 000 L) in the U.S., with a typical capacity range between 1,500 to 10,000 gallons (5 678 L to 37 854 L).

Figure 5.45 Typically, low pressure chemical tank trucks are involved in spills and leaks rather than violent ruptures. *Courtesy of Barry Lindley.*

Nonpressure cargo tanks almost always carry flammable/combustible liquids such as gasoline, fuel oil, alcohol, or other nonhazardous liquids. Individual compartments may carry different products. Fire control will be a primary concern at incidents involving these vehicles. **(Figure 5.46)**.

Figure 5.46 Nonpressure cargo tanks usually carry flammable and combustible liquids such as gasoline. *Courtesy of Rich Mahaney.*

MC-306 or DOT/TC-406 cargo tanks have the following features:

- Oval shape
- Manways located in overturn protection areas
- Bottom valves
- Longitudinal rollover protection
- Valve assembly and unloading control box under tank **(Figure 5.47, p. 248)**
- Vapor-recovery system on curb side and rear, if present
- Multiple compartments
- Manway assemblies and vapor-recovery valves on top for each compartment
- Emergency shut-off systems

Figure 5.47 Nonpressure cargo tanks may have an unloading control box under the tank.

Figure 5.48 Nonpressure cargo tanks may be breached by splits, tears, and damage to attachments.

Nonpressure cargo tanks can be breached by punctures, splits, tears, or damage to attachments **(Figure 5.48)**. When involved in fire, steel tanks can rupture violently; aluminum tanks will melt. More commonly, these cargo tanks are involved in liquid spills or leaks.

Corrosive Liquid Tank Trucks

Maximum Allowable Working Pressure (MAWP) — A percentage of a container's test pressure. Can be calculated as the pressure that the weakest component of a vessel or container can safely maintain.

Corrosive liquid tank trucks are also known as MC-312 or DOT/TC-412 (or equivalent) cargo tank trucks depending on the specifications to which they were built. These tank trucks typically have a pressure range of 35 to 55 psi (241 kPa to 379 kPa) and may have a much higher **maximum allowable working pressure (MAWP)**. Typical tank capacities are from 3,300 to 6,300 gallons (12 492 L to 23 848 L). Aluminum, mild steel, stainless steel, and fiberglass reinforced plastics (FRP) tanks can be rubber or polymer lined. The outer jacket may be aluminum or stainless steel and often covers a layer of insulation. Usually, these tank trucks only have one compartment.

CAUTION
Liquid coming out of drain hoses may indicate leaks on the top of the tanks.

Corrosive liquid tank trucks carry corrosives, typically acids such as acetyl chloride, hydrochloric acid, and sodium hydroxide **(Figure 5.49)**. Avoid contact with corrosive liquids and vapors. Be aware that corrosives can damage tools and equipment, including firefighter protective clothing.

MC-312 or DOT/TC-412 corrosive liquid tank trucks may have the following features **(Figure 5.50)**:

- Small-diameter round shape

- Exterior stiffening rings (may be visible on uninsulated tanks)

- Top unloading on the rear of the tank with exterior piping extending to the bottom of the tank.

- Rollover protection around the valve assembly

- A pressure relief device (PRD) typically located in turnover protection.

- Discolored loading/unloading area

- An area painted or coated with corrosive-resistant material

Figure 5.49 Corrosive liquid tank trucks carry corrosive liquids such as hydrochloric acid and sodium hydroxide.

WARNING!
Most DOT/TC-412s do not have a shut-off valve.

Corrosive liquid tank trucks can be breached by damage to attachments, punctures, splits or tears **(Figure 5.51, p. 250)**. Most commonly, these cargo tanks are involved in liquid spills or leaks, but on rare occasions chemical reactions can cause violent ruptures.

Figure 5.50 This corrosive liquid tank truck has an unloading area painted/coated in a different color. *Courtesy of Rich Mahaney.*

Figure 5.51 Releases from corrosive liquid tank trucks may involve spilled/leaking liquids and vapors/fumes. Rarely, these tank trucks can rupture violently due to chemical reactions. *Courtesy of Barry Lindley.*

Compressed-Gas/Tube Trailers

Compressed-gas/tube trailers transport individual steel cylinders stacked and mounted together. Typical pressures in the tubes range from 2,400 to 5,000 psi (16 547 kPa to 34 474 kPa) (gas only). Each cylinder typically has an overpressure device.

Compressed-gas/tube trailers carry helium, hydrogen, methane, oxygen, and other gases. Often, they are parked or located at the facility where the gas is used, much like a semi-permanent storage tank **(Figure 5.52)**.

Compressed-gas/tube trailers typically have the following features:

- A pressure relief device (PRD) for each cylinder **(Figure 5.53)**
- Bolted manway at front or rear
- Valves in a protected housing **(Figure 5.54)**
- Valves manifolded together **(Figure 5.55)**
- Permanent markings for the material or ownership that is locally identifiable, including proper shipping name

Figure 5.52 Compressed-gas/tube trailers may be permanently parked at a facility.

Figure 5.53 The numbers on this compressed-gas/tube trailer correspond to the pressure relief device for each cylinder. *Courtesy of Barry Lindley.*

Figure 5.54 Compressed-gas/tube trailer cylinder valves are protected in a boxlike housing. *Courtesy of Rich Mahaney.*

Figure 5.55 Compressed-gas/tube trailer cylinder valves are manifolded together.

Compressed-gas/tube trailers may experience disintegration, runaway cracking, damage to attachments, punctures, splits or tears. They may release via violent rupture, rapid relief, or leaks. When exposed to heat or flames, they may BLEVE. Flammable gases may explode/ignite when coming into contact with an ignition source. Because of the high pressures in the tubes, accidental releases from these trailers can be violent, and released gases will expand rapidly.

Dry Bulk Cargo Trailers

Dry bulk cargo trailers transport solids, including hazardous solids such as oxidizers, corrosive solids, cement, plastic pellets, and fertilizers **(Figure 5.56)**. While contents are not usually under pressure, low pressures between 15 to 20 psi (103 to 138 kPa) may be used to discharge or transfer the product from the container. These cargo trailers are constructed to transport heavy loads, but damage to attachments, punctures, splits, or tears may occur if they are involved in an accident.

Dry bulk cargo trailers have the following features:

Figure 5.56 Dry bulk cargo trailers may transport oxidizers, corrosive solids, and other materials. *Courtesy of Rich Mahaney.*

- Typically not under pressure

- Varying shapes that often include bottom valves with *V-* or *W*-shaped bottom-unloading compartments **(Figure 5.57, p. 252)**

- Rear-mounted, auxiliary-engine-powered compressor or tractor-mounted power-take-off air compressor

- Air-assisted, exterior loading and bottom unloading pipes

- Top manway assemblies **(Figure 5.58, p. 252)**

Figure 5.57 Dry-bulk cargo trailers may have V- or W-shaped bottom-unloading compartments. *Courtesy of Rich Mahaney.*

Figure 5.58 Dry-bulk cargo trailers may have top manway assemblies, as indicated by the arrows on this picture. *Courtesy of Barry Lindley.*

Bulk Transportation Containers: Tank Cars

Tank cars carry the bulk of the hazardous materials transported by rail. Some railroad tank cars have capacities of 4,000 to 34,000 gallons (15 142 L to 128 704 L) **(Figure 5.59)**. Because these cars carry large quantities, an accidental release of gases or liquids can cause many difficulties for responders. By recognizing distinctive railroad cars, first responders can begin the identification process from the greatest possible distance. The type of car gives clues as to what material may be within as well as the material's weight and volume.

Tank cars are divided into the following three main categories:

- Low pressure tank cars, also known as *general service* tank cars and *nonpressure* tank cars
- Pressure tank cars
- Cryogenic liquid tank cars

NOTE: Most of the following information on railroad tank cars is courtesy of *A General Guide to Tank Cars*, prepared by the Union Pacific Railroad, April 2003.

Figure 5.59 Railroad tank car capacities are much greater than cargo tank trucks. *Courtesy of Rich Mahaney.*

Hazardous Materials Transport by Railroad

In addition to the types of cars previously described, other types of railroad cars may also carry hazardous materials:

- Hopper cars (including pneumatically unloaded hopper cars)
- Boxcars
- Flat cars transporting other containers of hazardous materials (including intermodal containers, see Intermodal Container section)
- Well cars
- Spine cars (a type of flat car)
- Special service (or specialized) cars

Pressure Tank Cars

Pressure tank cars typically transport flammable, nonflammable, and poisonous gases at pressures greater than 25 psi (172 kPa) at 68°F (20°C) **(Figure 5.60)**. They also transport flammable liquids and liquified compressed gases. Tank test pressures from these tank cars range from 100 to 600 psi (689 kPa to 4 137 kPa). Pressure tank car capacities range from 4,000 to 34,000 gallons (15 142 L to 128 704 L).

Figure 5.60 Pressure tank cars transport flammable, nonflammable, and poisonous gases. *Courtesy of Rich Mahaney.*

NOTE: Though less common, some older jumbo cars can have a capacity of up to 50,000 gal (189 271 L).

Pressure tank cars often have the following features:

- Cylindrical.
- Noncompartmentalized.
- Metal (steel or aluminum).
- Rounded ends (heads).
- Top-loading cars.
- Out-of-sight fittings (loading/unloading, pressure-relief, and gauging) located inside the protective housings mounted on the manway cover plates in the top center of the tanks. Pressure tank cars typically have all fittings out of sight under the single protective housings on top of the tanks **(Figure 5.61)**.
- May be insulated and/or thermally protected **(Figure 5.62, p. 254)**.

Figure 5.61 Pressure tank cars have fittings protected in a single housing on the top of the tank. *Courtesy of Walter Schneider.*

New pressure tank cars have greater accident protection features and will withstand greater damage without leaking. They feature thicker walls, a lower profile of protective housing, and higher tank test pressures. New pressure tank cars are significantly heavier than old cars, and they may also be equipped with GPS tracking devices and anti-tampering mechanisms **(Figure 5.63, p. 254)**.

NOTE: Several highly hazardous liquids are shipped in pressure cars that have little to no vapor pressure.

Pressure tanks are subject to thermal, mechanical, and chemical damage that will release expanding gases or vapors. Because capacities are so large, affected areas and evacuation zones may be quite large. When exposed to heat or flame, pressure tank cars may BLEVE. Per the *ERG*, the initial isolation zone for a pressure tank car involved in fire is one mile (1.6 km) **(Figure 5.64)**.

Figure 5.62 Pressure tank cars may have thermal insulation. *Courtesy of Barry Lindley.*

Figure 5.63 Pressure tank cars may be equipped with GPS tracking devices. *Courtesy of Rich Mahaney.*

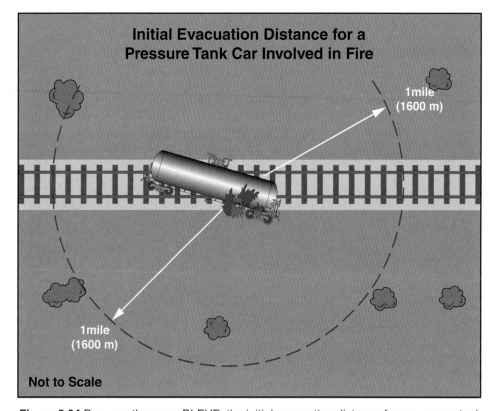

Figure 5.64 Because they may BLEVE, the initial evacuation distance for a pressure tank car involved in fire is one mile (1 600 meters).

Cryogenic Liquid Tank Cars

Cryogenic liquid tank cars carry low pressure (usually below 25 psi [172 kPa]) refrigerated liquids (-130°F and below [-90°C and below]). Materials found in these tanks include argon, hydrogen, nitrogen, and oxygen. Liquefied natural gas (LNG) and ethylene may be found in these containers at somewhat higher pressures. Cryogenic liquid tank cars often have fittings for loading/unloading, pressure relief, and venting in ground-level cabinets on the sides of the car or the end of the car **(Figure 5.65)**.

A cryogenic liquid tank car is in the tank-within-a-tank category with a stainless steel inner tank supported within a strong outer tank. The space between the inner tank and outer tank is filled with insulation. This space is also kept under a vacuum. The combination of insulation and vacuum protects the contents from ambient temperatures for only 30 days. The shipper tracks these time-sensitive shipments. Per the *ERG*, the initial isolation zone for a cryogenic liquid tank car is ½ mile (1 km).

Figure 5.65 Cryogenic liquid tank cars carry low pressures, and they may be recognized by ground-level cabinets on the side or end of the car. *Courtesy of Rich Mahaney.*

Low Pressure Tank Cars

Low pressure tank cars transport hazardous and nonhazardous solids and liquids with vapor pressures below 25 psi (172 kPa) at 105°F to 115°F (41°C to 46°C) **(Figure 5.66, p. 256)**. Tank test pressures for low pressure tank cars are 60 and 100 psi (414 kPa and 689 kPa). Capacities range from 4,000 to 34,000 gallons (15 142 L to 128 704 L) in newer tanks made of aluminum, mild steel, or stainless steel.

NOTE: Though less common, some older jumbo cars can have a capacity of up to 45,000 gal (170 344 L).

Low pressure tank cars transport hazardous materials, such as:

- Flammable liquids
- Flammable solids
- Reactive liquids
- Reactive solids
- Oxidizers **(Figure 5.67, p. 256)**
- Organic peroxides

Figure 5.66 Low pressure tank cars transport hazardous and nonhazardous liquids and solids.

Figure 5.67 Oxidizers may be transported in low pressure tank cars. *Courtesy of Rich Mahaney.*

- Poisons
- Irritants
- Corrosive materials

  They also transport nonhazardous materials, such as:

- Fruit and vegetable juices
- Wine and other alcoholic beverages
- Tomato paste
- Other agricultural products

  Low pressure tank cars have the following features:

- Cylindrical with rounded ends (heads)
- At least one manway for access to the tank's interior
- Compartmentalized with up to six compartments constructed as distinct tanks, each with its own set of fittings, capacity, and ability to transport a different commodity **(Figure 5.68)**.
- Fittings for loading/unloading, pressure and/or vacuum relief, gauging, and other purposes visible at the top and/or bottom of the car **(Figure 5.69)**.

  For many years, one method for identifying low pressure tank cars was to look for multiple fittings and equipment on top of the tank car. However, some new DOT/TC 111 tank cars enclose some or all of those fittings inside a protective housing similar to a pressure car (see previous section). First responders must now look at the top of the car and, if a single protective housing is present, they must verify whether it is a high-pressure tank car or a DOT 111 tank car by identifying the DOT/TC specifications stenciled on the right-hand side of the car **(Figures 5.70a and b)**.

Trains transporting multiple low pressure tank cars containing ethanol, crude oil, and other Class 3 products may be called **High-Hazard Flammable Trains (HHFT)(Figures 5.71a and b, p. 258)**. When involved in accidents, these tank cars can release their products, ignite, and violently rupture. These types of accidents are primarily attributed to human error. In addition to DOT-/TC-111 tank cars, new DOT-/TC-117 and DOT-/TC-120 tanks cars may be encountered **(Figures 5.72a, b, and c, p. 258)**.

High-Hazard Flammable Trains (HHFT) — Trains that have a continuous block of twenty or more tank cars loaded with a flammable liquid or thirty-five or more cars loaded with a flammable liquid dispersed through a train.

Figure 5.68 This low pressure tank car has two tanks. These may carry separate products. *Courtesy of Rich Mahaney.*

Figure 5.69 Many low pressure tank cars have multiple fittings visible on the top and/or bottom of the car.

Figures 5.70a and b Low pressure DOT/TC 111 tank cars have fittings protected in a single housing, much like pressure tank cars. Check the specification markings if in doubt. *Courtesy of Rich Mahaney.*

Figures 5.71a and b High-Hazard Flammable Trains (HHFT) often carry ethanol and crude oil. When involved in accidents, they can release their products, ignite, and rupture violently. *Courtesy of Rich Mahaney.*

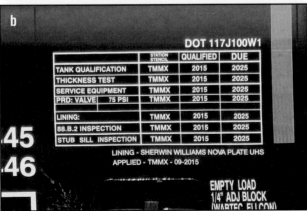

Figures 5.72a, b, and c Responders may encounter new DOT/TC 117 tank cars. The fittings for these cars are also in a protective housing. **a)** and **c)** *Courtesy of Barry Lindley*; **b)** *Courtesy of Rich Mahaney.*

Lac Mégantic Train Derailment (2013)

HHFT trains have been involved in several high-profile incidents, including a 2013 incident in which a 72-car train carrying around 2 million gallons (7.7 million liters) of petroleum crude oil derailed in Lac Mégantic, Quebec, Canada, killing 47 people and destroying much of the city's downtown.

After a Montreal, Maine and Atlantic Railway (MMA) train arrived in the city of Nantes on the evening of July 5, 2013, the engineer applied the train's hand brakes and left on its air brakes. Parked on a descending grade, the train gave the false impression that it was meeting railway rules that require a train to be kept in place by only its hand brakes.

Soon afterward, the Nantes Fire Department received a 9-1-1 call about a fire on the train. Firefighters responded and quickly handled the problem, shutting off the locomotive's fuel supply and switching off the electrical breakers inside the cab. After speaking with railway officials, firefighters left the station with the impression that the threat had been handled and the train was in fine condition.

That was not the case, though. When firefighters shut off the locomotives, the air compressor stopped supplying air to the air brakes and left the train unable to hold still on the main track. The train started rolling early in the morning, reaching 65 miles per hour as it headed for Lac Mégantic, a little more than seven miles away.

Sixty-three of the train's tank cars derailed in downtown Lac Mégantic, killing dozens of people, leveling buildings, and creating massive fires when the train's crude oil ignited. Fire departments responded to the large accident scene, and firefighters used a well-coordinated plan to extinguish the fires and protect the site.

An investigation of the derailment determined that many factors, not one in particular, contributed to the deadly accident. Amongst the factors, investigators found that the crude oil being transported was more volatile than described in the train's shipping documents. MMA increased testing for its trains and stopped having trains being operated by a one-person crew of just an engineer.

Other Railroad Cars

Other railroad cars include hopper cars and miscellaneous cars such as boxcars and gondolas. Descriptions of these cars are as follows:

- **Covered hopper cars** — Often transport dry bulk materials such as grain, calcium carbide, ammonium nitrate, and cement **(Figure 5.73, p. 260)**.

- **Uncovered (or open top) hopper cars** — May carry coal, sand, gravel, or rocks **(Figure 5.74, p. 260)**.

- **Pneumatically unloaded hopper cars** — Unloaded by air pressure and used to transport dry bulk loads such as ammonium nitrate fertilizer, dry caustic soda, plastic pellets, and cement. Pressure ratings during unloading range from 20 to 80 psi (69 kPa to 552 kPa) **(Figures 5.75a and b, p. 260)**.

- **Miscellaneous cars** — Boxcars and gondolas, well cars, and spine cars often used to carry containers of hazardous materials. These cars can include mixed cargos of a variety of products in different types of packaging **(Figure 5.76, p. 261)**.

NOTE: Cars may be fumigated, presenting additional hazards.

Figure 5.73 Covered hoppers carry solids, including oxidizers. *Courtesy of Rich Mahaney.*

Figure 5.74 Uncovered hoppers often carry coal. *Courtesy of Rich Mahaney.*

Figures 5.75a and b Pneumatically unloaded hopper cars transport ammonium nitrate fertilizer, dry caustic soda, and other solids. *Courtesy of Rich Mahaney.*

Figure 5.76 Boxcars carry a variety of mixed cargo, including packages of hazardous materials. *Courtesy of Barry Lindley.*

North American Railroad Tank Car Markings

Responders can receive valuable information from the markings on railroad tank cars as well as its contents:

- Reporting marks (**railcar initials and numbers**)
- Capacity stencil
- Specification marking

The *ERG* provides a key to these markings in the railcar identification chart, and more information is provided in sections that follow. Additionally, manufacturers' names on cars may provide some contact information. Railcars are normally dedicated to transporting a single material. **Dedicated tank cars** may have the name of that material painted on the car. DOT/TC only requires a finite number of shipping names to be stenciled on the car. Some companies may also choose to include that information as a courtesy.

Reporting Marks

Tank cars, like all other freight cars, are marked with their own unique sets of reporting marks. Reporting marks (also called *initials and numbers*) may be used to obtain information about the car's contents from the railroad's computer, the shipper, CHEMTREC, CANUTEC, or SETIQ. Reporting marks should match the initials and numbers provided on the shipping papers for the car. They are stenciled on both sides (to the left when facing the side of the car) and both ends (upper center) of the tank car tank **(Figures 5.77a and b, p. 262)**. Some shippers and car owners also stencil the top of the car with the car's reporting marks to help identify the car.

Capacity Stencil

The **capacity stencil** shows the volume of the tank car tank. The volume in gallons (and sometimes liters) is stenciled on both ends of the car under the car's reporting marks. The capacity in pounds (and sometimes kilograms) is stenciled on the sides of the car under the car's reporting marks. The term *load limit* may be used to mean the same thing as capacity in pounds or kilograms. For certain tank cars, the water capacity (water weight) of the tank, in pounds (and typically kilograms) is stenciled on the sides of the tank near the center of the car **(Figure 5.78, p. 262)**.

Railcar Initials and Numbers — Combination of letters and numbers stenciled on rail tank cars that may be used to get information about the car's contents from the railroad's computer or the shipper. *Also known as* Reporting Marks.

Dedicated Tank Car — Rail tank car that is specked to meet particular parameters unique to the product including pressure relief device, linings, valves, fittings, and attachments. This type of car is often used for a single specified purpose for the life of the car, and may be marked to indicate that exact purpose.

Capacity Stencil — Number stenciled on the exterior of a tank car to indicate the volume of the tank.

Figures 5.77a and b
The reporting marks are highlighted on these tank cars. *Courtesy of Rich Mahaney.*

Figure 5.78 The load limit and capacity stencil are highlighted on this tank car. These are located under the reporting marks. *Courtesy of Rich Mahaney.*

Specification Marking — Stencil on the exterior of a tank car indicating the standards to which the tank car was built; may also be found on intermodal containers and cargo tank trucks.

Specification Marking

The **specification marking** indicates the standards to which a tank car was built. The marking is stenciled on both sides of the tank. When facing the side of the car, the marking will be to the right (opposite from the reporting marks) **(Figure 5.79)**. First responders can also get specification information from the railroad, shipper, car owner, or the Association of American Railroads by using the car's reporting marks. **Figure 5.80** provides a brief explanation of tank car specification markings.

Federal Railroad Administration (FRA) Regulation Update

After June 25, 2012, FRA regulations require stamping the tank specifications on a metal plate located on the frame of the car to certify that the tank complies with all specification requirements. The specification marking is also stamped into the tank heads where it is not readily visible.

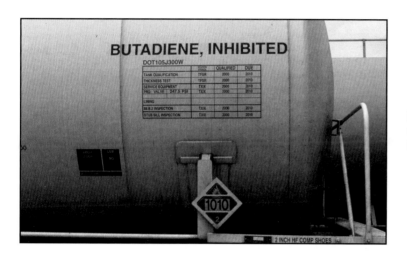

Figure 5.79 Specification markings will be on the opposite end of the tank from the reporting marks. *Courtesy of Rich Mahaney.*

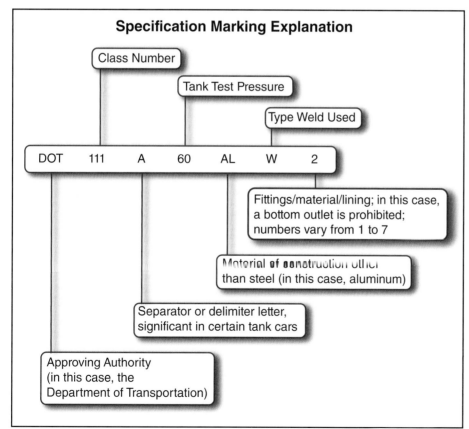

Figure 5.80 Key to tank car specification markings. First responders are most likely to need the DOT/TC class number.

Bulk Transportation Containers: Intermodal Tanks

An **intermodal container** is a freight container that is used interchangeably in multiple modes of transport such as rail, highway, and ship. The various types of intermodal containers can be divided into the following two main categories:

1. **Freight containers** — Transport a wide range of products, from foodstuffs to dry goods. They come in a variety of types and sizes, most commonly in 20, 40, 45, 48, and 53 foot (6 m, 12 m, 14 m, 15 m, and 16 m) lengths. Several common types of freight containers are **(Figure 5.81)**

   — Dry van intermodal containers (sometimes called *box containers*)

   — **Refrigerated intermodal containers** (also called *reefers*)

   — Open top intermodal containers

   — Flat intermodal containers

2. **Tank containers** — Also called *intermodal tanks* **(Figure 5.82)**. Three general classifications of intermodal tank containers are:

Intermodal Freight Containers

Dry Van *(box containers)*

Refrigerated *(reefers)*

Open Top

Flat

Figure 5.81 Many types of intermodal freight containers exist.

Intermodal Container — Freight containers designed and constructed to be used interchangeably in two or more modes of transport. *Also known as* Intermodal Tank, Intermodal Tank Container, *and* Intermodal Freight Container.

Refrigerated Intermodal Container — Cargo container having its own refrigeration unit. *Also known as* Reefer.

— Pressure intermodal tanks

— Specialized intermodal tanks such as cryogenic intermodal tanks and tube modules

— Low pressure intermodal tanks (also called *nonpressure intermodal tanks*)

Some intermodal freight containers may contain hazardous materials **(Figures 5.83a and b)**. Other intermodal freight containers may contain mixed loads that include both hazardous and nonhazardous materials. With many freight containers, you will not be able to determine by the shape of the container alone whether it contains hazardous materials. Instead, responders should use intermodal container markings or shipping papers (described later in this chapter) to identify the contents of these containers. Shipping papers may not be accurate, and hazardous materials may be shipped illegally in intermodal containers without proper identification.

Figure 5.82 Intermodal tank containers transport hazardous materials worldwide. *Courtesy of Rich Mahaney.*

Figures 5.83a and b Intermodal freight containers may transport hazardous materials in a variety of packaging. The freight containers are not always properly placarded. *Courtesy of Barry Lindley.*

Intermodal tank containers generally have a cylinder enclosed at both ends. First responders may also see tube modules, cryogenic tanks, compartmentalized tanks, or other shapes. **Table 5.7** provides examples of the most common types of intermodal tanks. The tank container is placed in frames to protect it and to allow stacking, lifting, and securing. The capacities of these containers ordinarily do not exceed 6,340 gallons (24 000 L) **(Table 5.8)**.

Table 5.7 Intermodal Tanks	
Tank Type	**Descriptions**
	**Nonpressure Intermodal Tank** • IM-101: 25.4 to 100 psi (175 kPa to 689 kPa) • IM-102: 14.5 to 25.4 psi (100 kPa to 175 kPa) **Contents:** Liquids or solids (both hazardous and nonhazardous)
	**Pressure Intermodal Tank** 100 to 500 psi (689 kPa to 3 447 kPa) **Contents:** Liquefied gases, liquefied petroleum gas, anhydrous ammonia, and other liquids
	**Cryogenic Intermodal Tank** **Contents:** Refrigerated liquid gases, argon, oxygen, helium
	**Tube Module Intermodal Container** **Contents:** Gases in high-pressure cylinders (3,000 or 5,000 psi [20 684 kPa or 34 474 kPa]) mounted in the frame

Table 5.8
Intermodal Tank Container Descriptions

Specification	Materials Transported	Capacity	Design Pressure
IM 101 Portable Tank	Hazardous and nonhazardous materials, including toxics, corrosives, and flammables with flash points below 32°F (0°C)	Normally range from 5,000 to 6,300 gallons (18 927 L to 23 848 L)	25.4 to 100 psi (175 kPa to 689 kPa) {1.75 to 6.89 bar}
IM 102 Portable Tank	Whiskey, alcohols, some corrosives, pesticides, insecticides, resins, industrial solvents, and flammables with flash points ranging from 32°F to 140°F (0° to 60°C)	Normally range from 5,000 to 6,300 gallons (18 927 L to 23 848 L)	14.5 to 25.4 psi (100 kPa to 175 kPa) {1 to 1.75 bar}
Spec. 51 Portable Tank	Liquefied gases such as LPG, anhydrous ammonia, high vapor pressure flammable liquids, pyrophoric liquids (such as aluminum alkyls), and other highly regulated materials	Normally range from 4,500 to 5,500 gallons (17 034 L to 0 820 L)	100 to 500 psi (689 kPa to 3 447 kPa) {6.89 to 34.5 bar}

CAUTION
Intermodal freight containers can contain almost anything.

Intermodal Container Specifications

Starting in 2003, new intermodal container specifications include "T" Codes in place of the older IMO types **(Figure 5.84)**. Although DOT/TC Specification 51, IM 101, or IM 102 portable tanks may not be manufactured after January 1, 2003, such tanks may continue to be used for the transportation of a hazardous material, as subject to provisions including material containment minimum requirements and periodic inspections of the particular tanks.

NOTE: "T" codes are explained in detail in the IFSTA manual, Hazardous Materials Technician.

Figure 5.84 New intermodal container specifications include "T" Codes. This tank meets T14 specifications. *Courtesy of Barry Lindley.*

Pressure Intermodal Tank

A pressure intermodal tank container is less common in transport. DOT/TC classifies this tank as Spec. 51; internationally it is known as an IMO Type 5 tank container **(Figure 5.85)**. This type of container is designed for MAWPs of 100 to 500 psi (689 kPa to 3 447 kPa) and usually transports liquefied gases under pressure.

Pressure intermodal tanks may be damaged during transport, loading, and unloading. Leaks frequently involve the fittings, with the release of rapidly expanding gases or vapors. Pressure intermodal tanks exposed to heat and/ or flame may BLEVE **(Figure 5.86)**.

Figure 5.85 Pressure intermodals are not very common. They are also known as *Spec 51* or *IMO Type 5* tanks. *Courtesy of Rich Mahaney.*

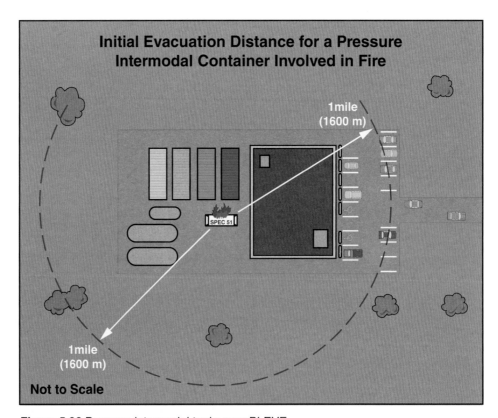

Figure 5.86 Pressure intermodal tanks may BLEVE.

Low Pressure Intermodal Tank

This tank is the most common intermodal tank used in transportation. Even though they are often called nonpressure intermodal tanks, these tanks may have pressures as high as 100 psi (689 kPa). They are also called *intermodal portable tanks* or *IM portable tanks*. The two common groups of low pressure/nonpressure intermodal tank containers are:

1. **IM 101 portable tanks** — Internationally, these are called International Maritime Organization (IMO) Type 1 tank containers **(Figure 5.87)**. They are built to withstand maximum allowable working pressures (MAWP) of 25.4 to 100 psi (175 kPa to 689 kPa). They transport both hazardous and nonhazardous materials.

2. **IM 102 portable tanks** — Internationally, these are called IMO Type 2 tank containers. They are designed to handle maximum allowable working pressures (MAWP) of 14.5 to 25.4 psi (100 kPa to 175 kPa). These containers are gradually being removed from service. They transport materials such as alcohols, pesticides, resins, industrial solvents, and flammables with flashpoints between 32 and 140°F (0°C to 60°C). Most commonly, they transport nonregulated materials (those not specifically covered by regulations) such as food commodities.

Low pressure intermodal tanks may be damaged during transport, loading, and unloading, including damage to fittings and container walls. Releases are commonly in the form of spilled liquids, often flammable or combustible.

Figure 5.87 IM 101 portable tanks transport hazardous and nonhazardous liquids and solids. *Courtesy of Rich Mahaney.*

Specialized Intermodal Tank or Container

There are several types of specialized intermodal tank containers. Cryogenic-type containers are built to IMO Type 7 specifications **(Figure 5.88, p. 270)**. The tube module transports gases in high-pressure cylinders with MAWPs of 2400 to 5,000 psi (16 547 kPa to 34 474 kPa). Cryogenic liquid tank containers carry refrigerated liquid gases, argon, oxygen, and helium. Dry bulk intermodal containers carry materials such as fertilizer, cement, and plastic pellets **(Figure 5.89, p. 270)**.

Figure 5.88 IMO Type 7 intermodals transport cryogenic materials. *Courtesy of Rich Mahaney.*

Figure 5.89 Dry bulk intermodals may carry fertilizer, cement and other solids. *Courtesy of Rich Mahaney.*

Plastic Pellet Spill, 2014

When firefighters in Texas arrived at the scene of an eight-car train derailment, they saw plastic polyethylene pellets covering the ground like snow. Several railcars carrying the pellets had overturned during the early morning derailment, which caused a highway to be closed in both directions. A tank car carrying around 1,000 gallons (4 000 L) of a chemical used in the production of shampoo and cosmetics also spilled during the accident, creating the need for a major cleanup.

Personnel rerouted vehicles and other trains were directed away from the area, as contractors hired by the train company assisted with the cleanup. Workers used bulldozers and flatbed trucks to move railcars and clean up the spilled chemicals and plastic pellets, which had fallen in large amounts from an overpass.

Local officials praised the quick response by firefighters and the implementation of an emergency plan. After workers cleared the debris, the railroad was reopened one day after the derailment.

International Intermodal Markings

In addition to required placards, identifying markings on intermodal tanks and containers include reporting marks **(Figure 5.90)**. Reporting markings are generally found on the tank or container, on the right-hand side as you face it from either the sides or the ends. As with tank car reporting marks, you can use this information in conjunction with shipping papers or computer data to identify and verify the contents of the tank or container. Other markings on intermodal containers can also provide specification information **(Figure 5.91)**. Intermodal containers carrying hazardous materials must have proper shipping names stenciled on two sides.

Figure 5.90 Intermodal reporting marks identify the specific container. *Courtesy of Rich Mahaney.*

Figure 5.91 Intermodal specification information identifies this as an IM 101 tank.

⚠ CAUTION

Read the intermodal container markings and understand all of the information provided.

Bulk Transportation Containers: Ton Containers

Some freight shippers and facilities use especially large capacity containers for bulk transportation and bulk storage. These large capacity containers may be classified as ton containers and Y cylinders/Y ton containers.

Ton Containers

Ton containers are pressure tanks that have capacities of 1 short ton or approximately 2,000 pounds (907 kg or 0.91 tonnes). They are typically stored on their sides **(Figure 5.92, p. 272)**. The ends (heads) of the containers are convex or concave, and they have two valves in the center of one end, one above the other. One valve connects to a tube that extends into the liquid space; the other valve connects to a tube that extends into the vapor space above **(Figure 5.93, p. 272)**. Ton containers may have pressure-relief devices or fusible plugs in case of fire or exposure to elevated temperatures. Ton containers holding

Figure 5.92 Ton containers are typically stored and transported on their sides. *Courtesy of Rich Mahaney.*

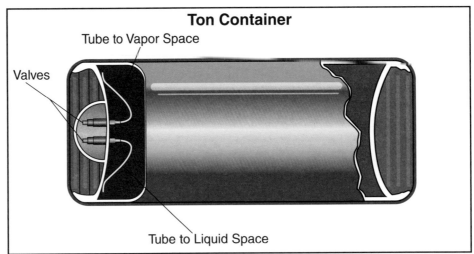

Figure 5.93 Ton containers have one valve connected to the vapor space and one valve connected to the liquid space.

Figure 5.94 Incidents involving ton containers typically require chemical protective clothing because these containers release highly toxic and/or corrosive gases.

Ton Container

Tube to Vapor Space

Valves

Tube to Liquid Space

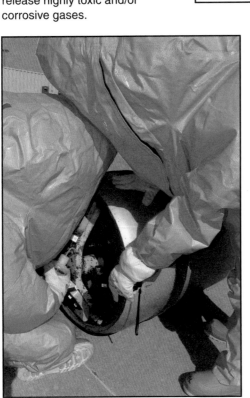

chlorine are often found at locations such as water treatment plants and commercial swimming pools. Ton containers may also contain products such as sulfur dioxide, anhydrous ammonia, or Freon® refrigerant.

As pressure containers, ton containers typically release their contents as gases or vapors. Always evacuate responders and civilians to a safe distance to avoid the vapor cloud that escapes from these containers. Because products stored in ton containers are often highly toxic and/or corrosive, evaluate available PPE for effectiveness during incidents involving these containers **(Figure 5.94)**.

WARNING!
Structural fire fighting gear does not provide adequate protection against the hazardous materials commonly stored in ton containers.

Y Cylinders/Y Ton Containers

Y cylinders are a type of compressed gas cylinder that can be bulk or nonbulk, but they are typically classified as bulk **(Figure 5.95)**. A typical Y ton container will have a specification such as DOT/TC 3AA-2400 or DOT/TC 3AA-480 (pressure is dependent on product). These containers are typically 7 feet (2 m) long, 2 ft (0.6 m) in diameter, have a wall thickness of about 0.6 inches (15 mm), and, when empty, weigh about 1,200 lbs (544 kg). These containers have a water capacity of approximately 120 gallons (454 liters), or 16 cubic feet (0.5 m³). Often used for refrigerants, they typically operate in a cascade system.

Figure 5.95 Y cylinders are a type of compressed gas cylinder. *Courtesy of Rich Mahaney.*

Y cylinders/Y ton containers have two specifications depending on size:

- **DOT/TC–3AA** — A seamless steel cylinder with a water capacity (nominal) of not over 1,000 lbs (454 kg) and a service pressure of at least 150 psig (1 034 kPaG).

- **DOT/TC–3AAX** — A seamless steel cylinder with a water capacity of not less than 1,000 lbs (454 kg) and a service pressure of at least 500 psig (3 447 kPaG).

Other Bulk and Nonbulk Containers

Not every type of container that a first responder will encounter can be classified according to the previously mentioned systems. The following section introduces the following additional types of bulk and nonbulk containers:

- Radioactive material containers
- Pipelines
- Intermediate bulk containers
- Nonbulk containers

Radioactive Material Containers

All shipments of radioactive materials (sometimes called *RAM*) must be packaged and transported according to strict regulations. These regulations protect the public, transportation workers, and the environment from potential exposure to radiation. The type of packaging used to transport radioactive materials is determined by the activity, type, and form of the material to be shipped. Depending upon these factors, radioactive material is shipped in one of the following five basic types of container listed in order of increasing level of radioactive hazard:

1. **Excepted** — This packaging is used to transport materials that have limited radioactivity, such as articles manufactured from natural or depleted uranium or natural thorium. **Excepted packagings** are only used to transport materials with low levels of radioactivity that present no risk to the public or environment. Empty packaging is excepted. Excepted packaging is not marked or labeled as such. Because of its low risk, excepted packaging is exempt from several labeling and documentation requirements.

Excepted Packaging — Container used for transportation of materials that have very limited radioactivity.

2. **Industrial** — This container design retains and protects its contents during normal transportation activities. Industrial packages are not identified as such on the packages or shipping papers. Industrial packages contain materials that present a limited hazard to the public and the environment. Examples of these materials include:

   — Slightly contaminated clothing

   — Laboratory samples

   — Smoke detectors

3. **Type A** — This container design protects its contents and maintains sufficient shielding under conditions normally encountered during transportation. These packages must demonstrate their ability to withstand a series of tests without releasing their contents. The package and shipping papers will have the words *Type A* on them. Radioactive materials with relatively high specific activity levels are shipped in Type A packages. Examples of these materials include:

   — Radiopharmaceuticals (radioactive materials for medical use)

   — Certain regulatory qualified industrial products

4. **Type B** — These packages must not only demonstrate their ability to withstand tests simulating normal shipping conditions, but they must also withstand severe accident conditions without releasing their contents. Type B packages are identified as such on the package itself as well as on shipping papers. The size of these packages range from small containers to those weighing over 100 tons (91 tonnes). These large, heavy packages provide shielding against radiation. Radioactive materials that exceed the limits of Type A package requirements must be shipped in Type B packages. Examples of these materials include:

   — Materials that would present a radiation hazard to the public or the environment if there were a major release

   — Materials with high levels of radioactivity such as spent fuel from nuclear power plants

Figures 5.96a-c Type C containers are built to withstand extreme stress and severe airplane accidents. *Courtesy of the National Nuclear Security Administration.*

5. **Type C** — These are rare packages used for high-activity materials (including plutonium) transported by aircraft **(Figures 5.96a-c)**. They are designed to withstand severe accident conditions associated with air transport without

loss of containment or significant increase in external radiation levels. The Type C package performance requirements are significantly more stringent than those for Type B packages.

Pipelines and Piping

According to PHMSA, as of the year 2015 there are over 2.5 million miles (4 million km) of pipelines in North America **(Figure 5.97)**. These pipelines transport a variety of flammable and nonflammable hazardous gases and liquids, including:

- Natural gas
- Propane
- Hydrogen
- Crude oil
- Diesel
- Gasoline
- Jet fuel
- Home heating oils
- Carbon dioxide
- Anhydrous ammonia

Figure 5.97 There are millions of miles of underground pipelines in North America. *Courtesy of Rich Mahaney.*

Pipelines are usually buried, but they may be located aboveground as well, especially in cold climates where the ground is often frozen. In some cases, multiple products may be pushed through the same pipeline at the same time, or separated by a pipeline pig. Hydrocarbons are often comingled **(Figure 5.98)**.

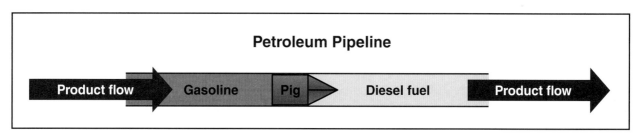

Petroleum Pipeline

Product flow | Gasoline | Pig | Diesel fuel | Product flow

Figure 5.98 A single pipeline may carry more than one type of product, separated by a pipeline pig.

Pipelines also come in a variety of sizes and pressures depending on the product and the function of the line. For example, large natural gas transmission pipelines operate under extreme pressure, while smaller distribution lines typically operate under much lower pressures. **Figure 5.99, p. 276**, provides an overview of a basic pipeline system for crude oil and natural gas.

Pipeline breaches can be caused by:

- Excavation
- Corrosion
- Equipment, material, joint, or weld failures
- Operation errors
- Natural disasters such as earthquakes and floods
- Vehicle collisions

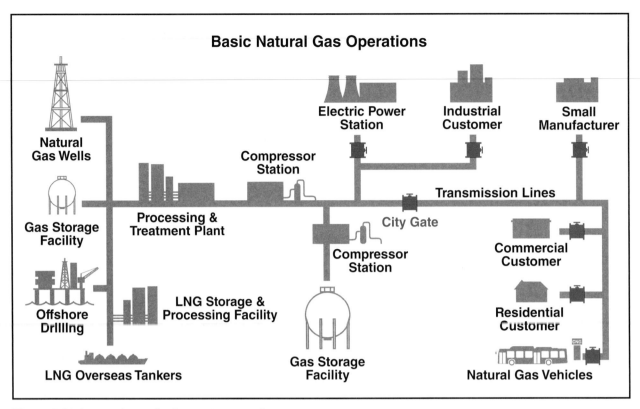

Figure 5.99 A natural gas pipeline system overview.

Emergency responders can view the transmission pipelines in their area via the National Pipeline Mapping System, available online. This system does not provide data on gathering and distribution lines.

Pipeline releases can be violent, particularly if gases and/or high pressure transmission lines are involved. Spills and leaks can involve large quantities of product. Since most products transported by pipelines are flammable or combustible, fire control should always be a priority consideration at pipeline incidents. Some indications of pipeline leaks and ruptures include:

- Visible vapor clouds or liquid spills
- Hissing, roaring, or explosive sounds
- Strong odors such as rotten eggs or petroleum smells
- Liquids bubbling up from water or soil with no obvious source
- High pressure gas blowing out of water or soil
- Dead vegetation or discolored snow above a pipeline right-of-way

Pipeline markers in the U.S. and Canada include the signal words *Caution*, *Warning*, or *Danger*, representing an increasing level of hazard. They also contain information describing the transported commodity, the carrier's (operator's) name, and an emergency telephone number.

Establishing good communication and cooperation with pipeline operators before an emergency occurs is an important element of emergency response preparedness. Pipeline operators are required to provide a wealth of information to emergency responders that can help reduce the impact of an actual release, including:

- Location of transmission pipelines that cross their area
- Name of the pipeline operator and emergency contact information
- Products carried and their hazards
- Location of pipeline emergency response plans
- How to contact the pipeline operator regarding questions, concerns, or emergencies
- How to safely respond to a pipeline emergency

NOTE: General considerations for responding to a pipeline emergency are provided on pages 23-25 of the *2016 Emergency Response Guidebook*.

Many industrial, commercial, and institutional facilities have piping carrying everything from water and steam to hazardous materials. Pipes carrying hazardous materials should be appropriately marked and labeled. Many facilities in the U.S. and Canada follow ANSI's A13.1-1981, *Scheme for Identification of Piping System*s, to mark and label pipes.

Intermediate Bulk Containers

Per the U.S. DOT, an intermediate bulk container (IBC) is either rigid or flexible portable packaging (other than a cylinder or portable tank) designed for mechanical handling **(Figure 5.100)**. Design standards for IBCs in the U.S., Canada, and Mexico are based on United Nations Recommendations on the Transportation of Dangerous Goods. The maximum capacity of an IBC is not more than 3 m³ (3,000 L, 793 gal, or 106 ft³). The minimum capacity is not less than 0.45 m³ (450 L, 119 gal, or 15.9 ft³) or a maximum net mass of not less than 400 kilograms (882 lbs).

NOTE: These measurements were established by the UN and primarily use the metric measurement. There is no weight limit on solid products.

Figure 5.100 Intermediate bulk containers (IBCs) contain hazardous and nonhazardous liquids and solids. *Courtesy of Rich Mahaney.*

IBCs are authorized to transport a wide variety of materials and hazard classes, including:

- Aviation fuel (turbine engine)
- Gasoline
- Hydrochloric acid
- Methanol
- Toluene
- Corrosive liquids
- Solid materials in powder, flake, or granular forms

IBCs are divided into two types: flexible intermediate bulk containers (FIBCs) and rigid intermediate bulk containers (RIBCs). Both types are often called *totes*.

Flexible Intermediate Bulk Containers (FIBC)

FIBCs are sometimes called *bulk bags, bulk sacks, supersacks, big bags,* or *tote bags.* They are flexible, collapsible bags or sacks that are used to carry both solid materials and fluids **(Figure 5.101)**. The designs of FIBCs are as varied as the products they carry. Often the bags used to transport wet or hazardous materials are lined with polypropylene or some other high-strength fabric. Others may be constructed of multiwall paper or other textiles. A common-sized supersack can carry the equivalent of four to five 55-gallon (208 L) drums and (depending on design and the material inside) can be stacked one on top of another. Sometimes FIBCs are transported inside a rigid exterior container made of corrugated board or wood.

Rigid Intermediate Bulk Containers (RIBC)

RIBCs are typically made of steel, aluminum, wood, fiberboard, or plastic; and they are often designed to be stacked. RIBCs can contain both solid materials and liquids. Some liquid containers may look like smaller versions of intermodal nonpressure tanks with metal or plastic tanks inside rectangular box frames. Other RIBCs may be large, square or rectangular boxes or bins **(Figure 5.102)**. Rigid portable tanks may be used to carry liquids, fertilizers, solvents, and other chemicals; and they may have capacities up to 400 gallons (1 514 L) and pressures up to 100 psi (689 kPa).

Figure 5.101 A single flexible intermediate bulk container may carry the equivalent of four to five 55-gallon (200 liter) drums. *Courtesy of Rich Mahaney.*

Figure 5.102 RIBCs may be square or rectangular boxes or bins and carry liquids, fertilizers, solvents, or other chemicals. They are often designed to stack. *Courtesy of Rich Mahaney.*

Nonbulk Packaging

Containers that are used to transport smaller quantities of hazardous materials than bulk or IBCs are called *nonbulk packaging.* They are usually used during highway transport or other routine transportation. **(Table 5.9)** shows common types of nonbulk packaging including the following types of containers:

- Bags
- Carboys and "jerricans"

- Cylinders
- Drums
- Dewar flasks (cryogenic liquids)

Table 5.9
Nonbulk Packaging

Package Type	Descriptions
	**Bags** • Made of paper, plastic, film, textiles, woven material, or others • Sizes vary **Contents:** Explosives, flammable solids, oxidizers, organic peroxides, fertilizers, pesticides, and other regulated materials.
	**Carboys and Jerricans** • Made of glass or plastic • Often encased in a basket or box • Sizes vary **Contents:** Flammable and combustible liquids, corrosives.
	**Cylinders** • Presssures higher than 40 psi (276 kPa) {2.76 bar} but vary • Sizes range from lecture bottle size to very large **Contents:** Compressed gases.

Continued

Table 5.9 (concluded)

Package Type	Descriptions
	**Drums** • Made of metal, fiberboard, plastic, plywood, or other materials • May have open heads (removable tops) or tight (closed) heads with small openings • Sizes vary from 55 gallons (208 L) to 100 gallons (379 L) **Contents:** Hazardous and nonhazardous liquids and solids.
	**Dewar Flasks** • Vacuum insulated • Made of glass, metal, or plastic with hollow walls from which the air has been removed • Sizes vary **Contents:** Cryogenic liquids; thermoses may contain nonhazardous liquids.

Bags

A bag is a flexible packaging made of paper, plastic film, textiles, woven material, or other similar materials. Bags may transport:

- Explosives
- Flammable solids
- Oxidizers
- Organic peroxides
- Fertilizers
- Pesticides
- Other regulated materials

Bags can be sealed in a variety of ways, including ties, stitching, gluing, heat sealing, and crimping with metal. Typically, bags are stored and transported on pallets.

Carboys and Jerricans

A *carboy* is a large glass or plastic bottle encased in a basket or box, primarily used to store and transport corrosive liquids, although its use has expanded to nonhazardous materials (such as water) as well. The outer packaging may be made of such materials as polystyrene or wood, and carboys may be round or rectangular. Their capacities may exceed 20 gallons (76 L), but 5-gallon (19 L) containers are more common.

Jerrican is another name for a rectangular plastic carboy and is the term used in UN regulations. Some organizations differentiate between carboys and jerricans, defining jerricans as rectangular metal containers typically transporting flammable and combustible liquids and carboys as transporting corrosives.

Cylinders

A *cylinder* is a pressure vessel designed for pressures higher than 40 psi (276 kPa) and has a circular cross section, but it does not include any of the containers, tanks, or vessels described in previous sections. Cylinders are used to store, transport, and dispense large volumes of gaseous materials. Compressed-gas cylinders range in size from small lecture bottles (small bottles used for classroom demonstrations) to large cylinders and have varying pressures.

All approved cylinders, with the exception of some that store poisons, are equipped with safety-relief devices. These devices may be spring-loaded valves that reclose after operation, heat-fusible plugs, or pressure-activated bursting disks that completely empty the container. All fittings and threads are standardized according to the material stored in the cylinder.

As yet, there is no nationally regulated color code that permits visual identification of cylinder materials by color. Some manufacturers use a single color for all their cylinders, while other manufacturers have their own color-coding system. If local manufacturers and distributors use an identification system, it should be identified in emergency response plans.

Drums

A *drum* is a flat-ended or convex-ended cylindrical packaging made of the following materials:

- Metal
- Fiberboard
- Plastic
- Plywood
- Other suitable materials

Drum capacities range up to 119 gallons (450 L), but 55-gallon (208 L) drums are the most common. Drums may contain a wide variety of hazardous and nonhazardous materials in both liquid and solid form. Typically, metal drums carry flammables and solvents, and plastic/poly drums carry corrosives.

Drums have the following two types of tops:

- **Open heads** — Removable tops
- **Tight (or closed) heads** — Nonremovable tops with small openings plugged by bungs (stoppers)

Dewar

A **dewar** flask (*vacuum flask*) is a nonpressurized, insulated container that has a vacuum space between the outer shell and the inner vessel. The dewar flask is designed for the storage and dispensing of cryogenic materials such as liquid nitrogen, liquid oxygen, and helium **(Figure 5.103)**. Dewars have a

Figure 5.103 Dewar flasks are vacuum insulated.

Dewar — All-metal container designed for the movement of small quantities of cryogenic liquids within a facility; not designed or intended to meet Department of Transportation (DOT) requirements for the transportation of cryogenic materials.

bulky appearance due to the insulation used to keep the cryogenic material at the desired temperature. The volume of dewar flasks is often between 4 gallons to 125 gallons (15 L to 500 L). Some dewar flasks may be as large as 1,250 gallons (5 000 L).

Chapter Review

Answer the following questions to review the information provided in this chapter.

1. What are some questions that should be asked to help you identify potential outcomes?

2. How does the General Hazardous Materials Behavior Model help predict hazards at hazmat incidents?

3. What are the major types of containers that hold hazardous materials?

4. List the types of bulk facility storage tanks and the hazards that they may present to first responders.

5. List the types of cargo tank trucks and the hazards that they may present to first responders.

6. List the types of tank cars and the hazards that they may present to first responders.

7. List the types of intermodal tanks and the hazards that they may present to first responders.

8. What is the difference between ton containers and Y Cylinder/Y Ton Containers?

9. List other types of containers that first responders may encounter and what kinds of hazardous materials they may carry.

5-1
Analyze a hazardous materials scenario to identify potential hazards.

Step 1: Identify type of container.

ID No.	Guide No.	Name of Material	ID No.	Guide No.	Name of Material
1184	131	Ethylene dichloride	1204	127	Nitroglycerin, solution in alcohol, with not more than 1% Nitroglycerin
1185	131P	Ethyleneimine, stabilized			
1188	127	Ethylene glycol monomethyl ether	1206	128	Heptanes
			1207	130	Hexaldehyde
1189	129	Ethylene glycol monomethyl ether acetate	1208	128	Hexanes
1190	129	Ethyl formate	1208	128	Neohexane
1191	129	Ethylhexaldehydes	1210	129	Ink, printer's, flammable
1191	129	Octyl aldehydes	1210	129	Printing ink, flammable
1192	129	Ethyl lactate	1210	129	Printing ink related material
1193	127	Ethyl methyl ketone	1212	129	Isobutanol
1193	127	Methyl ethyl ketone	1212	129	Isobutyl alcohol
1194	131	Ethyl nitrite, solution	1213	129	Isobutyl acetate
1195	129	Ethyl propionate	1214	132	Isobutylamine
1196	155	Ethyltrichlorosilane	1216	128	Isooctenes
1197	127	Extracts, flavoring, liquid	1218	130P	Isoprene, stabilized
1197	127	Extracts, flavouring, liquid	1219	129	Isopropanol
1198	132	Formaldehyde, solution, flammable	1219	129	Isopropyl alcohol
			1220	129	Isopropyl acetate
1198	132	Formalin (flammable)	1221	132	Isopropylamine
1199	132P	Furaldehydes	1222	130	Isopropyl nitrate
1199	132P	Furfural	1223	128	Kerosene
1199	132P	Furfuraldehydes	1224	127	Ketones, liquid, n.o.s.
1201	127	Fusel oil	1228	131	Mercaptan mixture, liquid, flammable, poisonous, n.o.s.
1202	128	Diesel fuel			
1202	128	Fuel oil	1228	131	Mercaptan mixture, liquid, flammable, toxic, n.o.s.
1202	128	Gas oil			
1202	128	Heating oil, light	1228	131	Mercaptans, liquid, flammable, poisonous, n.o.s.
1203	128	Gasohol			
1203	128	Gasoline	1228	131	Mercaptans, liquid, flammable, toxic, n.o.s.
1203	128	Motor spirit			
1203	128	Petrol	1229	129	Mesityl oxide
			1230	131	Methanol
			1230	131	Methyl alcohol

Page 31

Step 2: Identify materials involved.

Step 3: Identify any leaks.

Step 4: Identify location of release.

Step 5: Identify conditions surrounding the incident.

Step 6: Communicate with applicable operators or representatives.

Step 7: Collect and determine hazard information using approved reference sources.

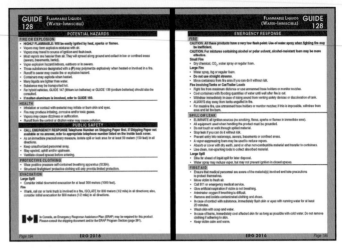

Step 9: Describe potential hazards, harm, and outcomes based on potential behavior.

Step 8: Describe any potential behavior of material and containers.

Planning the Response: Identifying Action Options

Courtesy of Barry Lindley.

Chapter Contents

Key Terms

NFPA Job Performance Requirements

This chapter provides information that addresses the following job performance requirements of NFPA 1072, *Standard for Hazardous Materials/Weapons of Mass Destruction Emergency Response Personnel Professional Qualifications (2017)*.

5.3.1

Planning the Response: Identifying Action Options

Learning Objectives

After reading this chapter, students will be able to:

1. Explain predetermined procedures. (5.3.1)

2. List incident priorities for hazardous materials incidents. (5.3.1)

3. Describe the process of size-up and hazard and risk assessment. (5.3.1)

4. Define hazardous materials incident levels. (5.3.1)

5. Explain the three modes of operations at hazardous materials incidents. (5.3.1)

6. Identify methods for planning the initial response. (5.3.1)

7. Distinguish common response objectives and action options at hazardous materials incidents. (5.3.1)

8. Identify actions available at a hazardous materials incident. (Skill Sheet 6-1, 5.3.1)

Chapter 6
Planning the Response: Identifying Action Options

This chapter introduces the following topics related to incident action planning during a hazmat incident:

- Predetermined procedures
- Incident priorities
- Size-up and hazard and risk assessment
- Incident levels
- Modes of operation
- Planning the initial response
- Common response objectives and options at hazmat incidents

Predetermined Procedures

During a hazmat incident, developing an appropriate Incident Action Plan (IAP) is vital because mistakes made in the initial response can make the difference between solving the problem and becoming part of it **(Figure 6.1)**. While the Incident Commander (IC) may be responsible for developing the Incident Action Plan and identifying the response objectives and options, all responders must understand the process and know the tasks they may be asked to perform.

Despite obvious variations, hazmat incidents typically share some similarities. These similarities are the basis for an organization's predetermined standard operating procedures (SOPs), standard operating guidelines (SOGs), or operating instructions (OIs), collectively referred to as SOP/Gs throughout the rest of this manual. They may also include emergency response plans. SOP/Gs should include considerations for the following:

- Chemical responses
- Biological responses
- Radiological/nuclear responses
- Explosives/explosive materials responses

Figure 6.1 The consequences can be severe if responders make mistakes in the initial stages of the incident. *Courtesy of the U.S. Air Force.*

- WMD responses

- Significant incident responses

Standard guidelines have a built-in flexibility that allows adjustments, with reasonable justification, when unforeseen circumstances occur. The first units that reach the scene usually initiate the predetermined actions. The initial response actions supplement, but do not replace, incident size-up and other decisions based on professional judgment, evaluation, or command. In addition, several predetermined procedures may be available from which to choose, depending on incident severity, location, and the ability of first-in units to achieve control.

Following predetermined procedures reduces chaos on the hazmat scene. All resources can be used in a coordinated effort to rescue victims, stabilize the incident, and protect the environment and property. Operational procedures that are standardized, clearly written, and mandated to each department/ organization member, establish accountability and increase command and control effectiveness.

Predetermined procedures also help prevent duplication of effort and uncoordinated operations because all positions are assigned and covered. In addition, predetermined procedures describe assumption and transfer of command, communications procedures, and tactical procedures.

SOP/Gs should define your role according to your training level at emergency incidents, including those involving hazardous materials. While the procedures may vary considerably in different localities, the principles are usually the same. You must know the location of your agency's emergency response plan and written SOP/Gs **(Figure 6.2)**.

Figure 6.2 Responders should be familiar with their emergency response plan and SOPs/SOGs. *Courtesy of the U.S. Air Force.*

Incident Priorities

All hazmat incidents have three incident priorities (which are the same for all emergency services organizations — law enforcement, fire, emergency medical services [EMS], or other). The three priorities (in order) for hazmat incidents are as follows:

1. Life safety

2. Incident stabilization

3. Protection of property and the environment

NOTE: A fourth priority, societal restoration, is sometimes added to this list to ensure that the recovery phase of major incidents is considered from the beginning.

All plans must be made with these priorities in mind. However, incidents are dynamic and priorities may change according to the situation.

NOTE: Never risk your life to save property that is replaceable or cannot be saved.

Chlorine Incident in Apex, NC

At 9:38 p.m. on Thursday, October 5, 2006, the Apex (NC) Fire Department (AFD) was dispatched to a reported chlorine odor near a business that collects, processes, and repackages industrial waste for transport and disposal. AFD dispatched a standard response of two engines and a chief officer (shift commander). Later reports indicated that the fire involved pesticides, oxidizers, contaminated metals, flammable and combustible materials, lead, and sulfur.

Upon arrival, the AFD reported a large vapor cloud and requested a second-alarm assignment. Crews initiated a reconnaissance to determine the cloud's source and the Incident Commander (IC) ordered a community evacuation. The daily manifest could not be located inside the structure to ascertain what chemicals were burning and what the cloud might contain.

The IC decided not to fight the fire and ordered companies to contain the liquid runoff near the property's edge and continue evacuations. Changing winds and the threat of explosions resulted in the relocation of the Incident Command Post (ICP), shelter locations, and the Emergency Operations Center. A mobile command vehicle was requested from Raleigh, NC, through mutual-aid agreements.

Exposures affected by the incident included:

- Police equipment, including some vehicles that were assigned to off-duty personnel

- CSX freight rail and Amtrak passenger service

- Airspace over the fire. The Federal Aviation Administration (FAA) received a request to restrict the airspace.

A joint information center and media site were established near the ICP. Media briefings were held hourly. The media provided the public with essential emergency public information regarding hazards, evacuation orders, and evacuation routes.

Evacuations continued throughout the night of October 5 using school buses and shelters for the evacuees. Apex EMS coordinated with public transportation, schools, and the area EMS to evacuate one hundred nursing home residents. Wheelchairs had to be used to remove all of the residents. The patients were transported to three area hospitals. This evacuation was completed in four hours with no injuries.

The Medical Branch ultimately included sixteen EMS units, two buses, and two engine companies. Mutual-aid fire departments set up decontamination stations at the three area receiving hospitals. Three schools were used as shelters.

Offensive operations were started around 9 a.m. on Friday, October 6, 2006. By 5 p.m., Apex Fire Command was terminated, and the site was turned over to the company's contract firefighters. The last of the fires was extinguished by 1 a.m. on Saturday, October 7, and by the time the incident demobilized:

- Approximately 17,000 people had been evacuated from their homes due to the threat posed by the chemical cloud.

- No fatalities occurred.

- Thirty civilians were treated for respiratory distress and skin irritation.

- Twelve police officers and one firefighter were treated for respiratory difficulties similar to tear gas exposure.

- More than 250,000 air and water samples showed no complications from runoff.

At 5 p.m. on October 6, the Apex police took responsibility for coordinating traffic control to allow reentry to the evacuated areas. The reentry was conducted in phases, with traffic controls that allowed only people with proper identification to enter their neighborhoods.

Size-Up and Hazard and Risk Assessment

Upon arrival at the incident, the IC assesses the incident's conditions to recognize clues indicating problems or potential problems. This process is called *size-up*, the mental process of the Incident Commander considering all available factors that will affect an incident during the operations' course. The information gained from the size-up is used to determine the response objectives (strategies) and action options (tactics) that are applied to the incident during the planning and implementation stages.

Hazard and Risk Assessment

Hazard and risk assessment is part of the size-up process, focusing particularly on the dangers, hazards, and risks present at the incident. Hazard and risk assessment is a continual evaluation. It starts with preincident planning and continues throughout the incident response operation. The first IC who arrives on the scene conducts an extensive size-up and then continues assessing hazards throughout the incident, altering the mitigation process to minimize risk and maximize benefit, as appropriate.

Hazard and Risk Assessment — Formal review of the hazards and risks that may be encountered by firefighters or emergency responders; used to determine the appropriate level and type of personal and respiratory protection that must be worn. *Also known as* Hazard Assessment.

During hazmat size-up, the IC must consider all sides of the incident (**Figure 6.3**). Hazmat size-up is frequently complicated by limited information or an inability to access the scene due to hazards present. The IC's view of the incident may be limited by the size of the hazard area or location of the release (such as inside a vehicle or structure). In addition, limited or conflicting information regarding the product or products involved is possible. Initial assessment is based on anticipated conditions and updated as additional information becomes available.

What This Means To You
Continual Risk Assessment

Size-up/hazard and risk assessment is a continuing process that is not done solely by the IC upon arrival at the scene. As a responder, be aware of the situation around you. Furthermore, use the appropriate channels to report this information to the IC. Conditions can change rapidly, and you must be continually alert. When a cloud of green gas starts drifting in your direction because the wind direction has changed unexpectedly, you need to notice it, react to it accordingly, and then report it!

The following information needed for hazard and risk assessment can be obtained at the time the incident is reported:

- Number and type of injuries
- Occupancy type
- Type of incident

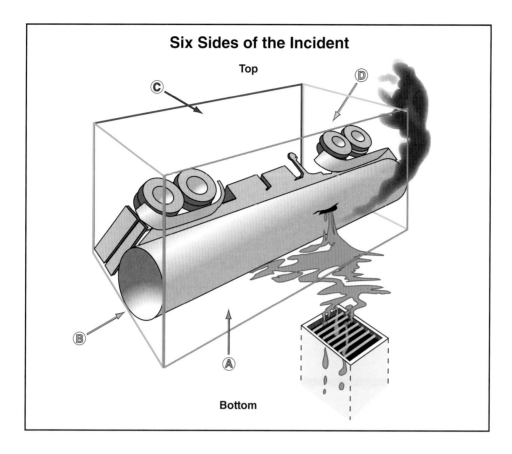

Six Sides of the Incident

Figure 6.3 Size-up must consider all six sides of an incident. Hazardous vapors and gases rise or sink depending on their vapor density.

- Product and container information if available
- Location of the incident
- Equipment and resources responding
- Time of day
- Weather

Once on the scene, additional pieces of the hazard and risk assessment are added to the information made available before arrival. The following factors may have an effect on the situation, such as:

- Wind direction
- Topography
- Land use
- Presence of victims
- Equipment access
- Available response personnel

The initial survey should consider the following questions, if appropriate:

- Where is the incident scene in relation to population, environmental, and property exposures?
- Is the incident scene inside a building or outside?
- What are the hazardous materials?
- Is it a terrorist attack or other criminal incident?

- What hazard classes are involved?
- What quantities are involved?
- What concentrations are involved?
- How could the material react?
- What containers are involved?
- How is the material likely to behave?
- Is it a liquid or solid spill or a gas release?
- Is something burning?
- What kind of container holds the material?
- What is the condition of the container?
- How much time has elapsed since the incident began?
- Can we anticipate where the product is going? Where will it be in 10 minutes, 30 minutes, 60 minutes?
- What personnel, equipment, and extinguishing agents are available?
- Is there private fire protection or other help available?
- What effect can the weather have?
- Are there nearby lakes, ponds, streams or other bodies of water?
- Are there overhead wires, underground pipelines, or other utilities?
- Where are the nearest storm and sewer drains?
- What has already been done?
- What would be the end result if no action was taken?

When the incident requires a rescue, consider the following variables before rushing into a potentially dangerous situation:

- Risk to rescuers
- Ability of rescuers to protect themselves
- Probability of rescue
- Difficulty of rescue
- Capabilities and resources of on-scene forces
- Possibilities of explosions or sudden material releases
- Available escape routes and safe havens
- Constraints of time and distance

NOTE: See Chapter 12 for more information concerning rescue at hazmat incidents.

After the material has been identified, responders can use references such as safety data sheets (SDSs), shipping papers with emergency response information, other written or computer references, and the generic information provided by the *ERG* to determine the health and physical hazards presented by the material. This will assist in determining the level of risk presented by the hazardous material itself.

Manufacturers, shippers, and carriers may provide additional response information such as hazards, behavior, and other recommendations when contacted. Emergency contact information may be provided on shipping papers, pipeline markings, or other container markings. Emergency response centers such as CHEMTREC, CANUTEC, and SETIQ will also contact manufacturers, shippers, and carriers.

First responders should be able to predict (or attempt to predict) where the hazardous material may be going by using the *ERG* and other sources (such as plume-modeling software, if available). Responders can also predict where the hazardous material is going given its physical state of matter (liquid, gas, or solid) and the environmental conditions present (night or day, wind or no wind, indoors or outdoors) **(Figure 6.4)**. Responders can determine the concentration and spread of the material by using monitoring and detection devices. **Skill Sheet 6-1** provides steps for using approved sources to collect response information.

Given this information, responders can estimate the size of the endangered area and predict potential exposures, including **(Figure 6.5)**:

- Number of people
- Buildings
- Property
- Environmental concerns in the area such as sewer drains, streams, lakes, ponds, and wells

Situational Awareness

Effective mitigation of any hazardous material incident requires that emergency responders establish and maintain situational awareness of the event. Situational awareness is more than just size-up of the incident. Situational awareness is a continuous process that includes:

- Size-up
- Interpreting signs **(Figure 6.6, p. 298)**
- Assessing what is happening over the life of the incident
- Predicting outcomes based on a plan of action

Maintaining situational awareness is one of the greatest challenges to emergency responders as the process is also met with barriers such as competing priorities, distractions, and information overload. Failure to establish and maintain situational awareness of the incident is likely to result in a failure to achieve the desired outcome.

Figure 6.4 Environmental conditions and the chemical and physical properties of the product will determine how the hazardous material behaves. *Courtesy of Steve Irby, Owasso (OK) Fire Department.*

Figure 6.5 Understanding where the hazardous material is going will help predict and protect potential exposures. *Courtesy of Rich Mahaney.*

Figure 6.6 Situational awareness includes interpreting the information available, assessing what is happening, and predicting potential outcomes. *Courtesy of MSA.*

Situational awareness is sometimes referred to as *a process working at three levels*:

Level 1: Perception — Perceive the situation around us.

Level 2: Comprehension — Apply our knowledge and past experiences to our perception and develop an understanding of the meaning of the situation.

Level 3: Application — Take our understanding of the situation and apply it to the future, thereby predicting how and when the situation will change and what action is appropriate on our part.

The loss of situational awareness creates an opportunity for errors to occur and improper decisions to be made. The following eight factors may lead to the loss of situational awareness:

1. **Ambiguity** — Information received is confusing or unclear.

2. **Distraction** — Loss of focus of the original mission without appropriate rationale.

3. **Fixation** — Too focused on a single element of the situation to the exclusion of all others. This indicator includes personal concerns such as financial or family problems.

4. **Overload** — Tasks or information overwhelm us, or we attempt to perform all the tasks ourselves.

5. **Complacency** — False sense of comfort based on a misconception of the hazard, risk, or situation sometimes based on past, seemingly similar experience.

6. **Improper procedure** — Policies or procedures are violated or ignored without justification.

7. **Unresolved discrepancy** — Two or more pieces of information do not agree.

8. **Lack of comprehensive hazard surveillance** — Crew members become so fixated on one detail that they ignore everything else.

Proper situational awareness depends on performing the following actions:

- Maintain effective communications.
- Recognize and make others aware of any deviation from standard operating procedures (SOP/Gs) or policies.
- Monitor crew member performance.
- Provide information in advance of an operation or mission.
- Identify any potential problems or existing hazards.
- Communicate the desired course of action.
- Communicate the mission's status continuously.
- Evaluate the situation for any changes continually.
- Clarify expectations of crew members continually.

What This Means To You

Incident Evaluation Skills

Understanding what is happening (and what has happened) at an emergency incident is not a linear process. Unfortunately, there is not a checklist of All the Things You Need to Know that you can follow, check off all that apply, and subsequently act upon. Every incident is going to be different. You may be bombarded with visual, audible, and sometimes conflicting information; yet you must be able to make sense of what you are given in order understand the situation. Skillful ICs are able to quickly identify relevant information and analyze it in order to form a clear picture of the incident.

Incident Levels

After the initial size-up has determined the scope of an incident, the level of the incident can be determined in accordance with the definitions in the Local Emergency Response Plan (LERP). Most incident level models define three levels of response graduating from Level I (least serious) to Level III (most serious). By defining the levels of response, an increasing level of involvement and necessary resources can be identified based on the severity of the incident. These levels are described as follows:

- **Level I** — Within the capabilities of the fire or emergency services organization or other first responders having jurisdiction. A Level I incident is the least serious and the easiest to handle. It may pose a serious threat to life or property, although this situation is not usually the case. Evacuation (if required) is limited to the immediate area of the incident. The following are examples of Level I incidents:

  — Small amount of gasoline or diesel fuel spilled from an automobile **(Figure 6.7, p. 300)**

  — Leak from domestic natural gas line on the consumer side of the meter

  — Broken containers of consumer commodities such as paint, thinners, bleach, swimming pool chemicals, and fertilizers (owner or proprietor is responsible for cleanup and disposal)

Figure 6.7 A small gasoline spill is a Level I incident. *Courtesy of Rich Mahaney.*

- **Level II** — Beyond the capabilities of the first responders on the scene and may be beyond the capabilities of the first response agency/organization having jurisdiction. Level II incidents may require the services of a formal hazmat response team. A properly trained and equipped response team could be expected to perform the following tasks:

  — Use chemical protective clothing **(Figure 6.8)**.

  — Dike and confine within the contaminated areas.

  — Perform plugging, patching, and basic leak control activities.

  — Sample and test unknown substances **(Figure 6.9)**.

  — Perform various levels of decontamination.

  The following are examples of Level II incidents:

  — Spill or leak requiring limited-scale evacuation

  — Any major accident, spillage, or overflow of flammable liquids

  — Spill or leak of unfamiliar or unknown chemicals

  — Accident involving extremely hazardous substances

  — Rupture of an underground pipeline

  — Fire that is posing a boiling liquid expanding vapor explosion (BLEVE) threat in a storage tank

- **Level III** — Requires resources from state/provincial agencies, federal agencies, and/or private industry and also requires Unified Command. A Level III incident is the most serious of all hazardous material incidents. A large-scale evacuation may be required. Most likely, the incident will not be concluded by any one agency **(Figure 6.10)**. Successful handling of the incident requires a collective effort from several of the following resources/procedures:

— Specialists from industry and governmental agencies

— Sophisticated sampling and monitoring equipment

— Specialized leak and spill control techniques

— Decontamination on a large scale

The following are examples of Level III incidents:

— An evacuation extending across jurisdictional boundaries

— Beyond the capabilities of the local hazardous material response team

— Activate (in part or in whole) the federal response plan

Figure 6.8 A Level II incident might require chemical protective clothing. *Courtesy of New South Wales Fire Brigades.*

Figure 6.9 Testing unknown substances might be required at Level II incidents.

Figure 6.10 To mitigate most Level III incidents, more than one agency will be needed. *Courtesy of the U.S. Department of Defense.*

NIMS Incident Types

Per NIMS, incidents may be categorized in order to make decisions about resource requirements. Incident types are based on the following five levels of complexity:

Type 5:

- The incident can be handled with one or two single resources with up to six personnel, including the Incident Commander (IC).
- Command and General Staff positions (other than the IC) are not activated.
- A written Incident Action Plan (IAP) is not required.
- The incident is contained within the first operational period and often within an hour to a few hours after resources arrive on scene.
- Examples include a vehicle fire, an injured person, or a police traffic stop.

Type 4:

- Command and General Staff functions are activated only if needed.
- Several resources are required to mitigate the incident.
- Incident is usually limited to one operational period in the control phase.
- Agency administrator may have briefings to update the complexity analysis and delegation of authority.
- A written IAP is not required, but a documented operational briefing will be completed for all incoming resources.
- The agency administrator's role includes operational plans, including objectives and priorities.

Type 3:

- When capabilities exceed initial attack, add the appropriate ICS positions to match the complexity of the incident.
- Some or all of the Command and General Staff positions may be activated, as well as Division/Group Supervisor and/or Unit Leader level positions.
- A Type 3 Incident Management Team (IMT) or Incident Command organization manages initial action incidents with a significant number of resources, an extended attack incident until containment/control is achieved, or an expanding incident until transition to a Type 1 or 2 team.
- Incident may extend into multiple operational periods.
- A written IAP may be required for each operational period.

Type 2:

- Extends beyond the capabilities for local control and is expected to go into multiple operational periods. A Type 2 incident may require the response of resources out of the area, including regional and/or national resources, to effectively manage the operations, command, and general staffing.
- Most or all of the Command and General Staff positions are filled.
- A written IAP is required for each operational period.
- Many of the functional units are needed and staffed.

Modes of Operation

Strategies are divided into three options that relate to modes of operation:

- **Nonintervention** — Allows the incident to run its course on its own.

- **Defensive** — Provides confinement of the hazard to a given area by performing diking, damming, or diverting actions.

- **Offensive** — Includes actions, such as plugging a leak, to control the incident **(Figure 6.11)**.

Figure 6.11 Leak control is an offensive action.

Selection of the strategic mode is based on the risk to responders, their level of training, and the balance between the resources required and those that are available. When selecting a mode of operation, the safety of first responders is the utmost consideration. The mode of operation may change during the course of an incident. Incident priorities will help determine which mode is used at the incident. The IC may decide to use different modes simultaneously at the same incident based on incident dynamics.

Nonintervention

Nonintervention operations are operations in which the responders do not take direct actions on the actual problem. Not taking any action is the only safe strategy in many types of incidents and the best strategy in certain types of incidents when mitigation is failing or otherwise impossible. An example of a situation for nonintervention is a pressure vessel that cannot be adequately

Nonintervention Operations — Operations in which responders take no direct actions on the actual problem.

cooled because it is exposed to fire. In such incidents, responders should evacuate personnel in the area and withdraw to a safe distance. The nonintervention mode is selected when one or more of the following circumstances exist:

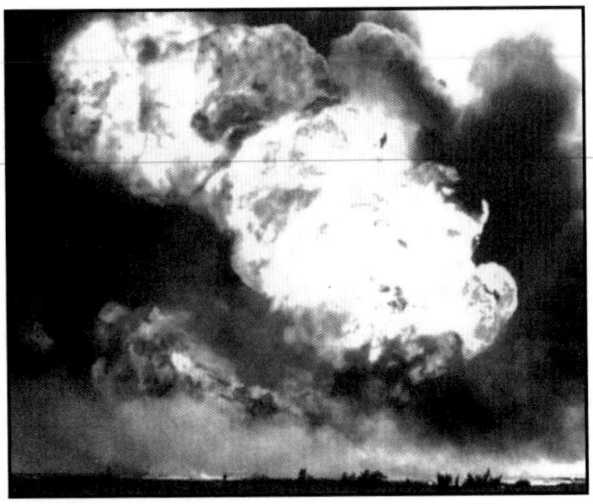

Figure 6.12 Nonintervention is an acceptable strategy at some incidents. *Courtesy of the U.S. Army Corps of Engineers.*

- The facility or Local Emergency Response Plan (LERP) calls for it based on a preincident evaluation of the hazards present at the site.

- The situation is clearly beyond the capabilities of responders (**Figure 6.12**).

- Explosions are imminent.

- Serious container damage threatens a massive release.

In such nonintervention situations, first responders should take the following actions:

- Withdraw to a safe distance.

- Report scene conditions to the telecommunications center.

- Initiate an incident management system.

- Call for additional resources as needed.

- Isolate the hazard area and deny entry.

- Begin evacuation where needed.

Defensive Operations — Operations in which responders seek to confine the emergency to a given area without directly contacting the hazardous materials involved.

Defensive

Defensive operations are those in which responders seek to confine the emergency to a given area without directly contacting the hazardous materials involved. The defensive mode is selected when one of the following two circumstances exists:

- The facility or LERP calls for it based on a preincident evaluation of the hazards present at the site.

- Responders have the training and equipment necessary to confine the incident to the area of origin.

In defensive operations, operations level first responders should take the following actions:

- Report scene conditions to the telecommunications center.

- Initiate an incident management system.

- Call for additional resources as needed.

- Isolate the hazard area and deny entry.

- Establish and indicate zone boundaries.

- Begin evacuation where needed.

- Control ignition sources.

- Use appropriate defensive control tactics (**Figure 6.13**).

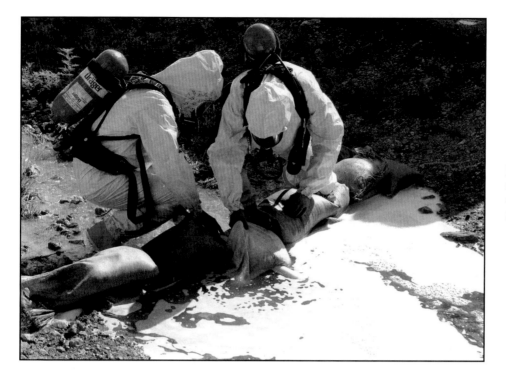

Figure 6.13 Defensive operations aim to confine the emergency without directly contacting the hazardous material involved.

- Protect exposures.
- Perform rescues when safe and appropriate.
- Evaluate and report incident progress.
- Perform emergency decontamination procedures.

Offensive

Offensive operations are those where responders take aggressive, direct action on the material, container, or process equipment involved in the incident **(Figure 6.14)**. These operations may result in contact with the material and therefore require responders to wear appropriate chemical-protective clothing

Offensive Operations — Operations in which responders take aggressive, direct action on the material, container, or process equipment involved in an incident.

Figure 6.14 Offensive operations may involve contacting the hazardous material while taking direct actions to mitigate the incident.

and respiratory protection. Some offensive operations are beyond the scope of responsibilities for first responders and are conducted by more highly trained hazardous materials personnel.

Offensive Tasks Allowed by U.S. OSHA and Public Safety Canada

The United States and Canadian governments recognize that Operations Level first responders who have appropriate training including certification, plus appropriate protective clothing and adequate resources can perform offensive operations involving flammable liquid and gas fire control of the following materials:

- Gasoline
- Diesel fuel
- Natural gas
- Liquefied petroleum gas (LPG)

Planning the Initial Response

Once the incident analysis is underway and there is a basic understanding of the hazards and products involved, the IC must use that information to plan the response. Thinking the situation through and developing a solid strategy with realistic response objectives will deliver a successful and safe outcome. The sections that follow will address aspects relevant to planning the initial response.

FEMA Incident Action Planning Process

FEMA identifies the following steps in the incident action planning process:

- Develop incident objectives and strategy.
- Develop tactics and resource assignments.
- Conduct incident and resource assessment.
- Identify required logistical support.
- Consider public information and interagency issues.
- Document assignments and required support.
- Monitor implementation.

Response Models

For an IC, the ability to make decisions and solve problems effectively is a necessary skill. Problem-solving and decision-making are fluid processes in hazmat incidents, so an IC's understanding of a problem (and consequent plans to address it) may change as more information becomes available and/ or conditions change. However, look at the issues present and move in an orderly manner to a successful mitigation strategy. Using a response model can simplify the problem-solving process because most response models incorporate an entire problem-solving process:

- An information gathering or input stage
- A processing or planning stage
- An implementation or output stage
- A review or evaluation stage

There are many models from which to choose. Most models move through each of the basic stages in problem solving and decision making, and the model used will often be dictated by departmental policy. As explained in Chapter 1, *APIE* is a simple response model containing the basic four-step problem-solving process model elements:

1. Analyze the incident.
2. Plan the initial response.
3. Implement the response.
4. Evaluate progress.

Other Hazmat Response Models

APIE is one response model used at hazmat/WMD incidents. Many other models exist, including the following:

GEDAPER

The GEDAPER response model, developed by David Lesak, has been adopted and embedded into the curriculum of the National Fire Academy. The acronym GEDAPER stands for:

G - Gather information.

E - Estimate potential course and harm.

D - Determine strategic goals.

A - Assess tactical options and resources.

P - Plan of action implementation.

E - Evaluate operations.

R - Review the process.

Eight Step Process©

Gregory G. Noll, Michael S. Hildebrand, and James G. Yvorra developed the Eight Step Incident Management Process©, which is a tactical decision making model that focuses on hazmat/WMD incident safe operating practices. The eight steps are:

1. Site management and control.

2. Identify the problem.

3. Hazard assessment and risk evaluation.

4 Select protective clothing and equipment.

5. Information management and resource coordination.

6. Implement response objectives.

7. Decontamination.

8. Terminate the incident.

HazMatIQ©

HazMatIQ© is a four-step decision-making response model that is used by major fire and law enforcement agencies across the United States. The HazMatIQ© system is a proprietary risk-based response system that will assist responders through a four-step process:

- A quick chemical size-up using supplied charts

- A streamlined chemical hazard research process

- Detection and monitoring prediction

- Selection of mission specific PPE

DECIDE

Ludwig Benner developed the DECIDE response model. The DECIDE mnemonic stands for:

D - Detect the presence of a hazardous material.

E - Estimate likely harm without intervention.

C - Choose response objectives.

I - Identify action options.

D - Do best option.

E - Evaluate progress.

Department of Transportation (DOT)

The Department of Transportation (DOT) has devised an eight-step response model approach to incident size-up and safety. This response model is good for lower levels of hazmat response but may not be practical for the Technician Level responder. The model can be found on page 4 of the 2016 edition of the *Emergency Response Guidebook*.

The following are the steps used in the DOT response model:

- Approach cautiously from upwind, uphill, or upstream.

- Secure the scene.

- Identify the hazards.

- Assess the situation.

- Obtain help.

- Decide on site entry.

- Respond.

- Above all, do not assume that gases or vapors are harmless because of a lack of smell.

Risk-Based Response
— Method using hazard and risk assessment to determine an appropriate mitigation effort based on the circumstances of the incident.

Risk-Based Response

A **risk-based response** uses information, science, and technology to mitigate a hazardous materials incident. The key is to equip responders with the critical information that is needed to make good decisions, while not overwhelming them with nice-to-know information.

Product identification is a vital element in successful mitigation of a hazmat incident. While this is true, the reality of the situation is that it may not always be possible. In either case, implement a risk-based response for all hazardous materials incidents.

Risk-based response is a hierarchy of decisions needed to protect responders. As with most successful incidents, the response starts with a thorough size-up, identifying the immediate hazards so that decisions can be made in a logical and educated manner. Too much time can be wasted attempting to research a property that may not exist, may not be present, or cannot be found. While a strong size-up helps with prediction, it is detection and monitoring equipment that will help protect the responders. A risk-based response is a quick and efficient way to "thin slice" information and make educated, life-saving decisions.

Developing the Incident Action Plan (IAP)

Incident Action Plans (IAPs) are critical to the rapid, effective control of emergency operations. An IAP is a well-thought-out, organized course of events developed to address all phases of incident control within a specified time. The timeframe specified is one that allows the least negative action to continue. Written IAPs may not be necessary for short-term, routine operations; however, large-scale or complex incidents require the creation and maintenance of a written plan for each operational period **(Figure 6.15)**.

Action planning starts with identifying the response objective (strategy) to achieve a solution to the confronted problems. A response objective is broad in nature and defines what has to be done. Once the strategy has been defined, the Command Staff needs to select the action options (tactics, the how, where, and when) to achieve the objective. Action options are measurable in both time and performance. An IAP also provides for necessary support resources such as water supply, utility control, or SCBA cylinder filling.

The IAP ties the entire problem-solving process together by stating what the analysis has found, what the plan is, and how it shall be safely implemented. Once the plan is established and resources are committed, it is necessary to assess its effectiveness. Gather and analyze information so that necessary modifications may be made to improve the plan if necessary. This step is part of a continuous size-up process. Elements of an IAP include the following:

- Strategies/incident objectives
- Current situation summary
- Resource assignment and needs
- Accomplishments
- Hazard statement
- Risk assessment

Figure 6.15 Large scale, complex incidents may require a written IAP. *Courtesy of Phil Linder.*

- Safety plan and message
- Protective measures
- Current and projected weather conditions
- Status of injuries
- Communications plan
- Medical plan

All incident personnel must function according to the IAP. Company officers or sector officers should follow predetermined procedures, and every action should be directed toward achieving the goals and objectives specified in the plan.

For practical purposes, all first responders should be familiar with the concept of IAPs and site safety plans because they have a direct effect on actions taken at an emergency incident scene. A first responder assuming the role of IC will need to develop and implement an IAP.

Common Response Objectives and Action Options at Hazmat Incidents

Once first responders have a basic understanding of the problem, they can begin to plan their solution by establishing **response objectives** (strategies) and **action options** (tactics). Response objectives are broad statements of what must be done to resolve an incident. Action options are specific operations that must be done in order to accomplish those goals.

Response objectives must be selected based on the following criteria:

- Their ability to be achieved
- Their ability to prevent further injuries and/or deaths
- Their ability to minimize environmental and property damage within the constraints of safety, time, equipment, and personnel

Risk-based response objectives are based upon the hazards present at the incident. For example, if materials with higher levels of toxicity are involved, a more cautious response using higher levels of personal protective equipment might be used. An incident involving a hazardous material in a gaseous or vapor form might dictate a different strategy for control than an incident involving a hazardous material in a solid or liquid form that is far easier (and safer) to contain.

Some additional risk-based response principles are as follows:

- Activities that present a significant risk to the safety of members shall be limited to situations where there is a potential to save endangered lives.
- Activities that are routinely employed to protect property shall be recognized as inherent risks to the safety of members, and actions shall be taken to reduce or avoid these risks.
- No risk to the safety of members shall be acceptable when there is no possibility to save lives or property.

Response Objective — Statement based on realistic expectations of what can be accomplished when all allocated resources have been effectively deployed that provide guidance and direction for selecting appropriate strategies and the tactical direction of resources.

Action Option — Specific operations performed in a specific order to accomplish the goals of the response objective.

Making the right strategic decision at a hazmat incident is critical because of the variety of developments that can occur. Poorly developed decision-making processes can lead to greater problems. References such as the *Emergency Response Guidebook's* orange-bordered pages, relevant Safety Data Sheet sections, shipping papers with emergency response information, and other resources may provide some response information and guidance.

Some common response objectives at hazardous materials incidents are:

- Isolation
- Notification
- Identification
- Protection (life safety)
- Rescue **(Figure 6.16)**
- Spill control/confinement
- Leak control/containment
- Crime scene and evidence preservation
- Fire control
- Recovery/termination

Some of these response objectives have already been described in this manual. Identification of hazardous materials was covered in earlier chapters. Other response objectives will be addressed in chapters that follow.

While these are some of the common response objectives, ICs can set whatever objectives they deem appropriate, using whatever terms they prefer. Rescue might be considered an important response objective at one incident but not at another. If conditions at an incident change suddenly, rapid evacuation might become a response objective that springs to the top of the priority list.

Figure 6.16 Rescue is a response objective at some hazmat incidents.

Response objectives are prioritized depending on available resources and the particular details of the incident. Some response objectives may not be needed if the hazard is not present at the incident. If the material involved is nonflammable, fire control may not be an issue. Some goals may require the use of specialized resources (such as chemical protective clothing or specific absorbent materials) that are not yet available and therefore must be postponed or eliminated. Others may require the use of so large a percent of the available resources that the ability to complete other objectives might be compromised. The response objectives listed in this chapter are very broad strategic categories, and mitigation at an actual incident may require a variety of response objectives based on the problems presented at the scene.

Action options are the specific tactics that are used to accomplish response objectives. These are the tasks you will be asked to perform in order to mitigate the incident. For example, if "isolation" is the action objective, evacuating people from the hazard area might be an appropriate response option. **Table 6.1, p. 312**, Typical Hazmat Problems with Potential Response Objectives and Options, provides common problems presented at hazmat incidents with more narrowly defined strategies and tactics.

Problem	Strategies	Tactics
Access: Access problems may be related to gaining access or denying access (to civilians or unprotected responders). Generally the first problem presented is limiting access to civilians and unprotected responders.	Isolate and deny entry	• Establish control zones (Hot and Cold) • Control traffic
Container Under Stress: The two types of container stress that responders can readily affect are generally thermal stress (heating) and mechanical stress (due to overpressure).	Ignore	Protect exposures (protective actions only)
	Cool	• Use master stream • Use hoseline
	Extinguish fire	• Remove fuel • Use master stream • Use hoseline • Use foam master stream • Use foam hoseline
	Release pressure	• Transfer product • Release product to atmosphere • Vent and burn
Container Breach/Release: Active strategies to manage a breach/release generally require operations inside the hazard area (Hot Zone).	Ignore	Protect exposures (protective actions only)
	Contain	• Close valve(s) • Tighten attachments • Plug • Patch • Transfer product • Decontaminate (required for entry)
Dispersion: Active strategies to control dispersion may be either offensive or defensive (depending on where they are performed). Dispersion control strategies are driven by the form of the material that has been (or is being) released.	Ignore	Protect exposures (protective actions only)
	Confine: Solid	Cover
	Confine: Liquid	• Adsorb or absorb • Dike (Circle or V-shape) • Divert • Retain • Dam (underflow or overflow) • Suppress vapor (foam)
	Confine: Energy	Shield
	Disperse: Gas	Disperse vapor (water fog or blower)

Continued

Table 6.1 (concluded)

Problem	Strategies	Tactics
Fire: The fire problem includes a direct threat to life safety and exposures, potential to affect container integrity, and release of toxic products of combustion. However, in some cases (pesticides), fire may present less threat than fire-control operations.	Ignore	Protect exposures (protective actions only)
	Extinguish	• Use master stream • Use hoseline • Use foam master stream • Use foam hoseline • Use dry chemical • Use specialized extinguishing agent
Possible Victims: Possible victims may be reported (definitely a known imminent life threat) or inferred based on incident conditions. Victims removed from the hazard area (Hot Zone) may require decontamination.	Determine	Ask
	Notify	• Use public address system • Use telephone
	Locate	• Perform primary search/extraction • Perform decontamination • Perform secondary search
Visible/Known Victims: Victims may be visible or known to be inside the hazard area. These victims may (or may not) be able to rescue themselves. First responders must use care in assessing their capability to effect a rescue (due to limitations in personal protective equipment and training. Victims removed from the hazard area (Hot Zone) may require decontamination.	Rescue	• Rescue themselves • Move to safe refuge • Perform extraction • Perform decontamination
Potential Life Exposure: Potential victims may become exposed due to dispersion (downhill or downwind). Responders must consider dispersion, time, and incident conditions in evaluating potential life exposure.	Protect in place	• Notify face to face • Notify by telephone • Notify media
	Evacuate	• Notify face to face • Notify by telephone • Notify media • Shelter • Control traffic • Perform security
Environmental/Property Exposure: Active strategies to minimize environmental/property damage are generally offensive in nature.	Ignore	Self-mitigate
	Control chemical	• Dilute • Neutralize
	Cool	• Use master stream • Use hoseline • Use foam master stream • Use foam hoseline

Determining the Suitability of Available Personal Protective Equipment

Due to the nature of hazmat incidents, responders must be able to use SOP/Gs and other resources to determine if their PPE is adequate to perform their assigned tasks at an incident. If PPE is determined to be inadequate, the IAP will need to be revised. PPE requirements may differ depending on the following:

- Responder's mission/assignment
- Product(s) involved
- Circumstances at the incident, for example, in a confined space

NOTE: PPE will be examined in greater detail in Chapter 9.

Identifying Emergency Decontamination Needs

If responders or the public come into contact with (or potentially contact) hazardous materials, it may be necessary to remove the hazardous material as quickly as possible. This process is called **emergency decontamination** (emergency decon). If responders or the public exhibit signs and symptoms of exposure to hazardous materials or product is visible on their skin or clothing, emergency decon should be considered. Emergency decontamination will be addressed in greater detail in Chapter 10.

Chapter Review

Answer the following questions to review the information provided in this chapter.

1. How do predetermined procedures assist a first responder at a hazmat incident?

2. What are the three priorities for hazmat incidents?

3. Explain the three levels of perceptual awareness.

4. Give examples of Levels I, II, and III hazmat incidents.

5. What factors are considered when determining the mode of operation?

6. List the elements of an IAP.

7. What is the difference between a response objective and an action option?

Emergency Decontamination — The physical process of immediately reducing contamination of individuals in potentially life-threatening situations, with or without the formal establishment of a decontamination corridor.

Step 1: Identify response objectives based on the scope of the incident and available resources.

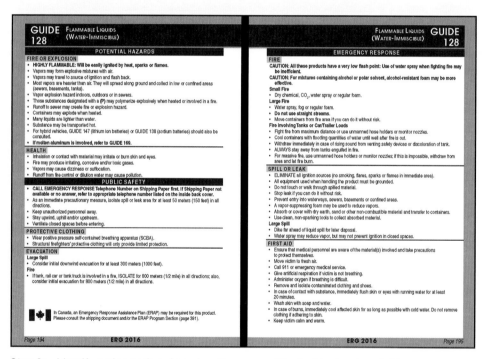

Step 2: Identify action options based on the scope of the incident and available resources.

Step 3: Identify safety precautions for the incident.

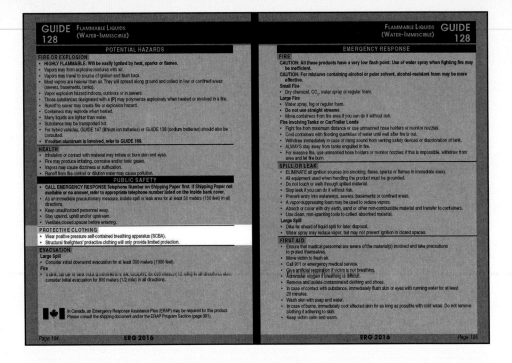

Step 4: Identify personal protective equipment appropriate for the incident.

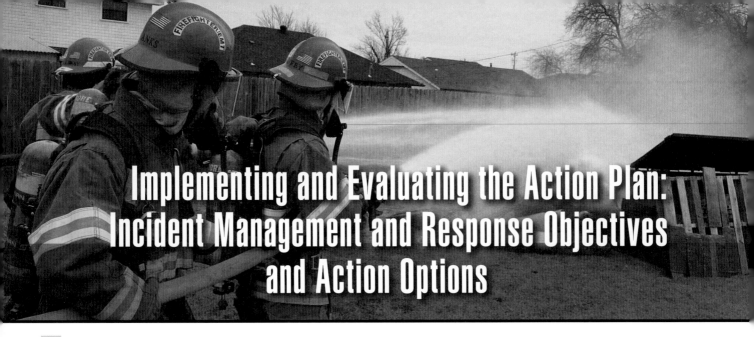

Implementing and Evaluating the Action Plan: Incident Management and Response Objectives and Action Options

Chapter Contents

Key Terms

NFPA Job Performance Requirements

This chapter provides information that addresses the following job performance requirements of NFPA 1072, *Standard for Hazardous Materials/Weapons of Mass Destruction Emergency Response Personnel Professional Qualifications (2017)*.

5.2.1

5.4.1

5.6.1

Implementing and Evaluating the Action Plan: Incident Management and Response Objectives and Action Options

Learning Objectives

After reading this chapter, students will be able to:

1. Describe the NIMS-ICS organizational functions that help initiate incident management. (5.4.1)

2. Describe secondary NIMS-ICS organizational functions. (5.4.1)

3. Explain ways of implementing response objectives and action options. (5.2.1, 5.3.1, 5.4.1)

4. Identify processes for evaluating progress. (5.4.1, 5.6.1)

5. Provide scene control at a hazardous materials incident. (5.4.1, Skill Sheet 7-1)

6. Evaluate and report progress made at a hazardous materials incident. (5.4.1, Skill Sheet 7-2)

Chapter 7
Implementing and Evaluating the Action Plan: Incident Management and Response Objectives and Action Options

This chapter will detail the following topics:

- Initiating the Incident Management System
- Implementing response objectives and action options
- Evaluating progress

Initiating Incident Management: NIMS-ICS Organizational Functions

Implementing the action plan is the third step in the APIE process, following analysis and planning. A crucial step in implementing the action plan is initiating the Incident Management System. An **Incident Management System (IMS)** is a management framework used to organize emergency incidents. As an emergency responder, you will initiate and operate under the IMS that your AHJ uses anytime that you respond to an emergency. The IMS provides the command structure and management terminology you will use at all emergency incidents.

By mandate, all emergency service organizations in the U.S. use the National Incident Management System - Incident Command System (NIMS-ICS). NIMS-ICS is designed to be applicable both to small, single-unit incidents that may last a few minutes and complex, large-scale incidents involving several agencies and mutual-aid units that possibly may last days or weeks. NIMS-ICS combines command strategy with organizational procedures. It provides a functional, systematic organizational structure that clearly shows the lines of communication and chain of command. NIMS-ICS provides the following to an incident:

- Modular organization
- Manageable span of control
- Organizational facilities such as a command post and staging areas
- Standardized position titles
- Integrated communication
- Accountability of resources

> **Incident Management System (IMS)** — System described in NFPA 1561, *Standard on Emergency Services Incident Management System and Command Safety,* that defines the roles, responsibilities, and standard operating procedures used to manage emergency operations. Such systems may also be referred to as Incident Command Systems (ICS).

Due to the nature of hazardous materials incidents, not all NIMS-ICS organizational functions may be required at an incident. As with fire incidents, the command structure at a hazmat incident will be dictated by the size and complexity of the emergency. NFPA 1026 and NFPA 1561 both include more information on ICS structure and applications.

NIMS-ICS involves five major organizational functions:

- Command
- Operations
- Planning
- Logistics
- Finance/Administration

Command Section

Command has the delegated authority to direct, order, and control resources **(Figure 7.1)**. Lines of authority must be clear to all involved, and lawful commands should be followed immediately without question. Responders should follow the chain of command and use correct radio protocols. To help avoid confusion during rapidly evolving situations, responders should not address anyone by name, rank, or job title; therefore, it does not matter who answers their radio messages. The basic Command organization configuration includes the following three levels:

- **Strategic level** — Entails the overall direction and goals of the incident
- **Tactical level** — Identifies the objectives that the tactical level supervisor/officer must achieve to meet the strategic goals
- **Task level** — Describes the specific tasks needed to meet tactical-level requirements and assigns these tasks to operational units, companies, or individuals

Figure 7.1 Basic NIMS-ICS Command structure.

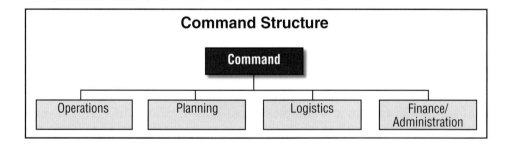

Incident Commander

The Incident Commander (IC) is the officer at the top of an incident chain of command and in overall charge of the incident. The Incident Commander's responsibilities include:

- Keeping an up-to-date report for the emergency scene
- Establishing the Command Post (CP) and formulating the Incident Action Plan
- Coordinating and directing all incident resources to implement the plan and meet its goals and objectives

- Informing the telecommunicator and other responders when Command is assumed or transferred

ICs at hazmat incidents have specific responsibilities in addition to standard IC functions. The IC does not have to actually perform or supervise each function but may choose to delegate them to others. The IC is required to perform the following functions at hazmat incidents:

- Establish the site safety (scene safety) plan.
- Implement a site security and control plan to limit the number of personnel operating in the control zones.
- Designate a Safety Officer.
- Identify the materials or conditions involved in the incident.
- Implement appropriate emergency operations.
- Ensure that all emergency responders (not just those of their own organizations) wear appropriate personal protective equipment (PPE) in restricted zones.
- Establish a decontamination plan and operation.
- Implement post-incident emergency response procedures (incident termination).

An aggressive plan should not be undertaken unless sufficient information is available to make logical decisions and the safe coordination of operations can be accomplished. If the incident is large and/or complex, the IC may delegate authority to the following command staff positions:

- Safety Officer (OSHA requires the appointment of a Safety Officer at hazmat incidents)
- Liaison Officer
- Public Information Officer

NOTE: Appendix F provides ICS color codes as recommended in NFPA 1561.

Safety Officer

The **Safety Officer**'s responsibilities include:

- Identifying and monitoring hazardous and unsafe situations
- Ensuring operational and personnel safety

Although the Safety Officer may exercise emergency authority to stop or prevent unsafe acts when immediate action is required, the officer generally chooses to correct them through regular lines of authority. The Safety Officer must be trained to the level of operations conducted at the incident and is required to perform the following duties:

- Obtain a briefing from the IC.
- Review IAPs for safety issues.
- Identify hazardous situations at the incident scene.
- Participate in the preparation and monitoring of incident safety considerations, including medical monitoring of entry team personnel before and after entry **(Figure 7.2)**.
- Maintain communications with the IC, and advise the IC of deviations from the incident safety considerations and of any dangerous situations.

Safety Officer — Member of the IMS command staff responsible to the Incident Commander for monitoring and assessing hazardous and unsafe conditions and developing measures for assessing personnel safety on an incident. *Also known as* Incident Safety Officer.

Figure 7.2 The Safety Officer monitors the scene for unsafe conditions.

- Alter, suspend, or terminate any activity that is judged to be unsafe.
- Conduct safety briefings.

The Safety Officer must ensure that safety briefings are conducted for entry team personnel before entry **(Figure 7.3)**. Safety briefings include the following information about the incident:

- Identification of hazards
- Description of the site
- Tasks to be performed
- Anticipated duration of the tasks
- PPE requirements
- Monitoring requirements
- Notification of identified risks
- Additional pertinent information

At incidents involving potential criminal or terrorist activities, the safety briefing should cover the following guidelines.

- Be alert for secondary devices.
- Do not touch or move any suspicious-looking articles (such as bags, boxes, briefcases, or soda cans).
- Do not touch or enter any damp, wet, or oily areas.
- Wear full protective clothing, including self-contained breathing apparatus (SCBA).
- Limit the number of personnel entering the crime scene.
- Document all actions.
- Do not pick up or take any souvenirs.
- Photograph or videotape anything suspicious.
- Do not destroy any possible evidence.
- Seek professional crime-scene assistance.

Figure 7.3 The Safety Officer conducts safety briefings with entry team members before they enter hazardous areas.

Command Post (CP)

The CP should be established at a safe location (uphill, upwind, and upstream from the incident, if possible). The IC must be accessible (either directly or indirectly), and a CP ensures this accessibility. A CP can be a predetermined location at a facility, a conveniently located building, or a radio-equipped vehicle located in a safe area **(Figure 7.4)**. Ideally, the CP location will allow the IC to observe the scene, although such a location is not absolutely necessary. The location of the CP is relayed to the telecommunicator/dispatcher and emergency responders. A CP needs to be readily identifiable with common identifiers such as:

- Custom designed command vehicles, or removable vehicle signage
- Marked building or tent
- Pennants, flags, or signs
- Marking lights, such as vehicle hazard lights

Operations Section

The Operations Section directly manages all incident tactical activities, the tactical priorities, as well as the safety and welfare of personnel working in the Operations Section **(Figure 7.5)**. The Operations Section Chief reports directly to the IC and is responsible for managing all operations that directly affect the primary mission of eliminating a problem incident. The Operations Section Chief directs the tactical operations to meet the IC's strategic goals.

One of the functions of the Operations section is the establishment and maintenance of the Staging Area. The Staging Area is where personnel and equipment awaiting assignment are held. This practice keeps the responders and their equipment a short distance from the scene until they are needed and minimizes confusion at the scene.

Figure 7.4 Many departments have mobile command posts; however, a command post can be a predetermined location at a facility, a conveniently located building, or a radio-equipped vehicle located in a safe area. *Courtesy of Ron Jeffers.*

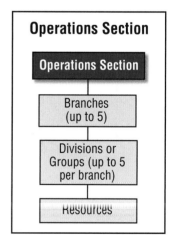

Figure 7.5 The Operations Section can be divided into Branches, Divisions, or Groups. ICs can designate Groups and/or Divisions without establishing Branches, depending on the needs of the incident.

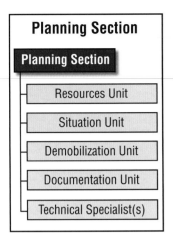

Figure 7.6 Specific units under the Planning Section include the Resource Unit, Situation Unit, Demobilization Unit, Documentation Unit, and Technical Specialists

Planning Section

The Planning Section gathers, assimilates, analyzes, and processes information needed for effective decision-making **(Figure 7.6)**. Information management is a full-time task at large incidents. The Planning Section serves as the IC's clearinghouse for incidents, allowing the IC's staff to provide information instead of having to deal with dozens of information sources. Command uses the information compiled by Planning to develop strategic goals and contingency plans. Specific units under Planning include:

- Resources Unit
- Situation Unit
- Documentation Unit
- Demobilization Unit
- Technical specialists as needed

Logistics Section

The Logistics Section is the support mechanism for the organization. It provides services and support systems to all the organizational components involved in the incident including the following:

- Facilities
- Transportation needs
- Supplies
- Equipment
- Maintenance
- Fueling supplies
- Meals
- Communications
- Responder medical services

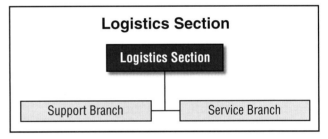

Figure 7.7 The Support Branch and the Service Branch are two branches within Logistics.

Support Branch and Service Branch are two branches within the logistics section **(Figure 7.7)**. The Service Branch includes medical, communications, and food services. The Support Branch includes supplies, facilities, and ground support (vehicle services).

Finance/Administration Section

The Finance/Administration Section is established when agencies responding to incidents have a specific need for financial services **(Figure 7.8)**. Not all agencies require the establishment of a separate

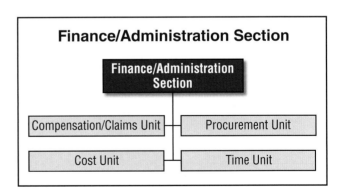

Figure 7.8 At large-scale, long-term incidents, the Finance/Administration Section is often activated.

Finance/Administration Section. In some cases, such as cost analysis, that function could be established as a Technical Specialist in the Planning Section. Specific units under the Finance/Administration Section include:

- Time Unit
- Procurement Unit
- Compensation Claims Unit
- Cost Unit

Hazmat teams may address cost recovery after the incident; therefore, an individual may not need to be staff on site. The AHJ will be a factor on how this section may be addressed at or after the incident, based on relevant ordinances.

Other NIMS-ICS Organizational Functions

In addition to the NIMS-ICS organizational functions previously described, other components and functions of NIMS-ICS include:

- Intelligence and Information Section
- Incident Command establishment and transfer
- Unified Command
- Hazmat Branch

Intelligence and Information Section

The Intelligence and Information Section is established at incidents when WMDs or criminal activities are suspected. It can be placed at any of several different organizational levels. This Section may be a function of Command, Operations, Planning, or other places in the organizational structure as the IC determines. The Intelligence and Information Section ensures that all intelligence/investigations operations and activities are properly managed, coordinated, and directed. These operations and activities help authorities:

- Prevent/deter potential unlawful activity, incidents, and/or attacks.
- Collect, process, analyze, secure, and appropriately disseminate information and intelligence.
- Identify, document, process, collect, create a chain of custody, safeguard, examine, analyze, and store probative evidence.
- Conduct a thorough and comprehensive investigation that leads to the identification, apprehension, and prosecution of the perpetrators.
- Serve as a conduit to provide situational awareness (local and national) pertaining to an incident.
- Inform and support life safety operations, including the safety and security of all response personnel.

Incident Command Establishment and Transfer

Under the IMS, the first person on the scene or the ranking individual of the first company on the scene assumes Command of the incident. That individual maintains Command until a higher ranking or more extensively trained responder arrives on the scene and assumes Command. The IC must have IMS training and be trained at the hazardous materials Operations

Figure 7.9 The Incident Commander (IC) has overall responsibility for managing the incident and must be trained at the hazardous materials Operations Level.

Level (**Figure 7.9**). Before transferring Command, the IC must ensure that the new IC is both capable of assuming Command (that is, have the necessary qualifications) and willing to accept Command. Command can be transferred face-to-face or over the radio, but only to someone who is on scene. As an incident grows larger, Command may be transferred several times before the situation is brought under control. A smooth and efficient transfer of Command contributes greatly to bringing an incident to a timely and successful conclusion.

The person relinquishing Command must provide the person assuming Command with as clear of a picture of the situation as possible. Provide a briefing or a situation status report, which is an updated version of the incident evaluation performed on arrival. The person assuming Command acknowledges receipt of the information by repeating it back to the current IC. If the reiteration is accurate, the recipient is ready to accept control of and responsibility for the management of the incident. The former IC can then be reassigned to an operating unit or retained at the Command Post as an aide or a member of the Command Staff.

When Command is transferred, the former IC must announce the change to avoid any possible confusion caused by others hearing a different voice acknowledging messages and issuing orders. Announcing the change of command keeps all responders informed of essential information during a rapidly evolving situation.

NIMS Incident Command Transfer Steps

Per NIMS, the following steps should be taken to transfer command:

Step 1: The incoming Incident Commander should, if at all possible, personally perform an assessment of the incident situation with the existing IC.

Step 2: The incoming IC must be adequately briefed. This briefing must be by the current IC, and take place face-to-face if possible. The briefing must cover the following:

- Incident history (what has happened)
- Priorities and objectives
- Current plan
- Resource assignments
- Incident organization
- Resources ordered/needed
- Facilities established
- Status of communications
- Any constraints or limitations
- Incident potential
- Delegation of Authority

Step 3: After the incident briefing, the incoming IC should determine an appropriate time for transfer of command.

Step 4: At the appropriate time, notice of a change in incident command should be made to:

- Agency headquarters (through dispatch)
- General Staff members (if designated)
- Command Staff members (if designated)
- All incident personnel

Step 5: The incoming IC may give the previous IC another assignment on the incident. There are several advantages of this:

- The initial IC retains first-hand knowledge at the incident site.
- This strategy allows the initial IC to observe the progress of the incident and to gain experience.

The ICS Form 201 is especially designed to assist in incident briefings (Step 2). It should be used whenever possible because it provides a written record of the incident as of the time prepared. The ICS Form 201 contains:

- Incident objectives
- A place for a sketch map
- Summary of current actions
- Organizational framework
- Resources summary

Source: NIMS

Unified Command

A multijurisdictional incident involves services (such as fire, law enforcement, and EMS) that are beyond the jurisdiction of one organization/agency. In these situations, the chain of command must be clearly defined. Control of an incident involving multiple agencies with overlapping authority and responsibility is accomplished through the use of Unified Command. When working under a Unified Command structure, several individuals may be working in Command, but only one person will ultimately answer for the operation. The concept of Unified Command simply means that all agencies that have a jurisdictional responsibility at a multijurisdictional incident contribute to the process by taking the following actions:

- Determine overall incident objectives.
- Select strategies.
- Accomplish joint planning for tactical activities.
- Ensure integrated tactical operations.
- Use all assigned resources effectively.

Proactive organizations identify target hazards in their areas of jurisdiction and also identify any other agencies with authority and responsibility for those target hazards. Ideally, those agencies meet, identify differences in agency IMS practices, and establish a **Memorandum of Understanding (MOU)** for Unified Command: a written agreement defining roles and responsibilities within a Unified Command structure. The lead officials of the agencies sign the MOU, and it becomes policy governing the personnel within those agencies.

Memorandum of Understanding (MOU) — Form of written agreement created by a coalition to make sure that each member is aware of the importance of his or her participation and cooperation.

Controlling hazardous material incidents may require the coordinated efforts of several agencies/organizations, such as the following:

- Fire service
- Law enforcement
- EMS
- Private concerns
  — Material's manufacturer
  — Material's shipper
  — Facility manager
- Government agencies (local, state/provincial, federal) with mandated interests in health and environmental issues
- Privately contracted cleanup and salvage companies
- Specialized emergency response groups, organizations, and technical support groups
- Utilities and public works

Before an incident occurs, agencies should perform the following to avoid jurisdictional and command disputes:

- Identify the specific agency/organization responsible for handling and coordinating response activities.
- Know what your mutual-aid contracts cover.
- Plan your preincident coordination at the local level. Document the identities and capabilities of nearby support sources.

Proper planning and preparation lead to safe and successful responses to hazardous materials incidents. The occurrence of a serious hazmat incident is not the time to discover that a neighboring fire department or industry cannot provide desperately needed equipment, personnel, or technical expertise. When emergency services organizations work together to develop their hazmat preincident surveys, they can meet the following objectives:

- Share vital resource information.
- Develop rapport among participating emergency services organizations.
- Identify and pool needed resources.

Hazmat Branch

Hazmat functions will be based on the AHJ and the needs of the IC at the actual scene **(Figure 7.10)**. For states that have implemented OSHA 1910.120, responsibilities of hazmat positions are defined.

Types of Command Structures

Your organization may use NIMS-ICS or some other Incident Management System. As a first responder, you must understand your role and responsibilities within the Command structure, whatever the system may be. Freelancing or taking action on your own without consent or knowledge of the Incident Commander (IC) is unacceptable and potentially dangerous.

At large and/or complicated hazmat/WMD incidents, you may see unfamiliar teams or personnel who arrive at the scene to assist. National response and/or Incident Management teams may arrive to manage different aspects of the incident. You may have to work with these individuals, and you need to understand your role in the big picture. Even though your agency was first on the scene, you may end up working with people with wide-ranging expertise, from a variety of organizations.

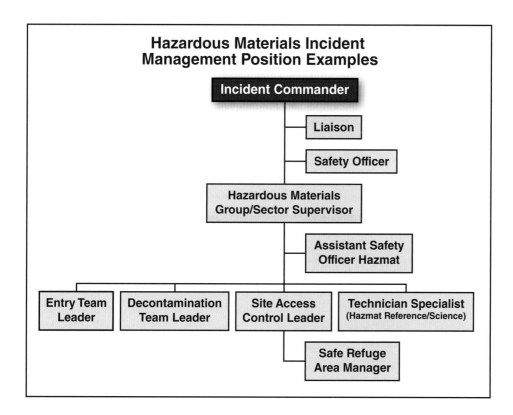

Hazardous Materials Incident Management Position Examples

Figure 7.10 An example of how a hazmat branch might be staffed. The AHJ will determine hazmat functions at an incident.

Implementing Response Objectives and Action Options

The purpose of the IAP is to develop the necessary strategy and tactics to affect a positive and safe outcome. Once the strategies are in place, tactics are developed. Tactics are the operational tasks that are used to accomplish the strategies. Tactics should be measureable in both time and performance. They should be evaluated to ensure that they will meet the strategic goals that were put in place.

Some typical hazmat strategies and tactics that you may be assigned to implement are presented in the following sections, including:

● Notification

● Isolation and scene control

● Hazard-control zones

● Protection of responders

● Protection of the public

● Protection of the environment and property

- Product control
- Fire control
- Emergency decontamination
- Evidence preservation

Notification and Requests for Assistance

Emergency response plans must ensure that responders understand their role in notification processes and predetermined procedures such as standard operating procedures (SOPs). Notification may include actions such as incident-level identification and public emergency information/notification. It is better to dispatch more resources than necessary in an initial response to ensure appropriate weight of attack to combat incident conditions. Responders should be familiar with the assets available in their jurisdictions.

Because hazmat incidents have the potential to overwhelm local resources, responders must know the procedure to request additional assets. This process should be described in local, district, regional, state, and national emergency response plans as well as in mutual/automatic aid agreements.

Notification also involves contacting law enforcement whenever a terrorist or criminal incident is suspected, as well as notifying other agencies (such as public works and the local emergency operations center) that an incident has occurred. Procedures will differ depending on the AHJ. Always follow SOPs and emergency plans for notification procedures.

In the U.S., the notification process is detailed in the National Response Framework (NRF), and all local, state, and federal emergency response plans must comply with these provisions. While incidents are handled at the lowest geographic, organizational, and jurisdictional level, when local agencies need additional assistance beyond the local level, the hazmat IC or AHJ may request help.

The first responders in the U.S. should turn to the local emergency response plan (LERP) if they need to request outside assistance for an incident **(Figure 7.11)**. Per the NRF, the local response agency should be closely tied with the community's Emergency Operations Center (EOC). If local assets are insufficient to manage the emergency, requests for additional assistance (such as activation of National Guard units) will be made to the state EOC. States may then request federal assistance through the Department of Homeland Security. The proper authorities (local, state, and federal) must be informed that an incident has occurred, even if additional assistance is not required for an incident.

The following are resources that may be requested to help at hazmat/WMD incidents in the U.S.:

- **Weapons of Mass Destruction-Civil Support Teams (WMD-CST)** — These teams support civil authorities at domestic chemical, biological, radiological, nuclear, or high-yield explosive incident sites by identifying CBRNE agents/substances. The National Guard Bureau fosters the development of WMD-CSTs. There are plans for at least one CST in each state. Their duties are to:
  — Assess current and projected consequences
  — Advise on response measures

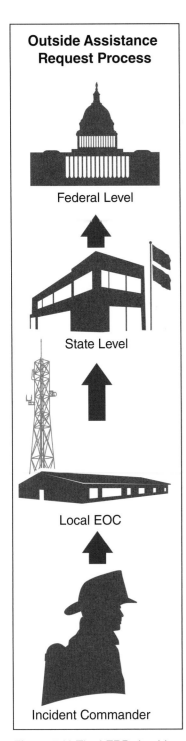

Outside Assistance Request Process

Federal Level

State Level

Local EOC

Incident Commander

Figure 7.11 The LERP should spell out the process for requesting outside help at incidents.

— Assist with appropriate requests for state support

— Provide an extensive communications capability

- **Disaster Medical Assistance Teams (DMAT)** — These are groups of professional and paraprofessional medical personnel (supported by a cadre of logistical and administrative staff) that provide emergency medical care during a disaster or other event **(Figure 7.12)**. The National Disaster Medical System (NDMS), through the U.S. Public Health Service (PHS), promotes and fosters the development of DMATs.

- **Disaster Mortuary Operational Response Teams (DMORT)** — These teams work under the guidance of local authorities and provide technical assistance and personnel to recover, identify, and process deceased victims. The teams are composed of private citizens, each with a particular field of expertise, who are activated in the event of a disaster. The NDMS, through the PHS and the National Association for Search and Rescue (NASAR), promotes and fosters the development of DMORTs.

- **National Medical Response Team-Weapons of Mass Destruction (NMRT-WMD)** — These are specialized response forces that provide medical care following a nuclear, biological, and/or chemical incident. Four NMRT-WMD teams are geographically dispersed throughout the U.S. The NDMS, through the PHS, fosters the development of NMRT-WMD. These units are capable of providing the following services:

— Mass-casualty decontamination

— Medical triage

— Primary and secondary medical care to stabilize victims for transportation to tertiary-care facilities in a hazardous material environment

- **National Guard Chemical, Biological, Radiological, Nuclear and High Yield Explosive (CBRNE) Enhanced Response Force Package (CERFP)** — The CERFPs and Civil Support Teams (CSTs) provide a phased capability. The CSTs detect and identify CBRNE agents/substances, assess their effects, advise the local authorities on managing response to attacks, and assist with requests for other forces. The CERFPs locate and extract victims from a contaminated environment, perform mass patient/casualty decontamination, and provide treatment as necessary to stabilize patients for evacuation.

Figure 7.12 DMAT teams provide emergency medical care during disasters or other events. *Courtesy of FEMA News Photos; photo by Andrea Booher.*

- **Urban Search and Rescue (US&R) Task Forces** — These highly trained teams provide search-and-rescue operations in damaged or collapsed structures and stabilization of damaged structures. They can also provide emergency medical care to the injured. Currently there are 28 federal US&R teams and numerous state teams that follow the DHS-FEMA US&R model regarding training, equipment, and personnel. The task forces are a partnership among the following entities:
  — Local fire departments
  — Law enforcement agencies
  — Federal and local governmental agencies
  — Private companies
- **Incident Management Teams (IMT)** — These teams of highly trained, experienced individuals are organized to manage large and/or complex incidents. They provide full logistical support for receiving and distribution centers. The Geographic Area Coordination Centers host and manage the National IMTs. During wildland fires, the U.S. Forest Service (USFS) hosts the teams. Both states and regions can have IMTs. The following features distinguish Incident Management Teams (IMTs):
  — Many fire and emergency services want to develop local and regional/metropolitan IMTs, which would be based on USFS models.
  — IMTs train to support Command and general staff functions of the Incident Command System (ICS).
  — Many states have organized IMTs at the state and local level. For example, Tualatin Valley Fire and Rescue, Oregon, maintains five IMTs, rotating on-call status on a weekly basis. These teams provide strategic incident management and support for incidents involving large areas, long durations, technical or political complexities, or any other aspects extending beyond routine response capabilities. Incidents involving deployment of IMTs rarely occur, in part, because these resources are rare. An example that proves the exception is the West, Texas, fertilizer plant explosion. Incident response included an IMT because of many factors that complicated the scope of the incident.

Isolation and Scene Control

The isolation perimeter may be comprised of an inner and outer perimeter, and it may be expanded or reduced in size as needed. In most cases, the outcomes of an on-site risk assessment determine the initial isolation perimeter established.

Once resources have been committed to an incident, it is easier to reduce the isolation perimeter in size than it is to extend it. If resources have arrived and have been tasked at an incident, it may be difficult to disengage and relocate them should the initial perimeter be inadequate.

The IC must undertake a risk assessment or size-up of the incident in order to determine an appropriate size for the isolation perimeter. To determine the perimeter size, the IC should consult with other onsite agency commanders to ensure that the spatial requirements and tactical objectives can be met.

NOTE: From a risk-management perspective, it is better to encompass a larger area that can be reduced in size once incident-site conditions have been assessed for risks such as secondary devices, unidentified hazardous materials, and atmospheric monitoring.

The isolation perimeter is also used to control both access and egress from the incident site. Unauthorized personnel may be kept out, while witnesses and persons with information about the incident may be directed to a safe location until being interviewed and released. Another important aspect of scene control at hazmat/WMD incidents is the establishment of hazard-control zones and staging areas, explained in the following sections. **Skill Sheet 7-1** provides basic steps for providing scene control and performing various assigned tasks at a hazardous materials incident.

Hazard-Control Zones

Hazard-control zones provide the scene control required at hazmat and terrorist incidents to:

- Prevent interference by unauthorized persons.
- Help regulate first responders' movements within the zones.
- Minimize contamination (including secondary contamination from exposed or potentially exposed victims).
- Help ensure accountability of all personnel operating at large, multiagency response incidents.

Hazard-control zones divide the levels of hazard of an incident, and what a zone is called generally depicts this level. These zones are often referred to as **hot**, **warm**, and **cold (Figure 7.13)**. While typically represented as concentric

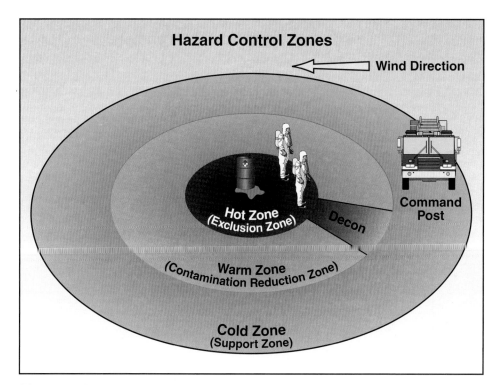

Figure 7.13 Hazard-control zones divide the levels of hazard at an incident into hot, warm, and cold zones, with the hot zone indicating the highest degree of danger.

Hazard-Control Zones — System of barriers surrounding designated areas at emergency scenes, intended to limit the number of persons exposed to a hazard and to facilitate its mitigation. A major incident has three zones: Restricted (Hot) Zone, Limited Access (Warm) Zone, and Support (Cold) Zone. EPA/OSHA term: Site Work Zones. *Also known as* Control Zones and Scene Control Zones.

Hot Zone — Potentially hazardous area immediately surrounding the incident site; requires appropriate protective clothing and equipment and other safety precautions for entry. Typically limited to technician-level personnel. *Also known as* Exclusion Zone.

Warm Zone — Area between the hot and cold zones that usually contains the decontamination corridor; typically requires a lesser degree of personal protective equipment than the Hot Zone. *Also known as* Contamination Reduction Zone *or* Contamination Reduction Corridor.

Cold Zone — Safe area outside of the warm zone where equipment and personnel are not expected to become contaminated and special protective clothing is not required; the Incident Command Post and other support functions are typically located in this zone. *Also known as* Support Zone.

circles, control zones take whatever shape is needed, often dictated by the features of the location and incident. Control zones are not necessarily static and can be adjusted as the incident changes.

The U.S. Occupational Safety and Health Administration (OSHA) and the U.S. Environmental Protection Agency (EPA) refer to these zones collectively as site work zones. They are sometimes called *scene-control zones*. Other countries may use different terminology for these zones.

Different agencies may have different control zone needs. At incidents involving crimes, law enforcement may designate a zone to incorporate the entire crime scene, and this zone may not correspond to traditional fire service activities. For example, at terrorist incidents in the U.S., the FBI establishes an evidence search perimeter 1.5 times the distance of the farthest known piece of evidence **(Figure 7.14)**. These law enforcement zones might change as evidence is processed and the crime scene is released. When establishing these zones, the crime scene dynamics may create a need for flexibility on the part of all agencies in the Unified Command.

Incidents involving bombs are an example of where traditional control zones and the operations that are usually conducted within those zones may vary. Because of the blast effects, there may be multiple buildings in danger of collapse, which will require the designation of a much larger hot zone. In order to preserve evidence in a bombing incident, law enforcement may require that the hot zone be extended out as far as the perimeter of the debris field. In these cases, there will likely be tight perimeter control as well as a large hot zone **(Figure 7.15)**. Due to these logistics, responders may need to conduct operations, such as triage, treatment, and transportation, in an area designated as the hot zone.

Figure 7.14 In the U.S., the FBI will establish a control perimeter at 1.5 times the distance of the farthest known piece of evidence.

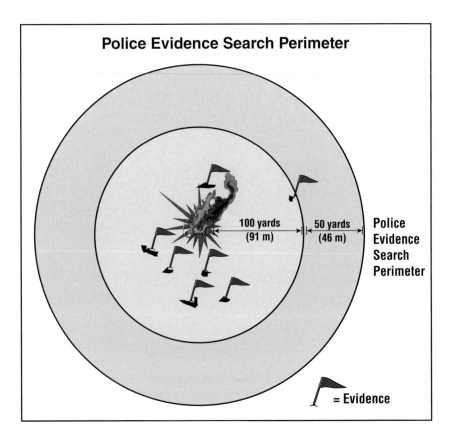

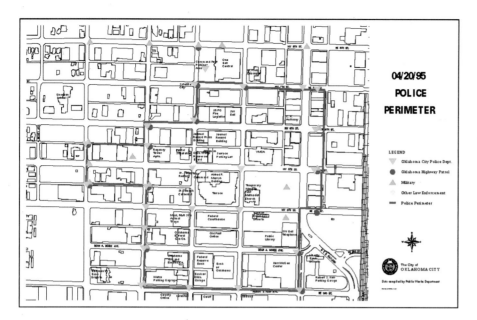

Figure 7.15 If evidence is widespread, as it was at the Oklahoma City bombing, the evidence perimeter may encompass a very large area.

An event that has multiple devices or release points is unique to terror events and may require a nontraditional scene management plan. In these cases, there may be more than one hot zone for a given incident. Whether the incident involves bombs or multiple devices, it is necessary for the Incident Commander to remain flexible and establish a scene management plan and control zones to meet the needs of all of the responders to the incident. Responders at the incident must be made aware of control zones as they are established.

Hot Zone

Traditionally, the hot zone (also called *exclusion zone*) is an area surrounding an incident that is potentially dangerous either because it presents a threat in the form of a hazardous material or the effects thereof. The area may be contaminated by chemical warfare agents, or it may have the potential to become contaminated by a released hazardous material. The area has been or could be exposed to the gases, vapors, mists, dusts, or runoff of the material. Responders must have proper training and wear appropriate personal protective equipment (PPE) to work in the hot zone or to support work being done inside the hot zone. There will be established access and egress points to ensure both accountability and designated PPE prior to entry.

The hot zone extends far enough to prevent people outside the zone from suffering ill effects from the released material, explosion, or other threat. Work performed inside the hot zone is often limited to highly trained personnel such as SWAT teams, US&R teams, hazardous materials technicians, Joint Hazard Assessment Teams (JHAT), mission specific operations, and bomb technicians.

WARNING!
Responders must have proper training and wear appropriate personal protective equipment (PPE) to work in the hot zone or to support work being done inside the hot zone.

Warm Zone

The warm zone (also called *contamination reduction zone* or *corridor*) is an area adjoining the hot zone and extending to the cold zone (see following section). The warm zone serves as a buffer between the hot and cold zones and as the decontamination location for personnel and equipment exiting the hot zone. Decontamination usually takes place within a corridor (decon corridor) located in the warm zone **(Figure 7.16)**. At incidents involving crimes, parts of the warm zone may be part of the crime scene; responders should create minimal disturbance. PPE will normally be required in this zone, although in some circumstances it may be at a reduced level from the hot zone. Monitoring and detection may be conducted around the perimeter of the warm zone to determine the extent of the hazards. The Unified Commander or the IC will approve the level of PPE required for work within this zone after input from others.

Figure 7.16 Decontamination typically takes place within the warm zone.

Cold Zone

The cold zone (also called *support zone*) surrounds the warm zone and is used to carry out all logistical support functions of the incident. Workers in the cold zone are not required to wear PPE because the zone is considered safe. However, some personnel may still be wearing PPE (such as body armor) in case of secondary devices and/or attacks to ensure safe evacuation in the case of rapid expansion of the hot zone.

The cold zone is the site for the following:

- Multiagency Command post (CP)
- Staging area
- Donning/doffing area
- Backup teams
- Research teams
- Logistical support

- Criminal investigation teams
- Triage/treatment/rehabilitation (rehab)
- Transportation areas

Staging

The **staging area** needs to be located in an isolated spot in a safe area where occupants cannot interfere with ongoing operations. Staging minimizes confusion and freelancing at the scene. Staging areas should be located at spots in the cold zone where occupants cannot interfere with ongoing operations. A safe direction of travel to the staging area should be broadcast to all resources responding to the incident.

Ideally, emergency responders and equipment at terror incidents should be staged between multiple locations in case staging areas are attacked. Some departments use the concept of a cornering/quartering staging procedure **(Figure 7.17)**. This has two basic purposes.

- Spreads out emergency response personnel from one another to limit their exposure as a target and minimizes the effects of a secondary type of attack/device.

- Allows personnel to envelop the scene and provide multiple treatment areas or operation function points.

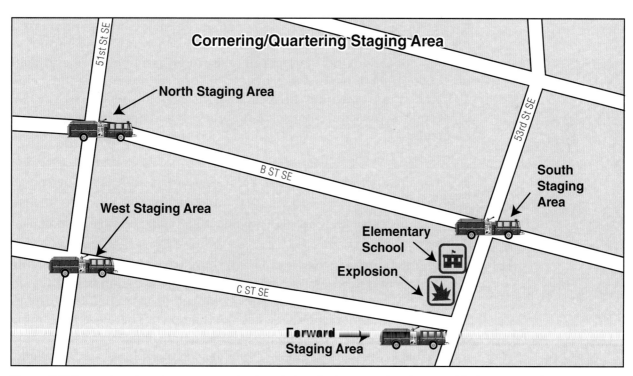

Figure 7.17 Some departments use a cornering/quartering staging procedure to spread their resources between multiple points.

Protection of Responders

The first priority at any incident is the protection and safety of emergency responders. Injured or incapacitated responders are unable to assist in mitigation efforts or protection of the public.

Measures to protect responders include the following:

- Staying uphill, upstream, and upwind of hazardous materials
- Wearing appropriate PPE
- Using time, distance, and shielding for protection
- Decontaminating responders when necessary
- Ensuring accountability of all personnel
- Tracking and identifying all personnel working at an incident
- Working as part of a team or buddy system
- Assigning safety officers
- Putting evacuation and escape procedures in place

The following sections describe some of these measures in greater detail.

Wearing PPE

During a risk-based response, responders must wear appropriate PPE to protect against hazards. The AHJ issues personal protection equipment, and responders must be trained in its selection, use, and maintenance. They must be aware of thermal issues such as heat stress. PPE is addressed in more detail in Chapter 9.

Accountability Systems

All personnel and equipment assigned to the incident must be tracked via an accountability system **(Figure 7.18)**. Most units responding to an incident arrive fully staffed and ready to be assigned an operational objective; other personnel may have to be formed into units at the scene. To handle these and other differences in the available resources, the IAP must contain a tracking and accountability system that has the following elements:

- Procedure for checking in at the scene
- Way of identifying and tracking the location of each unit and all personnel on scene
- Procedure for releasing people, equipment, and apparatus that are no longer needed

Accountability systems are especially important for incidents involving multiple agencies and organizations. All these agencies may have different levels of PPE and training. The agency/organization in command is responsible for tracking all responders, not just their own. Therefore, determine methods for tracking accountability in preplans and implement them as soon as possible at the incident scene. Types of accountability systems include:

- Fire service passport systems
- T-card systems for wildland incidents
- GPS and GIS

NOTE: NFPA 1500 and 1561 address accountability system requirements.

Figure 7.18 All personnel assigned to the incident must check in and out via the established accountability system.

Buddy Systems and Backup Personnel

NFPA and OSHA mandate the use of buddy systems and backup personnel at hazmat incidents. A buddy system organizes personnel into workgroups containing at least two members so that nobody works alone. The buddy system's primary benefit is to provide rapid help if there is an emergency; if one person becomes incapacitated, the other can call for help and provide immediate assistance.

In addition to using the buddy system, backup personnel must be in place and prepared to enter the hot zone with appropriate equipment to provide assistance or rescue if necessary. Backup personnel must be dressed in the same level of personal protective clothing as entry personnel **(Figure 7.19)**.

Figure 7.19 Backup personnel must dress in the same level of PPE as entry personnel, and they must be ready to enter the Hot Zone quickly, if needed.

///////////////////////////

CAUTION

At a minimum, four appropriately trained and equipped responders are needed to perform tasks in the hot zone — two working in the area itself and two standing by as backup.

Time, Distance, and Shielding

Using time, distance, and shielding is an effective strategy to protect first responders at hazmat incidents. The following describes the ways in which these strategies protect responders:

- **Time** — Limiting the time to which responders are exposed (or potentially exposed) to hazards and hazardous materials reduces the likelihood they will suffer serious harm. To limit the time of hazard exposure, restrict work times in the hot zone, and frequently rotate personnel on work groups.

- **Distance** — Maximizing distance from potential hazards will often prevent or reduce harm. The closer a responder is to the source of an explosion, the greater the harmful effects will be. Staying away from hazardous areas will also prevent harmful exposures. Distance may be controlled by implementing hazard-control zones.

- **Shielding** — Shielding places a physical barrier between a responder and the hazard. Shielding may consist of wearing PPE or positioning personnel so that another object, such as a wall, building, or apparatus, is between the responder and the hazard, thereby minimizing the chance of contact or harmful effect.

Evacuation/Escape Procedures

At hazmat incidents, a signaling system should be used that will advise personnel inside the danger area when to evacuate. The FEMA US&R Task Force program has developed a system for evacuating rescuers from dangerous areas. Notification can be made using devices such as:

- Handheld CO_2 boat air horns
- Air horns on fire apparatus
- Vehicle horns

Other communication methods in the event of an emergency can include portable radios, voice, hand signals, and the use of other predetermined signals. The US&R designated signals and their meanings are as follows:

- Cease Operations/All Quiet: one long blast (three seconds)
- Evacuate the Area: three short blasts (one second each)
- Resume Operations: one long and one short blast

Responders must also plan multiple escape procedures. If the primary means of egress becomes blocked, rescuers should determine the possibility of using the alternate route.

Protection of the Public

Measures to protect the public include such operations as conducting rescues, performing mass decontamination, and providing emergency medical care and first aid. Additional measures include evacuation, sheltering in place, and protecting/defending in place. The IC selects the best option (or combination of options) based on the incident.

Rescue

Based on the nature of the incident, victims may be found in a variety of locations, such as out in the open or within a structure or a confined space. Before attempting a rescue, evaluate the location and viability of the victim as well as available tools and equipment. If the decision is made to attempt a rescue, safety should be a paramount concern.

The IC makes decisions about rescue based on a variety of factors at the incident. The following factors affect the ability of personnel to perform a rescue:

- Nature of the hazardous material and incident severity
- Training
- Availability of appropriate PPE

- Availability of monitoring equipment

- Number of victims and their conditions

- Time needed to complete a rescue

- Tools, equipment, and other devices needed to effect the rescue

NOTE: Chapter 12 addresses information needed by Ops-Level responders who will enter the hot zone to conduct rescues. Chapter 4 provides information about the hazards associated with each DOT hazard class so that first responders can assess potential risks at incidents involving these materials (for example, at an incident involving corrosives, they can determine that chemical burns are probably one of the major hazards).

First responders without Mission-Specific training should avoid contact with hazardous materials. Viable victims who are contaminated should be moved as carefully as possible, and transferred to the care of medical responders located within the initial-isolation zone or hot zone.

Actions that can be taken without risk of contamination include:

- Directing people to an area of safe refuge or evacuation point located in a safe place within the hot zone that is upwind and uphill of the hazard area.

- Instructing victims to move to an area that is less dangerous before moving them to an area that offers complete safety.

- Directing contaminated or potentially contaminated victims to an isolation point, safe refuge area, safety shower, eyewash facility, or decontamination area **(Figure 7.20)**.

- Giving directions to a large number of people for mass decontamination.

- Conducting searches during reconnaissance or defensive activities.

- Conducting searches on the edge of the hot zone.

If there are injured victims at the scene, first responders must always be aware of the potential dangers of contamination and the need to decontaminate as part of the treatment process (see Chapter 10, Implementing the Response: Decontamination). They must follow local procedures for determining prioritization of emergency medical care and decontamination.

Figure 7.20 First responders can direct contaminated or potentially contaminated victims to safety showers or areas of safe refuge. *Courtesy of the U.S. Marine Corps; photo by Sgt J.A. Lee II.*

Evacuation

To evacuate means to move all people from a threatened area to a safer place. To perform an evacuation, there must be enough time to warn people, for them to prepare to leave, and for them to leave the area by a safe route (uphill, upwind, etc.). Generally, if there is enough time for evacuation, it is the best protective action. Emergency responders should begin evacuating people who are most threatened by the incident in accordance with distances recommended by the ERG, preincident surveys, or other sources. Even after people move these recommended distances, they are not necessarily completely safe from harm. Do not permit evacuees to congregate at the scene. Instead, send them to a designated place (or area of safe refuge) along a specific route.

The number of responders needed to perform an evacuation varies with the size of the area and number of people to evacuate. Evacuation can be an expensive, labor-intensive operation; therefore, it is important to assign enough personnel resources to conduct it.

Evacuation and traffic-control activities on the downwind side could cause responders and evacuees to become contaminated and, consequently, need decontamination. Responders may also need to wear PPE to safely conduct the evacuation. The local emergency response plan should include a preplan for evacuation (including casualties) for likely terrorist targets such as stadiums and other public gathering places.

The IC must address the following factors regarding large-scale evacuations:

- **Notification** — Alert the public of the need to evacuate and tell them where they should go. The local emergency response plan details the notification methods. Relay clear and concise information to avoid confusion or additional panic. Notification methods include:
  — Knocking on doors
  — Public address systems
  — Radio
  — TV
  — Sirens
  — Building alarms
  — Short message service (SMS) through cell phones (text messages)
  — Reverse 9-1-1
  — Emergency Alerting System (EAS, in the United States)
  — Loudspeakers mounted on helicopters or emergency vehicles
  — Electronic billboards
- **Transportation** — Plan, in advance, alternate means of transportation, such as school buses, public transit systems, planes, trains, boats, barges, and ferries **(Figure 7.21)**.
- **Relocation facilities and temporary shelters** — Designate appropriate evacuation shelters in the local emergency response plan. Determine staffing arrangements in advance. Shelters must be able to provide food, water,

medicine, bathroom and shower facilities, and places to sleep (for evacuations of long duration) **(Figure 7.22)**. Establish an information/registration system to track the whereabouts of evacuees so their friends and relatives can find them.

- **Reentry** — Consider how people will be allowed to return to evacuated areas.

Figure 7.21 Some people will not be able to self-evacuate. Therefore, plans must be made in advance to provide transportation to individuals without a means to leave the hazard area. *Courtesy of FEMA News Photos; photo by Win Henderson.*

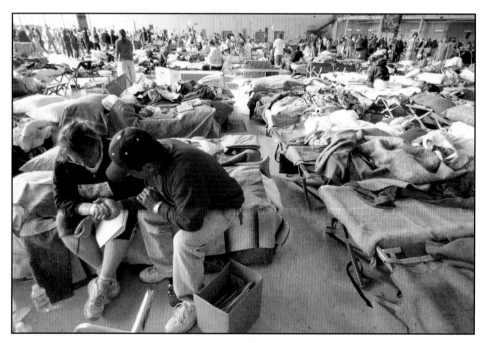

Figure 7.22 Evacuees must have some place to stay. *Courtesy of FEMA News Photos; photo by Andrea Booher.*

Evacuation of Contaminated Victims

Individuals who have been exposed or potentially exposed to chemical, biological, or radiological agents must be decontaminated (see Chapter 10). While it may be impossible to keep these individuals at the scene, efforts to keep them in place should be made in order to prevent the spread of harmful or potentially deadly materials to other locations. Evacuate contaminated or potentially contaminated individuals to an area of safe refuge (or a triage and treatment area as appropriate) within the isolation perimeter to await decontamination. Because victims may leave the scene before emergency responders arrive (or ignore requests to stay in order to undergo decon), shelters, hospitals, and other public health care facilities must be prepared to conduct decon of people who walk in on their own or with help from others at their facilities.

Sheltering in Place

Sheltering in place means to direct people to go quickly inside or to stay inside a room or a building and remain inside until danger passes. Some situations may make sheltering in place preferable to evacuation. The decision to shelter in place may be guided by the following factors:

- The population is unable to initiate evacuation because of health care, detention, or educational occupancies.

- The material is spreading too rapidly to allow time for evacuation.

- The material is too toxic to risk any exposure.

- Vapors are heavier than air, and people are in a high-rise or multi-level structure **(Figure 7.23)**.

When protecting people inside a structure, close all doors, windows, heating, ventilation and air-conditioning systems. Vehicles are not as effective as buildings for sheltering in place, but they can offer temporary protection if windows are closed and the ventilation system is turned off.

First responders should also pay attention to the condition of surrounding buildings before ordering sheltering in place. Some areas may have old and dilapidated structures without air-conditioning or with openings between floorboards. Sheltering in place might not provide sufficient protection in such cases, making evacuation the better option.

Similarly, evacuation may be a better option than sheltering in place when explosive vapors or gases are involved, for two reasons:

- These vapors or gases may take a long time to dissipate from the surrounding environment.

- Vapors or gases may permeate into any building that cannot be sealed from the outside atmosphere.

Whether using evacuation or shelter in place, inform the public needs as early as possible and provide additional instructions and information throughout the course of an emergency. Sheltering in place may be more effective if public education has been initiated ahead of the incident via emergency planning.

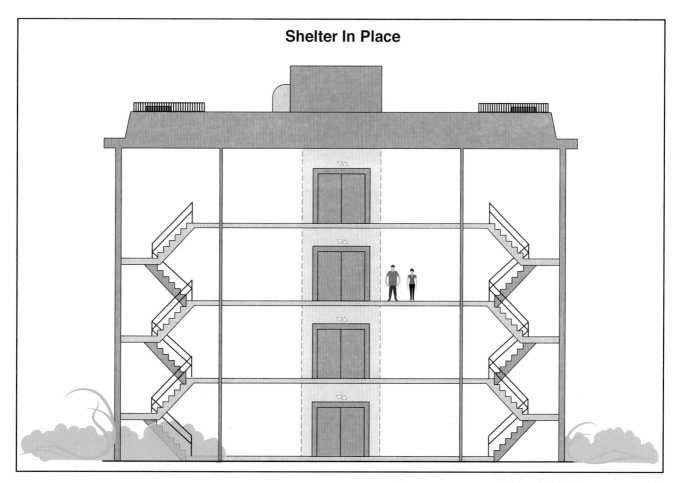

Shelter In Place

Figure 7.23 When vapors and gases are heavier than air, it may be safest to shelter in place in high-rise or multi-level structures.

Protecting/Defending in Place

Protecting/defending in place is an active (offensive) role or aggressive posture to physically protect those individuals in harm's way. When appropriate and safe to do so, defending in place eliminates the need for unnecessary evacuations, which, if initiated, will require additional logistical support to ensure the health and safety of civilians who are protected in place. Actions that may be taken during this kind of operation include:

- Using hose streams to diffuse a plume **(Figure 7.24, p. 348)**
- Securing a neighborhood or area
- Turning off HVAC systems to minimize spread of contaminants

Protection of the Environment and Property

Exposure protection is a defensive control tactic. Most firefighters should be familiar with the concept of protecting exposures in fire situations, usually in terms of protecting property that is exposed to a fire in order to keep it from spreading. However, at the hazmat scene, the same concept includes protecting the environment and protecting property that is threatened by an expanding incident (including closed containers and piping).

Protecting environment and property includes both protecting exposures from fires involving hazardous materials and protecting the environment from the harmful effects of hazardous materials that are not burning. For example, diking a storm drain is a tactic that protects the environment from being exposed to (and harmed by) potentially toxic materials **(Figure 7.25, p. 348)**.

Figure 7.24 Defending in place uses aggressive tactics to physically protect individuals in harm's way.

Figure 7.25 Diking a storm drain can protect the environment from harm.

Protecting the Environment

Environmental damage is also an important concern. The air, surface water, wildlife, water table, and land surrounding an incident may be seriously affected by released materials. Water used during fire-control activities may become contaminated with pollutants or other hazardous materials. The nonbiodegradable nature of many materials means that the consequences of contamination may take years for the full effect to be realized. The result of contamination may also require large sums of money to repair. All released materials and runoff need to be confined and held until their effect on the environment can be determined.

Protecting Property

The property risk at hazardous materials incidents is similar to that created by other fire hazards except that the threatening material may not always be readily evident. Flammable and toxic gases, mists, and vapors can contaminate and pose an ignition threat with no visible signs. Protective actions must be tailored to the material, its properties, and any reactions to the proposed protective medium. ICs may appropriately decide not to save property when operations pose a risk. Lives or the environment must not be unduly compromised to save property.

Product Control

When hazardous materials escape their containers, emergency responders may need to perform product control at incidents. There are two main product control strategies, one primarily defensive, and one offensive. Spill control is a defensive strategy that attempts to confine a hazardous material that has been released from its container. Leak control is an offensive strategy that attempts to contain a material in its original container, or transfer it to another container **(Figure 7.26)**. Product control is addressed in greater detail in Chapter 13.

Figure 7.26 Leak control attempts to keep a material in its original container.

Fire Control

Most hazmat incidents involve flammable materials. Fire control is the strategy used to prevent ignition and/or extinguish the fire when hazardous materials are involved. Tactics may include using fire fighting foam or water depending on the situation and the product involved. Fire control is addressed in greater detail in Chapter 13.

Evidence Recognition and Preservation

Incidents involving WMDs or other illegal activities are crimes, and the locations where they occur are crime scenes. Notify law enforcement as soon as a crime is suspected. Fire service first responders should not collect **evidence**.

Evidence — Information collected and analyzed by an investigator.

First responders need to identify and preserve evidence so that the investigator can collect and properly document it per the AHJ. Local emergency response plans should spell out responsibilities of individual agencies at such incidents as well as detail the acceptable procedures and techniques to be used.

First responders should preserve evidence so that investigators can identify and successfully prosecute guilty parties. The more the scene is disturbed, the more difficulties investigators encounter when attempting to develop a clear and accurate picture of what actually occurred. Law enforcement must gather accurate, acceptable information about the crime to be used in court. Even seemingly irrelevant things can have tremendous significance to forensic experts and other law enforcement investigators, including:

- Footprints
- Wrapping paper
- Containers
- Debris placement
- Victim locations
- Vehicles in the vicinity
- Location of witnesses and bystanders

Evidence can take many forms. Items that look like trash may be pieces of a bomb or an incendiary device. Evidence can include everything from body fluids to tire tracks to cigarette butts. The pattern of scattered debris may tell investigators about the force of an explosion (and consequently, how big the bomb was). Residue on debris can help identify what explosive materials were used. Clothing and jewelry removed from victims are considered evidence. At illegal clandestine labs, evidence may include fingerprints, weapons, chemical containers, notes, letters, and papers. Evidence can be anything; therefore, responders must — to the degree possible — avoid disturbing a scene.

The preservation of life is more important than the preservation of evidence. Lifesaving operations take precedence.

As soon as it is known or suspected that criminal or terrorist activity is involved at an incident, first responders should do the following to help preserve evidence and assist law enforcement:

- DO NOT touch anything unless it is necessary.
- Avoid disturbing areas not directly involved in rescue activities.
- Remember what the scene looked like upon first arrival as well as details about the progression of the incident. Note as many of the W's as possible: who, what, when, where, and why. If possible, pay attention to the following:
  — Who was present (including victims, people running from the scene, people acting suspiciously, bystanders, and potential witnesses).
  — What happened.
  — When important events occurred.
  — Where objects/people/animals were located.
  — Why events unfolded as they did.
- Document observations as quickly as possible. While it may be quite some time before responders have the opportunity, the sooner information is written down, the more accurate it will probably be. This documentation may be used as evidence for legal proceedings.
- Take photographs and videos of the scene as soon as possible (**Figure7.27**).

- Remember and document when something was touched or moved. Document in the report where it was and where it was placed. Photograph the item before doing anything if possible. DO NOT try to recreate the scene as it looked before something was touched or moved. In other words, do not move something back into the position you found it in after you have already moved it.

- Minimize the number of people working in the area if possible. Establish travel routes that minimize disturbance **(Figure 7.28)**.

- Leave fatalities and their surroundings undisturbed.

- Isolate and secure areas where evidence is found, and report findings to law enforcement authorities.

- Identify witnesses, victims, and the presence of evidence. Investigators will want to interview witnesses and victims as part of their investigations. Advise witnesses to remain at the scene in a safe location until they have been interviewed and released.

- Preserve potentially transient physical evidence (evidence present on victims or evidence such as chemical residue, body fluids, or footprints) that may be compromised by weather conditions.

- Have evidence collection points (such as ground tarps) located near decontamination corridors and hot zone exit locations to gather evidence during decon or doffing operations.

- At chemical or biological incidents, secure and isolate restaurants or food vendors near the incident area in the event contaminated food can be used as evidence.

- Follow predetermined procedures regarding operations at crime scenes.

Additional information about evidence preservation and sampling is provided in Chapter 14, Implementing the Response: Mission-Specific Evidence Preservation and Public Safety Sampling. The information found there is directed at personnel who may be charged with evidence collection, sampling, and documenting chain of custody.

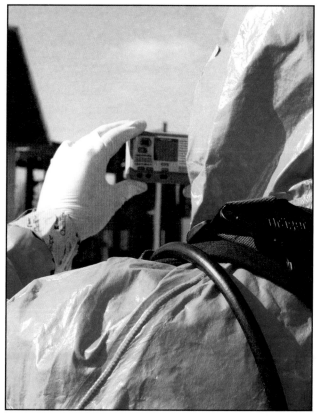

Figure 7.27 If a crime is suspected, take pictures of the incident scene as soon as possible and try to minimize how much the scene is disturbed.

Figure 7.28 Establishing travel routes through the scene can minimize disturbance. *Courtesy of FEMA News Photos, photo by Jocelyn Augustino.*

Evaluating Progress

The final aspect of the APIE process is looking back or evaluating progress. This evaluation is conducted throughout the incident and continues until termination. **Skill Sheet 7-2** provides basic steps for evaluating and reporting progress at a hazardous materials incident. The following sections address:

- Progress reports
- When to withdraw
- Recovery
- Termination

Progress Reports

If an IAP is effective, the IC should receive favorable progress reports from tactical and/or task supervisors, and the incident should begin to stabilize. As new information becomes available and circumstances change, the IC should also reevaluate the plan. If the initial plan is not working, it must be changed either by selecting new strategies or by changing the tactics used to achieve them. In accordance with predetermined communication procedures, first responders should communicate the status of the planned response and the progress of their actions to the IC. Responders must be trained to use the communication equipment they are issued and be familiar with the AHJ's communication procedures.

When to Withdraw

The IAP must be reevaluated and possibly revised should mitigation efforts fail or the situation worsen (or intensify). If the threat of BLEVE or other dangerous situation develops, it may become necessary to withdraw immediately. Some indicators to withdraw include (**Figure 7.29**):

- Sudden change in temperature
- Sudden change in pressure
- Audible indicators of a pressure relief device activating
- Sudden increase in flames

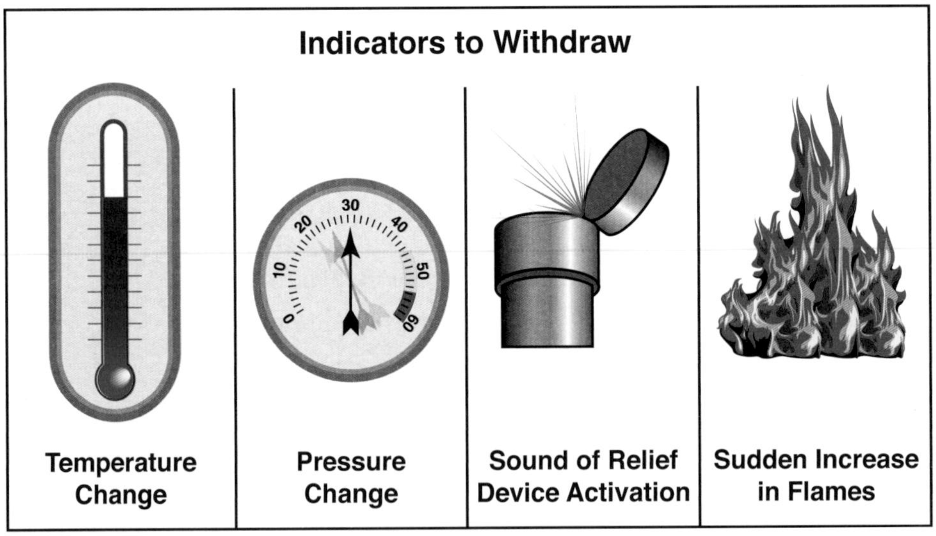

Figure 7.29 Consider withdrawing if you notice a sudden change in temperature, a container's pressure changes suddenly, a relief device activates, or there is a sudden increase in flames.

Indicators to Withdraw

Temperature Change | Pressure Change | Sound of Relief Device Activation | Sudden Increase in Flames

Recovery

Normally, the last strategic goals at a hazardous materials emergency are recovery and termination efforts. Recovery deals with returning the incident scene and responders to a preincident level of readiness. Termination involves documenting the incident and using this information to evaluate the response. The major goals of the recovery phase are as follows:

- Return the operational area to a safe condition.
- Debrief personnel before they leave the scene.
- Return the equipment and personnel of all involved agencies to the condition they were in before the incident.

On-Scene Recovery

On-scene recovery efforts aim to return the scene to a safe condition. These activities may require the coordinated effort of numerous agencies, technical experts, and contractors. Generally, fire and emergency services organizations do not conduct remedial cleanup actions unless those actions are absolutely necessary to eliminate conditions that present an imminent threat to public health and safety. If such imminent threats do not exist, contracted remediation firms under the oversight of local, state/provincial and federal environmental regulators generally provide for these cleanup activities. In these situations, the fire and emergency services organization may also provide control and safety oversight according to local SOPs.

On-Scene Debriefing

On-scene debriefing, conducted in the form of a group discussion, gathers information from all operating personnel, including law enforcement, public works, and EMS responders. During the debriefing stage, responders should obtain the following information:

- Important observations
- Actions taken
- Timeline of those actions

During the hazardous communication briefing (required by OSHA in the U.S.), provide personnel with information concerning the signs and symptoms of overexposure to the hazardous materials involved in the incident. It is important that this debriefing process be thoroughly documented. Each person attending must receive and understand the instructions and sign a document certifying that the information was both received and understood. Provide the following information to responders before they leave the scene:

- Identity of material involved
- Potential adverse effects of exposure to the material
- Actions to be taken for further decontamination
- Signs and symptoms of an exposure
- Mechanism by which a responder can obtain medical evaluation and treatment
- Exposure documentation procedures

Operational Recovery

Operational recovery involves actions necessary to return resource forces to preincident readiness. These actions involve:

- Release of units
- Resupply of materials and equipment
- Decontamination of equipment and PPE
- Preliminary actions necessary for obtaining financial restitution

The financial effect of hazardous materials emergencies can greatly exceed that of any other activity conducted by the fire and emergency services. Normally, a fire and emergency services organization's revenues obtained from taxes or subscriber fees are calculated based upon the equipment and personnel necessary to conduct fire suppression and other emergency activities. Communities should have in place the necessary ordinances to allow for the recovery of costs incurred from such emergencies. A vital part of this process is to document costs through the Unit Log and other tracking mechanisms.

Termination

In order to conclude an incident, the IC must ensure that all strategic goals have been accomplished and the requirements of laws have been met. Documentation, analysis, and evaluation must be completed. The termination phase involves two procedural actions: **postincident critiques** and **postincident analysis (PIA)**.

Postincident Critique

OSHA Title 29 *CFR* 1910.120 mandates that incidents are critiqued for the purposes of identifying operational deficiencies and learning from mistakes. As with all critiques performed by fire and emergency services, hazardous materials incident critiques need to occur as soon as possible after the incident and involve all responders, including law enforcement, public works, and EMS responders. As with other administrative and emergency-response functions, documentation of the critique lists those individuals in attendance as well as any operational deficiencies that were identified.

Postincident Analysis

The postincident analysis process compiles the information obtained from the debriefings, postincident reports, and critiques to identify trends regarding operational strengths and weaknesses. Once trends have been identified, recommendations for improvements are made.

Recommendations during this analysis may include several categories:

- Operational weaknesses
- Training needs
- Necessary procedural changes
- Required additional resources
- Necessary updates and/or required changes

    The postincident analysis also includes:

- Completion of necessary reporting procedures required to document personal exposures
- Equipment exposures
- Incident reports
- Staff analysis reports
- Change or improvement benchmarked for further consideration
- Follow-up analysis or training

Postincident analysis forms the basis for improved response. Therefore, schedule follow-up analysis or training to ensure successful implementation.

Chapter Review

Answer the following questions to review the information provided in this chapter.

1. List five major organizational functions of NIMS-ICS that help initiate incident management and identify their major responsibilities or activities.

2. List four secondary organizational functions of NIMS-ICS and identify their major responsibilities or activities.

3. What are common hazmat response objectives and options that may be assigned?

4. Explain how incident recovery and incident termination can aid in incident evaluation.

Step 1: Take protective actions.

Step 2: Establish an incident management system.

Step 3: Develop and implement an IAP.

Step 4: Establish and ensure scene control.

Step 5: Select and use personal protective equipment .

Step 6: Protect exposures and personnel.

Step 7: Ensure safety procedures are followed.

Step 8: Minimize or avoid hazards.

Step 9: Identify potential evidence.

Step 10: Preserve potential evidence.

Step 11: Complete assignment(s).

Step 1: Determine incident status.

Step 2: Evaluate IAP.

Step 3: Determine action objectives are being met.

Step 5: Communicate progress to supervisor using approved communication tools and equipment.

Step 4: Evaluate effectiveness of assigned tasks.

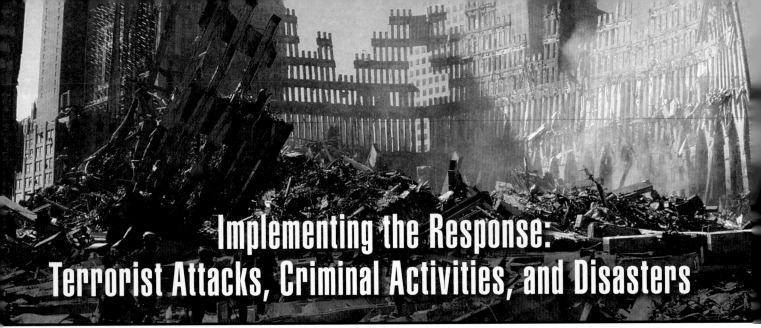

Implementing the Response: Terrorist Attacks, Criminal Activities, and Disasters

Chapter Contents

Courtesy of FEMA News Photo, photo by Michael Rieger.

Key Terms

NFPA Job Performance Requirements

This chapter provides information that addresses the following job performance requirements of NFPA 1072, *Standard for Hazardous Materials/Weapons of Mass Destruction Emergency Response Personnel Professional Qualifications (2017).*

5.2.1   5.3.1

Implementing the Response:
Terrorist Attacks, Criminal Activities, and Disasters

Learning Objectives

After reading this chapter, students will be able to:

1. Define terrorism. (5.2.1)

2. Explain ways of identifying terrorist attacks. (5.2.1)

3. Describe the range of tactics that may be used in a terrorist attack. (5.2.1)

4. Identify indicators and types of explosive attacks and devices. (5.2.1)

5. Identify indicators and types of chemical attacks. (5.2.1)

6. Identify indicators and types of biological attacks. (5.2.1)

7. Identify indicators and types of possible radiological attacks. (5.2.1)

8. Identify general hazards at illicit laboratories. (5.2.1)

9. Recognize illegal hazmat dumps. (5.2.1)

10. Describe hazmat operations after disasters. (5.2.1)

Chapter 8
Implementing the Response: Terrorist Attacks, Criminal Activities, and Disasters

This chapter explains the following:

- What is terrorism?
- Terrorism and emergency response
- Terrorist tactics and types of attacks
- Explosive attacks
- Chemical attacks
- Biological attacks
- Radiological and nuclear attacks
- Illegal hazmat dumps
- Hazmat during and after disasters

What Is Terrorism?

Terrorists have the knowledge and the capability to strike anywhere in the world, and they deliberately target locations where civilians are present. All societies are vulnerable to incidents involving terrorism, although countries in conflict are especially vulnerable.

The U.S. Federal Bureau of Investigation (FBI) heads U.S. government agencies in investigating and attempting to prevent terrorist attacks on U.S. soil. There are many different definitions of terrorism; however, the FBI defines *terrorism* as the unlawful use of force against persons or property to intimidate or coerce a government, the civilian population, or any segment thereof, in the furtherance of political or social objectives. Under this definition, all terrorist activities share the following three commitments:

1. Using force that involves illegal activities
2. Intimidation or coercion
3. Supporting political or social objectives

Under other definitions, terrorism may *not* require the use of force. Terrorism may be defined as the unlawful or threatened use of force or violence against individuals or property to coerce and intimidate governments or societies, often to achieve political, religious, or ideological objectives. The decision to engage in criminal, intimidating activities to achieve goals separates a terrorist organization from a legitimate organization. However, any organization, legitimate or not, can resort to terrorist means to achieve its political or social agenda. Terrorists can operate as a group or act alone.

Terrorist organizations plan activities that will have an emotional effect on the target population. They desire the target population to react to their attacks and demands in a manner that furthers their goals. Terrorism is designed to cause disruption, fear, and panic. Terrorists may want to draw attention to their cause, coerce or intimidate governments into granting their demands, or provoke governments into repressive actions that may inspire oppressed masses to revolt.

Terrorists are difficult to stop, even when security precautions are taken and attacks are expected. An act of terrorism can occur anywhere, at any time. Terrorists will attack targets on land, sea, or air **(Figure 8.1)**. No jurisdiction — urban, suburban, or rural — is immune from terrorist acts.

Figure 8.1 Terrorists will attack anywhere they detect vulnerability — on land, in the air, or at sea. Terrorists attacked the USS Cole while it was refueling in the port of Aden, Yemen. *Courtesy of the U.S. Department of Defense.*

Terrorism and Emergency Response

During the initial part of the response, emergency responders may not know if the incident is a terrorist act or something else. All incidents (terrorist or other) have the same priorities of life safety, incident stabilization, and protecting property and the environment. The same Incident Management System applies to all operations. Emergency responders will still respond and be among the first on the scene and consider traditional strategies and tactics to manage the incident. Responders will use the same risk-based response procedures to ensure safety and protection of themselves and the public. The size and type of incident plays a key role in how the response is managed. It may take some time before responders identify an incident as a terrorist attack.

Many emergency response organizations may have supplies, equipment, and emergency response plans for other disasters that will work for terrorist incidents. PPE used for illicit drug lab responses and other hazmat incidents may also provide protection at a terrorist incident, depending on the materials involved. Similarly, decontamination tents, trailers, and equipment will serve the same purpose at terrorist incidents as at hazmat incidents. Responders can adapt existing evacuation plans to suit conditions created during a terrorist incident.

Targeted Versus Nontargeted Incidents

Despite their many similarities, important differences exist between terrorist incidents and other emergencies. Differences between nontargeted emergencies and a targeted attack are explained in the following:

- **Intent** — An act of terrorism is intended to cause damage, inflict harm, and kill. Terrorists specifically target the public, first responders, or both. Most other emergency incidents are not criminal in nature. A secondary attack is a deliberate release of hazardous materials that may cause widespread harm, regardless of any intent to target any audience.

- **Severity and Complexity** — Terrorist events may involve large numbers of casualties. They may involve materials, such as radioactive materials, with which first responders have little experience. Secondary contamination from handling patients may present a hazard. Structural collapse and other dangers may occur significantly after the initial attack. Issues, such as securing the scene and managing the incident, may be especially complex and difficult because of the large area involved **(Figure 8.2)**.

- **Crime Scene Management** — At terrorist incidents, responders must preserve evidence and notify law enforcement as soon as possible **(Figure 8.3)**. Responders must quickly recognize a terrorist attack. Failure to act quickly could result in the loss or accidental destruction of valuable information. Evidence preservation is explained in Chapter 14.

- **Command Structure** — A Unified Command Structure is required at most terrorist incidents. Law enforcement will have jurisdiction over all incidents involving terrorism.

Figure 8.2 Terrorist attack incident scenes may be very dangerous to responders. *Courtesy of FEMA News Photos; photo by Mike Rieger.*

Figure 8.3 Responders must recognize potential evidence and avoid disturbing it if possible. *Courtesy of the U.S. Navy; photo by Journalist 1st Class Mark D. Faram.*

Identification of Terrorist Attacks

The following are a few examples of situations that can cue the responder to consider the possibility of a terrorist attack:

- Report of two or more medical emergencies in public locations such as a shopping mall, transportation hub, mass transit system, office building, assembly occupancy, or other public buildings

- Unusually large number of people with similar medical signs and symptoms arriving at physicians' offices or medical emergency rooms

- Reported explosion at a movie theater, department store, office building, government building, or a location with historical or symbolic significance

CBRNE — Abbreviation for Chemical, Biological, Radiological, Nuclear, and Explosive. These categories are often used to describe WMDs and other hazardous materials characteristics.

Additional information can provide clues as to the type of attack. **CBRNE** attacks (chemical, biological, radiological, nuclear, and explosive attacks) each have their own unique indicators **(Table 8.1)**. Monitoring and detection devices also play an important role in determining which of these materials may be present at the incident scene. If criminal or terrorist activity is suspected at an incident, first responders must quickly forward that information to law enforcement representatives.

Terrorist Tactics and Types of Attacks

Agroterrorism — Terrorist attack directed against agriculture, such as food supplies or livestock.

Cyber Terrorism — Premeditated, politically motivated attack against information, computer systems, computer programs, and data which result in violence against noncombatant targets by subnational groups or clandestine agents.

Traditional terrorist tactics include assassination, armed assault, and bombings (including suicide bombings). Some conventional attacks may produce devastating effects equal to or exceeding those produced by the use of weapons of mass destruction (WMD). For example, assassination of a political leader could affect regime stability, or the use of conventional weapons could produce mass casualties and destruction exceeding the response capability of the community. New tactics such as **cyber terrorism** and **agroterrorism** (also called agricultural terrorism) present threats to computer/network security and food supplies.

Table 8.1
Terrorist Attacks at a Glance

Chemical Attack	Biological Attack
• Victims in a concentrated area • Symptoms immediate (seconds to hours after exposure) • Symptoms very similar (SLUDGEM) • May have observable features such as chemical residue, dead foliage, dead animals/insects, and pungent odor	• Victims dispersed over a wide area • Symptons delayed (days — weeks after exposure) • Symptoms most likely vague and flu-like • No observable features
Explosive Attack	**Radiological Attack**
• Explosion self-evident (debris field, fire, etc.) • Victims in a concentrated area • Mechanical and thermal injuries • Potential radiation and chemical agent risk — monitoring for both is necessary	• Explosion self-evident (debris field, fire, etc.) • Victims in a concentrated area • Mechanical and thermal injuries initially, radiological symptoms (if any) will likely be delayed • Radiation detected through monitoring

Potential Terrorist Targets

Certain occupancies are more likely to be terrorist targets than others. Terrorists are likely to target locations where an attack has the potential to do the greatest harm, such as:

• Killing or injuring persons

• Causing panic and/or disruption

• Damaging the economy

• Destroying property

• Demoralizing the community

When the goal is to kill as many people as possible, any location or occupancy that has large public gatherings, such as football stadiums, sports arenas, theaters, and shopping malls, might become a potential target. Terrorists might also target places with historical, economic, or symbolic significance such as local monuments, high-profile buildings, or high-traffic bridges. Examples of potential terrorist targets include:

• **Mass transportation** — Airports, ferry terminals and buildings, maritime port facilities, planes, subways, buses, commuter trains, and mass transit stations

• **Critical infrastructure** — Dams, water treatment facilities, power plants, electrical substations, nuclear power plants, trans-oceanic cable landings, telecommunication switch centers (telecom hotels), financial institutions, rail and road bridges, tunnels, levees, liquefied natural gas (LNG) terminals, natural gas (NG) compressor stations, petroleum pumping stations, and petroleum storage tank farms

- **Areas of public assembly and recreation** — Convention centers, hotels, casinos, shopping malls, stadiums, and theme parks

- **High profile buildings and locations** — Monuments, buildings/structures of historic or national significance, and high-rise buildings

- **Industrial sites** — Chemical manufacturing facilities, shipping facilities, and warehouses

- **Educational sites** — Colleges, universities, community colleges, vocational/training facilities, and primary and secondary schools

- **Medical and science facilities** — Hospitals, clinics, nuclear research labs, other research facilities, nonpower nuclear reactors, and national health stockpile sites

Scrutinize reported incidents at these occupancies closely for potential terrorist involvement. If you suspect terrorism, notify law enforcement authorities immediately.

Expansion of Terrorism

Experts fear that terrorists have the means to broaden their tactics to include the use of weapons of mass destruction (WMD). According to the U.S. Government (United States Code, Title 50, Chapter 40, Section 2302, and Title 18, Part I, Chapter 113B, Section 2332a), the term *weapon of mass destruction* means any weapon or device that is intended to or has the capability to cause death or serious bodily injury to a significant number of people through the release, dissemination, or impact of one of the following means:

- Toxic or poisonous chemicals or their precursors

- A disease organism

- Radiation or radioactivity

Other parts of the U.S. Code also include certain explosive and incendiary devices in the definition.

Other Acronyms for WMDs

While this chapter explains types of attacks based on the CBRNE acronym, other organizations may use other acronyms to indicate essentially the same thing. These terms include COBRA (chemical, ordinance, biological, radiological agents); B-NICE (biological, nuclear, incendiary, chemical, explosive); and NBC (nuclear, biological, chemical) in addition to others. This manual will often refer to *CBR materials* when describing chemical, biological, and radiological attacks.

Many WMDs present significant hurdles that must be overcome before they can be successfully deployed. Some are difficult to store and must be used very quickly. Other materials, such as botulism toxin, may have to be kept in temperature-controlled environments. The barriers to deployment also present barriers in the production of WMDs. For example, producing these highly sophisticated weapons requires a high level of resources and knowledge. There-

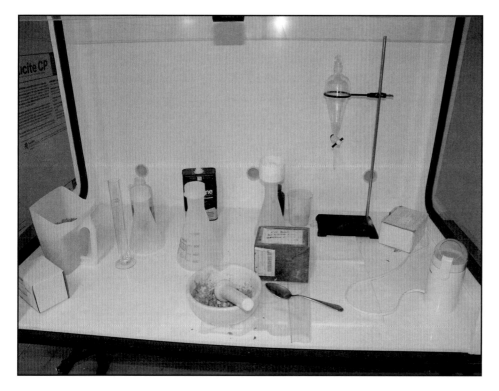

Figure 8.4 Ricin can be made from recipes available on the Internet.

fore, the greatest threat of mass-produced WMD agents comes from nations or organizations with the infrastructure, finances, and scientific knowledge to produce them and not from isolated terrorist groups. Even organizations with the needed resources cannot easily produce WMDs. For many decades, WMDs were difficult to make in uncontrolled conditions. Unfortunately, technologies are evolving to make this kind of hazard more prevalent. Some WMDs are relatively readily produced and/or acquired, such as:

- **Ricin** — A biological toxin made from castor beans. Recipes for making the toxin are available on the Internet **(Figure 8.4)**.

- **Triacetone triperoxide (TATP)** — An explosive that can be made from common household products without expensive laboratory equipment.

- **Foot-and-mouth disease** — Biological agricultural threats that have natural reservoirs in nature.

- **Radiological materials** — Materials can be stolen and/or acquired from a variety of accessible medical and construction sources.

- **Toxic industrial chemicals** — Chemicals are available in every jurisdiction.

WMD Threat Spectrum

Experts have not reached consensus on which types of WMDs first responders are most likely to encounter. However, given the availability of parts, relative ease of production, and ease of deployment, the following list is a probable WMD threat spectrum from most likely to least likely:

1. **Explosives** — Such as IEDs, vehicle bombs, and suicide bombers. Also explosives potentially combined with other materials such as industrial chemicals, biological materials, or radiological materials **(Figure 8.5, p. 368)**

2. **Biological toxins** — Such as ricin

3. **Industrial chemicals** — Such as chlorine and phosgene

Triacetone Triperoxide (TATP) — Triacetone triperoxide (TATP) is typically a white crystalline powder with a distinctive acrid (bleach) smell and can range in color from a yellowish to white color. *Similar to* Hexamethylene triperoxide diamine (HMTD).

Figure 8.5 Explosives are more commonly used than other WMDs.

4. **Biological pathogens** — Such as contagious diseases

5. **Radiological materials** — Such as those used in a radiological dispersal device

6. **Military-grade chemical weapons** — Such as nerve agents

7. **Nuclear weapons** — Such as nuclear bombs

NOTE: Conventional attacks such as hijackings, sniper attacks, and/or shootings are also highly likely, but not considered a WMD threat for purposes of this list.

Explosives and conventional attacks have been terrorists' weapons of choice throughout history, and most experts agree that explosives are the greatest WMD threat today. This manual will present this spectrum in the following order:

- Explosive attacks

- Chemical attacks

- Biological attacks

- Radiological and nuclear attacks

Ranking of WMD Threats

There is no way to predict which types of terrorist or WMD attacks will occur with what frequency. Several organizations have developed a framework to guide how they teach these topics, but no single framework is necessarily more accurate than any other.

Secondary Attacks and Booby Traps

The use of secondary devices at terrorist attacks or illicit laboratories is always a possibility. Secondary devices are often designed to affect an ongoing emergency response in order to create more chaos and injure responders and bystanders. Booby traps may be set to protect illicit laboratories **(Figure 8.6)**. Usually, secondary devices are explosives of some kind, most likely an improvised explosive device (IED). Booby traps utilizing other weapons are also possible, and some may use chemical, biological, or radiological materials. Some may use animals such as snakes or guard dogs. Secondary devices may also be deployed as a diversionary tactic to route emergency responders away from the primary attack area.

Secondary devices will be hidden or camouflaged. A time delay may detonate the devices, but other devices are also used, such as radio-controlled and cell-phone-activated devices. In some cases, an obvious IED may be used to lure personnel to a specific area where a less obvious IED is hidden. If any time booby traps or secondary devices are found or suspected, contact law enforcement and/or explosive ordnance disposal (EOD)/bomb squad personnel.

Figure 8.6 Booby traps, like this acid jar, might be set to protect illicit labs. Secondary devices might be set to kill or injure responders at terrorist attacks.

CAUTION

If one device has been found or detonated, always expect another.

To protect against possible secondary devices, use the following guidelines:

- Anticipate the presence of a secondary device at any suspicious incident.

- Conduct a visual search for a secondary device (or anything suspicious) before moving into the incident area.

- Limit the numbers of emergency response personnel to those performing critical tasks (rescue) until the area has been checked and confirmed that no additional devices are present.

- Avoid touching or moving anything that may conceal an explosive device (including items such as backpacks and purses).

- Manage the scene with cordons, boundaries, and scene control zones.

- Evacuate victims and nonessential personnel as quickly as possible.

- Preserve the scene as much as possible for evidence collection and crime investigation.

WARNING!

NEVER approach or move suspicious objects. Notify appropriate personnel (law enforcement/Explosive Ordnance Disposal/bomb squad personnel) and evacuate the area immediately.

While secondary devices and booby traps can be disguised as almost anything, responders should look for things that may seem out of place **(Figure 8.7)**. If anything suspicious is found, responders should note the item, treat the item with appropriate caution, notify appropriate personnel (law enforcement/Explosive Ordnance Disposal/bomb squad personnel), and evacuate the area immediately. Be cautious of any item(s) that arouse(s) curiosity, including the following:

- Containers with unknown liquids or materials

- Unusual devices or containers with electronic components, such as:
  — Wires
  — Circuit boards
  — Cellular phones
  — Antennas
  — Other items attached or exposed

- Devices containing quantities of the following:
  — Fuses
  — Fireworks
  — Match heads
  — Black powder
  — Smokeless powder
  — Incendiary materials
  — Other unusual materials

- Materials, such as nails, bolts, drill bits, and marbles, attached to or surrounding an item that could be used for shrapnel **(Figure 8.8).**

- Ordnance such as blasting caps, detonation cord (detcord), military explosives, commercial explosives, and grenades

- Devices, such as razor blades and trip wires, on containers or other items on handles, valves, ladders, or other locations

- Energized bare electrical wiring or exposed metal surfaces connected to an electrical system

- Any combination of the previously described items

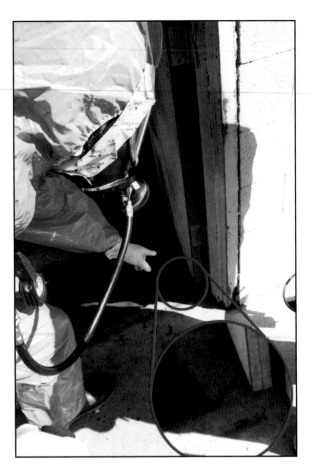

Figure 8.7 Booby traps and secondary devices may be concealed. Therefore, responders should look for things that seem out of place, like this wire leading under the floor mat.

Figure 8.8 This replica shows the nuts and broken glass used as shrapnel in a guitar IED detonated in Israel.

Explosive Attacks

Explosive devices can be anything from homemade pipe bombs to sophisticated military ordnance. For example, the truck bomb that exploded April 19, 1995, outside the Murrah Federal Building in Oklahoma City, killing 168 people and injuring many others, is testimony to the potential destructive power of simple resources. The following sections describe explosive attacks.

Explosive/Incendiary Attack Indicators

The majority of terrorist attacks involves the use of explosive materials and incendiary devices, and typically are considered conventional attacks **(Figure 8.9)**. However, explosives, such as with a car or truck bomb, destroying an occupied building may be classified as *weapons of mass destruction* when used to inflict high casualties. Explosives may also be used to disseminate chemical, biological, and radiological materials.

Figure 8.9 The majority of terrorist attacks utilize conventional weapons such as explosives and incendiary devices like the IEDs pictured here. *Courtesy of the U.S. Department of Defense.*

Explosive/incendiary attack indicators include:

- Warning or threat of an attack or received intelligence
- Reports of an explosion
- Explosion
- Accelerant odors (such as gasoline)
- Multiple fires or explosions
- Incendiary device or bomb components (such as broken glass from a Molotov cocktail or wreckage of a car bomb)
- Unexpectedly heavy burning or high temperatures
- Unusually fast burning fires
- Unusually colored smoke or flames
- Presence of propane or other flammable gas cylinders in unusual locations
- Unattended packages, backpacks, or objects left in high traffic/public areas
- Fragmentation damage/injury
- Damage that exceeds the level usually seen during gas explosions, including shattered reinforced concrete or bent structural steel **(Figure 8.10)**
- Crater(s)
- Scattering of small metal objects such as nuts, bolts, and/or nails used as shrapnel

Figure 8.10 Car and truck bombs can do greater damage than accidental gas explosions. Indicators include shattered reinforced concrete and bent structural steel. *Courtesy of U.S. Air Force; photo by Senior Airman Sean Worrell.*

Anatomy of an Explosion

An explosive (or energetic) material is any material or mixture that will undergo an extremely fast, self-propagating reaction when subjected to some form of energy **(Figures 8.11a and b)**. As described in Chapter 4, explosive materials react when they combine an oxidizing component with a fuel component.

An explosion results when a material undergoes a physical or chemical reaction that releases rapidly expanding gases. These gases form almost instantaneously. The pressure from the expanding gases compresses the surrounding atmosphere into a shock front that is sometimes visible, expanding outward from the point of detonation. The pressure wave formed can demolish almost anything in its path.

There are actually two phases to a blast-pressure wave: the positive-pressure phase and the negative-pressure phase (sometimes called the *suction phase*). Both phases cause damage.

The following occurs in the positive-pressure phase:

- The shock front leads the positive-pressure wave, striking anything in its path with a hammering force.

- The positive-pressure wave continues outwards in an expanding radius until its energy diminishes.

- The energy dissipates due to distance or because it transfers to objects (such as buildings) standing in its path.

After the initial expansion energy dissipates, a negative pressure or suction phase is created when the following occurs:

- Displaced atmosphere rushes back in to fill the vacuum left at the center of the explosion. This rush of air also has destructive power although not to the same degree as the positive-pressure wave.

- Structures damaged in the initial blast can be further damaged in the negative-pressure phase.

- An explosion's negative-pressure phase lasts about three times longer than the positive-pressure phase.

Figures 8.11a and b Black powder **(a)** and TNT **(b)** are examples of low and high explosives, respectively.

As explained in Chapter 4, multiple components of explosions cause destruction. The blast pressure phase is only one of those components. The rapid release of energy may throw debris and shrapnel outwards. The shock wave may travel through the ground, creating a seismic disturbance. The explosion may release thermal heat energy in the form of a fireball.

The quantity and type of explosives determine the size of an explosion. The hot zone may be limited to a small area or it could extend for blocks. An explosion can also create secondary hazards from breaking glass and any degree of structural collapse.

Classification of Explosives

Most commonly, explosives are categorized by chemical reaction or rate of decomposition. Only Division 1.1 and Division 1.4 explosives are used in explosive attacks. In general, high explosives create a larger effect, in sound and size, than low explosives.

High Explosives

High explosives decompose rapidly (almost instantaneously) in a **detonation** that can include velocities faster than the speed of sound. High explosives are placarded as DOT Division Number 1.1.

Examples of high explosives that are available for legal purchase include:

- Plastic explosives, such as C4 and C3 **(Figure 8.12)**
- Nitroglycerin
- TNT **(Figure 8.13, p. 874)**
- Blasting caps
- Dynamite
- **Ammonium nitrate and fuel oil (ANFO)** and other blasting agents **(Figure 8.14, p. 374)**

High Explosive — Explosive that decomposes extremely rapidly (almost instantaneously) and has a detonation velocity faster than the speed of sound.

Detonation — Explosion with an energy front that travels faster than the speed of sound.

Ammonium Nitrate and Fuel Oil (ANFO) — High explosive blasting agent made of common fertilizer mixed with diesel fuel or oil; requires a booster to initiate detonation.

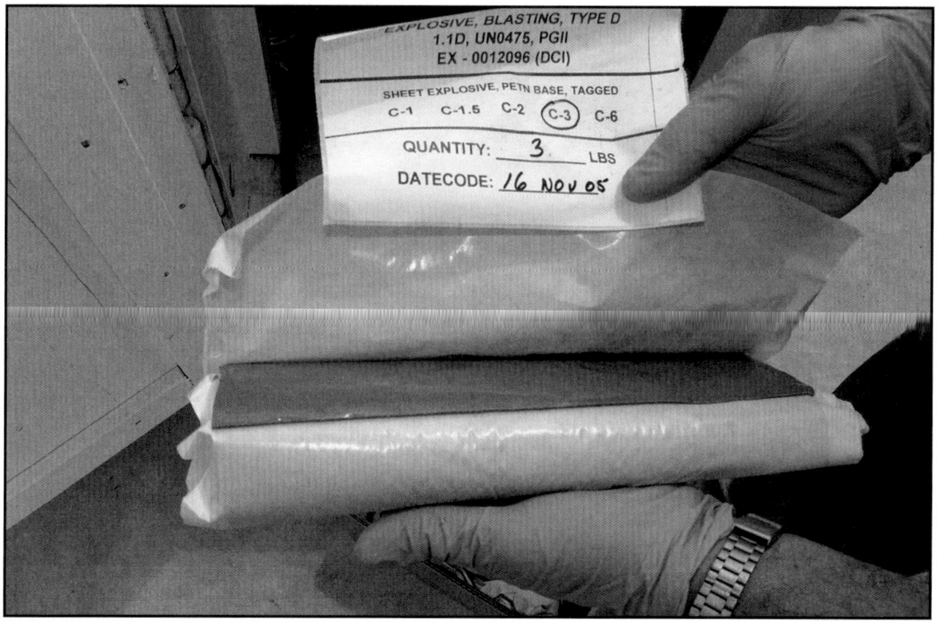

Figure 8.12 C3 is a plastic explosive.

Low Explosive — Explosive material that deflagrates, producing a reaction slower than the speed of sound.

Deflagrate — To explode (burn quickly) at a rate of speed slower than the speed of sound.

Incendiary Device — (1) Contrivance designed and used to start a fire. (2) Any mechanical, electrical, or chemical device used intentionally to initiate combustion and start a fire. *Also known as* Explosive Device.

Primary Explosive — High explosive that is easily initiated and highly sensitive to heat; often used as a detonator. *Also known as* Initiation Device.

Detonator — Device used to trigger less sensitive explosives, usually composed of a primary explosive; for example, a blasting cap. Detonators may be initiated mechanically, electrically, or chemically.

Secondary Explosive — High explosive that is designed to detonate only under specific circumstances, including activation from the detonation of a primary explosive. *Also known as* Main Charge Explosive.

Tertiary Explosive — High explosive that require initiation from a secondary explosive. Tertiary explosives are often categorized with secondary explosives. *Also known as* Blasting Agents.

Low Explosives

Low explosive materials decompose rapidly but do not produce an explosive effect unless they are confined. In other words, they **deflagrate** at a speed slower than the speed of sound. Low explosives confined in small spaces or containers are commonly used as propellants. Low explosives are placarded as DOT Division Number 1.4.

Black powder, a low explosive, is used to propel bullets and fireworks. Other examples of low explosives are the pyrotechnic substances used in fireworks and road flares. Some agencies may refer to unconfined low explosives as *incendiary materials*. Many experts do not distinguish **incendiary devices**/materials from other low explosives.

Primary and Secondary Explosives

Primary explosives are generally more sensitive than secondary explosives **(Figure 8.15)**. Emergency responders should also be familiar with the following classifications based on high explosives' susceptibility to initiation (or sensitivity):

Figure 8.13 TNT detonates faster than the speed of sound.

Figure 8.14 ANFO can be purchased legally. *Courtesy of David Alexander; Texas Commission on Fire Protection.*

- **Primary explosives** — Easily initiated and highly sensitive to heat and usually used as **detonators**. Small amounts of primary explosives, even a single grain or crystal, can detonate. Examples of primary explosives are lead azide, mercury fulminate, and lead styphnate.

- **Secondary explosives** — Designed to detonate only under specific circumstances usually by activation energy from a primary explosive. Secondary explosives are less sensitive than primary explosives to initiating stimuli, such as heat or flame. TNT is an example of a secondary explosive.

- **Tertiary explosives (blasting agents)** — Insensitive materials based on ammonium nitrate (AN); they usually require initiation from a secondary explosive. Not all experts recognize this category and would consider blasting agents to be secondary explosives.

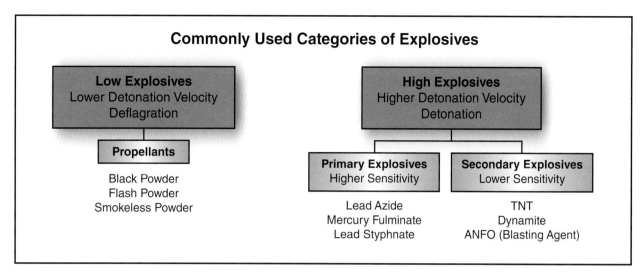

Figure 8.15 Commonly used categories of explosives.

Commercial/Military Explosives

Commercial and military explosives are normally used for such legitimate purposes as mining, demolition, excavation, construction, and military applications **(Table 8.2, p. 376)**. Unfortunately, criminals and terrorists may also attempt to steal and use explosives.

Terrorists may also use military **munitions** in explosive attacks, which may include:

- Mortars
- Grenades
- Antipersonnel mines
- Surface-to-air missiles
- Rocket propelled grenades
- Other types of military explosives

Munitions — Military reserves of weapons, equipment, and ammunition.

Homemade/Improvised Explosive Materials

Nonmilitary first responders are more likely to encounter homemade or improvised explosive materials than military weapons in their day-to-day response activities. Responders typically stage 300 meters (1 000 feet) away from a suspected explosive material incident.

Improvised explosive materials are typically made by combining an oxidizer with a fuel **(Figure 8.16, p. 377)**. Many of these materials are fairly simple to make and require very little technical expertise or specialized equipment. However, the explosive materials created are often highly unstable. More than one potential terrorist has died trying to make **homemade explosives (HME)**.

The following sections describe peroxide-based explosives, chlorate-based explosives, and nitrate-based explosives. These categories do not represent a comprehensive list; many oxidizers and fuels can be combined to form improvised explosive materials.

NOTE: Page 374 of the *2016 Emergency Response Guidebook* provides an IED Safe Stand-Off Distance table.

Homemade Explosive (HME) — Explosive material constructed using common household chemicals. The finished product is usually highly unstable.

Table 8.2
Commercial Explosives

Ammonium Nitrate

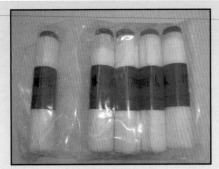

Binary Explosives

Black Powder

Blasting Caps

C-4

C-3 Sheet Explosive

DET Cord–RDX

DET Cord–PETN

Dynamite

Semtex

PETN

TNT

HMX

Composition B (60% RDX, 40% TNT

WARNING!
Never handle commercial or military explosives.

Components of Improvised Explosives

🔥 Potential Fuels   +   🔥 Potential Oxidizers   =   💥 Explosive Blends (Oxidizer + Fuel)

Hydrocarbons:
Alcohol
Carbon Black
Charcoal
Dextrin
Diesel
Ethylene Glycol
Gas
Kerosene
Naphtha
Rosin
Sawdust
Shellac
Sugar
Vaseline
Wax/Paraffin

Energetic Hydrocarbons:
Nitrobenzene
Nitromethane
Nitrocellulose

Elemental "Hot" Fuels:
Powdered Metals
- Aluminum
- Magnesium
- Zirconium
- Copper
Phosphorus
Sulfur
Antimony Trisulfide

Oxidizers:
Perchlorate
Chlorate
Hypochlorite
Nitrate
Peroxide
Iodate
Chromate
Dichromate
Permaganate
Sodium Chlorate
Potassium Chlorate
Ammonium Nitrate
Potassium Nitrate
Hydrogen Peroxide
Barium Peroxide
Ammonium Perchlorate
Calcium Hypochlorite
Nitric Acid
Lead Iodate
Sodium Chlorate
Potassium Permanganate
Lithium Chromate
Potassium Dichromate

Nitrate Blends:
ANFO (Ammonium Nitrate + Diesel Fuel)
ANAI (Ammonium Nitrate + Aluminum Powder)
ANS (Ammonium Nitrate + Sulfur Powder
ANIS (Ammonium Nitrate + Icing Sugar)
Black Powder (Potassium Nitrate + Charcoal + Sulfur)

Chlorate/Perchlorate Blends:
Flash Powder (Potassium Chlorate/Perchlorate + Aluminum Powder + Magnesium Powder + Sulfur)
Poor Man's C-4 (Potassium Chlorate + Vaseline)
Armstrong's Mixture (Potassium Chlorate + Red Phosphorus)

Liquid Blend:
Hellhoffite (Nitric Acid + Nitrobenzene)

Common Precursors Used To Make Explosives

🦅 **Precursors:**
Hydrogen Peroxide
Sulfuric Acid (battery acid)
Nitric Acid
Hydrochloric Acid (muriatic acid)
Urea
Acetone
Methyl Ethyl Ketone
Alcohol (Ethyl or Methyl)
Ethylene Glycol (antifreeze)
Glycerine
Hexamine (camp stove tablets)
Citric Acid (sour salt)

💥 **Nitrated Explosives:**
Nitroglycerine (Glycerine + Mixed Acid [Nitric Acid + Sulfuric Acid])
Ethylene Glycol Dinitrate (EGDN) (Ethylene Glycol + Mixed Acid [Nitric Acid + Sulfuric Acid])
Methyl Nitrate (Methyl Alcohol [methanol] + Mixed Acid [Nitric Acid + Sulfuric Acid])
Urea Nitrate (Urea + Nitric Acid)
Nitrocotton (Gun Cotton) (Cotton + Mixed Acid [Nitric Acid + Sulfuric Acid])

Peroxide Explosives:
Triacetone Triperoxide (TATP) (Acetone + Hydrogen Peroxide + Strong Acid [Sulfuric, Nitric, or Hydrochloric])
Hexamethylene Triperoxide Diamine (HMDT) (Hexamine + Hydrogen Peroxide + Citric Acid)
Methyl Ethyl Ketone Peroxide (MEKP) (Methyl Ethyl Ketone + Hydrogen Peroxide + Strong Acid [Sulfuric, Nitric, or Hydrochloric])

Figure 8.16 Most homemade explosives are made by combining an oxidizer with a fuel.

Peroxide-Based Explosives

Peroxide-based explosives are made by mixing concentrated hydrogen peroxide, acetone, and either hydrochloric or sulfuric acid. Peroxide-based explosives include acetone peroxide (triacetone triperoxide or TATP) and **hexamethylene triperoxide diamine (HMTD) (Figure 8.17)**. Both TATP and HMTD are unstable during the manufacturing process and also as a finished product. This lack of stability makes them dangerous to make and handle. On the other hand, specialized equipment is not required for the manufacture of TATP and HMTD, so they can be produced almost anywhere **(Figures 8.18a and b)**.

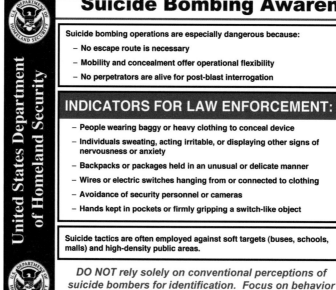

Figure 8.17 TATP looks like many common, white, powdery substances, but it may be very unstable.

Figures 8.18a and b (a) Peroxide-based explosives labs may have quantities of acetone and hydrogen peroxide. **(b)** Responders need to know the common indicators of suicide bombers. *Courtesy of the U.S. Department of Homeland Security.*

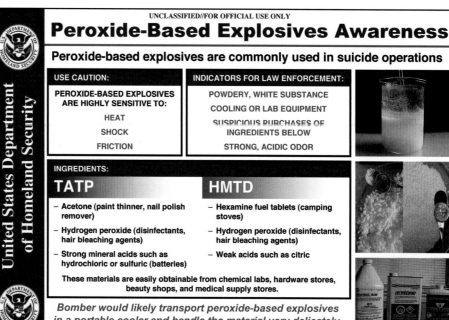

a

UNCLASSIFIED//FOR OFFICIAL USE ONLY

Peroxide-Based Explosives Awareness

Peroxide-based explosives are commonly used in suicide operations

United States Department of Homeland Security

USE CAUTION:
PEROXIDE-BASED EXPLOSIVES ARE HIGHLY SENSITIVE TO:
HEAT
SHOCK
FRICTION

INDICATORS FOR LAW ENFORCEMENT:
POWDERY, WHITE SUBSTANCE
COOLING OR LAB EQUIPMENT
SUSPICIOUS PURCHASES OF INGREDIENTS BELOW
STRONG, ACIDIC ODOR

INGREDIENTS:

TATP
- Acetone (paint thinner, nail polish remover)
- Hydrogen peroxide (disinfectants, hair bleaching agents)
- Strong mineral acids such as hydrochloric or sulfuric (batteries)

HMTD
- Hexamine fuel tablets (camping stoves)
- Hydrogen peroxide (disinfectants, hair bleaching agents)
- Weak acids such as citric

These materials are easily obtainable from chemical labs, hardware stores, beauty shops, and medical supply stores.

Bomber would likely transport peroxide-based explosives in a portable cooler and handle the material very delicately.

UNCLASSIFIED//FOR OFFICIAL USE ONLY

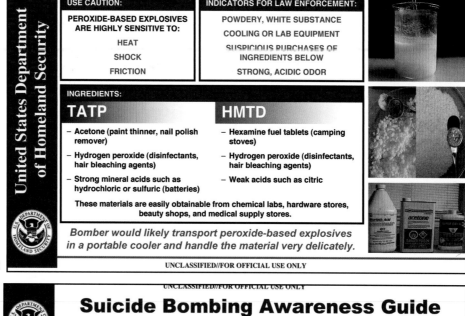

b

UNCLASSIFIED//FOR OFFICIAL USE ONLY

Suicide Bombing Awareness Guide

United States Department of Homeland Security

Suicide bombing operations are especially dangerous because:
- No escape route is necessary
- Mobility and concealment offer operational flexibility
- No perpetrators are alive for post-blast interrogation

INDICATORS FOR LAW ENFORCEMENT:
- People wearing baggy or heavy clothing to conceal device
- Individuals sweating, acting irritable, or displaying other signs of nervousness or anxiety
- Backpacks or packages held in an unusual or delicate manner
- Wires or electric switches hanging from or connected to clothing
- Avoidance of security personnel or cameras
- Hands kept in pockets or firmly gripping a switch-like object

Suicide tactics are often employed against soft targets (buses, schools, malls) and high-density public areas.

DO NOT rely solely on conventional perceptions of suicide bombers for identification. Focus on behavior and situation context.

UNCLASSIFIED//FOR OFFICIAL USE ONLY

Chlorate-Based Explosives

Improvised Explosive Devices (IEDs) may contain chlorate-based oxidizers. Chlorate-based oxidizers commonly take the form of a white crystal or powder that must be mixed with a fuel source. Chlorates are a common ingredient in some fireworks and can be purchased in bulk from fireworks and chemical supply houses. Many manufacturing processes and many products use chlorates, including printing, dying, steel, weed killer, matches, and explosives.

Nitrate-Based Explosives

Some IEDs may contain nitrate-based oxidizers, and some may already have a fuel source included, such as black powder and smokeless powder. Other explosives may require the addition of a separate fuel source. Nitrates are commonly found in ammonium nitrate, and fertilizers **(Figure 8.19)**.

Figure 8.19 Ammonium nitrate is a source of nitrates. *Courtesy of Texas Commission on Fire Protection.*

Improvised Explosive Devices (IEDs)

Depending on sophistication, IEDs are relatively easily to make and can be constructed in almost any location or setting. IEDs are not commercially manufactured; they are homemade. They are usually constructed for a specific target and can be contained within almost any object **(Figures 8.20a-d, p. 000)**.

Inexperienced designers may create IEDs that fail to detonate or, in some cases, detonate during the building process or when being moved or placed. Some bomb makers specialize in IED manufacture and make more sophisticated varieties. These sophisticated devices may be constructed with components scavenged from conventional munitions and standard consumer electronics components, such as speaker wire, cellular phones, or garage door openers **(Figure 8.21, p. 381)**. IEDs often include nails, tacks, broken glass, bolts, and other items that will cause additional shrapnel damage and fragmentation injuries.

IEDs In Different Objects

Figures 8.20a-d IEDs can be concealed as almost anything, as these IED replicas show.

What This Means To You

Targeted Attacks

Targeted attacks in the modern world routinely use unique combinations of common materials. As in any other kind of threat, first responders should consider the whole scene while evaluating whether an item may be a disguised danger. Be suspicious of items out of place.

Identification of IEDs

The bomber's imagination is the only limitation to the design and implementation of IEDs. Be cautious of any item(s) that attract attention because they seem out of place, anomalous, out of the ordinary, curious, suspicious, out of context, or unusual.

Figure 8.21 Ordinary objects, such as garage door openers, wires, and other electronic components, can be used to build IEDs. *Courtesy of the U.S. Army; photo by Spc. Ben Brody.*

IEDs may be placed anywhere. Usually, bombers try to avoid detection when placing IEDs. The level of security and awareness of the public, security forces, and employees affect where and how a terrorist will place an IED.

WARNING!
An IED can look like ANYTHING!

IED Types Categorized by Containers

IEDs are typically categorized by their container and the way in which they are initiated. Bomb types, based on the outer container, can include features similar to the following descriptions. You can identify types of IEDs by their method of transportation or delivery:

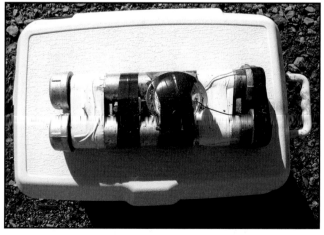

Figure 8.22 Steel or PVC pipes can be used to make pipe bombs. *Courtesy of August Vernon.*

* **Pipe bombs** — The most common type of IED found in the United States **(Figure 8.22)**. Characteristics of the pipe bomb and those of the pipe bomb manufacturers include the following:

    — Length ranges from 4 to 14 inches (102 mm to 356 mm).

    — Steel or polyvinyl chloride (PVC) pipe sections are filled with explosives and the ends are capped or sealed.

- Easily obtained materials such as black powder or match heads.

- Filled or wrapped with nails or other materials that will become shrapnel when the bomb detonates.

- Can throw shrapnel up to 300 feet (90 m) with lethal force.

- Detonate with a homemade fuse or with commercially available fuses.

- Explosive filler can get into the pipe threads making the device extremely sensitive to shock or friction **(Figure 8.23)**.

- **Satchel, backpack, knapsack, duffle bag, briefcase, or box bombs** — Some terrorists may fill these bags with explosives or an explosive device **(Figure 8.24)**. Terrorists use these devices because it is common to see people carrying backpacks or other types of bags. These IEDs may include electronic timers or radio-controlled triggers so there may be no external wires or other items visible. These bombs come in any style, color, or size (even as small as a cigarette pack).

- **Plastic bottle bombs** — Manufacturers of these bombs fill plastic soda bottles (or any size of plastic drink bottles) with a material (such as dry ice) or combination of reactive materials that will expand rapidly, causing the container to explode. The Internet lists many variations of plastic bottle bombs. Be careful around plastic containers containing multilayered liq-

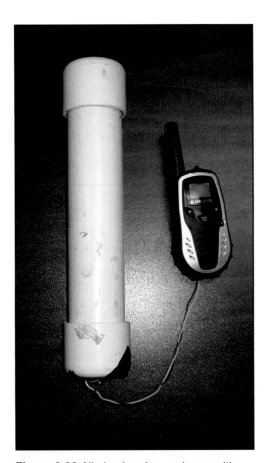

Figure 8.23 All pipe bombs can be sensitive to shock or friction. *Courtesy of August Vernon.*

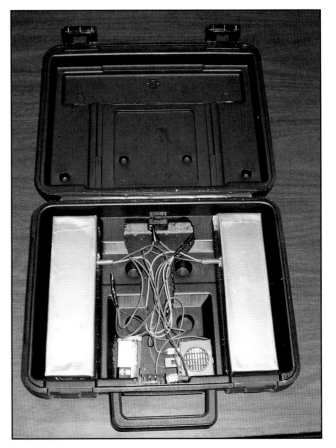

Figure 8.24 Suicide bombers can detonate backpack and briefcase bombs. The suicide bombers can also leave the backpack and briefcase bombs in crowded locations where the timer or remote signal detonates them. *Courtesy of August Vernon.*

uids and containers with white or gray liquids with cloudy appearance. Do not attempt to move or open plastic bottle bombs. Once initiated, they can detonate at any time. Manufacturers may use materials such as:

— Pool chemicals

— Dry ice

— Alcohol

— Acid

— Aluminum

— Toilet bowl cleaner

— Drain cleaner

— Driveway cleaners

- **Fireworks** — Some manufacturers may modify and/or combine legally obtained fireworks to form more dangerous explosive devices.

- **M-Devices** — Small devices constructed of cardboard tubes (often red) filled with flash powder and sealed at both ends and are ignited by fuses. The most common are M-80s, which measure $5/8 \times 1\frac{1}{2}$ inches (16 mm by 38 mm). At one time, M-80s were available in the U.S. as commercial fireworks. However, for safety reasons, they were made illegal in 1966.

- **Carbon Dioxide (CO_2) Grenades** — These devices are made by drilling a hole in used CO_2 containers (such as those used to power pellet pistols) and filling with an explosive powder; they are usually initiated by a fuse. Shrapnel may be added to the outside of the container. These devices, also known as *crickets,* have a small range but will create a great deal of destruction within that range.

- **Tennis ball bombs** — A tennis ball may be filled with an explosive mixture and ignite using a simple fuse.

- **Other existing objects** — Items that seem to have an ordinary purpose, such as fire extinguishers, propane bottles, trash cans, gasoline cans, and books, can be substituted or used as the bomb container **(Figure 8.25)**.

Figure 8.25 Replica of a fire extinguisher bomb used in Israel.

WARNING!
Do not move, handle, or disturb an IED when found!

Mail, Package, or Letter Bombs

A package or letter may be used to conceal the explosive device or material. Opening the package or letter usually triggers the bomb.

Package or letter bomb indicators include **(Figure 8.26, p. 384)**:

- Package or letter has no postage, non-cancelled postage, or excessive postage.

- Parcels may be unprofessionally wrapped with several combinations of tape to secure them and endorsed: *Fragile, Handle with Care, Rush,* or *Do Not Delay.*

- Sender is unknown, no return address is available, or the return address is fictitious.

- Mail may bear restricted endorsements such as *Personal* or *Private*. These endorsements are particularly important when addressees do not usually receive personal mail at their work locations.

- Postmarks may show different locations than return addresses.

- Common words are misspelled on the mail.

- Mail may display distorted handwriting, or the name and address may be prepared with homemade labels or cut-and-paste lettering.

- Package emits a peculiar or suspicious odor.

- Mail shows oily stains or discoloration.

- Letter or package seems heavy or bulky for its size and may have an irregular shape, soft spots, or bulges.

- Letter envelopes may feel rigid or appear uneven or lopsided.

- Mail may have protruding wires or aluminum foil.

- Package makes ticking, buzzing, or whirring noises.

- Unidentified person calls to ask if a letter or package was received.

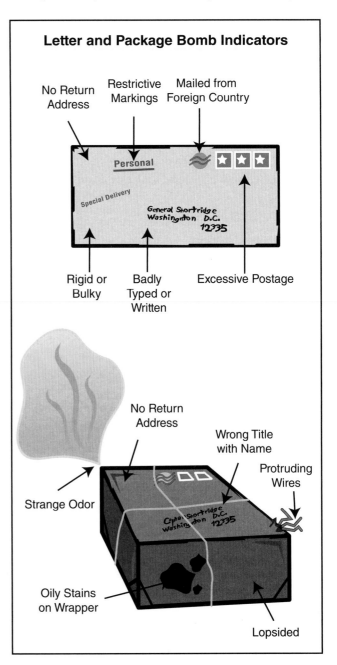

Figure 8.26 Common letter and package bomb indicators.

Person-Borne Improvised Explosives Devices (PBIEDs)

Person-Borne Improvised Explosives Devices (PBIEDs) typically consist of bombs worn or carried by a suicide bomber. Suicide bombers wear the PBIED in the form of vests with many pockets sewn into them to hold explosive materials **(Figures 8.27)**. Terrorists may also carry PBIEDs or they may be attached to coerced or unwilling victims.

Individuals carrying briefcases or packages are inherently suspicious to security forces, particularly in locations where personnel have to pass through a security checkpoint. In addition to bags, packages, and cases, clothing can conceal a bomb. The contours of bulky suicide vests or belts may be visible prior to detonation **(Figure 8.28)**. Terrorists might also wear unseasonable or atypical attire, such as a coat (which may conceal a suicide belt) during warm weather. Wires or other materials exposed on or around the body could also be an indication of a bomb.

Behavioral indicators of potential suicide bombers include the following:

- Fear, nervousness, or overenthusiasm
- Profuse sweating
- Keeping hands in pockets
- Repeated or nervous touching or patting of clothing
- Slow-paced walking while constantly shifting eyes to the left and right

Person-Borne Improvised Explosives Device (PBIED) — Improvised explosive device carried by a person. This type of IED is often employed by suicide bombers, but may be carried by individuals coerced into carrying the bomb.

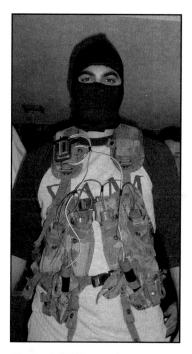

Figure 8.27 Person-borne bombs include suicide bombers and individuals coerced into carrying explosives. *Courtesy of August Vernon.*

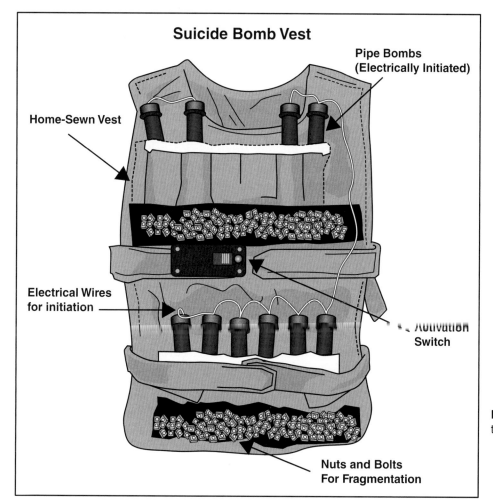

Suicide Bomb Vest

Pipe Bombs
(Electrically Initiated)

Home-Sewn Vest

Electrical Wires
for initiation

Activation
Switch

Nuts and Bolts
For Fragmentation

Figure 8.28 Components of a typical suicide bomb vest.

- Major attempts to avoid security personnel
- Obvious or awkward attempts to blend in with a crowd
- Obvious disguising of appearance
- Actions indicating a strong determination to get to a target
- Repeated visits to a high-risk location during the recon/ target acquisition phase
- Placing items in locations that seem out of place or arouse curiosity

ALERT

The FBI uses the acronym ALERT to designate indicators of a possible suicide bomber:

- **A**lone and nervous
- **L**oose and/or bulky clothing
- **E**xposed wires (possibly through a sleeve)
- **R**igid mid-section (explosives device or a rifle)
- **T**ightened hands (may hold a detonation device)

Explosive Ordnance Disposal (EOD) — Emergency responders specially trained and equipped to handle and dispose of explosive devices. *Also called* Hazardous Devices Units *or* Bomb Squad.

Never approach a suspected or confirmed suicide bomber who is injured or deceased. If there are several strong indicators of a suicide bomber, the first priority is to clear and isolate the area and observe the bomber with binoculars or spotting scopes. Trained personnel from an equipped **explosive ordnance disposal (EOD)** unit must conduct the first approach. These units may use a bomb disposal robot **(Figure 8.29)**.

Vehicle-Borne Improvised Explosive Devices (VBIEDs)

Vehicle-Borne Improvised Explosive Device (VBIED) — An improvised explosive device placed in a car, truck, or other vehicle. This type of IED typically creates a large explosion.

Vehicle-Borne Improvised Explosive Devices (VBIEDs) may contain many thousands of pounds (kilograms) of explosives that can cause massive destruction **(Figure 8.30)**. The explosives can be placed anywhere in a vehicle. When using small vehicles, such as passenger cars, the explosives are often concealed in the trunk.

Indicators of a possible VBIED include:

- Preincident intelligence or 9-1-1 calls leading to the suspected vehicle
- Vehicle parked suspiciously for a prolonged amount of time in a strategic or central location
- Vehicle abandoned in a public assembly, tourist area, pedestrian area, retail area, or transit facility
- Vehicle parked between, against, or close to the columns of a multistory building
- Vehicle that appears to be weighted down or sits unusually low on its suspension
- Vehicle with stolen plates, nonmatching plates, or no plates at all

Figure 8.29 Never approach a suspected suicide bomber who is injured, deceased, or has surrendered. Let EOD personnel orchestrate the first approach. *Courtesy of the U.S. Marine Corps; photo by Sgt. Lukas M. Atwell.*

Figure 8.30 VBIEDs can cause massive destruction. *Courtesy of the U.S. Air Force; photo by Master Sgt. Robert R. Hargreaves Jr.*

- Wires, bundles, electronic components, packages, unusual containers, liquids or materials visible in the vehicle

- Unknown liquids or materials leaking under vehicle

- Unusually screwed, riveted, or welded sections located on the vehicle's bodywork

- Unusually large battery or extra battery found under the hood or elsewhere in the vehicle

- Blackened windows or covered windows

- Taped, sealed, or otherwise inaccessible hollows of front or rear bumpers

- Tires that seem solid, instead of air-inflated

- Bright chemical stains or unusual rust patches on a new vehicle

- Chemical odor present or unusual chemical leak beneath vehicle

- Wiring protruding from the vehicle, especially from trunk or engine compartment

- Wires or cables running from the engine compartment, through passenger compartment, to the rear of vehicle

- Wires or cables leading to a switch behind sun visor

- Appearance or character of the driver does not match the use or type of vehicle

- Driver seems agitated, lost, and unfamiliar with vehicle controls

- Anything about a vehicle that seems out of place, unusual, abnormal, or arouses curiosity

WARNING!
Never approach a suspicious vehicle once an indicator of possible VBIED has been noticed.

Response to Explosive/IED Events

All operations must be conducted within an Incident Command System and determined by the risk/benefit analysis. In addition, do the following:

- Follow designated SOP/Gs.

- ALWAYS proceed with caution, especially if an explosion has occurred or it is suspected that explosives may be involved in an incident.

- Understand that secondary devices may be involved.

- Request EOD (bomb squad) personnel, hazmat, and other specialized personnel as needed (**Figure 8.31**).

- Treat the incident scene as a crime scene until proven otherwise.

Figure 8.31 Only certified EOD technicians should touch, move, defuse, or otherwise handle explosive devices. *Courtesy of the U.S. Air Force; photo by Airman Matthew Flynn.*

- NEVER touch or handle a suspected device, even if someone else already has. Only certified, trained bomb technicians should touch, move, defuse, or otherwise handle explosive devices.

- Do not use two-way radios, cell phones, or **mobile data terminals (MDT)** within a minimum of 300 feet (90 m) of any device or suspected device. The larger the suspicious device, the larger the standoff distance should be.

- Use intrinsically safe communications equipment within the isolation zone.

- Note unusual activities or persons at the scene and report observations to law enforcement.

- Limit personnel exposure until the risk of secondary devices is eliminated.

Mobile Data Terminal (MDT) — Mobile computer that communicates with other computers on a radio system.

Train with Specialized Bomb Disposal Personnel

If possible, ask local EOD (bomb squads) for their assistance with training and planning. Most bomb technicians will gladly provide your agency with training on their procedures and equipment, since they will require your support during an incident. Fire and EMS departments should become familiar with local bomb squad operations and entry suits; responders should know how to remove this specialized gear from an injured bomb technician in case of an emergency.

Responders must complete a primary search when responding to a bombing incident. In these situations, follow local protocols. Regardless of who completes the primary search and subsequent rescue operations, the Incident Commander should limit the number of personnel in the blast area to the minimum number of personnel required to carry out critical lifesaving operations. When determining risk, evaluate both the potential for additional explosions and structural stability.

WARNING!

Avoid staging near gardens, garbage bins, or other vehicles that could conceal explosive or incendiary devices. Limit exposure until the secondary device risk is eliminated.

Chemical Attacks

A **chemical attack** is the deliberate release of a toxic gas, liquid, or solid that can poison people and the environment. Attackers may use **chemical agents** or toxic industrial materials (TIMs). Chemical agents are intended for use in warfare or terrorist activities to kill, seriously injure, or seriously incapacitate people through their physiological effects. TIMs are particularly poisonous hazardous materials that are normally used for industrial purposes, but they could be used by terrorists to deliberately kill, injure, or incapacitate people.

This section explains the following types of chemical agents:

- Nerve agents
- Blister agents (vesicants)
- Blood agents (cyanide agents)
- Choking agents (pulmonary or lung-damaging agents)
- Riot control agents (irritants)
- Toxic industrial materials (common hazardous materials used for terrorist purposes)

Table 8.3, p. 390, provides the UN/DOT identification number and hazard class for some of the common chemical warfare agents, as well as the *ERG* where responders can obtain additional information to manage the initial response phases of the incident.

Chemical Attack — Deliberate release of a toxic gas, liquid, or solid that can poison people and the environment.

Chemical Agent — Chemical substance that is intended for use in warfare or terrorist activities to kill, seriously injure, or incapacitate people through its physiological effects. *Also known as* Chemical Warfare Agents.

Table 8.3
UN/DOT ID Number, Hazard Class, and *ERG* Guide Number
for Selected Chemical Agents

Agent	UN/DOT ID #	UN/DOT Class	*ERG* Guide U.S.	Military Symbol
Nerve Agents				
Tabun (GA)	2810	6.1	153	GA
Sarin (GB)	2810	6.1	153	GB
Soman (GD)	2810	6.1	153	GD
V Agent (VX)	2810	6.1	153	VX
Blister Agents/Vesicants				
Mustard (H)	2810	6.1	153	H
Distilled mustard (HD)	2810	6.1	153	HD
Nitrogen mustard (HN)	2810	6.1	153	
Lewisite (L)	2810	6.1	153	L
Phosgene Oxime (CX)	2811	6.1	154	
Blood Agents				
Hydrogen Cyanide (AC)	1051	6.1	117	
Cyanogen Chloride (CK)	1589	2.3	125	
Choking Agents				
Chlorine (CL)	1017	2.3	124	
Phosgene (CG)	1076	2.3	125	
Riot Control Agents / Irritants				
Tear Gas (CS)	1693	6.1	159	
Tear Gas (CR)	1693	6.1	159	
Mace (CN)	1697	6.1	153	
Pepper Spray (OC)		2.2 (6.1)*	159	
Adamsite (DM)	1698	6.1	154	DM

Hazard class can be 2.2 or 6.1, depending on how the pepper spray is packaged.

NOTE: Letters in parentheses next to the name represent military designations, not chemical formulas.

Chemical Attack Indicators

Chemical attacks usually result in readily observable features including signs and symptoms that develop very rapidly. Chemical attack indicators include:

- Warning or threat of an attack or received intelligence

- Presence of hazardous materials or laboratory equipment that is not relevant to the occupancy

- Intentional release of hazardous materials

- Unexplained patterns of sudden onset of similar, nontraumatic illnesses or deaths (the pattern could be geographic, by employer, or associated with agent dissemination methods)

- Unexplained odors or tastes that are out of character with the surroundings

- Multiple individuals exhibiting unexplained skin, eye, or airway irritation

- Unexplained bomb or munition-like material, especially if it contains a liquid

- Unexplained vapor clouds, mists, and plumes, particularly if they are not consistent with their surroundings

- Multiple individuals exhibiting unexplained health problems such as:
  — Nausea
  — Vomiting
  — Twitching
  — Tightness in chest
  — Sweating
  — Pinpoint pupils (miosis) **(Figure 8.32)**
  — Runny nose (rhinorrhea)
  — Disorientation
  — Difficulty breathing
  — Convulsions

- Unexplained deaths and/or mass casualties

- Casualties distributed downwind (outdoors) or near ventilation systems (indoors)

- Multiple individuals experiencing blisters and/or rashes

- Trees, shrubs, bushes, food crops, and/ or lawns that are dead (not just a patch of dead weeds), discolored, abnormal in appearance, or withered (not under drought conditions) **(Figure 8.33)**

- Surfaces exhibiting oily droplets or films and unexplained oily film on water surfaces

- Abnormal number of sick or dead birds, animals, and/or fish

- Unusual security, locks, bars on windows, covered windows, and barbed-wire enclosures

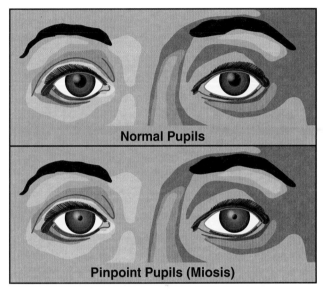

Figure 8.32 Exposure to chemical agents may cause miosis (pinpoint pupils).

Figures 8.33 Like this acid release, chemical agents may kill or wither trees and vegetation. *Courtesy of Barry Lindley.*

SLUDGEM or DUMBELS

Some people teach the symptoms of exposure to chemical warfare agents with acronyms, specifically SLUDGEM and DUMBELS. These symptoms do not present in victims in a chronological order, so both acronyms are appropriate.

SLUDGEM indicators are:

- **S**alivation (drooling)
- **L**acrimation (tearing)
- **U**rination
- **D**efecation
- **G**astrointestinal upset/aggravation (cramping)
- **E**mesis (vomiting)
- **M**iosis (pinpointed pupils) or Muscular twitching/spasms

DUMBELS indicators are:

- **D**efecation
- **U**rination
- **M**iosis or Muscular twitching
- **B**ronchospasm (wheezing)
- **E**mesis
- **L**achrimation
- **S**alivation

TIMs used as chemical weapons may be identified through traditional methods such as:

- Identification of occupancy types and locations
- Container shapes
- Hazardous materials placards, labels, and markings
- Written resources
- Sensory indicators
- Use of monitoring and detection devices

Nerve Agents

Nerve Agent — A class of toxic chemical that works by disrupting the way nerves transfer messages to organs.

Nerve agents are the most toxic chemical warfare agents. Exposure to even minute quantities can kill quickly by attacking the nervous system. Stable, easily dispersed, and highly toxic, nerve agents have rapid effects when absorbed through the skin or respiratory system. Although nerve agents are generally clear and colorless, colors and odors can vary with impurities. Impure "G" agents may have a slight fruity odor. VX is odorless. Although people sometimes use the term *nerve gas*, the term is a misnomer. Nerve agents are liquids at ambient temperatures and dispersed as an aerosolized liquid (vapor, not gas).

First responders should be familiar with the following nerve agents (military designations are provided in parentheses):

- **Tabun (GA)** — Usually low **volatility**, **persistent chemical agent** that is absorbed through skin contact or inhaled as a vapor.

- **Sarin (GB)** — Usually volatile, **nonpersistent chemical agent** that is mainly inhaled.

- **Soman (GD)** — Usually moderately volatile chemical agent that can be inhaled or absorbed through skin contact.

- **Cyclohexyl sarin (GF)** — Low-volatility persistent chemical agent that is absorbed through skin contact and inhaled as a vapor.

- **V-agent (VX)** — Low-volatility persistent chemical agent that can remain on material, equipment, and terrain for long periods; it is usually absorbed through the skin but can be inhaled. First responders may also see reference to other V-agents including VE, VG, and VS, but the most common is VX. VX is primarily a contact exposure hazard.

Nerve agents' volatilities vary widely. For example:

- **G-series agents** tend to be nonpersistent (vaporize and disperse quickly) unless manufacturers thicken them with some other agent to increase their persistency.

- V-Agents are persistent (remain effective in the open for a significant amount of time). For example, the consistency of VX is similar to motor oil. Its primary route of entry is through direct contact with the skin.

- GB is an easily volatile liquid that is primarily an inhalation hazard.

- The volatilities of GD, GA, and GF are between those of GB and VX, and their vapors are heavier than air.

Considering the low vapor pressures, nerve agent vapors will not travel far under normal conditions. Therefore, the size of the endangered area may be relatively small. However, the vapor hazard can significantly increase if the liquid is exposed to high temperatures, spread over a large area, or aerosolized. **Table 8.4, p.394**, provides information about nerve agents including descriptions and symptoms of exposure.

Speed is the most important factor in medical management of individuals who have been exposed to nerve agents because of their extremely rapid effects. Effective treatment is best achieved by immediate use of **autoinjectors** containing **antidotes (Figure 8.34, p. 395)**.

Blister Agents

Blister agents (vesicants) burn and blister the skin or any other part of the body they contact. They act on the eyes, mucous membranes, lungs, skin and blood-forming organs. These agents damage the respiratory tract when in-

Volatility — Ability of a substance to vaporize easily at a relatively low temperature.

Persistent Chemical Agent — Chemical agent that remains effective in the open (at the point of dispersion) for a considerable period of time, usually more than 10 minutes.

Nonpersistent Chemical Agent — Chemical agent that generally vaporizes and disperses quickly, usually in less than 10 minutes.

G-Series Agents — Nonpersistent nerve agents initially synthesized by German scientists.

Antidote — Substance that counteracts the effects of a poison or toxin.

Autoinjector — Spring loaded syringe filled with a single dose of a lifesaving drug.

Blister Agent — Chemical warfare agent that burns and blisters the skin or any other part of the body it contacts. *Also known as* Vesicant *and* Mustard Agent.

Table 8.4
Nerve Agent Characteristics

Nerve Agent (Symbol)	Descriptions	Symptoms (All Listed Agents)
Tabun (GA)	• Clear, colorless, and tasteless liquid • May have a slight fruit odor, but this feature cannot be relied upon to provide sufficient warning against toxic exposure • ***Probable Dispersion Method:*** Aerosolized liquid	***Low or moderate dose by inhalation, ingestion (swallowing), or skin absorption:*** Persons may experience some or all of the following symptoms within seconds to hours of exposure: • Runny nose • Diarrhea • Watery eyes • Increased urination • Small, pinpoint pupils • Confusion • Eye pain • Drowsiness • Blurred vision • Weakness • Drooling and excessive sweating • Headache • Nausea, vomiting, and/or abdominal pain • Cough • Slow or fast heart rate • Chest tightness • Abnormally low or high blood pressure • Rapid breathing
Sarin (GB)	• Clear, colorless, tasteless, and odorless liquid in pure form • ***Probable Dispersion Method:*** Aerosolized liquid	
Soman (GD)	• Pure liquid is clear, colorless, and tasteless; discolors with aging to dark brown • May have a slight fruity or camphor odor, but this feature cannot be relied upon to provide sufficient warning against toxic exposure • ***Probable Dispersion Method:*** Aerosolized liquid	
Cyclohexyl sarin (GF)	• Clear, colorless, tasteless, and odorless liquid in pure form • Only slightly soluble in water • ***Probable Dispersion Method:*** Aerosolized liquid	***Skin contact:*** Even a tiny drop of nerve agent on the skin can cause sweating and muscle twitching where the agent touched the skin ***Large dose by any route:*** These additional health effects may result: • Loss of consciousness • Convulsions • Paralysis • Respiratory failure possibly leading to death
V-Agent (VX)	• Clear, amber-colored odorless, oily liquid • Miscible with water and dissolves in all solvents • Least volatile nerve agent • Very slow to evaporate (about as slowly as motor oil) • Primarily a liquid exposure hazard, but if heated to very high temperatures, it can turn into small amounts of vapor (gas) • ***Probable Dispersion Method:*** Aerosolized liquid	***Recovery Expectations:*** • Mild or moderately exposed people usually recover completely • Severely exposed people are not likely to survive • Unlike some organophosphate pesticides, nerve agents have *not* been associated with neurological problems lasting more than 1 to 2 weeks after the exposure

Source: Information on symptoms provided by the Centers for Disease Control and Prevention (CDC).

haled and can cause vomiting and diarrhea when ingested. Blister agents are more likely to produce casualties than fatalities, although exposure to such agents can be fatal.

Blister agents are usually persistent and may be oily liquids ranging from colorless to pale yellow to dark brown, depending on purity. Blister agents may take several days or weeks to evaporate. It is more difficult to remove these agents during decontamination than less viscous products. **Table 8.5, p. 396,** provides information about blister agents.

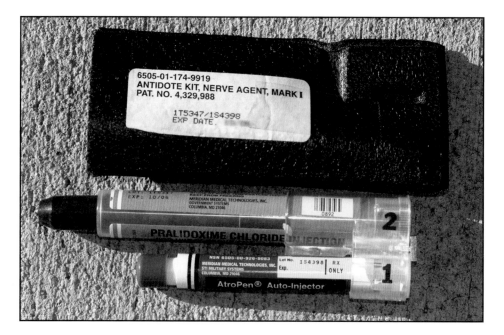

Figure 8.34 The most effective treatment for chemical agent exposure are autoinjectors containing antidotes.

Blister agents can be categorized into the following groups:

- Mustard agents
  - Sulfur mustards (H, HD [also called distilled mustard], and HT)
  - Nitrogen mustards (HN, HN-1, HN-2, and HN-3)
- Arsenical vesicants
  - Lewisite (L, L-1, L-2, and L-3)
  - Mustard/lewisite mixture (HL) (a mixture of lewisite [L] and distilled mustard [HD])
  - Phenyldichloroarsine (PD)
- Halogenated oximes — Phosgene oxime (CX)

Blood Agents

Blood agents are **chemical asphyxiants**. They interfere with the body's ability to use oxygen either by preventing red blood cells from carrying oxygen to other cells in the body or by inhibiting the ability of cells to use oxygen for producing the energy required for metabolism. Some sources may use the terms blood agents and cyanogen agents synonymously, but not all blood agents are cyanogens (for example, arsine is not). Similarly, not all cyanogens are necessarily blood agents. Blood agents are sometimes categorized as TIMs because they also have industrial applications.

First responders should be familiar with the following blood agents:

- **Arsine (SA)** — Arsine gas is formed when arsenic comes in contact with an acid. It is a colorless, nonirritating toxic gas that has a reported mild garlic odor. Most people can only detect this odor at levels higher than those necessary to cause poisoning **(Table 8.6, p. 397)**.

- **Hydrogen cyanide (AC)** — Hydrogen cyanide is a colorless, highly volatile liquid that is extremely flammable, highly soluble, and stable in water. Mixtures of the vapor and air may be explosive **(Table 8.7, p. 398)**. The vapor

Chemical Asphyxiant — Substance that reacts to prevent the body from being able to use oxygen. *Also known as* Blood Agent.

Table 8.5
Common Blister Agent Characteristics

Blister Agent (Symbol)	Description	Symptoms
Sulfur Mustard (H/HD)	• Can be clear to yellow or brown when in liquid or solid form • Sometimes smells like garlic, onions, or mustard; sometimes has no odor • Can be a vapor, an oily-textured liquid, or a solid • Vapors are heavier than air • **Probable Dispersion Method:** Aerosolized liquid	Symptoms include: • **Skin:** Redness and itching of the skin may occur 2 to 48 hours after exposure and change eventually to yellow blistering of the skin. • **Eyes:** Irritation, pain, swelling, and tearing may occur within 3 to 12 hours of a mild to moderate exposure. A severe exposure may cause symptoms within 1 to 2 hours and may include the symptoms of a mild or moderate exposure plus light sensitivity, severe pain, or blindness (lasting up to 10 days). • **Respiratory tract:** Runny nose, sneezing, hoarseness, bloody nose, sinus pain, shortness of breath, and cough within 12 to 24 hours of a mild exposure and within 2 to 4 hours of a severe exposure. • **Digestive tract:** Abdominal pain, diarrhea, fever, nausea, and vomiting. Other factors include: • Typically, signs and symptoms do not occur immediately. • Depending on the severity of the exposure, symptoms may not occur for 2 to 24 hours. • Some people are more sensitive than others. • Exposure is usually not fatal.
Nitrogen Mustard (HN)	• Comes in different forms that can smell fishy, musty, soapy, or fruity • Can be in the form of an oily-textured liquid, a vapor (the gaseous form of a liquid), or a solid • Is liquid at normal room temperature (70°F or 21°C) • Can be clear, pale amber, or yellow colored when in liquid or solid form • Vapors are heavier than air • **Probable Dispersion Method:** Aerosolized liquid	Symptoms include: • **Skin:** Redness usually develops within a few hours after exposure followed by blistering within 6 to 12 hours. • **Eyes:** Irritation, pain, swelling, and tearing may occur. High concentrations can cause burns and blindness. • **Respiratory tract:** Nose and sinus pain, cough, sore throat, and shortness of breath may occur within hours. Fluid in the lungs is uncommon. • **Digestive tract:** Abdominal pain, diarrhea, nausea, and vomiting. • **Brain:** Tremors, incoordination, and seizures are possible following a large exposure. Other factors include: • Typically, signs and symptoms do not occur immediately. • Depending on the severity of the exposure, symptoms may not occur for several hours.
Lewisite (L)	• Colorless liquid in its pure form; can appear amber to black in its impure form • Has an odor like geraniums • Vapors are heavier than air • **Probable Dispersion Method:** Aerosolized liquid	Signs and symptoms occurring immediately following exposure include: • **Skin:** Pain and irritation within seconds to minutes; redness within 15 to 30 minutes followed by blister formation within several hours. — Blister begins small in the middle of red areas and then expands to cover the entire reddened area of skin. — Lesions (sores) heal much faster than lesions caused by other blistering agents (sulfur mustard and nitrogen mustards). — Discoloring of the skin that occurs later is much less noticeable. • **Eyes:** Irritation, pain, swelling, and tearing may occur on contact. • **Respiratory tract:** Runny nose, sneezing, hoarseness, bloody nose, sinus pain, shortness of breath, and cough. • **Digestive tract:** Diarrhea, nausea, and vomiting. • **Cardiovascular:** *Lewisite shock* or low blood pressure.

Continued

Table 8.5 (concluded)

Blister Agent (Symbol)	Description	Symptoms
Phosgene Oxime (CX)	• Colorless in its solid form and yellowish-brown when liquid • Has a disagreeable, irritating odor • Vapors are heavier than air • **Probable Dispersion Method:** Aerosolized liquid	Signs and symptoms occur immediately following exposure: • **Skin:** Pain occurring within a few seconds, and blanching (whitening) of the skin surrounded by red rings occurring on the exposed areas within 30 seconds. — Within about 15 minutes, the skin develops hives. — After 24 hours, the whitened areas of skin become brown and die, and a scab is then formed. — Itching and pain may continue throughout the healing process. • **Eyes:** Severe pain and irritation, tearing, and possibly temporary blindness. • **Respiratory tract:** Immediate irritation to the upper respiratory tract, causing runny nose, hoarseness, and sinus pain. • Absorption through the skin or inhalation may result in fluid in the lungs (pulmonary edema) with shortness of breath and cough.

Source: Information on symptoms provided by the Centers for Disease Control and Prevention

Table 8.6
Arsine (SA)

Description	Symptoms
• Colorless, nonirritating toxic gas with a mild garlic odor that is detected only at levels higher than those necessary to cause poisoning • Is formed when arsenic comes in contact with an acid • **Probable Dispersion Method:** Vapor release	**Low or moderate dose by inhalation:** Persons may experience some or all of the following symptoms within 2 to 24 hours of exposure: • Weakness • Fatigue • Headache • Drowsiness • Confusion • Shortness of breath • Rapid breathing • Nausea, vomiting, and/or abdominal pain • red or dark urine • Yellow skin and eyes (jaundice) • muscle cramps **Large dose by any route:** These additional health effects may result: • Loss of consciousness • Convulsions • Paralysis • Respiratory failure, possibly leading to death Other factors: • Showing these signs and symptoms does not necessarily mean that a person has been exposed. • If people survive the initial exposure, chronic effects may include: — Kidney damage — Numbness and pain in the extremities — Neuropsychological symptoms such as memory loss, confusion, and irritability

Table 8.7
Blood Agent Characteristics for AC and CK

Blood Agent (Symbol)	Description	Symptoms
Hydrogen cyanide (AC)	• Colorless gas or liquid • Characteristic *bitter almond* odor • Slightly lighter than air • Miscible • Extremely flammable • Explosive gas/air mixtures • Reacts violently with oxidants and hydrogen chloride in alcoholic mixtures, causing fire and explosion hazard • ***Probable Dispersion Method:*** Aerosolized liquid	May be absorbed through skin and eyes. Symptoms include: • ***Inhalation:*** Headache, dizziness, confusion, nausea, shortness of breath, convulsions, vomiting, weakness, anxiety, irregular heartbeat, tightness in the chest, and unconsciousness. Effects may be delayed. • ***Skin:*** May be absorbed. See *Inhalation* for other symptoms. • ***Eyes:*** Redness; vapor is absorbed. See *Inhalation* for other symptoms. • ***Ingestion:*** Burning sensation. See *Inhalation* for other symptoms.
Cyanogen chloride (CK)	• Colorless gas • Pungent odor • Heavier than air • ***Probable Dispersion Method:*** Vapor release	Symptoms include: • ***Inhalation:*** Runny nose, sore throat, drowsiness, confusion, nausea, vomiting, cough, unconsciousness, edema with symptoms which may be delayed. • ***Skin:*** Readily absorbed through intact skin, causing systemic effects without irritant effects on the skin; frostbite may occur on contact with liquid; liquid may be absorbed; redness and pain. • ***Eyes:*** Frostbite on contact with liquid; redness, pain, and excess tears.

Source: Information on symptoms provided by the Centers for Disease Control and Prevention (CDC).

is less dense than air and has a faint odor that is reported to be similar to bitter almonds; about 25 percent of the population cannot smell hydrogen cyanide.

- **Cyanogen chloride (CK)** — Cyanogen chloride is a colorless, highly volatile liquid that dissolves readily in organic solvents but is only slightly soluble in water (also see Table 8.7). Its vapors are heavier than air. Cyanogen chloride has a pungent, biting odor. Normally, it is a nonpersistent hazard. Exposure effects to cyanogen chloride are similar to hydrogen cyanide but with additional irritation to the eyes and mucous membranes.

Choking Agents

Choking Agent — Chemical warfare agent that attacks the lungs, causing tissue damage.

Choking agents attack and cause tissue damage to the lungs. They are sometimes called *pulmonary* or *lung-damaging* agents. Like blood agents, choking agents have industrial applications, and responders may encounter them during normal hazmat incidents. Choking agents include chemicals such as diphosgene (DP), chloropicin (PS), ammonia, hydrogen chloride, phosphine, and elemental phosphorus. Two of the most common choking agents are chlorine and phosgene:

- **Chlorine** — Yellow-green in color, chlorine gas is usually pressurized and cooled to a liquid state for storage and transportation. It has a pungent, bleach-like, irritating odor. When liquid chlorine is released, it quickly turns into a gas that is heavier than air. Chlorine does not remain in its liquid form for long, so decon is usually not required. Exposure may cause:

— Coughing

— Chest tightness

— Burning eyes, nose, and throat

— Watering eyes

— Blurred vision

— Nausea and vomiting

• **Phosgene** — Phosgene is a colorless, nonflammable gas that has the odor of freshly cut hay. Its odor threshold is well above its permissible exposure limit, so it is already at a harmful concentration by the time someone smells it. Phosgene's boiling point is 47°F (8.2°C), but its vapor density is much heavier than air; therefore, it may remain for long periods of time in trenches and other low-lying areas. Phosgene does not remain in its liquid form very long, so decon is usually not required. Exposure symptoms are similar to chlorine, although phosgene may also cause burns and rash to skin.

Riot Control Agents

Riot control agents (sometimes called *tear gas* or *irritating agents*) are chemical compounds that cause immediate irritation to the eyes, mouth, throat, lungs, and skin, temporarily disabling people. Several different compounds are considered riot control agents.

All riot agents are solids and require dispersion as aerosolized particles, usually released by pyrotechnics (such as with an exploding tear gas canister) or a propelled spray with the particles suspended in a liquid. Some are sold in small containers as personal defense devices containing either a single agent or a mixture. Some devices also contain a dye to visually mark a sprayed assailant. When dispersed, riot control agents are usually heavier than air.

Table 8.8, p. 400, provides common riot control agent characteristics. Because the symptoms of exposure are very similar for all the agents, they are listed only once.

In addition to tear gas, mace, pepper spray, and other irritants, the following agents are sometimes categorized as riot control agents:

• **Incapacitant** — Produces a temporary disabling condition that persists for hours to days after exposure has occurred (unlike that produced by most riot control agents). Examples of incapacitants include:

— Central nervous system (CNS) depressants (anticholinergics)

— CNS stimulants (lysergic acid diethylamide or LSD)

• **Vomiting agent** — Causes violent, uncontrollable sneezing, coughing, nausea, vomiting, and a general feeling of bodily discomfort. It is dispersed as an aerosol and produces its effects by inhalation or direct action on the eyes. Principal vomiting agents include:

— Diphenylchlorarsine (DA)

— Diphenylaminearsine chloride (Adamsite or DM)

— Diphenylcyanarsine (DC)

Riot Control Agent — Chemical compound that temporarily makes people unable to function, by causing immediate irritation to the eyes, mouth, throat, lungs, and skin.

Table 8.8
Riot Control Agent Characteristics

Riot Control Agent (Symbol)	Descriptions	Symptoms (All Listed Agents)
Chlorobenzylidene malononitrile (CS)	• White crystalline solid • Pepper-like smell	**Immediately after exposure:** People exposed may experience some or all of the following symptoms: • **Eyes:** Excessive tearing, burning, blurred vision, and redness
Chloroacetophenone (CN, mace)	• Clear yellowish brown solid • Poorly soluble in water, but dissolves in organic solvents • White smoke smells like apple blossoms	• **Nose:** Runny nose, burning, and swelling • **Mouth:** Burning, irritation, difficulty swallowing, and drooling • **Lungs:** Chest tightness, coughing, choking sensation, noisy breathing (wheezing), and shortness of breath • **Skin:** Burns and rash
Oleoresin Capsicum (OC, pepper spray)	• Oily liquid, typically sold as a spray mist • **Probable Dispersion Method:** Aerosol	• **Other:** Nausea and vomiting Long-lasting exposure or exposure to a large dose, especially in a closed setting, may cause severe effects such as the following: • Blindness • Glaucoma (serious eye condition that can lead to blindness)
Dibenzoxazepine (CR)	• Pale yellow crystalline solid • Pepper-like odor • **Probale Dispersion Method:** Propelled	• Immediate death due to severe chemical burns to the throat and lungs • Respiratory failure possibly resulting in death Prolonged exposure, especially in an enclosed area, may lead to long-term effects such as the following:
Chloropicrin (PS)	• Oily, colorless liquid • Intense odor • Violent decomposition when exposed to heat	• Eye problems including scarring, glaucoma, and cataracts • May possibly cause breathing problems such as asthma **Recovery Expectations:** If symptoms go away soon after a person is removed from exposure, long-term health effects are unlikely to occur.

Source: Information on symptoms provided by the Centers for Disease Control and Prevention (CDC).

Toxic Industrial Materials (TIMs)

A toxic industrial material (TIM) is an industrial chemical that is toxic at a certain concentration and is produced in quantities exceeding 30 tons (30.5 tonnes) per year at one production facility. TIMs are not as lethal as highly toxic nerve agents. However, TIMs pose a far greater threat than chemical warfare agents because they are produced in very large quantities (multitons) and are readily available. For example, sulfuric acid is not as lethal as a nerve agent, but it is easier to disseminate large quantities of sulfuric acid because large amounts of it are manufactured and transported every day.

Based on a hazard index ranking (high, medium, or low hazard) that OSHA provides, toxic industrial materials (TIMs) are divided into three hazard categories (**Table 8.9**). The categories are defined as follows:

● **High hazard** — Indicates a widely produced, stored, or transported TIM that has high toxicity and is easily vaporized.

● **Medium hazard** — Indicates a TIM that may rank high in some categories but is lower in others such as number of producers, physical state, or toxicity.

Table 8.9
Toxic Industrial Materials Listed by Hazard Index Ranking

High Hazard	Medium Hazard	Low Hazard
Ammonia	Acetonic cyanohydrin	Allyl isothiocyanate
Arsine	Acrolein	Arsenic trichloride
Boron trichloride	Acrylonitrile	Bromine
Boron trifluoride	Allyl alcohol	Bromine chloride
Carbon disulfide	Allylamine	Bromine pentafluoride
Chlorine	Allyl chlorocarbonate	Bromine trifluoride
Diborane	Boron tribromide	Carbonyl fluoride
Ethylene oxide	Carbon monoxide	Chlorine pentafluoride
Fluorine	Carbonyl sulfide	Chlorine trifluoride
Formaldehyde	Chloroacetone	Chloroacetaldehyde
Hydrogen bromide	Chloroacelonitrile	Chloroacetyl chloride
Hydrogen chloride	Chlorosulfonic acid	Crotonaldehyde
Hydrogen cyanide	Diketene	Cyanogen chloride
Hydrogen fluoride	1,2-Dimethylhydrazine	Dimethyl sulfate
Hydrogen sulfide	Ethylene dibromide	Diphenylmethane-4,4'-diisocyanate
Nitric acid, fuming	Hydrogen selenide	Ethyl chloroformate
Phosgene	Methanesulfonyl chloride	Ethyl chlorothioformate
Phosphorus trichloride	Methyl bromide	Ethyl phosphonothioic dichloride
Sulfur dioxide	Methyl chloroformate	Ethyl phosphonic dichloride
Sulfuric acid	Methyl chlorosilane	Ethyleneimine
Tungsten hexafluoride	Methyl hydrazine	Hexachlorocyclopentadiene
	Methyl isocyanate	Hydrogen iodide
	Methyl mercaptan	Iron pentacarbonyl
	Nitrogen dioxide	Isobutyl chloroformate
	Phosphine	Isopropyl chloroformate
	Phosphorus oxychloride	Isopropyl isocyanate
	Phosphorus pentafluoride	n-Butyl chloroformate
	Selenium hexafluoride	n-Butyl isocyanate
	Silicon tetrafluoride	Nitric oxide
	Stibine	n-Propyl chloroformate
	Sulfur trioxide	Parathion
	Sulfuryl chloride	Perchloromethyl mercaptan
	Sulfuryl fluoride	sec-Butyl chloroformate
	Tellurium hexafluoride	tert-Butyl isocyanate
	n-Octyl mercaptan	Tetraethyl lead
	Titanium tetrachloride	Tetraethyl pyrophosphate
	Trichloroacetyl chloride	Tetramethyl lead
	Trifluoroacetyl chloride	Toluene 2,4-diisocyanate
		Toluene 2,6-diisocyanate

Source: "Summary of the Final Report of the International Task Force 25: Hazard from Industrial Chemicals," April 15, 1999.

- **Low hazard** — Indicates that this TIM is not likely to be a hazard unless specific operational factors indicate otherwise.

Emergency responders should attempt to identify the material involved just as they would at any other hazardous materials incident. Follow all predetermined procedures and the guidelines provided in the *ERG* and other sources when responding to emergencies involving TIMs.

What This Means To You
Toxic Industrial Materials as Weapons

You are far more likely to have to deal with a toxic industrial chemical used as a weapon than chemical warfare agents in part because TIMs are much cheaper and easier to obtain. Chemical warfare agents, such as sarin, are notoriously difficult (and expensive) to produce.

For example, it is estimated that the cult, Aum Shinrikyo, spent over $30 million to produce the sarin used in their attacks in Japan. In contrast, chlorine cylinders could potentially be stolen from your local public swimming pool. Additionally, some TIMs are nearly as dangerous and deadly as warfare agents. Phosgene, for example, has industrial applications, but is also listed as a chemical warfare agent. Some commercially available pesticides disrupt nerve impulses in the same way nerve agents do, and they are very similar chemically.

TIMs may be well suited for use in terrorist attacks due to a variety of factors. Some may have poor warning properties. Others may be lethal in extremely low doses. Some may have ideal dispersal properties. In addition to the use of TIMs in terrorist attacks, their prevalence, especially in large quantities, make them more likely to be involved in a hazardous materials incident than other chemical agents.

Pay attention to the TIMs in your community and determine what chemicals might be used for terrorist purposes in your jurisdiction. Then, prepare to respond to incidents involving those TIMs.

Operations at Chemical Attack Incidents

The primary operational objective at a chemical attack is to do the greatest good for the greatest number. Responders must be familiar with SOPs/SOGs for handling chemical terrorist attacks and hazardous materials incidents.

Chemical attacks may differ from other hazmat incidents in the following ways:

- Severity of hazards present (such as deadly nerve agents) and need for appropriate PPE to protect against them
- Possibility of secondary devices
- Mass casualties
- Need for rapid decon
- Administration of antidotes

Biological Attacks

The Centers for Disease Control and Prevention (CDC) defines biological terrorism as "an intentional release of viruses, bacteria, or their toxins for the purpose of harming or killing citizens." Four types of **biological agents** are:

1. **Viral agents** — Viruses are the simplest types of microorganisms that can only replicate in their host's living cells. Viruses do not respond to **antibiotics**, making them an attractive weapon.

2. **Bacterial agents** — Bacteria are microscopic, single-celled organisms. Most bacteria do not cause disease in people, but when they do, two different mechanisms are possible: invading the tissues or producing poisons (toxins).

3. **Rickettsia** — Rickettsia are specialized bacteria that live and multiply in arthropods' (ticks and fleas) gastrointestinal tracts. They are smaller than most bacteria, but larger than viruses. Like bacteria, they are single-celled organisms with their own metabolisms, and they are susceptible to broad-spectrum antibiotics. However, like viruses, rickettsias only grow in living cells. Most rickettsias spread only through the bite of infected arthropods and not via human contact.

4. **Biological toxins** — Biological toxins are poisons produced by living organisms; however, the biological organism itself is usually not harmful to people. Some biological toxins have been manufactured synthetically and/or genetically altered in laboratories **(Figure 8.35)**. They are similar to chemical agents in the way they are disseminated and in their effectiveness as biological weapons.

Biological agents can be transmitted via:

- Aerosolization
- Food
- Water
- Insects

An attack using a biological weapon may not be as immediately obvious as an attack using a bomb or industrial chemical. Generally, biological weapons agents do not cause immediate health effects. Most biological agents take hours, days, or weeks to make someone ill, depending on the agent's incubation period. Because of this delay, the cause of illness may not be immediately evident, and the source of the attack may be difficult to trace.

In the beginning of a biological weapon attack, only a few patients may exhibit symptoms. The number of infected individuals will increase as the disease continues to transmit from person to person (such as might happen with smallpox). The scope of the problem may not be evident for days or even weeks. However, certain biological toxins (such as saxitoxin, a neurotoxin produced by marine organisms) could potentially act more quickly (in minutes to hours).

Antibiotic — Antimicrobial agent made from a mold or a bacterium that kills or slows the growth of bacteria; examples include penicillin and streptomycin. Antibiotics are ineffective against viruses.

Biological Agent — Viruses, bacteria, or their toxins which are harmful to people, animals, or crops. When used deliberately to cause harm, may be referred to as a Biological Weapon.

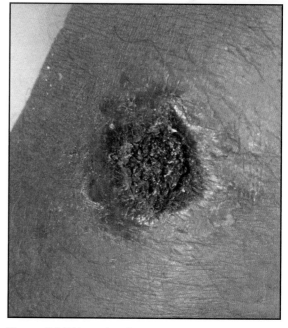

Figure 8.35 Natural anthrax is more of a skin hazard, causing lesions such as this. Weaponized anthrax is a more dangerous inhalation hazard. *Courtesy of the CDC.*

Biological Attack Indicators

Biological attacks utilize viruses, bacteria, and/or biological toxins. The effects of biological attacks may not be readily noticeable. Signs and symptoms may take many days to develop.

Biological attack indicators include:

- Warning or threat of an attack or received intelligence
- Presentation of specific unusual diseases such as smallpox
- Unusual number of sick or dying people or animals (often of different species)
- Multiple casualties with similar signs or symptoms
- Dissemination of unscheduled or unusual spray
- Abandoned spray devices (devices may have no distinct odors)
- Nonendemic illness for the geographic area (for example, Venezuelan equine encephalitis in Europe)
- Casualty distribution aligned with wind direction
- Electronic tracking of signs and symptoms (syndromic surveillance) reported to hospitals, pharmacies, and other health care organizations
- Illnesses associated with a common source of food, water, or location
- Large numbers of people exhibiting flu-like symptoms during non-flu months

Depending on the agent used and the scope of an incident, emergency medical services (EMS) responders and health care personnel may be the first to realize that there has been a biological attack. In some cases, there may be reliable evidence, such as a witness to an attack, to implicate terrorist activity, or the discovery of a delivery system, such as finding a contaminated dissemination device, from which an infectious agent is subsequently isolated and identified. If a biological attack is suspected, first responders should immediately notify their local health care agency.

Disease Transmission

Specific methods of infectious disease transmission include (**Figure 8.36, p. 406**):

- **Airborne transmission (inhalation of airborne organisms or toxins)** — Diseases remain suspended in air for a long time and when inhaled may penetrate deep into the respiratory tract. Airborne-transmitted diseases, such as influenza, pneumonia, and polio, can typically survive outside the body for long periods of time.

- **Contact with infected droplets** — Infected droplets, such as rubella, tuberculosis, and SARS, transmit diseases through contact with mucous membranes of the eyes, nose, and mouth. Droplets generally do not stay airborne for long periods of time.

- **Direct contact (such as touching or kissing an infected person)** — Most sexually transmitted diseases, such as HIV, fall into this category; other diseases, such as Ebola, transmitted in this way typically do not survive outside the human body for long.

- **Indirect contact (such as touching contaminated surfaces)** — Indirect contact diseases can generally survive on exposed surfaces for extended periods of time. The Norwalk Virus is an example of a disease transmitted through indirect contact.

- **Ingestion of contaminated food or water** — Normally this occurs due to contact with infected fecal material. Examples of diseases transmitted through contaminated food or water include amoebic dysentery and cholera.

- **Vectors** — Some diseases, such as Lyme Disease and the bubonic plague, are spread by insects (fleas, flies, and mosquitoes) and animals (**vectors**) such as rodents (mice and rats) and livestock.

    NOTE: Many diseases have more than one route of transmission.

When developing a biological weapon, the method of transmission is an important consideration. A disease that is spread by airborne transmission (such as smallpox) has the potential to infect a large number of people more quickly than one that is only transmitted through direct contact (such as HIV or Ebola).

> **Vector** — An animate intermediary in the indirect transmission of an agent that carries the agent from a reservoir to a susceptible host.

Contagiousness

An infectious disease is one that is caused by a microorganism with the potential to transfer to another person. A contagious disease is one that can spread rapidly from person to person. An attack with a contagious agent, such as smallpox or SARS, has the potential to become an epidemic. Noncontagious diseases will only affect those individuals who have direct exposure to the disease agent itself. Noncontagious diseases will not spread to other people except by contact with the disease agent. Biological attacks with noncontagious agents, such as anthrax and biological toxins, are not contagious.

Operations at Biological Attack Incidents

Bioterrorism incidents will most likely cross jurisdictional boundaries. Planning efforts must include provisions for sharing resources, critical information, and management responsibilities.

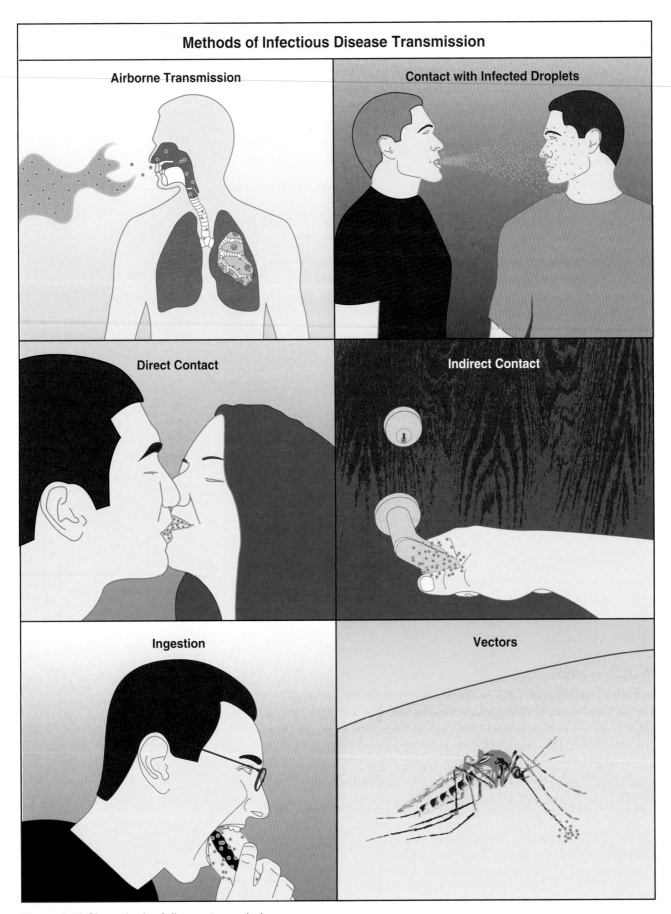

Figure 8.36 Six methods of disease transmission.

First responders should always adhere to universal precautions whenever they have contact with broken or moist skin, blood, or body fluids. These precautions will protect against many biological agents and/or infected individuals. Universal precautions include:

- Use disposable gloves **(Figure 8.37)**.
- Change gloves between patients to prevent transmitting the infection from patient to patient.
- Wash hands immediately after removing gloves.
- Use disposable PPE and a face shield if you anticipate any splashing.
- Contact the local health department for additional instructions for vaccinations, prophylactic antibiotic therapy, or other appropriate measures for a given disease.

In the event of an overt attack or incident, responders should focus on isolation and containment of the biological agent to prevent the spread of pathogens or toxins. Overt attacks could include white powder incidents (with a credible threat), discovery of a suspected biologic laboratory, or witnessed use of spray devices.

The following measures may contain indoor attacks:

- Turning off ventilation systems
- Closing doors and windows
- Turning off elevators
- Sealing ducts, windows, and doors using tape, plastic sheets, and expanding foams to restrict air flow

The following actions may help contain overt, outdoor attacks:

- Cover the device or dispersed agent with tarps or other physical barriers to prevent spreading.

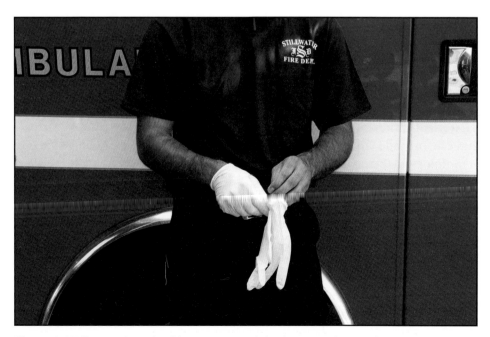

Figure 8.37 Responders should use common infection control procedures to protect themselves from biological agents.

- Decontaminate the dispersed agent with a light spray of water and bleach.

- Secure and place the suspicious item, package, object, or substance in a sealed hazmat recovery bin or container to mitigate spreading.

If possible, keep individuals who have been exposed to biological agents from leaving the scene until a thorough risk assessment has been conducted and appropriate measures taken (potentially in consultation with local health authorities). Whether victims exhibit signs or symptoms of illness, decontamination is recommended for any credible threat involving aerosols or contact with potentially harmful substances.

To ensure containment, use the following guidelines:

- Initially contain persons that may be affected.

- Decontaminate victims if circumstances indicate the need prior to treatment and transport to a medical facility.

- Register (record name and contact information) all persons potentially exposed at the incident in case follow-up is required.

Pandemic — Epidemic occurring over a very wide area (several countries or continents), usually affecting a large proportion of the population.

At biological attack incidents, isolation and containment issues will primarily involve managing infected victims; public health authorities will likely manage these issues. Local plans for handling a **pandemic** flu may translate to other contagious disease outbreaks.

What This Means To You

Ebola Outbreak

Though not a targeted attack, the hazmat community learned from the international response to the Ebola virus outbreak in 2014-2015. The outbreak demonstrated the need to focus on the appropriate use of PPE and the need to develop and use effective decon responses. These are global considerations across any number of risks, and the recent Ebola crisis demonstrates the consequences of not adequately following these procedures.

Radiological and Nuclear Attacks

Historically, there have been few attempts at radiological terrorism. Threats have been made, and plans to carry out nuclear attacks have been foiled, but a radiological terrorism attack has not yet occurred.

Response to a radiological incident is likely to be similar to the response to other emergency incidents. For example, a response to an attack on a shipment of radioactive materials might follow *ERG* guidelines for radiological materials, with additional consideration given to secondary devices and evidence preservation. Responders may not immediately detect the presence or involvement of radiological materials. Emergency response agencies must include radiation monitoring as a normal part of response to any fire and/or explosion incident. The only way to confirm if radiation is present at an incident is to use radiological monitoring equipment.

In the event of a nuclear attack, the scale and scope of the disaster facing the local first responders will probably overwhelm them. Responders will undoubtedly require outside assistance to successfully mitigate the incident. Communication, transportation, water supplies, and resources may be limited or nonexistent. The number of casualties and destruction may be overwhelming. When an organized response is possible, responders should apply the same framework for any emergency response with special consideration given to nuclear/radiological hazards. Because nuclear attacks are extremely unlikely, the following sections will only address radiological devices.

Radiological and Nuclear Attack Indicators

Radiological attacks utilize weapons that release radiological materials, most likely in the form of dust or powder. Dispersal may be accomplished by including the material in a bomb or explosive device, i.e., a radiological dispersal device (RDD). Radiological attack indicators include:

- Warning or threat of an attack or received intelligence

- Individuals exhibiting signs and symptoms of radiation exposure

- Radiological materials packaging left unattended or abandoned in public locations

- Suspicious packages that seem to weigh more than their appearance suggests they should (such packages may contain lead to shield a radiation source)

- Activation of radiation detection devices, with or without an explosion

- Material that is hot or seems to emit heat without any sign of an external heat source

- Glowing material (strongly radioactive material may emit or cause radio-luminescence)

Nuclear attacks are a little different from radiological attacks; nuclear attacks are the intentional detonation of a nuclear weapon. Indicators include:

- Warning or threat of an attack or received intelligence

- Mushroom cloud

- Exceptionally large/powerful explosion

- Electromagnetic pulse (EMP)

Radiological Devices

Several types of designs exist for radiological devices. All designs will expose people to radiation or to disperse radiological material. Radiological devices are sometimes referred to as *dirty bombs* because the contamination they spread could ruin property, crops, and livestock and cause large areas to become unusable. These devices include radiation-exposure devices (REDs), radiological-dispersal devices (RDDs), and radiological-dispersal weapons (RDWs).

Radiation-Exposure Devices

A **radiation-exposure device (RED)** is a powerful gamma-emitting radiation source. Terrorists may place it in a high-profile location, such as a high-traffic urban area, entertainment arena, or a shopping complex which could expose

Radiation-Exposure Device (RED) — Powerful gamma-emitting radiation source used as a weapon.

a large number of people to the intense radiation source **(Figure 8.38)**. Terrorists may also use REDs to target specific individuals and/or harm a limited number of people over a long period of time.

Radiological-Dispersal Devices

The U.S. Department of Defense defines a radiological-dispersal device (RDD) as any device, including weapons or equipment (other than a nuclear explosive device), specifically designed to disseminate radioactive material to cause destruction, damage, or injury by means of the radiation produced by the decay of such material. An RDD is intended to disperse radioactive material over a large area, but an RDD is incapable of producing a nuclear yield. Terrorists use RDDs to create fear and panic by exposing people to radioactive material or to contaminate areas and buildings, making them unusable until decontaminated. An RDD typically uses the force of conventional explosives to scatter radioactive material **(Figure 8.39)**.

Radiological-Dispersal Weapons

Radiological-dispersal weapons (RDWs) or Simple Radiological Dispersal Devices (SRDDs) are nonexplosive RDDs. RDWs can use inexpensive and common items such as pressurized containers, building ventilation systems, fans, and mechanical devices to spread radioactive contamination **(Figure 8.40)**. For example, radioactive material could be placed into a ventilation system and then dispersed throughout a building when the building's ventilation system is operated. Dispersal by these means would require putting the radioactive material into a dispersible form (powder or liquid) and would require large amounts of radioactive material to pose a hazard once dispersed.

Radiological Dispersal Weapons (RDW) — Devices that spread radioactive contamination without using explosives, instead, radioactive contamination is spread using pressurized containers, building ventilation systems, fans, and mechanical devices.

Operations at Radiological Attack Incidents

The Incident Command System (ICS) and local/jurisdictional procedures will establish priorities at radiological incidents. For most terrorism events, individual fire departments will eventually fold into a larger ICS structure. After multiple agencies with overlapping authority arrive, a Unified Command Structure will establish incident control. Until those agencies arrive, the AHJ's Incident Management System will provide the necessary structure for managing the incident at the lowest level. Regardless of the entity in Command, responders will need to gather essential information, including conducting a scene size-up, for use by incoming agencies.

Responders conducting scene size-up need to look for:

- Unusual or out-of-place incident-scene indicators
- Size and shape of smoke plumes
- Odors
- Large debris fields
- Craters from explosions

At the scene, hazard identification and characterization are important. Responders should always evaluate the area for new hazards, weather changes, or changing conditions.

Radiation-Exposure Device

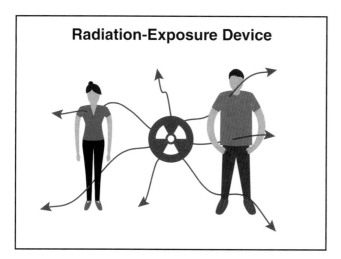

Figure 8.38 Radiation-exposure devices emit gamma radiation. They may be used to target specific individuals or harm a limited number of people over a long period of time.

Radiological-Dispersal Device

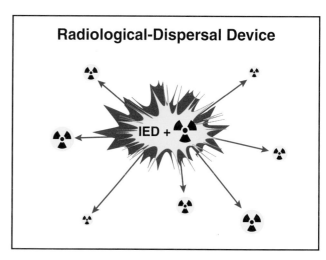

Figure 8.39 Radiological-dispersal devices use explosives to scatter radioactive materials over an area.

Radiological-Dispersal Weapon

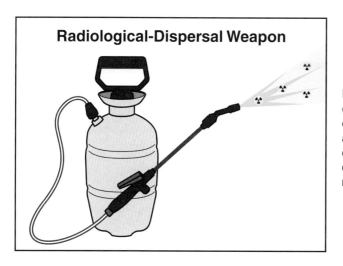

Figure 8.40 Radiological-dispersal weapons use common items, such as spray containers or ventilation fans, to disperse radioactive materials.

If responders suspect terrorism, they should proceed cautiously, evaluate the scene for radiation levels, and note potential locations of secondary devices. They should designate and enforce scene control zones. The following geographic and environmental factors can complicate a radiological terrorism incident:

- Prevailing winds that can carry airborne radioactive particulates
- Broken water mains
- Vehicle and/or pedestrian traffic flow
- Ventilation systems
- Air and rail corridors
- Natural or man-made influences

Tactics for radiological incidents include the following:

- Position apparatus upwind of the incident.
- Secure the area and prevent unauthorized entry.

- Stay alert and look for small explosive devices designed to disseminate an agent.

- Use time, distance and shielding as protective measures.

- Use full PPE, including SCBA.

- Avoid contact with any visible smoke or fumes.

- Monitor radiation and contamination levels.

- Establish background radiation levels outside the suspected contamination area.

- Detain or isolate uninjured people or equipment.

- Remove victims from high hazard areas.

- Assist the medical personnel as necessary to triage, treat, and decontaminate trauma victims.

- Call for expert guidance, following AHJ SOPs.

- Preserve possible evidence for subsequent criminal and forensic investigations.

- Do not conduct overhaul and clean-up operations; avoid disturbing the incident scene as much as possible.

The *ERG* provides response information for general radiological incidents involving low to high levels of radiation in Guide No.163. Radiological materials fall under UN/DOT Class 7.

Sources for Help

Sources for information and assistance at radiological terrorism incidents include:

- FBI WMD Coordinator
- REACTS, https://orise.orau.gov/reacts/
- State / local radiation safety office

Illicit Labs

Illicit laboratories may be used to manufacture drugs, explosives, biological weapons such as ricin, or chemical warfare agents. Drug labs, in particular, can be found virtually anywhere, from hotel rooms and private residences to portable labs in vehicles and campgrounds **(Figure 8.41)**. These labs can be extremely hazardous and, if you discover one, you should: **Stop, get out, and report it to authorities immediately!** The following are clues that may indicate an illicit laboratory:

- Laboratory glassware in unusual locations
- Large quantities of household chemicals and pharmaceuticals
- Hot plates
- Unusual odors in unexpected places such as hotel rooms

- Unusual uses of common materials such as coffee filters, water bottles, coffee grinders
- Increased security such as barred windows
- Unusual traffic patterns such as pedestrian and vehicular
- Unusual behavior and suspicious activity
- Personal protective equipment such as gloves, masks
- Disinfectants

Table 8.10 provides key indicators of different illicit lab types. While there may be some similarities between lab types, the clues may add up overall to indicate one type of lab.

NOTE: More information about, and clues to identify illicit laboratories are provided in Chapter 15.

Figure 8.41 Meth labs can be very portable. This is a small box lab. *Courtesy of MSA.*

Table 8.10 Illicit Lab Indicators			
Methamphetamine Lab Indicators	**Explosives Lab Indicators**	**Biological (Including Ricin) Lab Indicators**	**Chemical Weapons Lab Indicators**
Packages of cold medicine	Blasting caps	Agent samples: soil, blood, or organs; vials from commercial vendors	Chemical agent detection kits
Unusual numbers of matches	Detonation cord	Agar plates, petri dishes, liquid growth medium	Auto injector antidotes for nerve agents
Propane tanks with blue fittings	Wires, fuses, batteries, switches	Castor beans or plants	Cyanide salts
Hot plates, camp stoves, deep fryer, mantles	Tubes, pipes, potential shrapnel components (nails, bolts, broken glass)	Fermenters	Chemicals including: • Phosgene • Thiodiglycol • Thionyl chloride • Phosphorous trichloride
Capped two-liter or quart containers with clear (most commonly) to opaque liquids used in "one-pot" method	Ammonium nitrate	Drying and milling equipment	Commercial chemical glassware and chemical containers
Red phosphorous	Hexamine fuel tablets	Sterilization equipment	Chemistry textbooks
Lithium	Fuel oil	Incubators	Pressurized spray bottles for dissemination
Hydriodic acid	Urea nitrate	Animals in cages (alive or dead)	Animals in cages (alive or dead)

Illegal Hazmat Dumps

Illegal chemical dumps may occur in any jurisdiction. Some illegal disposers may consider lawful disposal too expensive or complicated. In other cases, the disposers may have used the materials in illicit labs or other illegal activities. Some chemical dumpsites may have existed years before any regulations prohibited such actions. Illicit labs are described in greater detail in Chapter 15.

Depending on the chemicals involved and the location of the site, discovery of an illegal dump may be considered an emergency. However, first responders may be the first person called to the scene. Illegal dumpsites may be expensive to clean up, and cleanup often requires state/provincial and/or federal/national involvement. Frequently, illegal dumps pose the following significant problems and hazards:

- **Unlabeled containers** — Disposers may have removed chemicals from their original containers or labels. Identification information may have been deliberately removed as well.

- **Mixed chemicals** — Containers and dump sites may have many different (and potentially incompatible) chemicals mixed together, making hazard and risk assessment difficult.

- **Aged chemicals** — Many chemicals become unstable when subjected to age and weathering in outside climates.

- **Environmental contamination** — When chemicals are dumped in ponds, streams, rivers, wetlands, and lakes, environmental contamination becomes a serious issue. Even if chemicals are not dumped in a body of water, leaking drums and other containers can pose a threat to groundwater sources.

Hazmat During and After Disasters

Natural disasters, such as floods, hurricanes, tornadoes, and earthquakes, can create hazmat incidents. For example, flood waters may move containers of all shapes, sizes, and contents (**Figure 8.42**). Floods can float tanks off foundations and sweep away entire chemical storage yards. Some containers may release their contents into the flood waters, and dead animals may also be present. Tornadoes and earthquakes may damage containers, rip up piping, or move containers around. Large-scale events, such as hurricanes, can cause problems with massive quantities of abandoned household hazardous waste (such as refrigerators with Freon cannot be taken to the local landfill) (**Figure 8.43**). Industries may be affected and experience chemical or oil spills. After the disaster, many hazardous materials containers may not be correctly placarded or labeled, resulting in identification difficulties.

Problems created by natural disasters may overwhelm local response capabilities. On-scene conditions may also limit defensive and offensive actions. Know who and how to call for assistance, and keep in mind the need to follow all hazardous waste rules for disposal of orphaned hazardous materials. In many cases, federal help will be needed and available. Include consideration of potential disasters in your preincident surveys and preplanning.

Figure 8.42 Natural disasters can cause hazmat containers to be moved far from their original locations. Often, they are damaged during this time. *Courtesy of FEMA News Photos; photo by Liz Roll.*

Figure 8.43 Massive quantities of household waste can be generated after a disaster. These propane containers were collected after one such incident in the U.S. *Courtesy of FEMA News Photos; photo by Greg Henshall.*

Flooding in Cedar Rapids, Iowa, 2008

From June 10, 2008, until about June 18, 2008, Cedar Rapids, Iowa, experienced a devastating flood FEMA called the flooding one of the worst disasters and most costly disasters in United States history. Approximately 9.2 square miles (15 square kilometers) experienced flooding, and the river depth was over 31 feet (9 m) deep on the morning of Friday, June 13, 2008. The flooding affected over 5,000 homes, businesses, industries and government buildings in Cedar Rapids. Other communities along the Cedar River, both up and down stream, experienced the same devastating damages and harm.

Many homes, businesses, industries, and government buildings had hazardous materials in them. The containers ranged in size and quantity from small cans of paint, to containers of gasoline, to compressed gas cylinders, 55 gallon (200 liter) drums, 500 gallon (1 893 liter) LPG tanks, underground gasoline storage tanks, and all kinds of containers in between. Hazardous materials containers were swept out of buildings and traveled with the flood waters and were deposited far from their original locations.

Once the flood waters receded, property owners began to clean out flood-damaged buildings and carry the debris to the side of the road for pick up by clean-up crews. Road sides were lined with different-sized containers waiting for pick up as well as refrigerators that needed to have the refrigerant removed before they could be disposed in a land fill. Other items included lawn mowers and other gasoline fueled equipment that needed to have the fuel tanks drained before being placed in a land fill. Additionally, other containers had floated into the community from upstream and now presented problems in the Cedar Rapids area of the Cedar River **(Figure 8.44)**.

The Cedar Rapids Fire Department Hazardous Materials Response Team and the Linn County Hazardous Materials Response Team were overwhelmed by hazardous material calls. They were also challenged with the following:

Figure 8.44 A bridge showing some of the drums and hazmat containers washed downstream during the flood. *Courtesy of Rich Mahaney*

- Collecting all hazardous material containers

- Identifying the products in the containers

- Transferring small amounts of products to larger collection containers

- Operating a hazard waste site collection operation

- Developing a plan to dispose of the entire mess

There were also issues of determining who would fund the cleanup. The Linn County Emergency Operations Center decided to contact and request help from the EPA Region 7 Hazardous Materials Response Team. The EPA Region 7 Team would take over most of the responsibilities for the hazmat problems with the assistance and support from the Cedar Rapids Fire Department and the Linn County hazardous Materials Response Teams and contractors that the EPA Region 7 Team brought in to help.

Figure 8.45 Thousands of sandbags contacted contaminated flood waters. *Courtesy of Rich Mahaney.*

Upon their arrival, the local hazmat teams briefed the EPA Team on the issues, problems, and challenges. The EPA Team made plans, assumed responsibility, located a site for collecting, storing, and identifying containers and products in a secure fenced location. The Team developed plans for container and product disposal. This location became a hazardous waste site where containers were stored in diked areas, separated by DOT hazard classes. The site met all of the rules of OSHA 1910.120. This operation continued for months until the EPA completed the mission.

Another interesting hazmat challenge was determining what to do with the thousands of sand bags that were used to help control the flood waters **(Figure 8.45)**. These sand bags were considered hazardous waste after the flood was over because they had been exposed to the dirty water from the Cedar River. They needed to be disposed of in a landfill.

Contributed by Rich Mahaney

Chapter Review

Answer the following questions to review the information provided in this chapter.

1. How is a terrorist organization different from a legitimate organization?

2. What are some of the cues that should prompt consideration of a terrorist attack?

3. What types of places are terrorists likely to target?

4. What types of WMDs may be readily available or easily made?

5. What should you do if you suspect a booby trap or secondary device?

6. What type of explosive are nonmilitary first responders most likely to encounter?

7. What are indicators of an explosive attack?

8. What are indicators of a chemical attack?

9. What types of chemical attacks agents are first responders most likely to encounter? Why?

10. How might chemical attacks differ from other hazmat incidents?

11. What four types of biological agents are likely to be used in a biological attack?

12. What are indicators of a probable biological attack and how do they differ from chemical or explosive attack indicators?

13. List types of radiological devices.

14. What are clues to an illicit laboratory?

15. What hazards are frequently encountered at illegal hazmat dumps?

16. What are some problems and hazards that disasters may create?

Implementing the Response: Personal Protective Equipment

Chapter Contents

Key Terms

Chemical Degradation 442
Emergency Breathing Support System
   (EBSS) 425
Encapsulating.......................................438
Flame-Resistant (FR) 436
Frostbite 458
Heat Cramps 455
Heat Exhaustion 455
Heat Rash..455
Heat Stroke...455
Hypothermia..458
Immediately Dangerous to Life and
   Health (IDLH).....................................423

Level A PPE..443
Level B PPE..443
Level C PPE..443
Level D PPE 443
Liquid Splash-Protective Clothing 438
Penetration..442
Permeation..441
Powered Air-Purifying Respirator
   (PAPR)..421
Supplied Air Respirator (SAR)...............425
Trench Foot 458
Vapor-Protective Clothing 439

NFPA Job Performance Requirements

This chapter provides information that addresses the following job performance requirements of NFPA 1072, *Standard for Hazardous Materials/Weapons of Mass Destruction Emergency Response Personnel Professional Qualifications (2017)*.

5.3.1	6.3.1	6.7.1
5.4.1	6.4.1	6.8.1
5.5.1	6.5.1	6.9.1
6.2.1	6.6.1	

Implementing the Response: Personal Protective Equipment

Learning Objectives

After reading this chapter, students will be able to:

1. Describe respiratory protection used at hazardous materials incidents. (5.3.1, 5.4.1, 6.2.1)

2. Explain varieties of protective clothing worn at hazardous materials incidents. (5.3.1, 5.4.1, 6.2.1)

3. Describe personal protective equipment ensembles used during hazardous materials incidents. (5.3.1, 6.2.1, 6.3.1, 6.4.1, 6.5.1, 6.6.1, 6.7.1, 6.8.1, 6.9.1)

4. Explain PPE related stresses. (5.4.1, 6.2.1)

5. Describe procedures for safely using PPE. (5.3.1, 5.4.1, 5.5.1, 6.2.1, 6.3.1, 6.4.1, 6.5.1, 6.6.1, 6.7.1, 6.8.1, 6.9.1)

6. Identify procedures for inspection, storage, testing, maintenance, and documentation of PPE. (5.4.1, 6.2.1)

7. Skill Sheet 9-1: Select appropriate PPE to address a hazardous materials scenario. (5.4.1, 6.2.1, 6.3.1, 6.4.1, 6.5.1, 6.6.1, 6.7.1, 6.8.1, 6.9.1)

8. Skill Sheet 9-2: Don, work in, and doff structural fire fighting personal protective equipment. (5.4.1, 6.2.1, 6.3.1, 6.4.1, 6.5.1, 6.6.1, 6.7.1, 6.8.1, 6.9.1)

9. Skill Sheet 9-3: Don, work in, and doff a Level C ensemble. (5.4, 6.2, 6.3, 6.4, 6.5, 6.6, 6.7, 6.8, 6.9)

10. Skill Sheet 9-4: Don, work in, and doff liquid splash protective clothing. (5.4, 6.2, 6.3, 6.4, 6.5, 6.6, 6.7, 6.8, 6.9)

11. Skill Sheet 9-5: Don, work in, and doff vapor protective clothing (5.4, 6.2, 6.3, 6.4, 6.5, 6.6, 6.7, 6.8, 6.9)

Chapter 9
Implementing the Response: Personal Protective Equipment

This chapter will explain the following topics:

- Respiratory protection
- Types of protective clothing
- PPE ensembles
- PPE related stress
- PPE use
- Classification, selection, inspection, testing, and maintenance of PPE

Respiratory Protection

Respiratory protection is a primary concern for first responders because inhalation is the most significant route of entry for hazardous materials. When correctly worn and used, protective breathing equipment protects the body from inhaling hazardous substances. Respiratory protection is, therefore, a vital part of any personal protective equipment (PPE) ensemble used at hazmat/WMD incidents **(Figure 9.1)**.

Responders use the following basic types of protective breathing equipment at hazmat/WMD incidents:

- Self-contained breathing apparatus (SCBA)
  - Closed circuit SCBA
  - Open circuit SCBA
- Supplied air respirators (SARs)
- Air-purifying respirators (APRs)
  - Particulate-removing
  - Vapor-and-gas-removing
  - Combination particulate- and vapor-and-gas-removing
- **Powered air-purifying respirators (PAPRs)**

Figure 9.1 Because inhalation is one of the most dangerous routes of entry for many hazardous materials, respiratory protection is extremely important.

Powered Air-Purifying Respirator (PAPR) — Motorized respirator that uses a filter to clean surrounding air, then delivers it to the wearer to breathe; typically includes a headpiece, breathing tube, and a blower/battery box that is worn on the belt.

WARNING!

You must wear your SCBA during emergency operations at terrorist/hazmat incidents until air monitoring and sampling determines other options are acceptable.

Each type of respiratory protection equipment has limits to its capabilities. For example, open-circuit self-contained breathing apparatus (SCBA) offers a limited working duration based upon the quantity of air (see SCBA section) contained within the apparatus' cylinder.

You may also need to be familiar with powered-air hoods, escape respirators, and combined respirators depending on what PPE you are issued. The sections that follow describe respiratory equipment (including basic limitations) as well as U.S. and international standards for respiratory protection.

Benicia, California, PPE Failure

On August 12, 1983, in Benicia, California, emergency response personnel in acid-resistant suits approached a leaking railroad tank car to conduct a size-up of the leak. Initial examinations of the tank car's placards indicated that the tank car contained anhydrous methylamine, a gas. However, the consist showed it was actually dimethylamine anhydrous. After initial assessment, mutual aid was requested from the San Francisco Fire Department, and a combined team of fire personnel and hazmat responders worked to cap the leak and transfer the gas into cylinders for safer transport.

During the incident, measures were taken to prevent air contamination, and all hazmat team members were decontaminated between trips back to the leaking tanker car. However, after three trips into the area without incident, the four-man team experienced equipment failure. The facepieces on the suits of all four members failed and shattered allowing vapors into the suits. Each firefighter reported to decontamination immediately and removed the damaged suits promptly. While an SCBA unit was able to shield each team member's respiratory system, each individual still suffered severe dermatitis because of skin exposure. Not only did the facepiece on each suit malfunction, but the chemical acted as a solvent and degraded the suit materials and adhesives at the seams to the extent that the heels of their fire boots had come unglued from the soles.

Along with the failure of the team's protective equipment, flashlights, and other tools being used were damaged and became unserviceable. Further support personnel were called in from chemical specialist companies and the leak was finally stopped without further equipment failure. Once the leak was under control, the Incident Commander elected to put a 24-hour hold on further operations due to a large influx of visitors expected in Benicia for a fair. In the end, the tanker leak required the work of five agencies to safely contain the gas.

Standards for Respiratory Protection at Hazmat/WMD Incidents

The U.S. Department of Homeland Security has adopted standards recommended by the National Institute for Occupational Safety and Health (NIOSH) and the NFPA for respiratory equipment to protect responders at hazmat/WMD incidents. These standards have been developed because of the extreme hazards associated with chemical (such as military nerve agents), biological, radioactive, and nuclear materials that could be used in terrorist attacks. NIOSH also certifies SCBA and recommends ways for responders to select and use protective clothing and respirators at biological incidents. Depending on their location, responders may also need to be familiar with standards regarding respiratory equipment issued by the ISO (International Standards Organization), the European Union, or other authorities. OSHA 29 *CFR* 1910.134 is the mandatory respiratory standard in the U.S.

Self-Contained Breathing Apparatus (SCBA)

Self-contained breathing apparatus (SCBA) is an atmosphere-supplying respirator for which the user carries the breathing-air supply. SCBA is perhaps the most important piece of PPE a responder can wear at a hazmat incident in terms of preventing dangerous exposures to harmful substances. The unit consists of the following:

- Facepiece
- Pressure regulator
- Air hoses
- Compressed air cylinder
- Harness assembly
- End-of-service-time indicators (low-air supply or low-pressure alarms)

In the U.S., NIOSH, and the Mine Safety and Health Administration (MSHA) must certify all SCBA for **immediately dangerous to life or health (IDLH)** atmospheres. Do not use SCBA that are not NIOSH/MSHA certified. The apparatus must also meet the design and testing criteria of NFPA 1981 in jurisdictions that have adopted that standard by law or ordinance. In addition, American National Standards Institute (ANSI) standards for eye protection apply to the facepiece lens design and testing.

NIOSH classifies SCBA as either closed-circuit or open-circuit. Two types of SCBA are currently being manufactured in closed- or open-circuit designs: pressure-demand, or positive-pressure. SCBA may also be either a high- or low-pressure type. However, use of only positive-pressure open-circuit or closed-circuit SCBA is allowed in incidents where personnel are exposed to hazardous materials **(Figure 9.2, p. 424)**.

Immediately Dangerous to Life and Health (IDLH) — Description of any atmosphere that poses an immediate hazard to life or produces immediate irreversible, debilitating effects on health; represents concentrations above which respiratory protection should be required. Expressed in parts per million (ppm) or milligrams per cubic meter (mg/m³), companion measurement to the permissible exposure limit (PEL).

Figure 9.2 Positive-pressure SCBA must be worn at hazmat incidents where personnel may be exposed to hazardous materials.

The advantages of using open-circuit SCBA-type respiratory protection are independence, maneuverability, and protection from toxic and/or asphyxiating atmospheres; however, several disadvantages are as follows:

- Weight of the units

- Limited air-supply duration

- Change in profile that may hinder mobility because of the configuration of the harness assembly and the location of the air cylinder

- Limited vision caused by facepiece fogging

- Limited communications if the facepiece is not equipped with a microphone or speaking diaphragm

NIOSH has entered into a Memorandum of Understanding with the National Institute of Standards and Technology (NIST), OSHA, and NFPA to jointly develop a certification program for SCBA used in emergency response to terrorist attacks. Working with the U.S. Army Soldier and Biological Chemical Command (SBCCOM), they developed a new set of respiratory protection standards and test procedures for SCBA used in situations involving WMD. Under this voluntary program, NIOSH issues a special approval and label identifying the SCBA as appropriate for use against chemical, biological, radiological, and nuclear agents. The SCBA certified under this program must meet the following minimum requirements:

- Approval under 42 *CFR* 84, Subpart H

- Compliance with NFPA 1981

- Special tests under 42 *CFR* 84.63(c):

  — *Chemical Agent Permeation and Penetration Resistance Against Distilled Sulfur Mustard (HD [military designation]) and Sarin (GB [military designation])*

  — *Laboratory Respirator Protection Level (LRPL)*

NIOSH maintains and disseminates a list of the SCBAs approved under this program and is entitled "'CBRN SCBA." This list contains the name of the approval holder, model, component parts, accessories, and rated duration. CBRN SCBA criteria are maintained as a separate category within the NIOSH Certified Equipment List.

NIOSH authorizes the use of an additional approval label on apparatus that demonstrate compliance to the *CBRN SCBA* criteria. This label is placed in a visible location on the SCBA backplate (on the upper corner or in the area of the cylinder neck) **(Figure 9.3)**. The addition of this label provides visible and easy identification of equipment for its appropriate use.

Supplied Air Respirators

The **supplied air respirator (SAR)** or airline respirator is an atmosphere-supplying respirator where the user does not carry the breathing air source. The apparatus usually consists of the following **(Figure 9.4)**:

- Facepiece
- Belt- or facepiece-mounted regulator
- Voice communications system
- Up to 300 feet (90 m) of air supply hose
- Emergency escape pack or **emergency breathing support system (EBSS)**
- Breathing air source (either cylinders mounted on a cart or a portable breathing-air compressor)

Supplied Air Respirator (SAR) — Atmosphere-supplying respirator for which the source of breathing air is not designed to be carried by the user; not certified for fire fighting operations. *Also known as* Airline Respirator System.

Emergency Breathing Support System (EBSS) — Escape-only respirator that provides sufficient self-contained breathing air to permit the wearer to safely exit the hazardous area; usually integrated into an airline supplied-air respirator system.

Figure 9.3 A NIOSH label showing compliance with CBRN criteria.

Figure 9.4 A typical SAR with EBSS. The EBSS should provide at least 5 minutes of air in case of emergency, enough to escape the hazard area into a safe atmosphere. *Courtesy of MSA.*

Because of the potential for damage to the air-supply hose, the EBSS provides enough air, usually 5, 10, or 15 minutes worth, for the user to escape a hazardous atmosphere. SAR apparatus are not certified for fire fighting operations because of the potential damage to the airline from heat, fire, or debris.

NIOSH classifies SARs as Type C respirators. Type C respirators are further divided into two approved types. One type consists of a regulator and facepiece only. The second type consists of a regulator, facepiece, and EBSS, and may also be referred to as a SAR with escape (egress) capabilities. The second type is used in confined-space environments, IDLH environments, or potential IDLH environments. Any type of SAR used at hazmat or CBR incidents must provide positive pressure to the facepiece.

SAR apparatus have the advantage of reducing physical stress to the wearer by removing the weight of the SCBA. The air supply line is a limitation because of the potential for mechanical or heat damage. In addition, the length of the airline (no more than 300 feet [90 m] from the air source) restricts mobility. Problems with hose entanglement must also be addressed. Other limitations are the same as those for SCBA: restricted vision and communications.

Air-Purifying Respirators

Air-Purifying Respirators (APRs) — Respirator that removes contaminants by passing ambient air through a filter, cartridge, or canister; may have a full or partial facepiece.

Air-purifying respirators (APRs) contain an air-purifying filter, canister, or cartridge that removes specific contaminants found in ambient air as it passes through the air-purifying element. Based on which cartridge, canister, or filter is being used, these purifying elements are generally divided into the three following types:

- Particulate-removing APRs

- Vapor-and-gas-removing APRs

- Combination particulate-removing and vapor-and-gas-removing APRs

APRs may be powered (PAPRs) or nonpowered. APRs do not supply oxygen or air from a separate source, and they protect only against specific contaminants at or below certain concentrations **(Figure 9.5)**. Combination filters combine particulate-removing elements with vapor-and-gas-removing elements in the same cartridge or canister.

Respirators with air-purifying filters may have either full facepieces that provide a complete seal to the face and protect the eyes, nose, and mouth or half facepieces that provide a complete seal to the face and protect the nose and mouth. Half-face respirators will NOT protect against CBR materials that can be absorbed through the skin or eyes and therefore are not recommended for use at hazmat/WMD incidents except in very specific situations (explosive attacks where the primary hazard is dust or particulates).

Figure 9.5 APRs should not be used where unknown atmospheric conditions exist. They do not provide oxygen from a separate source, and they only filter specific hazards. *Courtesy of U.S. Marine Corp, photo by Cpl. Alissa Schuning.*

Disposable filters, canisters, or cartridges are mounted on one or both sides of the facepiece. Canister or cartridge respirators pass the air through a filter, sorbent, catalyst, or combination of these items to remove specific contaminants from the air. The air can enter the system either from the external atmosphere through the filter or sorbent or when the user's exhalation combines with a catalyst to provide breathable air.

No single canister, filter, or cartridge protects against all chemical hazards. Therefore, you must know the hazards present in the atmosphere in order to select the appropriate canister, filter, or cartridge. Responders should be able to answer the following questions before deciding to use APRs for protection at an incident:

- What is the hazard?
- What is the oxygen level?
- Is the hazard a vapor or a gas?
- Is the hazard a particle or dust?
- Is there some combination of dust and vapors present?
- What concentrations are present?
- Does the material have a taste or smell?

WARNING!
Do not wear APRs during emergency operations where unknown atmospheric conditions exist. Wear APRs only in controlled atmospheres where the hazards present are completely understood and at least 19.5 percent oxygen is present.

APRs do not protect against oxygen-deficient or oxygen-enriched atmospheres, and they must not be used in situations where the atmosphere is immediately dangerous to life and health (IDLH). APRs can only be used if the hazardous material has a taste or smell. The three primary limitations of an APR are as follows:

- Limited life of its filters and canisters
- Need for constant monitoring of the contaminated atmosphere
- Need for a normal oxygen content of the atmosphere before use

Take the following precautions before using APRs:

- Know what chemicals/air contaminants are in the air.
- Know how much of the chemicals/air contaminants are in the air.
- Ensure that the oxygen level is between 19.5 and 23.5 percent.
- Ensure that atmospheric hazards are below IDLH conditions.

At hazmat/WMD incidents, APRs may be used after the hazards at the scene have been properly identified. In some circumstances, APRs may also be used in other situations (law enforcement working perimeters of the scene or EMS/medical personnel) and escape situations. APRs used for these CBRN situations should utilize a combination organic vapor/high efficiency particulate air (OV/HEPA) cartridge (see the sections that follow).

Particulate-Removing Filters

Particulate filters protect the user from particulates, including biological hazards, in the air. These filters may be used with half facepiece masks or full facepiece masks. Eye protection must be provided when the full facepiece mask is not worn.

Particulate-removing filters are divided into nine classes, three levels of filtration (95, 99, and 99.97 percent), and three categories of filter degradation. The following three categories of filter degradation indicate the use limitations of the filter:

● N — **N**ot resistant to oil

● R — **R**esistant to oil

● P — **P**resent when oil or nonoil lubricants are used.

Particulate-removing filters may be used to protect against toxic dusts, mists, metal fumes, asbestos, and some biological hazards **(Figure 9.6)**. High-efficiency particulate air (HEPA) filters used for medical emergencies must be 99.97 percent efficient, while 95 and 99 percent effective filters may be used depending on the health risk hazard.

Particle masks (also known as dust masks) are also classified as particulate-removing air-purifying filters **(Figure 9.7)**. These disposable masks protect the respiratory system from large-sized particulates. Particle masks provide very limited protection and should not be used to protect against chemical hazards or small particles such as asbestos fibers.

Vapor-and-Gas-Removing Filters

As the name implies, vapor-and-gas-removing cartridges and canisters are designed to protect against specific vapors and gases. They typically use some kind of sorbent material to remove the targeted vapor or gas from the air.

Figure 9.6 There may be high levels of contaminants in the air at hazmat/WMD incidents, including smoke, asbestos, and other particulates such as fiberglass. *Courtesy of FEMA News Photos, photo by Andrea Booher.*

Figure 9.7 A particle mask rated N-100, meaning it is not resistant to oil. Particle masks provide very limited protection and should not be used to protect against chemicals or small particles such as asbestos.

Figure 9.8 Most manufacturers color-code their filters/canisters for easy identification. *Courtesy of MSA.*

Individual cartridges and canisters are usually designed to protect against related groups of chemicals such as organic vapors or acid gases. Many manufacturers color-code their canisters and cartridges so it is easy to see what contaminant(s) the canister or cartridge is designed to protect against **(Figure 9.8)**. Manufacturers also provide information about contaminant concentration limitations.

Powered Air-Purifying Respirators (PAPR)

The PAPR uses a blower to pass contaminated air through a canister or filter to remove the contaminants and supply the purified air to the full facepiece. Because the facepiece is supplied with air under positive pressure, PAPRs offer a greater degree of safety than standard APRs in case of leaks or poor facial seals **(Figure 9.9)**. For this reason, PAPRs may be of use at hazmat/WMD incidents for personnel conducting decontamination operations and long-term operations. Air flow also makes PAPRs more comfortable to wear for many people.

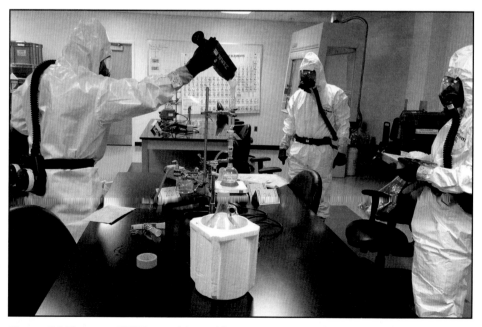

Figure 9.9 Because PAPRs provide positive pressure to the facepiece, they offer a greater degree of safety than APRs.

Several types of PAPR are available. Some units are supplied with a small blower and are battery operated. The small size allows users to wear one on their belts. Other units have a stationary blower (usually mounted on a vehicle) that is connected by a long, flexible tube to the respirator facepiece.

WARNING!
Do not use PAPRs in explosive or potentially explosive atmospheres.

As with all APRs, PAPRs should only be used in situations where the atmospheric hazards are understood and at least 19.5 percent oxygen is present. PAPRs are not safe to wear in atmospheres where potential respiratory hazards are unidentified, nor should they be used during initial emergency operations before the atmospheric hazards have been confirmed. Continuous atmospheric monitoring is needed to ensure the safety of the responder.

Combined Respirators

Combination respirators include SAR/SCBA, PAPR/SCBA, and SAR/APR **(Figure 9.10)**. These respirators can provide flexibility and extend work duration times in hazardous areas. SAR/SCBAs will operate in either SAR or SCBA mode, for example, using SCBA mode for entry and exit while switching to SAR mode for extended work. PAPR/SCBA is a bulky combination. When using PAPR/SCBA combinations, it is necessary to know the composition of the atmosphere. The PAPR mode allows a longer operational period if conditions are safe for use. SAR/APRs will also operate in either mode, but the same limitations that apply to regular APRs apply when operating in the APR mode. All combinations require specific training to use.

Supplied-Air Hoods

Powered- and supplied-air hoods provide loose fitting, lightweight respiratory protection that can be worn with glasses, facial hair, and beards **(Figure 9.11)**. Hospitals, emergency rooms, and other organizations use these hoods as an alternative to other respirators, in part, because they require no fit testing and are simple to use.

Respiratory Equipment Limitations

The following are equipment and air supply limitations:

- **Limited visibility** — Facepieces reduce peripheral vision, and facepiece fogging can reduce overall vision.

- **Decreased ability to communicate** — Facepieces hinder voice communication.

- **Increased weight** — Depending on the model, the protective breathing equipment can add 25 to 35 pounds (12.5 to 17.5 kg) of weight to the emergency responder.

- **Decreased mobility** — The increase in weight and splinting effect of the harness straps reduce the wearer's mobility.

- **Inadequate oxygen levels** — APRs cannot be worn in IDLH or oxygen-deficient atmospheres.

- **Chemical specific** — APRs can only be used to protect against certain chemicals. The specific type of cartridge depends on the chemical to which the wearer is exposed.

- **Psychological stress** — Facepieces may cause some users to feel confined or claustrophobic.

Additionally, open- and closed-circuit SCBA have maximum air-supply durations that limit the amount of time a first responder has to perform the tasks at hand. Non-NIOSH certified SCBAs may offer only limited protection in environments containing chemical warfare agents.

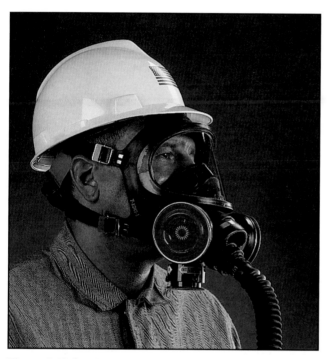

Figure 9.10 Combination respirators enable users to switch modes of operation between a combination of SAR, APR, PAPR, and SCBA depending on the equipment's design. *Courtesy of MSA.*

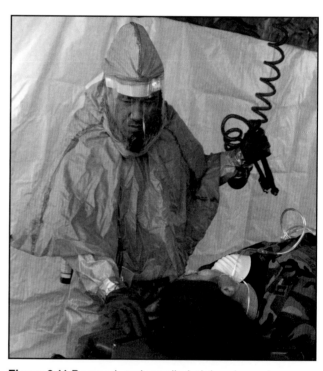

Figure 9.11 Powered- and supplied-air hoods can be worn with glasses, facial hair, and beards. *Courtesy of the U.S. Air Force, photo by Airman 1st Class Bradley A. Lail.*

CAUTION
Personnel wearing respiratory equipment must have good physical conditioning, mental soundness, and emotional stability due to the physiological and psychological stresses of wearing PPE.

Protective Clothing Overview

Protective clothing must be worn whenever an emergency responder faces potential hazards arising from thermal hazards and chemical, biological, or radiological exposure. Skin contact with hazardous materials can cause a variety of problems, including chemical burns, allergic reactions and rashes, diseases, and absorption of toxic materials into the body. Protective clothing is designed to prevent these problems. Body armor and bomb suits can be worn to protect against ballistic hazards and shrapnel from explosives **(Figure 9.12)**.

No single combination or ensemble of protective equipment (even with respiratory protection), can protect against all hazards. For example, fumes and chemical vapors can penetrate fire fighting turnout coats and pants, so the protection they provide is not complete. Similarly, chemical-protective clothing (CPC) offers no protection from fires **(Figures 9.13a and b)**.

An ensemble of appropriate PPE protects the skin, eyes, face, hearing, hands, feet, body, head, and respiratory system. While technological advances are being made to improve the versatility of all types of PPE (for example, developing more chemical-resistant turnouts and more fire-resistant CPC), you must understand your PPE's limitations in order to stay safe.

CAUTION

When you respond to hazmat/WMD incidents, you must have the appropriate personal protective equipment (PPE) to perform your mission safely and effectively.

The correct use of your PPE requires special training and instruction. When operating at the scene of an incident, use your PPE in accordance with standard operating procedures and manufacturer's recommendations, under the guidance of a hazardous materials technician, or under the supervision of an allied professional (someone with the knowledge, skills, and competence to provide correct guidance), as appropriate. The sections that follow explain the various standards that apply to protective clothing as well as the different clothing types that will commonly be used at hazmat/WMD incidents.

Standards for Protective Clothing and Equipment at Hazmat/WMD Incidents

As with respiratory protection, the U.S. Department of Homeland Security (DHS) has adopted NIOSH and NFPA standards for protective clothing used at hazmat/WMD incidents. Primarily, these apply to clothing worn at chemical and biological incidents in regards to chemical-protective clothing (CPC). However, you should be familiar with any standards pertaining

Figure 9.12 Body armor and bomb suits protect against ballistic hazards and shrapnel from explosives. *Courtesy of the U.S. Marine Corps, photo by Cpl. Antonio Rosas.*

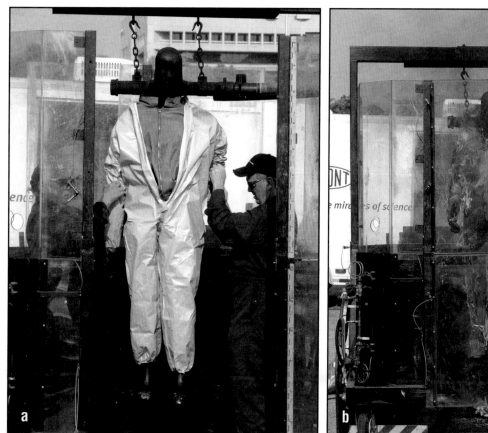

Figures 9.13a and b CPC is not always flame resistant. (a) A manikin is dressed in CPC and subjected to a very brief flash fire. (b) The CPC continues to burn and melt.

to design, certification, and testing requirements of any type of protective clothing, including body armor, structural fire fighting gear, and bomb suits. Depending on their location, responders may also need to be familiar with standards regarding respiratory equipment issued by the ISO, the European Union, or other authorities.

Structural Firefighters' Protective Clothing

Structural fire fighting clothing is not a substitute for chemical-protective clothing; however, it does provide some protection against many hazardous materials. The atmospheres in burning buildings, after all, are filled with toxic gases, and modern structural firefighters' protective clothing with SCBA provides adequate protection against some of those hazards **(Figure 9.14, p. 434)**. The multiple layers of the coat and pants may provide short-term exposure protection from such materials as liquid chemicals; however, there are limitations to this protection. For example, structural fire fighting clothing is neither corrosive-resistant nor vapor-tight. Liquids can soak through, acids and bases can dissolve or deteriorate the outer layers, and gases and vapors can penetrate the garment **(Figure 9.15, p. 434)**. Gaps in structural fire fighting clothing occur at the neck, wrists, waist, and the point where the pants and boots overlap.

Some hazardous materials can permeate (pass through at the molecular level) and remain in structural fire fighting clothing. Chemicals absorbed into the equipment can cause repeated exposure or a later reaction with another

Figure 9.14 Structural fire fighting protective clothing will provide limited protection against many hazardous materials.

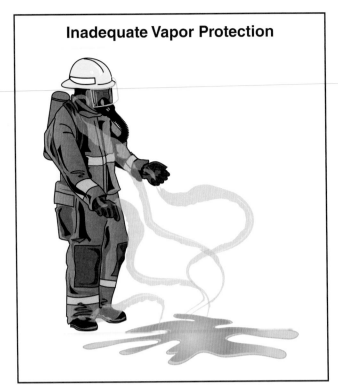

Inadequate Vapor Protection

Figure 9.15 Structural fire fighting protective clothing does not provide complete protection against chemical hazards. Toxic liquids can soak through the fabric; acids and bases can dissolve or deteriorate the outer layers; and vapors, fumes, and gases can penetrate through gaps in the material and ensemble.

Figure 9.16 Some jurisdictions allow the use of structural fire fighting protective clothing and SCBA to perform rescues.

chemical. In addition, chemicals can permeate the rubber, leather or neoprene in boots, gloves, kneepads, and SCBA facepieces making them unsafe to use. It may be necessary to discard equipment exposed to permeating chemicals.

While there is much debate among experts as to the degree of protection provided by structural fire fighting protective clothing (and SCBA) at hazmat/WMD incidents, there may be circumstances under which it will provide limited protection for short-term duration operations such as an immediate rescue **(Figure 9.16)**. Agency emergency response plans and SOPs should specify the conditions and circumstances under which it is appropriate for emergency responders to rely on firefighter structural protective clothing and SCBA during operations at hazmat/WMD incidents.

Structural fire fighting protective clothing will provide protection against thermal damage in an explosive attack, but it provides limited or no protection against projectiles, shrapnel, and other mechanical effects from a blast. It will provide adequate protection against some types of radiological materials, but not others. In cases where biological agents are strictly respiratory hazards, structural fire fighting protective clothing with SCBA may provide adequate protection. However, in any case where skin contact is potentially hazardous, it is not sufficient. Properly identify materials in order to make this determination. Any time a terrorist attack is suspected, but not positively identified, assume that responders are wearing only structural fire fighting protective clothing with SCBA and are at some level of increased risk from potential hazards such as explosives, radiological materials, and chemical or biological weapons.

High-Temperature Protective Clothing

High-temperature-protective clothing is designed to protect the wearer from short-term exposures to high temperature in situations where heat levels exceed the capabilities of standard fire fighting protective clothing. This type of clothing is usually of limited use in dealing with chemical hazards. The following are two types of high-temperature clothing that are available:

1. **Proximity suits** — Permit close approach to fires for rescue, fire-suppression, and property-conservation activities such as in aircraft rescue and fire fighting or other fire fighting operations involving flammable liquids **(Figure 9.17)**. Such suits provide greater heat protection than standard structural fire fighting protective clothing.

2. **Fire-entry suits** — Allow a person to work in total flame environments for short periods of time; provide short duration and close proximity protection at radiant heat temperatures as high as 2,000°F (1 100°C). Each suit has a specific use and is not interchangeable.

Figure 9.17 Proximity suits are frequently used in aircraft rescue and fire fighting. *Courtesy of U.S. Marine Corp, photo by Cpl. William Hester.*

Several limitations to high temperature-protective clothing are as follows:

- Contributes to heat stress by not allowing the body to release excess heat
- Is bulky
- Limits wearer's vision
- Limits wearer's mobility
- Limits communication
- Requires frequent and extensive training for efficient and safe use
- Is expensive to purchase
- Integrity of suit is designed for limited exposure time

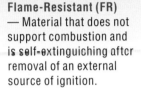

Flame-Resistant (FR) — Material that does not support combustion and is self-extinguishing after removal of an external source of ignition.

Flame-Resistant Protective Clothing

Many hazardous materials response personnel wear everyday **flame-resistant (FR)** work apparel. This apparel is designed for continuous wear during work activities in designated areas in which there is minimal risk for exposure to the following:

- Hot or molten materials
- Hot surfaces
- Radiant heat
- Flash fires
- Flame
- Electrical arc discharge

This protective apparel will not ignite or melt under exposure to fire or radiant heat **(Figure 9.18)**. Flame resistance in material can be achieved by using inherently flame-resistant fibers or by treating the material with a flame retardant chemical:

- **Inherently Flame-Resistant (IFR)** — Fibers that do not support combustion due to their chemical structure. They are flame-resistant without chemical additives. The high-temperature-resistant polymers in IFR fibers provide an inert barrier between the wearer and the hazard. Protective properties of the fabric are permanent and cannot be washed out or removed.

- **Flame Retardant** — A chemical compound that can be incorporated into a textile item during manufacture or applied to a fiber, fabric, or other textile item during processing to reduce its flammability. These fire retardants can be removed under some circumstances, such as washing.

Figure 9.18 Flame-resistant clothing will not ignite or melt when exposed to fire or extreme heat.

Chemical-Protective Clothing (CPC)

The purpose of chemical-protective clothing (CPC) is to shield or isolate individuals from the chemical, physical, and biological hazards that may be encountered during hazardous materials operations. CPC is made from a variety of different materials, none of which protects against all types of chemicals. Each material provides protection against certain chemicals or products, but only limited or no protection against others. The manufacturer of a particular suit must provide a list of chemicals for which the suit is effective. Selection of appropriate CPC depends on the specific chemical and on the specific tasks to be performed by the wearer.

> **WARNING!**
> CPC is not intended for fire fighting activities, nor for protection from hot liquids, steam, molten metals, welding, electrical arc, flammable atmospheres, explosive environments or thermal radiation.

CPC is designed to afford the wearer a known degree of protection from a known type, concentration, and length of exposure to a hazardous material, but only if it is fitted properly and worn correctly. Improperly worn equipment can expose and endanger the wearer.

Most protective clothing is designed to be impermeable to moisture, thus limiting the transfer of heat from the body through natural evaporation. This can contribute to heat disorders in hot environments. Other factors include the garment's degradation, permeation, and penetration abilities and its service life. A written management program regarding selection and use of CPC is required. Regardless of the type of CPC worn at an incident, it must be decontaminated. Responders who may be called upon to wear CPC must be familiar with (and comfortable going through) their local procedures for technical decontamination (see Chapter 10).

Design and testing standards generally recognize two types of CPC: liquid splash-protective clothing and vapor-protective clothing. The sections that follow describe these two types and, in addition, explain operations where CPC is required, written management programs that specify CPC use, the ways in which CPC can be damaged, and considerations for the service life of CPC.

> **WARNING!**
> No single type of CPC protects against all chemical hazards.

> **WARNING!**
> You must have sufficient training to operate in conditions requiring the use of chemical-protective clothing.

Liquid Splash-Protective Clothing

Liquid splash-protective clothing is designed to protect users from chemical liquid splashes but not against chemical vapors or gases **(Figure 9.19)**. NFPA 1992 sets the minimum design criteria for one type of liquid splash-protective clothing. Liquid splash-protective clothing can be **encapsulating** or nonencapsulating **(Figure 9.20)**.

An encapsulating suit is a single, one-piece garment that protects against splashes or, in the case of vapor-protective encapsulating suits, also against vapors and gases. Boots and gloves are sometimes separate, or attached and replaceable. Two primary limitations to fully encapsulating suits are as follows:

1. Impairs worker mobility, vision, and communication

2. Traps body heat which might necessitate a cooling system, particularly when SCBA is also worn

A nonencapsulating suit commonly consists of a one-piece coverall, but sometimes is composed of individual pieces such as a jacket, hood, pants, or bib overalls. Gaps between pant cuffs and boots and between gloves and sleeves are usually taped closed. Limitations to nonencapsulating suits include the following:

• Protects against splashes and dusts but not against gases and vapors

• Does not provide full body coverage: parts of head and neck are often exposed

• Traps body heat and contributes to heat stress

Figure 9.19 Liquid-splash protective clothing is not designed to be completely gas- and vapor-tight.

Figure 9.20 An encapsulating suit covers the entire body and the SCBA.

Encapsulating and nonencapsulating liquid splash-protective clothing are not resistant to heat or flames, nor do they protect against projectiles or shrapnel. Liquid splash-protective clothing is made from the same materials used for vapor-protective suits (see following section).

When used as part of a protective ensemble, liquid splash-protective ensembles may include an SCBA, an airline (supplied-air respirator [SAR]), or a full-face, air-purifying, canister-equipped respirator. Class 3 ensembles described in NFPA 1994 use liquid splash-protective clothing. This type of protective clothing is also a component of EPA Level B chemical protection ensembles.

Vapor-Protective Clothing

Vapor-protective clothing protects the wearer against chemical vapors or gases and offers a greater level of protection than liquid splash-protective clothing **(Figure 9.21)**. NFPA 1991 specifies requirements for a minimum level of protection for response personnel facing exposure to specified chemicals. This standard sets performance requirements for vapor-tight, totally encapsulating chemical-protective (TECP) suits and includes rigid chemical-resistance and flame-resistance tests and a permeation test against twenty-one challenge chemicals. NFPA 1991 also includes standards for performance tests in simulated conditions.

Vapor-protective ensembles must be worn with positive-pressure SCBA or combination SCBA/SAR. Vapor-protective ensembles are components of ensembles to be used at chemical and biological hazmat/WMD incidents.

Vapor-Protective Clothing — Gas-tight chemical-protective clothing designed to meet NFPA 1991, *Standard on Vapor-Protective Ensembles for Hazardous Materials Emergencies, 2016 Edition*; part of an EPA Level A ensemble.

Figure 9.21 Vapor-protective clothing provides the best protection against dangerous gases and vapors such as toxic and corrosive gases.

These suits are also primarily used as part of a Level A protective ensemble, providing the greatest degree of protection against respiratory, eye, or skin damage from hazardous vapors, gases, particulates, sudden splash, immersion, or contact with hazardous materials.

Vapor-protective suits have the following limitations:

- When exposed to fire, they melt and burn. They cannot be used in potentially flammable atmospheres.

- Do not protect the user against all chemical hazards.

- Impair mobility, vision, and communication **(Figure 9.22)**.

- Do not allow body heat to escape, so can contribute to heat stress, which may require the use of a cooling vest.

Vapor-protective ensembles are made from a variety of special materials. No single combination of protective equipment and clothing is capable of protecting a person against all hazards.

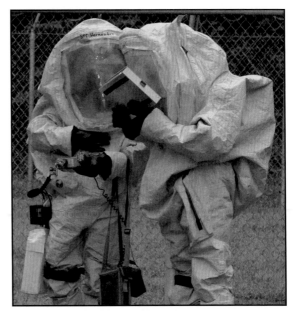

Figure 9.22 Vapor-protective clothing can significantly impair vision, mobility, and communication.

Mission Specific Operations Requiring Use of Chemical-Protective Clothing

Chemical-protective clothing must be worn in certain circumstances. Without regard to the level of training required to perform them, these are operations that may require the use of CPC:

- Site survey
- Rescue
- Spill mitigation
- Emergency monitoring
- Decontamination
- Evacuation

If responders are involved in any of these activities, consideration must be given to what type of protective equipment is necessary given the known and/or unknown hazards present at the scene. Always follow AHJ SOP/Gs for operations requiring use of chemical-protective clothing.

Written Management Programs

All emergency response organizations that routinely use CPC must establish a written Chemical-Protective Clothing Program and Respiratory Protection Management Program. A written management program includes policy statements, procedures, and guidelines. Copies must be made available to all personnel who may use CPC in the course of their duties or job.

The two basic objectives of any management program are protecting the user from safety and health hazards and preventing injury to the user from incorrect use or malfunction. To accomplish these goals, a comprehensive CPC management program includes the following elements:

- Hazard identification
- Medical monitoring
- Environmental surveillance
- Selection, care, testing, and maintenance
- Training

Permeation, Degradation, and Penetration

CPC's effectiveness can be reduced by three actions: permeation, degradation, and penetration. These are also characteristics that must be considered when choosing and using protective ensembles.

Permeation is a process that occurs when a chemical passes through a fabric on a molecular level **(Figure 9.23)**. In most cases, there is no visible evidence of chemicals permeating a material **(Figures 9.24a and b)**. The rate at which a compound permeates CPC depends on factors such as the chemical properties of the compound, nature of the protective barrier(s) in the CPC, and concentration of the chemical on the surface of the CPC. Most CPC manufacturers provide charts on breakthrough time (time it takes for a chemical to permeate the material of a protective suit) for a wide range of chemical compounds. Permeation data also includes information about the permeation rate (or the speed) at which the chemical moves through the CPC material after it breaks through.

> **Permeation** — Process in which a chemical passes through a protective material on a molecular level.

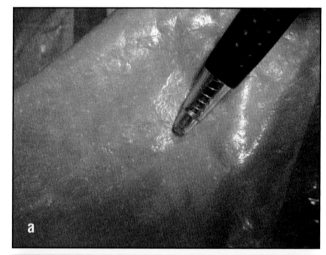

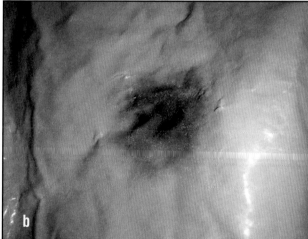

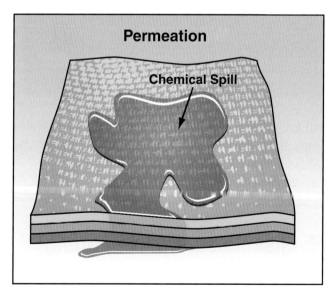

Figure 9.23 Permeation occurs when a chemical passes through a fabric or material on a molecular level.

Figures 9.24a and b A quick inspection of this suit's exterior might miss this small area of permeation (a). The damage is far more visible on the interior (b). *Courtesy of Barry Lindley.*

> **WARNING!**
> Decontamination will not stop permeation from occurring.

Chemical Degradation — Process that occurs when the characteristics of a material are altered through contact with chemical substances.

Penetration — Process in which a hazardous material enters an opening or puncture in a protective material. *See* Routes of Entry.

Chemical degradation occurs when the characteristics of a material are altered through contact with chemical substances. Examples include cracking, brittleness, and other changes in the structural characteristics of the garment **(Figure 9.25)**. The most common observations of material degradation are discoloration, swelling, loss of physical strength, or deterioration.

Penetration is a process that occurs when a hazardous material enters an opening or a puncture in a protective material **(Figure 9.26)**. Rips, tears, and cuts in protective materials — as well as unsealed seams, buttonholes, and zippers — are considered penetration failures. Often such openings are the result of faulty manufacture or problems with the inherent design of the suit.

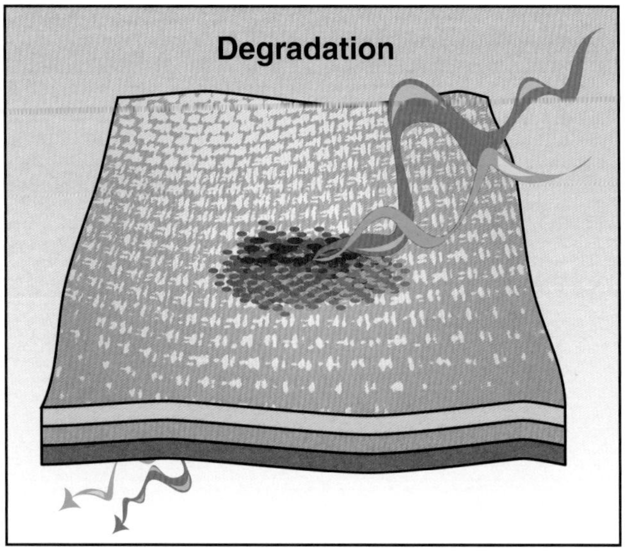

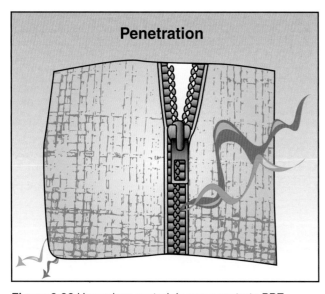

Figure 9.25 Contact with a chemical(s) can alter a material's characteristics.

Figure 9.26 Hazardous materials can penetrate PPE through gaps, tears, punctures, or other openings.

Service Life

Each piece of CPC has a specific service life over which the clothing is able to adequately protect the wearer. For example, a Saranex/Tyvek garment may be designed to be a coverall (covering the wearer's torso, arms, and legs) intended for liquid splash-protection and single use. If any garment is contaminated, remove it from service. Always follow AHJ SOP/Gs and manufacturer's specifications in regards to serviceability.

> **WARNING!**
> Never use CPC that is beyond its expiration date and/or has exceeded its service life.

All potentially contaminated CPC requires proper decontamination when the wearer leaves a potentially hazardous area. Chapter 10, Decontamination, provides more information about contamination and decontamination of CPC.

PPE Ensembles, Classification, and Selection

To achieve adequate protection, an ensemble of respiratory equipment and clothing is typically used. When determining the appropriate ensemble, consider the hazards present and the actions that need to be per-

formed. For example, simple protective clothing, such as gloves and a work uniform, in combination with a face shield or safety goggles may be sufficient to prevent exposure to biological hazards such as bloodborne pathogens. At the other end of the spectrum, a vapor-protective, totally-encapsulating suit combined with positive-pressure SCBA may be needed when dealing with extremely hazardous, corrosive, and/or toxic vapors or gases, especially if the hazardous materials can damage other types of PPE and readily be absorbed through the skin.

WARNING!
No single type of PPE protects against all hazards.

While the EPA has established a set of chemical-protective PPE ensembles providing certain protection levels that are commonly used by fire and emergency service organizations, other organizations such as law enforcement, industrial responders, and the military may have their own standard operating procedures or equivalent procedures guiding the choice and use of appropriate combinations of PPE. Law enforcement personnel may be equipped with a far different PPE ensemble than a firefighter, hazmat technician, civil support response team, or environmental cleanup person working at the same hazmat/WMD incident. The sections that follow describe a variety of factors concerning PPE ensembles.

CAUTION
Always follow your agency's SOP/Gs in determining the level of PPE necessary to perform a task.

Levels of Protection

Different levels of protective equipment are used at incidents involving hazardous materials/WMD: **Level A**, **Level B**, **Level C**, and **Level D (Figures 9.27 a-d, p. 444)**. They can be used as the starting point for ensemble creation; however, each ensemble must be tailored to the specific situation in order to provide the most appropriate level of protection.

Selecting protective clothing and equipment by how they are designed or configured alone is not sufficient to ensure adequate protection at hazmat incidents. Just having the right components to form an ensemble is not enough. The EPA levels of protection do not define or specify what performance (for example, vapor protection or liquid splash protection) the selected clothing or equipment must offer, and they do not identically mirror the performance requirements of NFPA performance standards.

Level A PPE — Highest level of skin, respiratory, and eye protection that can be given by personal protective equipment (PPE), as specified by the U.S. Environmental Protection Agency (EPA); consists of positive-pressure self-contained breathing apparatus, totally encapsulating chemical-protective suit, inner and outer gloves, and chemical-resistant boots.

Level B PPE — Personal protective equipment that affords the highest level of respiratory protection, but a lesser level of skin protection; consists of positive-pressure self-contained breathing apparatus, hooded chemical-protective suit, inner and outer gloves, and chemical-resistant boots.

Level C PPE — Personal protective equipment that affords a lesser level of respiratory and skin protection than levels A or B; consists of full-face or half-mask APR, hooded chemical-resistant suit, inner and outer gloves, and chemical-resistant boots.

Level D PPE — Personal protective equipment that affords the lowest level of respiratory and skin protection; consists of coveralls, gloves, and chemical-resistant boots or shoes.

Figures 9.27a-d EPA levels of protective ensembles: (a) Level A, (b) Level B, (c) Level C, (d) Level D.

Level A

The Level A ensemble provides the highest level of protection against vapors, gases, mists, and particles for the respiratory tract, eyes, and skin. Level A protection provides very little protection against fire. Operations Level responders do not typically operate in situations requiring Level A protection. However, you must be appropriately trained to wear Level A PPE if you are required to wear it.

The elements of Level A ensembles are:

- **Components** — Ensemble requirements include:
  - Positive-pressure, full facepiece, SCBA, or positive-pressure airline respirator with escape SCBA approved by NIOSH
  - Vapor-protective suits: Totally-Encapsulated Chemical Protective (TECP) suits constructed of protective-clothing materials that meet the following criteria:
    - o Cover the wearer's torso, head, arms, and legs
    - o Include boots and gloves that may either be an integral part of the suit or separate and tightly attached
    - o Enclose the wearer completely by itself or in combination with the wearer's respiratory equipment, gloves, and boots
    - o Provide equivalent chemical-resistance protection for all components of a TECP suit (such as relief valves, seams, and closure assemblies)
    - o May meet the requirements in NFPA 1991
  - Coveralls (optional)
  - Long underwear (optional)
  - Chemical-resistant outer gloves
  - Chemical-resistant inner gloves
  - Chemical-resistant boots with steel toe and shank
  - Hard hat (under suit) (optional)
  - Disposable protective suit, gloves, and boots (can be worn over totally encapsulating suit, depending on suit construction)
  - Two-way radios (worn inside encapsulating suit)
- **Protection provided** — Highest available level of respiratory, skin, and eye protection from solid, liquid, and gaseous chemicals.
- **Use** — Level A ensembles are used when risk analysis indicates it is appropriate. For example, Level A protection may be appropriate when site operations and work functions involve a high potential for splash, immersion, or exposure to unexpected vapors, gases, or particulates of material that are harmful to skin or capable of damaging or being absorbed through the intact skin.

Level B

Level B protection requires a garment that includes an SCBA or a supplied-air respirator and provides protection against splashes from a hazardous chemical. This ensemble is worn when the highest level of respiratory protection is necessary but a lesser level of skin protection is needed. Level B protection provides very little protection against fire. The Level B CPC ensemble may be encapsulating or nonencapsulating.

The elements of Level B ensembles are:

- **Components** — Ensemble requirements include:
  - Positive-pressure, full facepiece, SCBA, or positive-pressure airline respirator with escape SCBA approved by NIOSH
  - Hooded chemical-resistant clothing that may meet the requirements of NFPA 1992 (overalls and long-sleeved jacket, coveralls, one- or two-piece [encapsulating or nonencapsulating] chemical splash suit, and disposable chemical-resistant overalls)
  - Coveralls (optional)
  - Chemical-resistant outer gloves
  - Chemical-resistant inner gloves
  - Chemical-resistant boots with steel toe and shank
  - Disposable, chemical-resistant outer boot covers (optional)
  - Hard hat (outside or on top of nonencapsulating suits or under encapsulating suits)
  - Two-way radios (worn inside encapsulating suit or outside nonencapsulating suit)
  - Face shield (optional)
- **Protection provided** — Ensembles provide the same level of respiratory protection as Level A but have less skin protection. Ensembles provide liquid splash-protection, but no protection against chemical vapors or gases.
- **Use** — Ensembles may be used in the following situations:
  - Type and atmospheric concentration of substances have been identified and require a high level of respiratory protection but less skin protection.
  - Atmosphere contains less than 19.5 percent oxygen or more than 23.5 percent oxygen.
  - Presence of incompletely identified vapors or gases is indicated by a direct-reading organic vapor detection instrument, but the vapors and gases are known not to contain high levels of chemicals harmful to skin or capable of being absorbed through intact skin.
  - Presence of liquids or particulates is indicated, but they are known not to contain high levels of chemicals harmful to skin or capable of being absorbed through intact skin.

Level C

Level C protection differs from Level B in the area of equipment needed for respiratory protection. Level C is composed of a splash-protecting garment and an air-purifying device (APR or PAPR). Level C protection provides very little protection against fire. Level C protection includes any of the various types of APRs. Periodic air monitoring is required when using this level of PPE. Level C equipment is only used by emergency response personnel under the following conditions:

- The specific material is known.
- The specific material has been measured.
- This protection level is approved by the IC after all qualifying conditions for APRs and PAPRs have been met:
  — The product is known.
  — An appropriate filter is available.
  — The atmospheric oxygen concentration is between 19.5 to 23.5 percent.
  — The atmosphere is not IDLH.

  The elements of Level C ensembles are:
- **Components** — Ensemble requirements include:
  — Full-face or half-mask APRs, NIOSH approved
  — Hooded chemical-resistant clothing (overalls, two-piece chemical-splash suit, and disposable chemical-resistant overalls)
  — Coveralls (optional)
  — Chemical-resistant outer gloves
  — Chemical-resistant inner gloves
  — Chemical-resistant boots with steel toe and shank
  — Disposable, chemical-resistant outer boot covers (optional)
  — Hard hat
  — Escape mask (optional)
  — Two-way radios (worn under outside protective clothing)
  — Face shield (optional)
- **Protection provided** — Ensembles provide the same level of skin protection as Level B but have a lower level of respiratory protection. Ensembles provide liquid splash-protection but no protection from chemical vapors or gases on the skin.
- **Use** — Ensembles may be used in the following situations:
  — Atmospheric contaminants, liquid splashes, or other direct contact will not adversely affect exposed skin or be absorbed through any exposed skin.
  — Types of air contaminants have been identified, concentrations have been measured, and an APR is available that can remove the contaminants.
  — All criteria for the use of APRs are met.
  — Atmospheric concentration of chemicals does not exceed IDLH levels. The atmosphere must contain between 19.5 and 23.5 percent oxygen

Level D

Level D ensembles consist of typical work uniforms, street clothing, or coveralls. Level D protection can be worn only when no atmospheric hazards exist.

The elements of Level D ensembles are:
- **Components** — Ensemble requirements include:
  — Coveralls

— Gloves (optional)

— Chemical-resistant boots/shoes with steel toe and shank

— Disposable, chemical-resistant outer boot covers (optional)

— Safety glasses or chemical splash goggles

— Hard hat

— Escape device in case of accidental release and the need to immediately escape the area (optional)

— Face shield (optional)

- **Protection provided** — Ensembles provide no respiratory protection and minimal skin protection.

- **Use** — Ensembles are typically not worn in the hot zone and are not acceptable for hazmat emergency response above the Awareness Level. Level D ensembles are used when both of the following conditions exist:

  — Atmosphere contains no hazard.

  — Work functions preclude splashes, immersion, or the potential for unexpected inhalation of or contact with hazardous levels of any chemicals.

PPE Selection Factors

The risks and potential hazards present at an incident will determine the PPE needed. Many available sources can be consulted to determine which type and what level of PPE to use at hazmat incidents/terrorist attacks depending on the circumstances and hazards at the scene. SOP/Gs may also provide guidance for situations involving rescue and initial responses. At the Operations Level, responders must operate under the guidance of allied professionals, hazmat technicians, the emergency response plan, or SOP/Gs. Once the IAP is developed, the Site Safety Plan will spell out PPE requirements for tasks performed at the incident.

In general, the higher the level of PPE, the greater the associated risks will be. For any given situation, personnel should select equipment and clothing that provide an adequate level of protection. Overprotection, as well as underprotection, can be hazardous and should be avoided.

Determining the PPE level needed to enter the hot zone is ultimately the Incident Commander's responsibility, but all responders should understand the selection process. **Skill Sheet 9-1** provides steps for selecting appropriate PPE at a hazmat incident.

Consider the following general selection factors:

- **Chemical and physical hazards** —Consider and prioritize both chemical and physical hazards **(Table 9.1)**. Depending on what materials are present, any combination of hazards may need to be protected against.

  NOTE: Many types of PPE do not provide thermal protection.

- **Monitoring and detection readings** — Identify hazards that need to be addressed in PPE selection by monitoring and detecting readings.

- **Physical environment** —The ensemble components must be appropriate for whatever varied environmental conditions are present:

  — Industrial settings, highways, or residential areas

Table 9.1
Effectiveness of Typical Fire Service PPE Ensembles
for Specific Hazards

Fire Service Ensembles	Flammables/ Incendiaries*	Toxics/ Chemical Warfare Agents	Corrosives	Biological Hazards	Radiological Hazards	Explosives/ Ballistics
Standard Structural-Fire Fighting Ensemble Including SCBA**	Adequate	Inadequate for extended hot zone use***	Inadequate for extended hot zone use**	Varies Inadequate for incidents in which biological agents/ hazards or dissemination methods are unidentified or may still be occurring May be adequate in circumstances where agent/ hazard and dissemination methods are known	Adequate for Alpha and Beta radiation Inadequate for Gamma radiation	Inadequate for protection against explosives and ballistics Adequate for operations after an explosion not involving other CBRNE hazards
Chemical Protective Ensembles	Inadequate	EPA Level A, B, or C (NFPA 1994 Class 1, 2 and 3) as appropriate	EPA Level A, B, or C (NFPA 1994 Class 1, 2 and 3) as appropriate	EPA Level A, B, or C (NFPA 1994 Class 1, 2 and 3) as appropriate	Adequate for Alpha and Beta radiation Inadequate for Gamma radiation	Inadequate for protection against explosives or ballistics Adequate for operations after an explosion involving other CBR hazards as applicable
USAR Ensembles (without turnout gear)	Inadequate	Inadequate	Inadequate	Inadequate	Adequate for Alpha radiation with appropriate respiratory protection Inadequate for Beta and Gamma radiation	Inadequate for protection against explosives or ballistics Adequate for rescue and mitigation operations after an explosion not involving other CBR hazards

* Flammability should be given first priority when selecting PPE

**Not including turnout gear designed with improved CBR protection

***May be adequate for short duration exposures in certain situations (for example, during rescue operations, as determined by the Incident Commander, SOPs, or emergency response plan, etc.), and depending on the incident specifics.

— Indoors or outdoors

— Extremely hot or cold environments

— Uncluttered or rugged sites

— Required activities involving entering confined spaces, lifting heavy items, climbing ladders, or crawling on the ground

- **Exposure duration** — The ensemble components' protective qualities may be limited by many factors including exposure levels, material chemical resistance, and air supply. Assume the worst-case exposure so that appropriate safety margins can be added to the ensemble wear time.

- **Available protective clothing or equipment** — An array of different clothing or equipment should be available to personnel to meet all intended applications. Reliance on one particular clothing type or equipment item may severely limit the ability to handle a broad range of hazardous materials or chemical exposures. In its acquisition of equipment and protective clothing, the responsible authority should provide a high degree of flexibility while choosing protective clothing and equipment that is easily integrated and provides protection against each conceivable hazard.

- **Compliance with regulations** — Agencies responsible for responding to CBR incidents should select equipment in accordance with regulatory standards for response to such incidents, such as NIOSH standards and NFPA 1994.

Protective clothing selection factors include the following:

- **Clothing design** — Manufacturers sell clothing in a variety of styles and configurations.

  Design considerations include the following:

  — Clothing configuration

  — Seam and closure construction

  — Components and options

  — Sizes

  — Ease of donning and doffing

  — Clothing construction

  — Accommodation of other selected ensemble equipment

  — Comfort

  — Restriction of mobility

- **Material chemical resistance** — The chosen material(s) must resist permeation, degradation, and penetration by the respective chemicals. Mixtures of chemicals can be significantly more aggressive towards protective clothing materials than any single chemical alone. One permeating chemical may pull another with it through the material. Other situations may involve unidentified substances. Details:

  — Very little test data are available for chemical mixtures. If clothing must be used without test data, choose clothing that demonstrates the best chemical resistance against the widest range of chemicals.

  — In cases of chemical mixtures and unknowns, serious consideration must be given to selecting protective clothing.

- **Physical properties** — Clothing materials may offer wide ranges of physical qualities in terms of strength, resistance to physical hazards, and operation in extreme environmental conditions. Comprehensive performance standards (such as those from NFPA) set specific limits on these material properties, but only for limited applications such as emergency response. Users may also need to ask manufacturers the following questions:

  — Does the material have sufficient strength to withstand the physical demands of the tasks at hand?

  — Will the material resist tears, punctures, cuts, and abrasions?

  — Will the material withstand repeated use after contamination and decontamination?

  — Is the material flexible or pliable enough to allow users to perform needed tasks?

  — Will the material maintain its protective integrity and flexibility under hot and cold extremes?

  — Is the material subject to creation of a static electrical charge and discharge that could provide an ignition source?

  — Is the material flame-resistant or self-extinguishing (if these hazards are present)?

  — Are garment seams in the clothing constructed so they provide the same physical integrity as the garment material?

- **Ease of decontamination** — The degree of difficulty in decontaminating protective clothing may dictate whether disposable clothing, reusable clothing, or a combination of both is used.

- **Ease of maintenance and service** — The difficulty and expense of maintaining equipment should be considered before purchase.

- **Interoperability with other types of equipment** — Interoperability issues should be considered, for example, whether or not communications equipment can be integrated into the ensemble.

- **Cost** — Equipment needed to meet response requirements must be purchased within budget constraints.

Typical Ensembles of Response Personnel

The ensemble worn at an incident will vary depending on the mission of the responder. PPE for urban search and rescue (US&R) personnel will differ from that of hazmat response teams, and so forth. However, responders of any discipline must be aware of what hazards are present at the incident and what PPE is necessary to protect against the hazards to which they may be exposed. For example, if respiratory hazards exist at the incident, all personnel who might be exposed to these hazards must wear respiratory protection regardless of their mission. It is important that personnel who may need to use such PPE be trained to do so. The sections that follow will outline some of the ensembles used by emergency responders at hazmat/WMD incidents, keeping in mind that the nature of the incident will dictate the PPE requirements.

Fire Service Ensembles

Fire service personnel will wear ensembles appropriate for their mission at the incident, including typical fire fighting operations (such as fire extinguishment), hazardous materials response, and urban search and rescue **(Figure 9.28)**. **Table 9.2** shows a conservative estimate of the effectiveness of typical fire service PPE ensembles in the hot zone of hazmat/WMD incidents. EMS ensembles are described in a later, separate section.

The majority of responders will initially be wearing structural fire fighting protective clothing ensembles (turnout gear) that may offer limited protection against hazmat/WMD hazards. These ensembles may be appropriate for conducting

Figure 9.28 With appropriate training, fire service responders may use chemical protective ensembles at hazmat incidents.

some operations (such as rescue) at hazmat/WMD incidents given appropriate protective measures such as limited exposure times.

Responders trained to use CPC at hazmat events may don EPA Level A or B ensembles as described in previous sections. Chemical-protective ensembles must be designed to protect the wearer's upper and lower torso, head, hands, and feet. Ensemble elements must include protective garments, protective gloves, and protective footwear. Ensembles must accommodate appropriate respiratory protection.

Law Enforcement Ensembles

Law enforcement personnel typically wear ballistic protection and no respiratory protection. PPE may also be assigned for emergency situations such as terrorist attacks. Law enforcement personnel must be trained to use the PPE they are assigned, whatever it may be. **Table 9.3, p. 454**, shows a conservative estimate of the effectiveness of typical law enforcement PPE ensembles in the hot zone of hazmat/WMD incidents.

Body armor is designed to protect against ballistic threats. Body armor is commonly used by law enforcement personnel, but some fire service and EMS agencies use it, particularly when operating in dangerous situations or areas where attacks might be likely. Body armor should always be replaced if it has been impacted or damaged.

Table 9.2
Effectiveness of Typical Fire Service PPE Ensembles in the Hot Zone of CBRNE Incidents

Fire Service Ensembles	Incendiaries /Fires	Chemical Warfare Agents	TIMs	Biological Agents	Radiological Hazards	Explosives /Ballistics
Standard Structural-Fire Fighting Ensemble Including SCBA*	Adequate	Inadequate for extended hot zone use**	Inadequate for extended hot zone use**	Varies Inadequate for incidents in which agents or dissemination methods are unidentified or may still be occuring May be adequate in circumstances where agent and dissemination methods are known	Adequate for Alpha and Beta radiation Inadequate for Gamma radiation	Inadequate for protection against explosives and ballistics Adequate for operations after an explosion not involving other CBR hazards
Haz-Mat/ Chemical Protective Ensembles	Inadequate	EPA Level A and B (NFPA® 1994 Class 1, 2 and 3) as appropriate	EPA Level A, B, or C Class 1, 2 and 3 as appropriate	EPA Level A, B, or C (NFPA® 1994 Class 1, 2 and 3) as appropriate	Adequate for Alpha and Beta radiation Inadequate for Gamma radiation	Inadequate for protection against explosives and ballistics Adequate for operations after an explosion involving other CBR hazards as applicable
US&R Ensembles (without turnout gear)	Inadequate	Inadequate	Inadequate	Inadequate	Adequate for Alpha radiation with appropriate respiratory protection Inadequate for Beta and Gamma radiation	Inadequate for protection against explosives and ballistics Adequate for rescue and mitigation operations after an explosion not involving other CBR hazards

*Not including turnout gear designed with improved CBR protection

**May be adequate for short duration exposures in certain situations (for example, during rescue operations, as determined by the Incident Commander, SOPs, or emergency response plan, etc.), and depending on the agent.

Bomb disposal suits must provide full body protection against fragmentation, overpressure, impact, and heat. Normally designed to meet appropriate military specifications, they incorporate high-tech materials and ballistic plates in a head-to-toe ensemble **(Figure 9.29, p. 454).** Helmets are usually designed with built-in communications capabilities and forced-air ventilation systems, some of which also provide protection/filtration against CBR materials. Bomb suits are heavy and significantly impair dexterity and range of motion. New technology is being incorporated into next generation bomb suits to improve protection against CBR materials.

Table 9.3
Effectiveness of Typical Law Enforcement PPE in the Hot Zone of CBRNE Incidents

Law Enforcement PPE	Incendiaries/ Fires	Chemical Warfare Agents	TIMs	Biological Agents	Radiological Agents	Explosives/ Ballistics
Body Armor (w/ duty uniform)	Inadequate	Inadequate	Inadequate	Inadequate	Inadequate	Protection provided per type of armor
Haz Mat / Chemical Protective Ensembles	Inadequate	EPA Level A and B (NFPA® 1994 Class 1, 2 and 3) as appropriate	EPA Level A, B, or C (NFPA® 1994 Class 1, 2 and 3) as appropriate	EPA Level A, B, or C (NFPA® 1994 Class 1, 2 and 3) as appropriate	Adequate for Alpha and Beta radiation Inadequate for Gamma radiation	Inadequate for protection against explosives or ballistics Adequate for operations after an explosion involving other CBR hazards as applicable
Bomb Suits*	Adequate for flash fires	Inadequate	Inadequate	Varies Inadequate for incidents in which agents or dissemination methods are unidentified or may still be occurring	Adequate for Alpha and Beta radiation with appropriate respiratory protection Inadequate for Gamma radiation	Inadequate

*Not including bomb suits designed with improved CBR protection

Figure 9.29 Bomb suits incorporate high-tech materials and ballistic plates in a head-to-toe ensemble. *Courtesy of the U.S. Marine Corps, photo by Cpl. Brian A. Tuthill.*

EMS Ensembles

EMS PPE must provide blood- and body-fluid pathogen barrier protection. PPE ensembles should include outer protective garments, gloves, footwear, and face protection. The items might be configured to cover only part of the upper or lower torso such as arms with sleeve protectors, torso front with apron-styled garments, and face with face shields.

At hazmat incidents, EMS personnel not working in the hot zone may achieve some protection using a high-quality respirator, butyl rubber gloves, and a commercial chemical overgarment (elastic wrists and hood closures with built-in boots) that provides some liquid-droplet and vapor protection **(Figure 9.30)**. This level of protection may or may not be adequate for personnel conducting triage and decontamination operations in the warm zone, depending on circumstances.

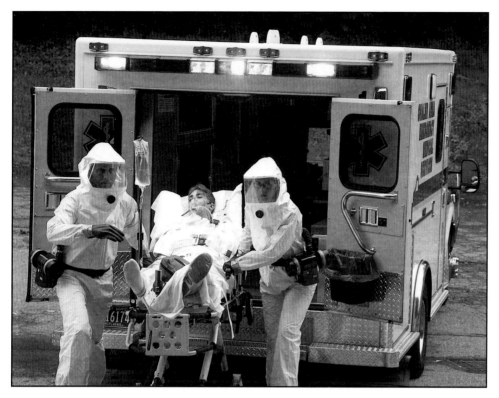

Figure 9.30 EMS ensembles consisting of high-quality respirators, butyl rubber gloves, and commercial chemical overgarments. *Courtesy of MSA.*

PPE-Related Stresses

Most PPE inhibits your body's ability to disperse heat and moisture, which is magnified because you are usually performing strenuous work while wearing the equipment. Thus, wearing PPE may increase your risk of heat-related disorders. However, when working in cold climates, you may also suffer cold-related disorders. CPC is not designed to provide insulation against the cold. Taking preventive measures will help protect you from these potential problems.

NOTE: Medical monitoring is required when environmental factors put you at risk.

Heat Emergencies

Wearing PPE or other special full-body protective clothing puts you at considerable risk of developing health effects ranging from transient heat fatigue to serious illness (heat stroke) or even death. Heat disorders include:

- **Heat stroke** (the most serious; see Safety Alert)
- **Heat exhaustion**
- **Heat cramps**
- **Heat rashes**

Heat Stroke — Heat illness in which the body's heat regulating mechanism fails; symptoms include (a) high fever of 105° to 106° F (40.5° to 41.1° C), (b) dry, red, hot skin, (c) rapid, strong pulse, and (d) deep breaths or convulsions. May result in coma or even death. *Also known as* Sunstroke.

Heat Exhaustion — Heat illness caused by exposure to excessive heat; symptoms include weakness, cold and clammy skin, heavy perspiration, rapid and shallow breathing, weak pulse, dizziness, and sometimes unconsciousness.

Heat Cramps — Heat illness resulting from prolonged exposure to high temperatures; characterized by excessive sweating, muscle cramps in the abdomen and legs, faintness, dizziness, and exhaustion.

Heat Rash — Condition that develops from continuous exposure to heat and humid air; aggravated by clothing that rubs the skin. Reduces the individual's tolerance to heat.

Heat Stroke

Heat stroke occurs when the body's system of temperature regulation fails and body temperature rises to critical levels. This condition is caused by a combination of highly variable factors, and its occurrence is difficult to predict. Heat stroke is a serious medical emergency and requires immediate medical treatment and transport to a medical care facility. The primary signs and symptoms of heat stroke are:

- Confusion
- Irrational behavior
- Loss of consciousness
- Convulsions
- Lack of sweating (usually)
- Hot, dry skin
- Abnormally high body temperature (for example, a rectal temperature of 105.8°F [41°C])

When the body's temperature becomes too high, it causes death. The elevated metabolic temperatures caused by a combination of workload and environmental heat load, both of which contribute to heat stroke, are also highly variable and difficult to predict. If you or any other responders show signs of possible heat stroke, obtain professional medical treatment immediately.

Heat-Exposure Prevention

Responders wearing protective clothing need to be monitored for the effects of heat exposure. Methods to prevent and/or reduce the effects of heat exposure include the following:

- **Fluid consumption** — Use water or commercial body-fluid-replenishment drink mixes to prevent dehydration. You should drink generous amounts of fluids both before and during operations. Drinking 7 ounces (200 ml) of fluid every 15 to 20 minutes is better than drinking large quantities once an hour. Balanced diets normally provide enough salts to avoid cramping problems. Details:

  — Before working, drinking chilled water is good.

  — After a work period in protective clothing and an increase in core temperature, drinking room-temperature water is better. It is not as severe a shock to the body.

- **Air cooling** — Wear long cotton undergarments, moisture-wicking modern fabrics, or similar clothing to provide natural body ventilation. Once PPE has been removed, blowing air can help to evaporate sweat, thereby cooling the skin. Wind, fans, blowers, and misters can provide air movement. However, when ambient air temperatures and humidity are high, air movement may provide only limited benefit.

- **Ice cooling** — Use ice to cool the body; however, use care not to damage skin with direct contact with ice, as well as to not cool off an individual too quickly. Ice will also melt relatively quickly. Ice cooling vests are available.

- **Water cooling** — Use water to cool the body. When water (even sweat) evaporates from skin, it cools. Provide mobile showers and misting facilities or evaporative cooling vests. Water cooling becomes less effective as air humidity increases and water temperatures rise.

- **Cooling vests** — Wear cooling vests beneath PPE **(Figure 9.31)**. Cooling vest technologies may use the technologies detailed in **Table 9.4**. Cooling vests may be cumbersome, bulky, and they may impair movement.

- **Rest/rehab areas** — Provide shade, humidity changers (misters), and air-conditioned areas for resting **(Figure 9.32, p. 458)**.

- **Work rotation** — Rotate responders exposed to extreme temperatures or those performing difficult tasks frequently.

- **Proper liquids** — Avoid liquids such as alcohol, coffee, and caffeinated drinks (or minimize their intake) before working. These beverages can contribute to dehydration and heat stress.

- **Physical fitness** — Encourage responders to maintain good physical fitness.

NOTE: NFPA 1584, *Standard on the Rehabilitation Process for Members During Emergency Operations and Training Exercises*, addresses many of these issues.

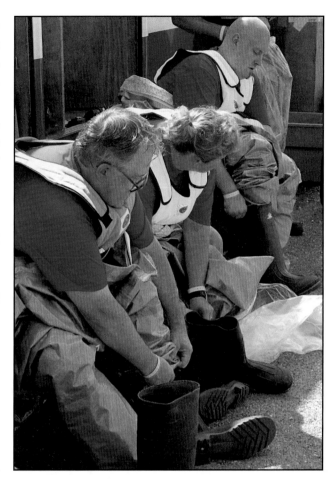

Figure 9.31 Some agencies may use cooling vests to combat heat illness when using CPC.

Table 9.4
Examples of Cooling Vest Technologies

Passive Technologies	Method of Cooling
Ice	Ice packs in vest provides cooling
Evaporation	Wetted vest material provides evaporative cooling
Gel	Cold gel in vest provides cooling
Phase Change	Phase change material in vest slowly solidifies to maintain a consistent, cool temperature
Active Technologies*	**Method of Cooling**
Circulating Fluids	Chilled liquid is circulated through small tubes in vest
Forced Air	Air is circulated through tubes and blown into an air-space above the body

* Require power to operate

Figure 9.32 Rehab can help prevent heat stress by allowing responders to cool off and rest. *Courtesy of Ron Jeffers.*

Trench Foot — Foot condition resulting from prolonged exposure to damp conditions or immersion in water; symptoms include tingling and/or itching, pain, swelling, cold and blotchy skin, numbness, and a prickly or heavy feeling in the foot. In severe cases, blisters can form, after which skin and tissue die and fall off.

Frostbite — Local tissue damage caused by prolonged exposure to extreme cold.

Hypothermia — Abnormally low body temperature.

Cold Emergencies

Cold temperatures may be caused by weather and/or other conditions such as exposure to cryogenic liquids. Prolonged exposure to freezing temperatures can result in health problems as serious as **trench foot**, **frostbite**, and **hypothermia**.

The primary environmental conditions that cause cold-related stress are low temperatures, high/cool winds, dampness, cold water, and standing/walking/working on cold, snowy and/or icy surfaces. Wind chill, a combination of temperature and velocity, is a crucial factor to evaluate when working outside. For example, when the actual air temperature of the wind is 40°F (4.5°C) and its velocity is 35 mph (55 km/h), the exposed skin experiences conditions equivalent to the still-air temperature of 11°F (-12°C) **(Table 9.5)**. Rapid heat loss may occur when exposed to high winds and cold temperatures.

You can prevent cold disorders with the following precautions:

- Being active
- Rehabbing in a warm area
- Wearing warm clothing/layers
- Dressing appropriately
- Avoiding cold beverages

Psychological Issues

The use of CPC can be a confining experience. Whether working in a Level A fully encapsulated suit or even a lower level suit, CPC will be much more confining than structural fire fighting clothing and equipment. This confinement may cause claustrophobia in responders. In addition to the confinement of the protective clothing, just knowing the hazards of the chemicals involved may be disconcerting to the responder.

Table 9.5
Wind Chill Chart

Temperature (°F)

Calm	40	35	30	25	20	15	10	5	0	-5	-10	-15	-20	-25	-30	-35	-40	-45
5	36	31	25	19	13	7	1	-5	-11	-16	-22	-28	-34	-40	-46	-52	-57	-63
10	34	27	21	15	9	3	-4	-10	-16	-22	-28	-35	-41	-47	-53	-59	-66	-72
15	32	25	19	13	6	0	-7	-13	-19	-26	-32	-39	-45	-51	-58	-64	-71	-77
20	30	24	17	11	4	-2	-9	-15	-22	-29	-35	-42	-48	-55	-61	-68	-74	-81
25	29	23	16	9	3	-4	-11	-17	-24	-31	-37	-44	-51	-58	-64	-71	-78	-84
30	28	22	15	8	1	-5	-12	-19	-26	-33	-39	-46	-53	-60	-67	-73	-80	-87
35	28	21	14	7	0	-7	-14	-21	-27	-34	-41	-48	-55	-62	-69	-76	-82	-89
40	27	20	13	6	-1	-8	-15	-22	-29	-36	-43	-50	-57	-64	-71	-78	-84	-91
45	26	19	12	5	-2	-9	-16	-23	-30	-37	-44	-51	-58	-65	-72	-79	-86	-93
50	26	19	12	4	-3	-10	-17	-24	-31	-38	-45	-52	-60	-67	-74	-81	-88	-95
55	25	18	11	4	-3	-11	-18	-25	-32	-39	-46	-54	-61	-68	-75	-82	-89	-97
60	25	17	10	3	-4	-11	-19	-26	-33	-40	-48	-55	-62	-69	-76	-84	-91	-98

WIND SPEED (mph)

Frostbite occurs in 15 minutes or less

Courtesy of NOAA

Psychological issues may be preventable through adequate training. As the responder works with the equipment and gains familiarity, confidence will build. Still, the mind is a very powerful organ and severe claustrophobia may be debilitating for a responder. If this is the case, emergency response in CPC may not be suitable for some responders.

Medical Monitoring

You should conduct medical monitoring before responders wearing PPE enter the warm and hot zones (pre-entry monitoring) as well as after leaving these zones (post-entry monitoring) as directed by the authority having jurisdiction. Check such things as vital signs, hydration, skin, mental status, and medical history. Each organization needs to establish written medical monitoring guidelines that establish minimum and maximum values for those evaluations. A post-medical monitoring follow-up is also recommended.

Keep exposure records in conjunction with any medical records for employees who have worked in proximity to the hazard. Because exposures to hazardous chemicals may not present any signs or symptoms for many years, it is a legal requirement to retain medical records per the AHJ. Exposure records should include the following information:

- Type of exposure
- Length of exposure

- Description of PPE used

- Type of decontamination used including any decontamination solutions

- On scene and follow-up medical attention and/or assistance

PPE Use

There is much more to the use of PPE ensembles than just putting on the ensemble. While it is imperative that you be proficient in donning the equipment, you must also be able to safely function in the suit while performing both simple and challenging tasks. As familiarity increases, so will the comfort levels. Increased comfort levels will help reduce your stress. This reduced stress can help increase both your proficiency and work time.

Pre-Entry Inspection

Check equipment before you enter into a hazardous atmosphere **(Figure 9.33)**. A thorough visual inspection should uncover any defects or deformities in the protective equipment. In addition to the visual inspection, confirm all pressure test completion dates, and conduct an operational check of the following items:

- Breathing apparatus

- All zippers and closures

- Valves

- Communications equipment

- Any equipment that will be taken or used in the hot zone

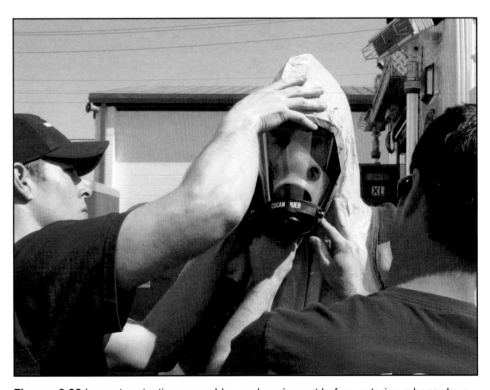

Figures 9.33 Inspect protective ensembles and equipment before entering a hazardous area.

READY

Components of the entry briefing can be summarized by the acronym, READY:

R Radio — Do you have a radio? Is it set on the correct channel? Perform a radio check.

E Equipment — What equipment is required, and do personnel know how to use it? What are the emergency signals?

A Air — Is your air cylinder full? What is the predetermined working time?

D Details — What is your team requested to do, typically no more than three items? The team should repeat this back to ensure accurate information transfer.

Y Yes — If all of the above steps are complete, the entry team should finish donning PPE and proceed to do entry.

Safety and Emergency Procedures

In addition to issues such as cooling, preventing dehydration, and medical monitoring, there are other safety and emergency issues involved with wearing PPE. For example, responders using PPE at hazmat incidents must be familiar with their local procedures for going through the technical decontamination process. Anytime emergency responders are to enter an IDLH atmosphere, they should always work in teams of two or more (buddy system), with a minimum of two equally trained and equipped personnel outside the IDLH atmosphere ready to rescue other emergency responders should the need arise (backup personnel). Responders should operate within their accountability systems, and know their evacuation and escape procedures.

Safety Briefing

A safety briefing will be conducted before responders enter the hot zone. The safety briefing will cover relevant information including:

- Incident status (based on the preliminary evaluation and subsequent updates)
- Identified hazards
- Description of the site
- Tasks to be performed
- Expected duration of the tasks
- Escape route or area of refuge
- PPE and health monitoring requirements
- Incident monitoring requirements
- Notification of identified risks
- Communication procedures, including hand signals

   NOTE: After using PPE at an incident, fill out any associated reports or documentation as required by the AHJ.

Air Management

Anytime a limited air supply such as SCBA is worn, air management is an important consideration. Emergency procedures should be developed for responder loss of air supply. These procedures may vary depending on the AHJ. To ensure adequate work time, calculate estimated times for the following tasks:

- Walk to the incident
- Return from the incident
- Decon
- Work time
- Safety time (extra time allocated for emergency use)

Air must be allocated for these estimated times. Responders should have a plan in place for dealing with air emergencies.

Many organizations have SOP/Gs that explain calculations for doing this and/or designate maximum entry times (such as 20 minutes) based on the air supply available. It may be necessary to stock SCBA cylinders of different sizes and volumes in an agency's cache of equipment **(Table 9.6)**.

NOTE: A cylinder's service pressure and rating are not a true indication of the overall work time. The one constant is the amount of air the cylinder will contain when it is full.

Table 9.6
Breathing Air Cylinder Capacities

Rated Duration	Pressure	Volume
30-minute	2,216 psi (15 290 kPa)	45 ft³ (1 270 L) cylinders
30-minute	4,500 psi (31 000 kPa)	45 ft³ (1 270 L) cylinders
45-minute	3,000 psi (21 000 kPa)	66 ft³ (1 870 L) cylinders
45-minute	4,500 psi (31 000 kPa)	66 ft³ (1 870 L) cylinders
60-minute	4,500 psi (31 000 kPa)	87 ft³ (2 460 L) cylinders

- Rated duration does not indicate the actual amount of time that the cylinder will provide air.

Contamination Avoidance

The terms *contamination* and *exposure* are sometimes used interchangeably, but the concepts are actually very different. Contamination can be defined as a condition of impurity resulting from contact or mixture with a foreign substance. In other words, the hazardous material has to touch or be touched by another object. In contrast, exposure means that a hazardous material has entered or potentially entered your body via the routes of entry, for example, by swallowing, breathing, or contacting skin or mucous membranes.

Most hazmat responses will likely include contamination, which can increase the risk of exposure. Because of this, avoid contamination as best as possible. As a responder, you should consider the following best practices:

- Always try to reduce any contact with the product. Avoid walking through and touching the product whenever possible.

- Do not kneel or sit on the ground in CPC, if possible. Contact avoidance is paramount, but allowing a suit to come in contact with the ground may cause chafing or abrasion on the suit allowing for faster suit degradation.

  NOTE: If avoidance is not possible and you need to protect the suit from damage, put something between your suit and the ground/contamination (options include: thick cardboard; rug; visqueen; absorbent pillows, pigs, booms, socks, and pads; knee pads).

- Protect monitoring instruments as best as possible.

Communications

Communication capabilities are required for all levels of personal protection. Communication devices may be integrated into PPE. Other nonemergency communication methods can include predesignated hand signals, motions, and gestures.

Signals for entry-team emergencies such as loss of air supply, medical emergency, or suit failure should also be designated. If possible, entry teams, backup personnel, and appropriate safety personnel at the scene should have their own designated radio channel.

Should responders lose radio communications or operate in an atmosphere not allowing radio communications, a backup system must be part of the operational plan. Hand signals used as the backup plan should be simple, easy to remember, and distinguishable from a distance. Hand signals should be designated for the following situations **(Table 9.7, p. 464)**:

- Loss of air supply

- Loss of suit integrity

- Responder down from injury or illness

- Emergency (waving hands above head)

- Loss of radio communications

- I am okay, or situation okay (tap on head with one hand or thumbs up)

  NOTE: Follow hand signals specified by AHJ.

All responders should follow local protocols for evacuation situations. Typically, these protocols will involve notifying the appropriate personnel (such as the Entry Team Leader and/or Hazmat Safety Officer), and exiting the hot zone as quickly as possible.

The remaining capabilities of equipment should also be communicated during evacuation situations. For example, if air supply is lost while wearing a vapor-protective suit, there is a limited amount of air in the suit itself that can be breathed if the SCBA facepiece or regulator is removed.

In addition to entry-team signals, include an emergency evacuation signal for all responders in the incident action plan. The emergency signal should indicate that an immediate exit from the hot zone is necessary. The signal should be audible (air horns) and also broadcast over the radio frequency.

Table 9.7
Hand Signals

Loss of
air supply

Loss of
suit integrity

Buddy down

Loss of radio
communications

Donning and Doffing of PPE

You should always train with the protective clothing that you will be using in the field. The donning process can be quite time consuming and confusing for a user who is not totally familiar with the garments. Instructions should be included with all PPE for the total donning and doffing process. **Skill sheets 9.2-5** provide steps for donning, working in, and doffing PPE.

Donning of PPE

While it is imperative that you follow the manufacturer and department recommendations for the donning of PPE, the following guidelines will outline generic donning procedures that may be included in an agency's procedures:

- Preselect the donning and doffing area in the cold zone as close to the entry point as possible. It should be clearly delineated.

- Ensure that the donning and doffing area is isolated from distractions and sheltered from the elements, if possible.

- Select an area that is large enough to accommodate all personnel involved in the donning and doffing procedures.

- Plan for as many people (including assistants) as needed to be involved in the donning procedures **(Figure 9.34)**.

- Before starting the donning process, each entry and backup team member should be medically evaluated based on AHJ procedures.

- Continue hydration per AHJ procedures.

- Conduct a mission briefing before the donning process to ensure that all members are attentive and there are no distractions. The mission briefing should include the specifics of the mission such as IAP and site safety plan.

- Deploy chemical-protective clothing in an organized manner.

- Check all equipment visually and operationally prior to donning to ensure proper working order.

- Ensure that the entry team members have removed all their personal effects such as rings, wallets, badges, watches, and pins.

- Don appropriate undergarments at this time, if applicable.

Seat the entry team so that their breathing apparatus is accommodated **(Figure 9.35)**. The physical activity of the donning process should be conducted by assistants to allow the entry and backup personnel the opportunity to rest and reduce stress levels. Once the donning process has begun, the donning supervisor should prepare both the entry team and the backup team at the same rate. The teams should remain ready and off air until the entry order has been given.

Once the entry order has been given, lead the entry teams to the entry access point. The safety officer should perform a final check of all equipment and closures before the teams are allowed to enter the hazard area. The backup team should be left off air and in a resting position until such a time that it may be called into service. Based on the hazards and chemicals involved, the backup team may be put on air and placed within the hot zone to reduce the travel time should the entry team need assistance with a rapid exit.

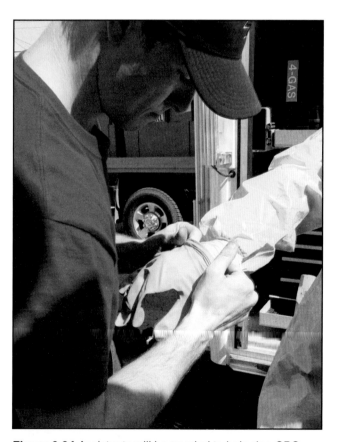

Figure 9.34 Assistants will be needed to help don CPC.

Figure 9.35 Backless chairs or benches will accommodate SCBAs.

Doffing of PPE

Many times, the donning supervisor may also serve as the doffing supervisor. This will be very helpful based on the person's knowledge of the members and equipment that were utilized for the entry.

Upon exit, it can be assumed that the entry team has either been contaminated or potentially contaminated by the hazard, thus needing decontamination prior to doffing. Based on the chemical hazards, it may be necessary to have the doffing personnel wear a lower level CPC. Any level of protection needed for doffing activities will be decided upon by the safety officer.

The personnel assisting in the doffing procedures should watch for signs and symptoms of heat stress. The entry personnel who will be doffing their protective equipment will most likely be hot, tired, and anxious to remove the clothing.

All doffing procedures should follow the manufacturer's directions, but these generic procedures may be included in any department's policies and guidelines:

- Personnel who are doffing equipment should allow the assisting personnel to perform the work.

- Entry team members should only touch the inside of the garments and never the outside. Likewise, assisting personnel should only touch the outside of the garments. It is critical that cross contamination be avoided.

- Once the garments are removed, zip or store them so that the inside and outside surfaces cannot touch.

- All entry garments should be placed in a containment bag and appropriately marked.

- The last item removed from the entry personnel should be the respirator facepiece. It should be removed by the user.

- The breathing apparatus should be isolated and marked for appropriate decontamination.

- All entry team and support team members must report immediately to rehab.

Inspection, Storage, Testing, Maintenance, and Documentation

You always want your PPE, tools, and equipment to perform as expected. An emergency incident is not the right place to discover problems with your protective clothing, respiratory protection, or other tools and equipment. The best way to ensure PPE, tools, and equipment always performs to expectation is by following a standard program for inspection, proper storage, maintenance, and cleaning. All inspections, testing, and maintenance must be conducted in accordance with manufacturer's recommendations.

Procedures are needed for both initial receipt of PPE, tools, and equipment and before and after use or exposure. PPE, tools, and equipment are initially inspected when purchased. Once the equipment is placed into service, the organization's personnel perform periodic inspections. Operational inspections of respiratory protection equipment occur after each use, daily or weekly, monthly, and annually. The organization must define the frequency and type of inspection in the respiratory protection policy, and they should follow manufacturer's recommendations. The care, cleaning, and maintenance schedules of respiratory protection and other equipment should be based on the manufacturer's recommendations, NFPA standards, or OSHA requirements.

PPE must be stored properly to prevent damage or malfunction from exposure to dust, moisture, sunlight, damaging chemicals, extreme temperatures (hot and cold), and impact **(Figure 9.36)**. Many manufacturers specify recommended procedures for storing their products. Follow these procedures to avoid equipment failure resulting from improper storage.

Keep records of all inspection, testing, and maintenance procedures. Reviewing these records periodically can show patterns about equipment that requires excessive maintenance or is susceptible to failure. Follow your agency's SOP/Gs for proper documentation.

After using PPE at an incident, it is important to fill out any associated reports or documentation as required by the AHJ. These reports may include PPE inspection forms, contaminated gear forms, deprovisioning forms, or any others as required by the AHJ.

Figure 9.36 Store CPC and other PPE so that it will not be damaged by sunlight or other potentially harmful exposures.

Chapter Review

Answer the following questions to review the information provided in this chapter.

1. What are the different types of respiratory protection used at hazmat/WMD incidents, and what types of incidents should each type be used?

2. What are the major categories of protective clothing worn at hazardous materials incidents?

3. What are the Levels of Protection for PPE Ensembles?

4. What are factors that must be considered when selecting PPE ensembles?

5. How do typical PPE ensembles vary between fire service, law enforcement, and EMS personnel?

6. What types of PPE-related stresses are hazmat responders likely to experience?

7. List general steps that should be performed during pre-entry inspection, donning, and doffing PPE.

8. Why is it so important for a responder to properly inspect, store, test, maintain, and document use of PPE and respiratory equipment?

9-1
Select appropriate PPE to address a hazardous materials scenario.

Step 1: Determine hazards.

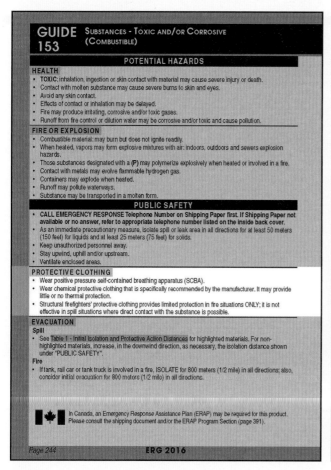

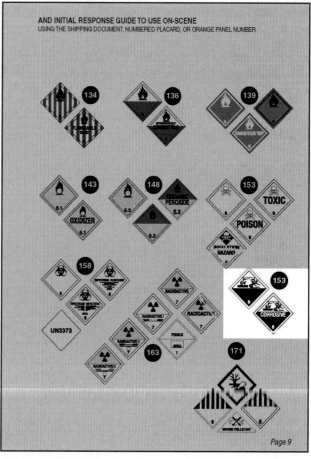

Step 2: Select an appropriate PPE ensemble for the mission-specific assignment.

9-2

Don, work in, and doff structural fire fighting personal protective equipment.

Step 1: Perform a visual inspection of PPE and SCBA for damage or defects.

Step 2: Don protective trousers and boots.

Step 3: Don protective hood, pulling hood down around neck and exposing head.

Step 4: Don protective coat.

Step 5: Don SCBA. Ensure that the cylinder valve is fully open and that all straps are secured.

Step 6: Don SCBA facepiece and ensure a proper fit and seal.

Step 7: Pull hood up completely so that facepiece straps and skin are not exposed.

Step 8: Don helmet and secure.

Step 9: Don inner gloves.

Step 10: Don gloves.

Step 11: Ensure that all fasteners, straps, buckles, etc., are fastened.

Step 12: Ensure that skin is not exposed.

Step 13: Attach SCBA regulators to facepiece and make sure SCBA is functioning properly.

Step 14: Perform preentry checks according to AHJ's SOPs.

Step 15: Perform work assignment.

Step 16: Undergo decontamination per AHJ's SOPs.

Step 17: Doff PPE in reverse order according to AHJ's SOPs, avoiding contact with outer ensemble or surfaces that may be contaminated.

Step 18: Conduct a post-entry inspection of PPE for damage or defects according to AHJ's SOPs and document finding.

Step 1: Perform a visual inspection of PPE for damage or defects.

Step 2: Don Level C PPE and secure closures.

Step 3: Don work boots.

Step 4: Pull ensemble leg opening over the top of the work boots.

Step 5: Don respirator.

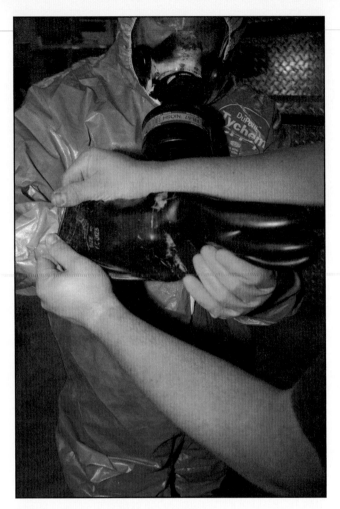

Step 6: Pull ensemble hood up completely so that facepiece straps and skin are not exposed.

Step 7: Don inner protective gloves.

Step 8: Don outer protective gloves.

Step 9: Pull ensemble sleeves over the outside of the gloves.

NOTE: Tape wrists and gaps if required by SOPs.

Step 10: Breathe through respirator and ensure that respirator is functioning properly.

NOTE: Don protective headgear per SOPs.

Step 11: Perform preentry checks as per AHJ's SOP.

Step 12: Perform work assignment.

Step 13: After assignment has been performed, proceed to decontamination line.

Step 14: Undergo decontamination as per AHJ's SOPs.

Step 15: Doff ensemble according to AHJ's SOPs, avoiding contact with outer ensemble or surfaces that may be contaminated.

Step 16: Doff SCBA according to AHJ's SOPs

Step 17: Conduct a post-entry inspection of PPE for damage or defects according to AHJ's SOPs and document finding.

Step 18: Return to proper storage as per manufacturer's instructions.

NOTE: If using an encapsulating ensemble, these steps will need to be modified according to AHJ's SOPs.

Step 1: Perform a visual inspection of PPE and SCBA for damage or defects.

Step 2: Don liquid splash PPE and secure closures.

Step 3: Don work boots according to AHJ's SOPs.

Step 4: Don the SCBA according to AHJ's.

Step 5: Don SCBA facepiece and ensure a proper fit and seal.

Step 6 Pull ensemble hood up completely so that facepiece straps and skin are not exposed.

Step 7: Don protective headgear (if required by AHJ).

Step 8: Don inner protective gloves.

Step 9: Don outer protective gloves.

NOTE: Tape wrists and gaps if required by SOPs.

Step 10: Attach SCBA regulator to facepiece and ensure proper operation.

Step 11: Perform preentry checks as per AHJ's SOP.

Step 12: Perform work assignment.

Step 13: Undergo decontamination as per AHJ's SOPs.

Step 14: To doff, remove PPE in reverse order of donning.

Step 15: Ensure medical monitoring is performed per AHJ's SOPs.

Step 16: Conduct a post-entry inspection of PPE for damage or defects according to AHJ's SOPs and document finding.

Step 17: Return to proper storage as per manufacturer's instructions.

Step 1: Perform a visual inspection of PPE and SCBA for damage or defects.

Step 2: Ensure the ensemble is the correct size.

Step 3: Ensure zipper is in good working order.

Step 4: Remove shoes, belts, and any objects that could damage ensemble.

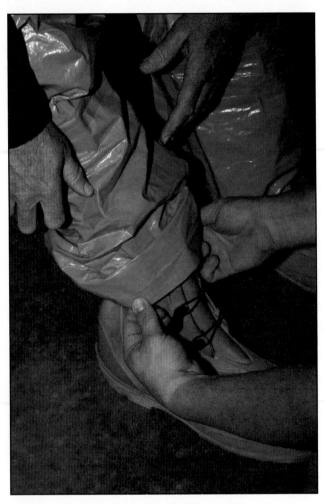

Step 7: Turn on air supply, don SCBA facepiece, check seal and breathe normally to ensure SCBA operates properly.

Step 5: Don ensemble according to AHJ's SOPs.

Step 6: Don the SCBA according to AHJ's SOPs.

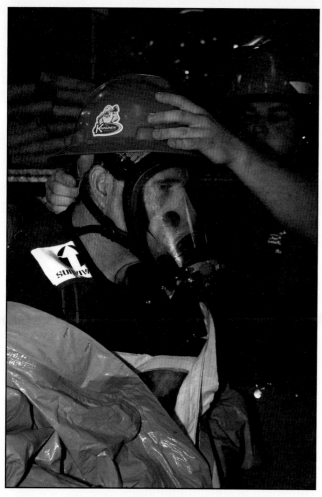

Step 8: Don protective headgear (if required by AHJ).

Step 9: Don outer protective gloves.

Step 10: Perform preentry checks as per AHJ's SOP.

Step 11: Perform work assignment.

Step 12: Undergo decontamination as per AHJ's SOPs.

Step 13: To doff, remove PPE in reverse order of donning.

Step 14: Ensure medical monitoring is performed per AHJ's SOPs.

Step 15: Conduct a post-entry inspection of PPE for damage or defects according to AHJ's SOPs and document finding.

Step 16 Return to proper storage as per manufacturer's instructions.

Implementing the Response: Decontamination

Chapter Contents

chapter 10

Key Terms

NFPA Job Performance Requirements

This chapter provides information that addresses the following job performance requirements of NFPA 1072, *Standard for Hazardous Materials/Weapons of Mass Destruction Emergency Response Personnel Professional Qualifications (2017)*.

5.3.1

5.4.1

5.5.1

6.2.1

6.3.1

6.4.1

Implementing the Response: Decontamination

Learning Objectives

After reading this chapter, students will be able to:

1. Define the different types of decontamination that may be used at a hazmat incident. (5.3.1, 5.5.1)

2. Identify decontamination methods. (6.3.1, 6.4.1)

3. Define gross decontamination. (5.4.1)

4. Explain processes for emergency decontamination. (5.3.1, 5.5.1, 6.2.1)

5. Explain processes for technical decontamination. (6.2.1, 6.3.1, 6.4.1)

6. Explain processes for mass decontamination. (6.3.1)

7. Identify victim management activities during decontamination operations.

8. Recognize general guidelines for decontamination operations. (6.3.1)

9. Describe decontamination implementation. (6.3.1, 6.4.1)

10. Explain decontamination termination activities. (6.3.1, 6.4.1)

11. Skill Sheet 10-1: Perform gross decontamination. (5.4.1)

12. Skill Sheet 10-2: Perform emergency decontamination. (5.5.1)

13. Skill Sheet 10-3: Perform technical decontamination on ambulatory people. (6.4.1)

14. Skill Sheet 10-4: Perform technical decontamination on nonambulatory victims. (6.4.1)

15. Skill Sheet 10-5: Perform mass decontamination on ambulatory people. (6.3.1)

16. Skill Sheet 10-6: Perform mass decontamination on nonambulatory victims. (6.3.1)

Chapter 10
Implementing the Response: Decontamination

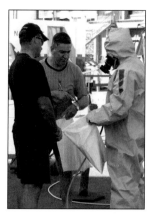

This chapter will address the following decontamination topics:

- Introduction to Decontamination
- Decontamination Methods
- Gross Decontamination
- Emergency Decontamination
- Technical Decontamination
- Mass Decontamination
- General Guidelines for Decontamination Operations
- Decontamination Implementation
- Evaluating Effectiveness of Decontamination Operations
- Termination of Decontamination Activities

The first four major sections of this chapter (Introduction to Decontamination, Decontamination Methods, Gross Decontamination, and Emergency Decontamination) address NFPA 1072 requirements at the Operations (core) level. The last six sections address NFPA 1072 requirements at the Operations Mission-Specific levels.

Introduction to Decontamination

Decontamination (decon) is an essential activity that must be considered at any hazardous materials or terrorism incident to ensure the safety of emergency responders and the public. Emergency decontamination should be established at all hazmat incidents.

Contamination is the transfer of a hazardous material to persons, equipment, and the environment in greater than acceptable quantities. There are two types of contamination:

- Direct contamination, when there is contact with the source of contamination

- Cross contamination (also called *secondary contamination*), when contamination occurs without contacting the direct source

Contaminates may be solids, liquids, or gases. Contaminate hazards vary depending on the material involved, but may be divided into the following types: chemical, physical, or biological. Contamination can be external (on the outside of the body or PPE) or internal (on the inside of the body).

Radiological Contamination

Radiological contamination is sometimes divided into:

- Loose surface

- Fixed

- Airborne

Externally, alpha contamination is relatively harmless and easy to protect against using appropriate PPE whereas internal contamination with alpha emitting material is much more dangerous. At high levels, beta contamination can present a hazard to the skin and lens of the eye and is also dangerous when deposited internally.

Decontamination (*decon*) or contamination reduction is the process of removing hazardous materials to prevent the spread of contaminants beyond a specific area and reduce contamination to levels that are no longer harmful. Decontamination prevents possible exposures to hazardous materials by removing contaminates.

Exposure is the process by which people, animals, or the environment are potentially subjected to, or come in contact with, a material; but, the material may not have been transferred. For example: If you smell perfume, you have been exposed to it because some of the perfume molecules have entered your nose in order to be smelled (*exposure route = inhalation*). However, you are not likely to carry the perfume (or its smell) around unless you have actually been contaminated by the perfume, which implies that you got enough of the material on you that it physically remains there. In the same way, decontamination may not be necessary if an individual has been exposed to a hazardous material rather than contaminated by it.

Decontamination is an essential activity that must be considered at any hazardous materials or terrorism incident to ensure the safety of emergency responders and the public. Decon operations minimize potentially harmful exposures and reduce or eliminate the spread of contaminants. Decontamination is performed at hazmat/WMD incidents to remove hazardous materials from responders, victims, PPE, tools, equipment, and anything else that has been contaminated **(Figure 10.1)**. Everyone and everything in the hot zone is subject to contact with the hazardous material and can become contaminated. Because of this potential, anything that goes into the hot zone passes through a decon area when leaving the zone. There are four types of decontamination addressed in the sections that follow:

- **Gross decontamination** — Decontamination phase where surface contamination is reduced as quickly as possible.

- **Emergency decontamination** — Decontamination to remove the threatening contaminant from the victim as quickly as possible without regard for the environment or property protection

- **Technical decontamination** — Decontamination using chemical or physical methods to thoroughly remove contaminants from responders (primarily entry team personnel) and their equipment; usually conducted within a formal decontamination line or corridor following gross decontamination.

Figure 10.1 Decontamination is performed to remove hazardous materials from victims, responders, and anything else that has been contaminated or potentially contaminated. *Courtesy of Boca Raton Fire Rescue.*

- **Mass decontamination** — Decontamination of large numbers of people in the fastest possible time to reduce surface contamination to a safe level, with or without a formal decontamination corridor or line.

Decontamination also provides victims with psychological reassurance. Some individuals who have been potentially exposed to hazardous materials may develop psychologically-based symptoms (such as shortness of breath, anxiety) even if they have not actually been exposed to harmful levels of contamination. Conducting decon can reduce or prevent these types of problems. It is important to continually assess the effectiveness of any decontamination operation. If monitoring determines that the selected method is not working, a different technique must be tried.

The type of decon operations conducted at an incident will be determined by a variety of factors, including **(Figure 10.2, p. 482)**:

- Number of persons requiring decon
- Type of hazardous materials involved
- Weather (washing off contaminants with a hose stream may not be a viable option in cold temperatures)
- Personnel and equipment available

Although emergency responders may have considerable experience with decon at hazmat incidents, performing decon at a terrorist incident may require some changes to the procedures used. Hazmat/WMD incidents may involve large numbers of people that have to be quickly assessed for injury or exposure and then passed through a decon corridor for treatment or safe sheltering away from the incident area (mass decon). Also, since a terrorist incident must be treated as a crime scene, any clothing, equipment, or contaminated materials have to be protected as evidence and handled in accordance with locally adopted policies and procedures.

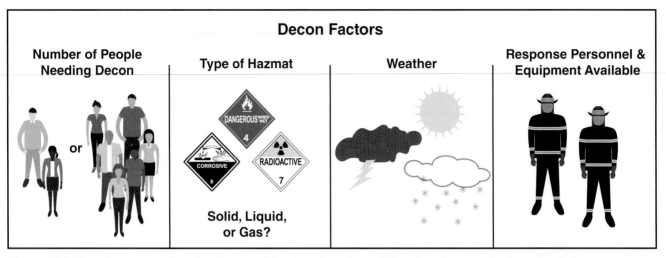

Decon Factors

Number of People Needing Decon

or

Type of Hazmat

DANGEROUS WHEN WET
4

CORROSIVE
8

RADIOACTIVE
7

Solid, Liquid, or Gas?

Weather

Response Personnel & Equipment Available

Figure 10.2 Many factors must be taken into consideration when determining what decon methods and techniques to use.

NOTE: Responders must be familiar with their organization's decon policies and procedures and how decon operations are implemented within the AHJ's incident command system.

Regardless of the many variables that may be encountered at the incident, the basic principles of any decontamination operation are easy to summarize:

1. Get it off.

2. Keep it off.

3. Contain it (prevent cross-contamination).

Before initiating any type of decontamination, the answers to the following questions should be considered:

- Do victims need to be decontaminated immediately or can they wait?

- Is it safe to conduct decon?

- Is there a safe place to conduct decon?

- What alternative decon methods are available?

- Are there adequate resources to conduct the operation? If not, can additional resources be obtained in a timely fashion?

- What is the time limit available to conclude decon before the victims deteriorate further?

- Is the equipment you are attempting to decontaminate going to be useable again and/or is it more cost effective to simply dispose of? Does decon save money or add value?

U.S. Anthrax Attacks

During the height of the U.S. anthrax attacks in 2001, a Michigan State University (MSU) police dispatcher received a call from an employee in Linton Hall, who reported that opening a letter had caused her to have a burning sensation in her throat. Unfortunately, the dispatcher confused this information with a call received minutes earlier from an employee in the

University Club (a different location on campus) who belatedly reported the receipt, during the previous month, of an envelope with a suspicious white powder in it. The dispatcher notified the FBI; the MSU Office of Radiation, Chemical, and Biological Safety; and the East Lansing Fire Department of a white powder incident at Linton Hall.

Because the two calls were confused, firefighters from East Lansing arrived at Linton Hall expecting a white powder incident, following procedures for a biological or chemical threat received in the mail. Even though the employee and her coworkers tried to explain that there had been no powder in the letter she reported, emergency responders decided to treat the situation as though there had been. Fifteen employees were forced to strip and be decontaminated in a makeshift cleansing station, being scrubbed down by male police officers and firefighters.

In the ensuing lawsuit brought by several of the women against the East Lansing FD, the women testified that they felt demeaned and traumatized by the experience. Those chosen to be decontaminated appeared to be at random, and police officers and firefighters not actively assisting in the decontamination process stood adjacent to the naked women, a fact that was highlighted as particularly disturbing. While precautions were taken to block doorways leading to the outside, none were taken to shield the windows in the hall from the stairwells leading to other levels of the building. This lack of privacy resulted in other employees, who were not being decontaminated, seeing the women naked during the cleaning process.

The women also were given conflicting information regarding the best way to prevent continued contamination. The emergency responders on the way to the hospital said to only take cold showers, while those at the hospital forced them to take hot showers. Despite the persistent nature of anthrax contamination, no further follow up was offered initially.

The confusion in the dispatcher's information, the conflicting advice from emergency personnel, and the disregard for the privacy of the women being decontaminated all contributed to a hazmat incident that ended badly. The lessons learned from this incident have led to changes in MSU's policies and procedures when dealing with situations like this. However, better communication and sensitivity to the personal rights of the women in the situation could have prevented the problems that did occur.

Decontamination Methods

Decontamination methods can be divided into four broad categories: Wet or dry methods and physical or chemical methods **(Figure 10.3)**. Decontamination methods vary in their effectiveness for removing different substances, and many factors may play a part in the selection decision, such as weather conditions and the chemical and physical properties of the hazardous material(s).

The most effective means of decontamination may be as simple as the removal of the outer clothing or PPE that has been contaminated by the hazardous mate-

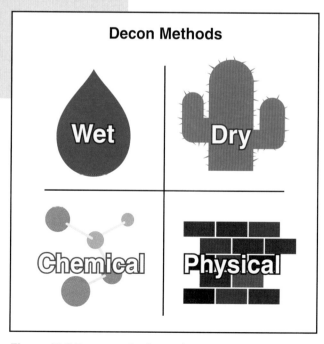

Figure 10.3 Decon methods can be wet or dry, chemical or physical.

rial. Additionally, flushing the contaminated surface with water is effective at removing the harmful substance or sufficiently diluting it to a safe level. For this reason, removal of contaminated clothing/PPE and flushing with water is usually sufficient for emergency and mass decon. Technical decon requires additional effort to meet the objective of thoroughly removing all contaminants and involves washing with water and some sort of soap, detergent, or chemical solution. The decision whether to perform emergency or technical decon is determined based on the hazardous material involved and the urgency in removing the victim from the contaminated environment.

Wet and Dry

As their names imply, wet and dry methods are categorized by whether they use water or other resources as part of the decon process. Wet methods usually involve washing the contaminated surface with solutions or flushing with water from a hose stream or safety shower; dry methods include scraping, brushing, and absorption.

Wet decon methods may necessitate the collection of runoff water in liquid-retaining (containment) devices such as wading pools **(Figure 10.4)**. Collected water may need to be analyzed for treatment and disposal. Disposal of runoff water and residue from decon operations must be properly accomplished in accordance with applicable laws and regulations. Proper authorities must be notified and consulted during this process. In some cases, wet methods may be difficult or impractical to use due to environmental or weather conditions. Life safety must take precedence over environmental considerations (for example, in mass decon situations).

Dry methods may be as simple as placing contaminated clothing into a suitable plastic bag (or recovery drum) or allowing the contaminant to evaporate. Other dry methods include vacuuming or brushing a powder or dust from a contaminated surface, scraping a material off, or using sticky tape (or a sticky pad) to clean or wipe off contamination. Dry methods have the advantage of not creating large amounts of contaminated liquid runoff (although absorption can result in a greater amount of contaminated material in general), and they may be accomplished through the systematic removal of disposable PPE while avoiding contact with any contaminants. Dry methods may be used during cold weather operations when wet methods are difficult to implement. When using dry methods, caution must be used to prevent the material becoming airborne.

Dry materials can also be used to remove liquid chemicals by absorption. Once used, these materials must be treated as contaminated waste and disposed of accordingly. Materials used for absorption may include **(Figure 10.5)**:

- Clay
- Sawdust
- Flour
- Dirt
- Fuller's earth
- Tissue paper
- Carbon

Figure 10.5 Dry decon can be conducted using sorbent powders or similar materials. *Courtesy of the U.S. Army, photo by Staff Sgt. Fredrick P. Varney, 133rd Mobile Public Affairs Detachment.*

Figure 10.4 To protect the environment, potentially contaminated runoff water from wet decon methods may need to be contained in wading pools or other catch basins. *Courtesy of the U.S. Marines, photo by Warren Peace.*

- Silica gel
- Paper towels
- Sponges

Physical and Chemical

Physical methods of decontamination remove the contaminant from a contaminated person without changing the material chemically (wet methods may dilute the chemical). The contaminant is then contained (when practical) for disposal. Examples of physical decontamination methods include:

- Absorption
- Adsorption
- Brushing and scraping
- Dilution
- Evaporation
- Isolation and disposal
- Washing
- Vacuuming

Chemical methods are used to make the contaminant less harmful by changing it through some kind of chemical process. For example, using bleach to sanitize tools and equipment that have been exposed to potentially harmful etiological agents is a form of chemical decontamination because the organisms are actually killed by the bleach. Another example would be using solvents to dissolve a contaminant. Examples of chemical decontamination methods include:

- Chemical degradation
- Sanitization
- Disinfection
- Sterilization
- Neutralization
- Solidification

When using chemical methods, hazardous materials technicians need to be careful to avoid creating additional hazards when introducing another chemical to the decon process. For example, using acetone as a solvent will add a high degree of flammability.

Gross Decontamination

Gross decontamination is a phase of decontamination where significant reduction of the amount surface contamination takes place as quickly as possible. Traditionally, gross decon was accomplished by mechanical removal of the contaminant or initial rinsing from handheld hose lines, emergency showers, or other nearby sources of water at hazmat incidents.

Because of increased awareness of firefighters' cancer risk, gross decon is now recommended at all emergency incidents involving exposure to potentially hazardous substances, including the toxic products of combustion. This may be accomplished by washing and/or doffing PPE at the scene and using wipes or other decon methods to remove soot from the face, head, and neck. PPE, tools, and equipment should be isolated, cleaned, and decontaminated according to SOPs, before reuse. It is recommended that structural firefighter protective clothing be machine washed in designated machines back at the station. Personnel should shower with soap and water thoroughly as soon as possible, even if wet methods of decon are used at the emergency incident scene.

Gross decontamination is performed in the following situations:

- Emergency responders exposed to smoke or products of combustion, before leaving the scene of the incident
- Responders before undergoing technical decontamination
- Victims during emergency decontamination
- Persons requiring mass decontamination

One advantage of gross decon is that it is conducted in the field, so the reduction of contaminates is immediate. A disadvantage is that, while it may remove the worst surface contamination, it may not remove *all* contaminates. Gross decon is not complete decon, and it should be followed by more thorough decontamination afterwards. **Skill Sheet 10-1** provides steps for conducting gross decon at an emergency scene involving toxic products of combustion.

Emergency Decontamination

The goal of emergency decontamination is to remove the threatening contaminant from the victim as quickly as possible — there is no regard for the environment or property protection. Emergency decon may be necessary for both victims and rescuers **(Figure 10.6)**. If either is contaminated, individuals must remove their clothing (or PPE) and wash quickly. Victims may need immediate medical treatment, and they cannot wait for the establishment of a formal decontamination corridor. The following situations are examples of instances where emergency decontamination is needed:

- Failure of protective clothing
- Accidental contamination of emergency responders
- Immediate medical attention is required by emergency workers or victims in the hot zone

Emergency decontamination has the following advantages:

- Fast to implement
- Requires minimal equipment (usually just a water source such as a hoseline)
- Reduces contamination quickly
- Does not require a formal contamination reduction corridor or decon process

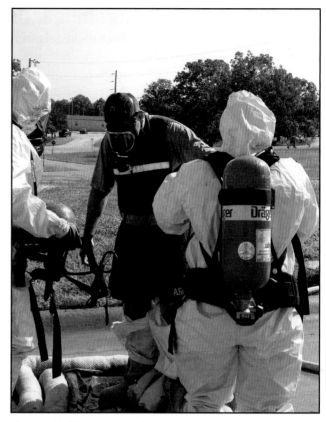

Figure 10.6 Emergency decon removes contamination as quickly as possible.

However, emergency decontamination has definite limitations. Removal of all contaminants may not occur, and a more thorough decontamination must follow. Emergency decontamination can harm the environment. If possible, measures must be taken to protect the environment, but such measures should not delay lifesaving actions. The advantage of eradicating a life-threatening situation far outweighs any negative effects that may result.

Seemingly normal incidents may involve hazardous materials. Emergency responders may become contaminated before they realize what the situation really is. When these situations occur, emergency responders need to withdraw immediately and follow local procedures for emergency decontamination. Should their air supply allow, responders should remain isolated until someone with the proper expertise and monitoring equipment can ensure that they have been adequately decontaminated.

Emergency decon should be conducted in a safe area. Responders conducting emergency decon should wear appropriate PPE, and they should always avoid contacting contaminates or potentially contaminated surfaces. If responders do contact contaminates, they may need to decontaminate themselves. Follow SOPs for conducting emergency decon. Emergency decontamination procedures may differ depending on the circumstances and hazards present at the scene. Refer to **Skill Sheet 10-2** for steps in conducting emergency decontamination.

Technical Decontamination
— Using chemical or physical methods to thoroughly remove contaminants from responders (primarily entry team personnel) and their equipment; usually conducted within a formal decontamination line or corridor following gross decontamination. *Also known as* Formal Decontamination.

Technical Decontamination

Technical decontamination uses chemical or physical methods to thoroughly remove or neutralize contaminants from responders' PPE (primarily entry team personnel) and equipment **(Figure 10.7)**. It may also be used on incident victims in non-life-threatening situations. Operations Level responders involved in technical decon operations must do so under the guidance of a hazmat technician, SOP/Gs, or an allied professional. Responders must be familiar with the AHJ's procedures for implementing technical decon within the incident command system, including decontamination team positions, roles, and responsibilities. During technical decon operations, Operations Level responders will typically act as follows:

- Protect themselves by dressing in appropriate PPE

- Establish a water supply

- Set up the decon corridor

- Establish perimeters

- Perform physical decontamination activities such as scrubbing, washing, and spraying

- Assist in the undressing/removal of PPE or clothing of individuals going through the decon line **(Figure 10.8)**

- Assist individuals going through the decon process

- Perform other duties per SOP/Gs and training

Figure 10.7 Technical decon uses chemical or physical methods to thoroughly remove or neutralize contaminants. *Courtesy of the U.S. Air Force, photo by Chiaki Iramina.*

Figure 10.8 Operations responders may assist with removing PPE or clothing of individuals undergoing decon.

Technical decon is usually conducted within a formal decon line or corridor. The contaminants involved at the incident determine the type and scope of technical decon. Resources for determining technical decon procedures may include the following:

- "First Aid" section of the safety data sheet(s) (SDS)
- Emergency response centers (such as CHEMTREC, CANUTEC, SETIQ)
- Container information labels
- Pre-incident plans
- Technical experts
- *ERG* (may tell responders if material is water soluble or water reactive)
- Poison control centers
- Other books, reference sources, computer programs, and/or data bases

In all types of decon, monitoring should be conducted to determine whether decon operations are effective. In some cases, equipment may need to be disposed of due to permeation of contaminants or other factors. Such equipment must be removed from service and properly contained before disposal according to established policies and procedures.

Technical Decontamination Techniques

Technical decon may use many techniques to remove a contaminant from a person or exposure. Emergency responders must know what to do when assigned to a decontamination corridor or line. They should be briefed before being assigned. The following sections explain several of the more common techniques used during technical decon.

NOTE: Table 10.1, p. 490, provides a list of advantages and disadvantages for each of these decon techniques.

Absorption

Absorption is the process of picking up liquid contaminants with absorbents **(Figure 10.9, p. 492)**. Some examples of absorbents used in decon are diatomaceous earth, baking powder, ashes, activated carbon, vermiculite, or other commercially available materials. Many absorbents are inexpensive and readily available, but expensive to dispose of once contaminated.

NOTE: The use of soil as an absorbent is not recommended by many experts.

Adsorption

Adsorption is the process in which a hazardous liquid interacts with (or is bound to) the surface of a sorbent material, such as activated carbon **(Figure 10.10, p. 492)**. Adsorbents tend to not swell like absorbents, and it is important to make sure that the adsorbent used is compatible with the hazardous material in order to avoid potentially dangerous reactions.

Brushing and Scraping

Brushing and scraping may remove large particles of contaminant or contaminated materials, such as mud from boots or other PPE. Generally, brushing and scraping alone is not sufficient decontamination. This technique is used before other types of decon **(Figure 10.11, p. 492)**.

Table 10.1
Advantages and Disadvantages of Technical Decon Techniques

Method	Advantages	Disadvantages
Absorption	• Many absorbent materials are inexpensive and readily available • Can be used as part of dry decon operations • Effective on flat surfaces	• Do not alter the hazardous material • Ineffective for decontaminating protective clothing and vertical surfaces • Disposal of contaminated absorbent materials may be problematic and expensive • Absorbent materials may increase in weight and/or volume as they absorb the hazmat • Absorbent materials must be compatible with the hazardous material
Adsorption	• Contains the hazardous material better than absorbent materials • Transportation of materials to disposal is simplified • Off-gasing (release of vapors/gases) is effectively reduced • Adsorptive materials do not swell	• Process can generate heat • Application typically limited to remediation of shallow liquid spills • Adsorptive materials are expensive • Adsorptive material must be compatible with the hazardous material (they are product specific)
Chemical Degradation	• Can reduce cleanup costs • Reduces risk posed to the first responder when dealing with biological agents • Often utilizes commonly available, inexpensive materials such as bleach, isopropyl alcohol, or baking soda • Utilizes products that are readily available	• Takes time to determine the right chemical to use (which should be approved by a chemist) and set up the decon process • Can cause violent reactions if done incorrectly and may create heat and toxic vapors • Rarely used to decontaminate people
Dilution	• Lessens the degree of hazard present by reducing the concentration of the hazardous material • Easy to implement (water is usually readily available) • Is very effective in many circumstances requiring decon • Can be used to decon large pieces of equipment/apparatus	• Can't be used on materials that react adversely to water • May be problematic in cold weather • May create large amounts of contaminated runoff • May be impractical because of the amount of water required for effective dilution
Disinfection	• Kills most of the biological organisms present • Can be used on site • Can be accomplished using a variety of chemical or antiseptic products • Disinfecting agent may be as simple as antibacterial soap or detergent	• Limited to biological decon only • May be difficult to decon large pieces of equipment/apparatus • Disinfecting agent may be toxic or harmful

Continued

Table 10.1 (concluded)

Method	Advantages	Disadvantages
Evaporation	• No additional materials necessary • No runoff collection necessary • No (or very limited) expense incurred	• Applicable for a very limited number of chemicals • Generally limited to decon of tools and equipment, not people • May be dramatically affected by weather conditions (including wind, temperature, humidity, and rain) • Hazardous vapors may travel and cause problems • May require a long time to complete • May not be acceptable method to use depending on applicable laws and regulations
Isolation and Disposal	• Isolation can be quick and effective • Easily achieved with containers such as isolation drums, heavy plastic bags, and other means of containment	• Disposal and transport costs may be extremely high • May require replacement of equipment and PPE that cannot be decontaminated and placed back in service
Neutralization	• Chemically alters the hazardous material to reduce the degree of hazard present • Effective on most corrosives and some poisons • Neutralizing agents are readily available (soda ash, vinegar)	• May be very difficult to successfully implement • Rarely done on living tissue • May require large quantities of neutralizing agents • May create violent chemical reaction including the release of heat and hazardous vapors • Preplanning is usually necessary
Solidification	• Solids are easier to contain than liquids and gases • Reduces the amount of vapor production and off gasing • Easier to clean up	• Requires specialized materials to implement
Sterilization	• Kills all microorganisms present	• Difficult or impossible to do onsite
Vacuuming	• Effective at removing dust and particulates • Effective indoors • Dry method, useful for cold weather operations in some situations	• Requires specialized vacuums equipped with hepa filters • May require high risk, negative air containment for decon area • Removing liquid chemical contamination requires special equipment • May require additional decon procedures to ensure complete decontamination (for example, washing) • Can't be used to decontaminate materials that react adversely to contact with water • May be problematic in cold weather • May create large amounts of contaminated runoff
Washing	• Quick and easy to implement (water is usually readily available) • Soap is readily available and inexpensive • Typically more effective than dilution alone • Is very effective in many circumstances requiring decon • Can be used to decon large pieces of equipment/apparatus	• Can't be used to decontaminate materials that react adversely to contact with water • May be problematic in cold weather • May create large amounts of contaminated runoff

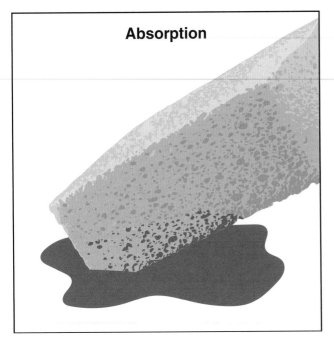

Figure 10.9 In absorption, liquid contaminants are absorbed into an absorbent.

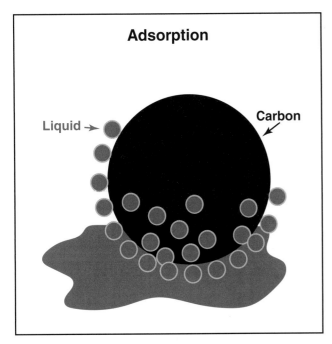

Figure 10.10 In adsorption, contaminants are bound to the surface of an adsorbent such as charcoal or carbon.

Figure 10.11 Brushing and scraping is used to remove large particles before other types of decon are employed. *Courtesy of Brian Canady/DFW-FRD.*

Chemical Degradation

The process of chemical degradation uses a material to change the chemical structure of a hazardous material. For example, household liquid bleach is commonly used to neutralize spills of etiological agents. The interaction of the bleach with the agent kills the dangerous germs and makes the material safer to handle.

Dilution

Dilution is the process of using water to flush contaminants from contaminated victims or objects and diluting water-soluble hazardous materials to safe levels. Dilution is advantageous because of the accessibility, speed, and economy of using water. However, there are disadvantages as well. Depending on the material, water may cause a reaction and create even more serious problems. Additionally, runoff water from the process is still contaminated and may have to be confined and disposed of properly. The amount of water needed for dilution may be impractical in some circumstances.

Evaporation

Evaporation is the process of a liquid turning into a gas. Some hazardous materials evaporate quickly and completely. In some instances, effective decontamination can be accomplished by simply waiting long enough for the materials to evaporate. Evaporation is used for decon of gaseous materials such as ammonia or other materials with high vapor pressures **(Figure 10.12)**. It can be used on tools and equipment when extending exposure time is not a safety issue.

Isolation and Disposal

In this process, contaminated items, such as clothing, tools, or equipment, are isolated by collecting them in some fashion and then disposed in accordance with applicable regulations and laws **(Figure 10.13)**. All equipment that cannot be sufficiently decontaminated must be disposed. All spent solutions and wash water must be collected and disposed. Disposal of equipment may be easier than decontaminating it; however, disposal can be very costly in circumstances where large quantities of equipment have been exposed to a material.

Neutralization

Neutralization changes the pH of a corrosive, raising or lowering it towards 7 (neutral) on the pH scale. Neutralization should not be performed on living tissue.

Sanitization, Disinfection, or Sterilization

Processes that render etiological contaminates harmless include:

- **Sanitization** – Reduces the number of microorganisms to a safe level (such as by washing hands with soap and water).

- **Disinfection** – Kills most of the microorganisms present. In a decon setting, a variety of chemical or antiseptic products may be used to accomplish disinfection. Most first responders are familiar with the disinfection procedures used to kill bloodborne pathogens such as wiping contaminated surfaces with a bleach solution.

- **Sterilization** – Kills all microorganisms present. Sterilization is normally accomplished with chemicals, steam, heat, or radiation. While sterilization of tools and equipment may be necessary before they are returned to service, this process is usually impossible or impractical to do in most onsite decon situations. Such equipment will normally be disinfected on the scene and then sterilized later.

Solidification

Solidification is a process that takes a hazardous liquid and treats it chemically so that it turns into a solid. Solidification is not used for personnel decontamination.

Figure 10.12 Evaporation can be used for decon of gaseous materials. *Courtesy of Rich Mahaney.*

Figure 10.13 Dispose of isolated items that cannot be decontaminated.

Figure 10.14 Washing typically involves soap or other surfactants mixed with water. *Courtesy of FEMA News Photos, photo by Jocelyn Augustino.*

Vacuuming

Vacuuming is a process using high efficiency particulate air (HEPA) filter vacuum cleaners to pull solid materials such as fibers, dusts, powders, and particulates from surfaces. Regular vacuums are not used for this purpose because their filters are not fine enough to catch all of the material.

Washing

Washing is a process similar to dilution in that they are both wet methods of decontamination. However, washing also involves using prepared solutions such as solvents, soap, and/or detergents mixed with water in order to make the contaminant more water-soluble before rinsing with plain water **(Figure 10.14)**. The difference is similar to simply rinsing a dirty dish in the sink versus washing it with a dishwashing liquid. In some cases, the former may be sufficient; in others, the latter may be necessary. Washing is an advantageous method of decontamination because of the accessibility, speed, and economy of using water and soap. As with the dilution process, runoff water from washing may need to be contained and disposed of properly.

Technical Decontamination for Ambulatory Victims

Ambulatory — People, often responders, who are able to understand directions, talk, and walk unassisted.

Victims who are able to understand directions, talk, and walk unassisted are considered **ambulatory**; technical decon corridors are typically designed for ambulatory persons such as emergency responders. Corridors may be set up for wet or dry decontamination methods. Technical decon corridors vary in the number of stations, depending on the needs of the situation. In some cases, technical decon may be as simple as washing one's hands and face with soap and water. **Table 10.2** is a sample technical decon checklist. Procedures for conducting technical decon of ambulatory victims are provided in **Skill Sheet 10-3**.

Technical Decontamination for Nonambulatory Victims

Nonambulatory victims are civilians or responders who are unconscious, unresponsive, or unable to move unassisted. These victims may be more seriously injured than ambulatory victims, and will need assistance being moved to a place where they undergo decon. They may have to remain in place if sufficient personnel are not available to remove them from the hot zone.

Technical decon personnel may need to do all or most of the decon process work for nonambulatory victims such as removing clothing and washing **(Figure 10.15)**.

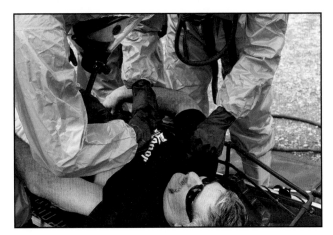

Figure 10.15 Responders may need to remove PPE or clothing from nonambulatory victims.

Table 10.2
Sample Decon Checklist

Date: **Location:**

- ☐ Initial briefing from the team leader
- ☐ Incident profile
- ☐ Decon solution and method
- ☐ PPE

Personnel Assignments

Decon Officer

[ ] Identified by vest

- ☐ All personnel monitored by Medical Branch

Decon Site Selection Criteria

- ☐ Decon is located in Warm Zone at exit from Hot Zone
- ☐ Decon area located uphill/upwind from Hot Zone
- ☐ Decon area level or sloped toward Hot Zone
- ☐ Water supply available

Decon Site Setup

- ☐ Area clearly marked with traffic cones and barrier tape to be secure against unauthorized entry
- ☐ Entry and exit points marked
- ☐ Emergency corridor established and clearly marked
- ☐ Runoff contained (tarp, plastic sheeting, dikes)
- ☐ Gross decon shower(s) setup
- ☐ Water supply established
- ☐ Containment basins and pools arranged in proper order
- ☐ Disposal containers in place for PPE and equipment drop
- ☐ Decon solutions mixed
- ☐ Brushes, hand sprayers, hoses and equipment in place
- ☐ Tool drop set up
- ☐ Spare SCBA cylinders available
- ☐ Relief personnel available

Branch Officers Briefing

- ☐ Preparation of branch status report
- ☐ Evaluation of branch readiness for mitigation plan

Entry/Decon Operations

- ☐ Decon and entry personnel briefed on hazards
- ☐ Emergency procedures and hand signals reviewed and understood
- ☐ Decon and entry personnel briefed on decon procedures
- ☐ Decon corridor complete
- ☐ Decon personnel on air
- ☐ Monitored for adequate relief personnel

Termination

- ☐ Disposable/contaminated materials isolated, bagged, and containerized
- ☐ All containers sealed, marked, and isolated
- ☐ All team equipment cleaned and accounted for

Source: Department of Fire Services, Office of Public Safety, Commonwealth of Massachusetts.

Mass Decontamination
— Process of
decontaminating large
numbers of people in
the fastest possible
time to reduce surface
contamination to a safe
level. It is typically a gross
decon process utilizing
water or soap and water
solutions to reduce the
level of contamination,
with or without a formal
decontamination corridor
or line.

Mass Casualty Incident
— Incident that results in a
large number of casualties
within a short time frame,
as a result of an attack,
natural disaster, aircraft
crash, or other cause that is
beyond the capabilities of
local logistical support.

Technical decon for nonambulatory victims is a more detailed process than mass decon for nonambulatory victims. The aim is to thoroughly decontaminate the individuals before transferring them to EMS. Procedures for conducting technical decon on nonambulatory victims are provided in **Skill Sheet 10-4**.

Mass Decontamination

Mass decontamination is the physical process of rapidly reducing or removing contaminants from multiple persons (victims and responders) in potentially life-threatening situations. Mass decon is initiated when the number of victims and time constraints do not allow the establishment of an in-depth decontamination process (such as technical decon) **(Figure 10.16)**. Put simply, the goal of mass decon is to do the greatest good for the greatest number of people.

All agencies should have a mass decon plan as part of their overall emergency response plan. Responders must be familiar with the AHJ's procedures for implementing mass decon within the incident command system, including decontamination team positions, roles, and responsibilities. To determine the correct mass decon procedures, responders must be familiar with established SOP/Gs, emergency response plans, training, and skills learned during drills/exercises, and preplans. Operations Level responders involved in mass decon operations must work under guidance from resources including:

- Hazmat technicians
- SOP/Gs
- Allied professionals

The scene of an incident requiring the use of mass decon may be quite chaotic and difficult to control, particularly if it is a **mass casualty incident**. To combat the chaos of the incident, responders should take the following actions:

- Communicate with victims by using hand signals, signs with pictures, apparatus public address systems, megaphones or other methods to direct them to decon gathering areas as well as through the decon process itself.

Figure 10.16 Mass decon is initiated to expedite decon of large numbers of people. *Courtesy of David Lewis.*

- Provide simple and specific directions that can be easily understood, since people may be traumatized and/or suffering from exposures.

- Use barrier tape, traffic cones, or other highly visible means to mark decon corridors (see Decontamination Corridor Layout section).

Mass decon methods include:

- Dilution

- Isolation

- Washing

As addressed in the technical decon section, each of these methods have their own advantages and limitations. Washing with a soap-and-water solution or universal decontamination solution will remove many hazardous chemicals and WMD agents; however, availability of such solutions in sufficient quantities cannot always be ensured. Therefore, mass decon can be most readily and effectively accomplished with a simple water shower system that merely dilutes the hazardous product and physically washes it away. Mass decon uses large volumes of low-pressure water, in a fog pattern, to quickly reduce the level of contamination **(Figure 10.17)**.

Figure 10.17 Large volumes of low-pressure water is often used in mass decon operations.

Mass decon showers should ensure that the process physically removes the hazardous material. The actual showering time is an incident-specific decision. When large numbers of potential victims are involved and queued for decon, showering time may be significantly shortened. This time may also depend upon the volume of water available in the showering facilities. Post-decon monitoring should be used to evaluate the effectiveness of decontamination operations.

WARNING!
Never delay decon while waiting for additional resources to arrive unless an assessment has been made that further injury or exposure will not occur.

Emergency responders should not overlook existing facilities when identifying means for rapid decontamination methods. For example, although water damage to a facility might result, the necessity of saving victims' lives could justify the activation of overhead fire sprinklers for use as showers. Similarly, having victims wade and wash in water sources such as public fountains, chlorinated swimming pools, or swimming areas, provides an effective, high-volume decontamination technique, although consideration must be given to the persistence of chemical agents in contained and contaminated water.

Figure 10.18 In most cases, disrobing and showering will provide effective removal of contaminants. Victims should be encouraged to remove as much clothing as possible, while remaining sensitive to privacy and modesty issues.

It is recommended that all victims undergoing mass decon remove clothing at least down to their undergarments before showering **(Figure 10.18)**. Removal of clothing can remove significant amounts of the contaminant materials. Victims should be encouraged to remove as much clothing as possible, proceeding from head to toe. Contaminated clothing should be isolated in drums, appropriate bags, or other containers for later disposal.

NOTE: You must have an accountability system for all valuables and personal items removed. See the Decontamination Implementation section for more details.

Removal of Clothing

There may be circumstances (such as with particulate contaminants) in which removal of clothing before showering actually increases the potential risk of exposure. These risks must be evaluated before decon methods are implemented. For example, with some radiological materials and biological agents, clothing should be dampened before removal ahead of showering to limit the potential of aerosolizing the agents.

Many innovations and products have been developed to assist in mass decon operations, from decon trailers and portable tents (that help alleviate privacy concerns) to portable water heaters, disposable coveralls, and "bagging and tagging" systems **(Figures 10.19a and b)**. The use of trailers and portable tents are best suited to long-duration incidents, incidents where weather may require them, and incidents where immediate decon is not as vital (such incidents involving biological hazards). Emergency responders should be familiar with the equipment and resources available as well as the mass decon procedures that their agency uses. **Figures 10.20a and b, p. 500**, provide examples of apparatus placement for generic mass decon schematics, however, many agencies have tents or trailers available for use.

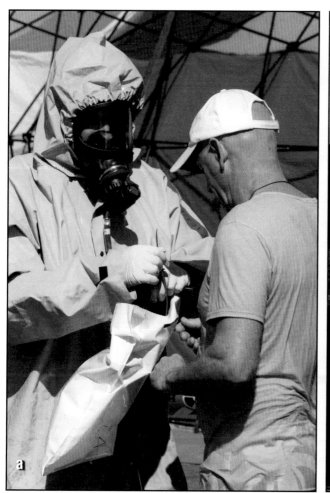

Figures 10.19a and b Pre-assembled decon kits may include everything from individual bags and ID tags to disposable garments, towels, and shoes. Bag and tag personal items so that they may be returned to their owners. *Courtesy of New South Wales Fire Brigades.*

To determine victim priority during decon, responders must consider factors related both to medical needs and decontamination. For maximum effectiveness, it is recommended that victims be divided into two groups: ambulatory and nonambulatory.

This division may require establishing separate decon areas for each group to avoid slowing down the progression of ambulatory victims through the decon area. Incidents involving a large number of incapacitated victims may require additional resources for separate decon corridors since nonambulatory victims will not be able to walk through the decon line **(Figure 10.21, p. 501)**. A separate decon line for emergency response personnel should also be provided.

If there are adequate resources available and the situation allows for additional time, it may be beneficial to separate victims by gender for privacy reasons; however, families should be allowed to stay together. Children, the elderly, and/or the disabled should not be separated from their parents or caretakers.

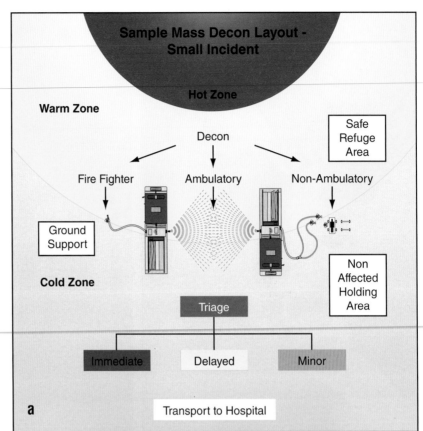

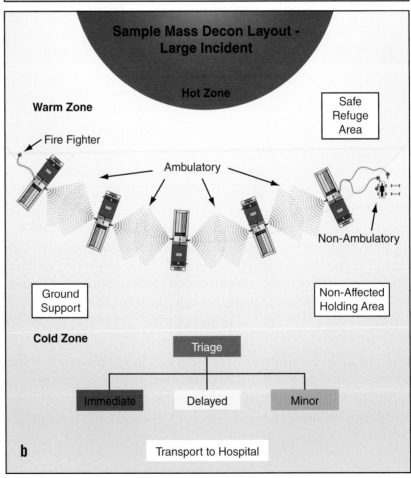

Figures 10.20a and b Sample mass decon schematics for small and large incidents. *Courtesy of Doug Weeks.*

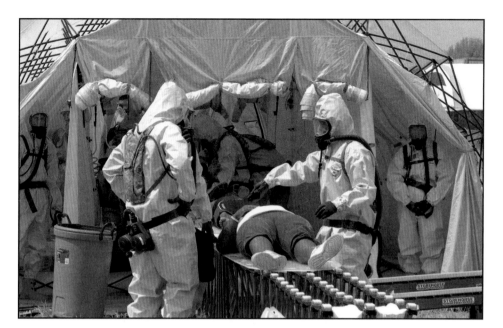

Figure 10.21 A system of rollers can be used to move nonambulatory victims through the decon corridor. *Courtesy of the U.S. Air Force, photo by Tech. Sgt. Todd Pendleton.*

Advantages and Limitations of Mass Decontamination

Advantages:

- Accommodates large numbers of people
- Can be implemented quickly using limited amount of personnel and equipment
- Reduces contamination quickly

Limitations:

- Does not always totally decontaminate the victim
- Relies on the cooperation of the victim
- Can create contaminated runoff that can harm the environment and other exposures

Mass Decontamination for Ambulatory Victims

Ambulatory victims should be directed to an area of safe refuge within the isolation perimeter to await prioritization for decontamination. **Figure 10.22, p. 502**, provides a sample layout for mass decon for ambulatory victims. Procedures for conducting mass decon on ambulatory victims are provided in **Skill Sheet 10-5**.

Several of the following factors may influence the priority of treatment for ambulatory victims:

- Victims with serious medical symptoms, such as shortness of breath or chest tightness
- Victims closest to the point of release
- Victims reporting exposure to the hazardous material
- Victims with evidence of contamination on their clothing or skin
- Victims with conventional injuries such as broken bones or open wounds

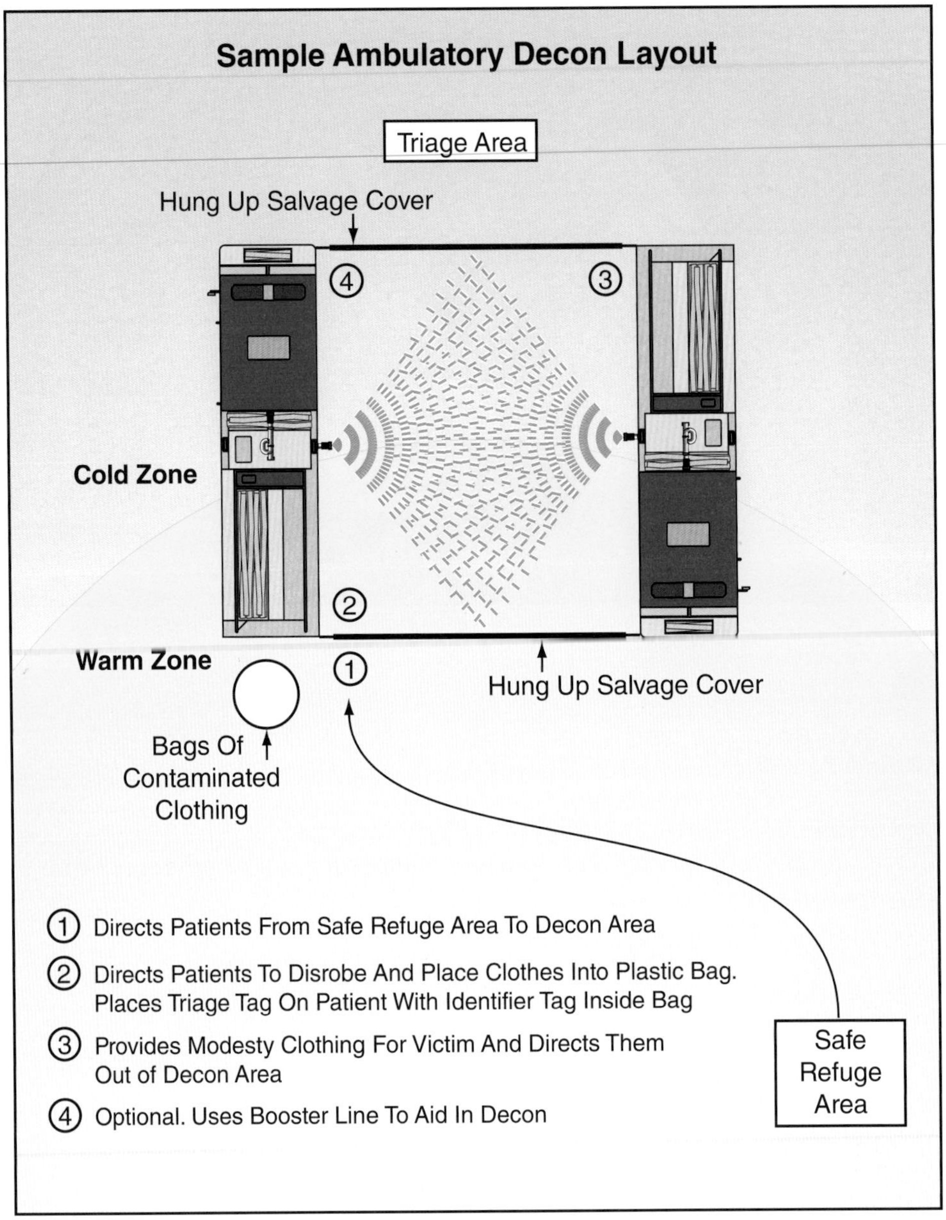

Sample Ambulatory Decon Layout

Triage Area

Hung Up Salvage Cover

④ ③

Cold Zone

② ①

Warm Zone

Hung Up Salvage Cover

Bags Of Contaminated Clothing

① Directs Patients From Safe Refuge Area To Decon Area

② Directs Patients To Disrobe And Place Clothes Into Plastic Bag. Places Triage Tag On Patient With Identifier Tag Inside Bag

③ Provides Modesty Clothing For Victim And Directs Them Out of Decon Area

④ Optional. Uses Booster Line To Aid In Decon

Safe Refuge Area

Figure 10.22 A sample layout for ambulatory decon. *Courtesy of Doug Weeks.*

Emergency Decontamination of Ambulatory Victims at Incidents Involving Chemical Agents

At incidents involving chemical agents, very little time may be available to successfully conduct decon. For example, after skin contact with nerve agents, industrial chemical agents, and vesicants, emergency decon should be conducted immediately (within minutes). It is therefore vital that rescuers conduct rapid extraction and prioritization of ambulatory victims as quickly and efficiently as possible.

Ambulatory victims showing visible signs of contamination or symptoms of exposure to chemical agents should be directed to undergo emergency decon. Emergency decon may precede additional mass decon procedures (for example, if portable decon tents or trailers are not yet set-up or available for use) and generally involves disrobing and flushing in high volume, low pressure showers (hand-held hose lines, or side-by-side apparatus

(Figure 10.23). Clothing and personal items may be bagged, but when victims require emergency decon, time should not be taken to tag items or conduct medical evaluations. The goal of emergency decon is to remove contaminants quickly. After undergoing emergency decon, victims may be directed to undergo additional mass decon, or they may be allowed to dry and redress (in provided clean clothing) as appropriate before proceeding for medical evaluation and treatment. In most cases where emergency decon and mass decon are needed, dilution is the solution.

Figure 10.23 An example of apparatus placement for emergency/mass decon.

Mass Decontamination for Nonambulatory Victims

Nonambulatory victims may be more seriously injured than ambulatory victims. They may have to remain in place if sufficient personnel are not available to remove them from the hot zone. **Figure 10.24, p. 504,** provides an example of a mass decon corridor layout for nonambulatory victims. The decon process for nonambulatory victims at mass casualty incidents will be more of a gross decon process, conducted quickly. Follow AHJ procedures for these incidents. Procedures for conducting mass decon on nonambulatory victims are provided in **Skill Sheet 10-6**.

Victim Management during Decontamination Operations

Victim management activities in an incident requiring decontamination include:

- Triage

- Handling deceased victims

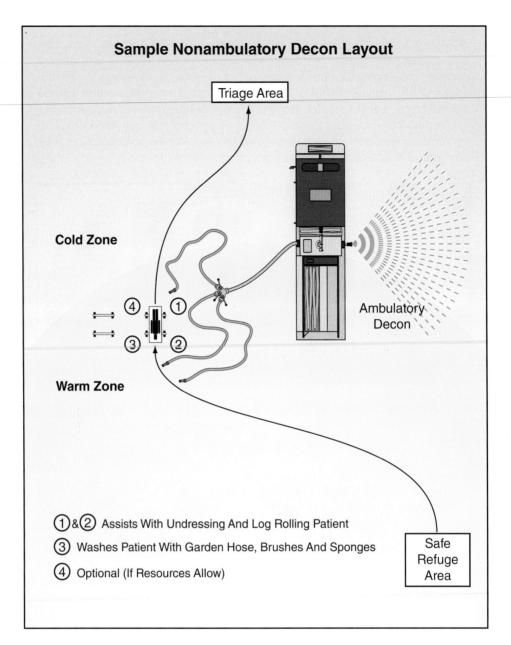

Sample Nonambulatory Decon Layout

Triage Area

Cold Zone

Ambulatory Decon

Warm Zone

Safe Refuge Area

① & ② Assists With Undressing And Log Rolling Patient

③ Washes Patient With Garden Hose, Brushes And Sponges

④ Optional (If Resources Allow)

Figure 10.24 Sample nonambulatory decon corridor. *Courtesy of Doug Weeks.*

Triage

Triage may be necessary when any victims are present at an incident and there is a need for medical assistance. All victims must undergo decon before being transferred to EMS. The type of exposure(s), products involved, injuries present, and other factors will determine if technical or mass decon is performed.

In most instances, triage will be conducted in the cold zone after decontamination has been performed. Prioritization of victims can be done using medical triage systems such as START (Simple Triage and Rapid Treatment/Transport) **(Figure 10.25)**. Procedures for conducting **triage** of victims should be predetermined within the local emergency response plan.

Triage at mass decon incidents is essentially the same as at technical decon incidents. Because there will typically be more individuals involved at mass decon incidents, there may be more EMS units and personnel involved. Triage, including skill steps, is covered in greater detail in Chapter 12, Implementing the Response: Mission-Specific Victim Rescue and Recovery.

Triage — System used for sorting and classifying accident casualties to determine the priority for medical treatment and transportation.

START Medical Triage System

START Category	Decon Priority	Classic Observations	Chemical Agent Observations
IMMEDIATE Red Tag	1	Respiration is present only after repositioning the airway. Applies to victims with respiratory rate >30. Capillary refill delayed more than 2 seconds. Significantly altered level of consciousness.	• Serious signs/symptoms • Known liquid agent contamination
DELAYED Yellow Tag	2	Victim displaying injuries that can be controlled/treated for a limited time in the field.	• Moderate to minimal signs/symptoms • Known or suspected liquid agent contamination • Known aerosol contamination • Close to point of release
MINOR Green Tag	3	Ambulatory, with or without minor traumatic injuries that do not require immediate or significant treatment.	• Minimal signs/symptoms • No known or suspected exposure to liquid, aerosol, or vapor
DECEASED/ EXPECTANT Black Tag	4	No spontaneous effective respiration present after an atempt to reposition the airway.	• Very serious signs/symptoms • Grossly contaminated with liquid nerve agent • Unresponsive to autoinjections

Figure 10.25 START or other triage systems can be used to triage patients at chemical agent incidents. *Courtesy of the U.S. Army Soldier and Biological Chemical Command (SBCCOM).*

Handling Deceased Victims

Responders must be prepared for the reality of handling deceased victims. As a general rule, deceased victims should remain untouched, but responders should always follow AHJ procedures. The AHJ's medical examiner will determine how and when bodies are handled. Normally, removal of deceased victims from the hot zone will be delayed until all viable victims have been removed.

Responders must consider ethical issues when removing the deceased and handle them with the utmost level of respect and dignity. Once deceased victims have been removed from the hot zone, decon operations must be completed before transferring the deceased to the medical examiner.

Emergency services personnel should be mindful of the need to preserve the incident scene and conduct operations with minimal disturbance and in consultation with those tasked with forensic evidence collection. At the appropriate time, the law enforcement agency (or designated authority) having jurisdiction will make a determination as to how victim remains will be managed.

Handling large numbers of deceased victims (mass casualty incident) may be beyond the capabilities of local emergency response personnel. Specialty response teams (i.e. disaster mortuary teams [DMORT]) may be requested to assist in these types of incidents. In the U.S. and Canada, these teams must be requested through the appropriate emergency management office in order to activate the assistance request. An on-scene morgue facility may have to be established if the incident involves large numbers of deceased victims.

General Guidelines for Decontamination Operations

General guidelines for decon operations include:

- Ensure technical decon setup is operational before entry personnel enter the hot zone.

- Begin emergency/mass decon operations quickly; the speed necessary will be determined by the material and type of incident involved (for example, chemical agents may need immediate removal whereas biological agents may not).

Figure 10.26 Victims may be traumatized, so communication must be clear and easily understood. *Courtesy of New South Wales Fire Brigades.*

- Always wear appropriate PPE.

- Avoid contacting hazardous materials, including contaminated victims.

- Decon operations may be coupled with an initial separation of victims into ambulatory/non-ambulatory and male/female.

- Assess all victims believed to have been in the hot zone to determine the need for decontamination before moving them to the cold zone. Decontaminate as necessary.

- Establish clearly designated decon entry points so that victims and responders both know where to go.

- When conducting decon of victims, the more clothing removed the better (disrobing is effective decon by itself). Unless a victim is soaked in something that would have penetrated outer clothing and into their underwear, there is no real need to have people disrobe completely.

- Decontaminate all emergency response personnel who have been in the hot zone before moving to the cold zone.

- Decon emergency responders separately from victims (establish separate decon lines when possible and practical).

- Establish a medical triage and treatment area just outside the decon zone so that victims exiting the decon area can be evaluated for injuries and exposure-related medical symptoms.

- Communicate with victims by using hand signals, signs with pictures, apparatus public address systems, megaphones or other methods to direct them to decon gathering areas as well as through the decon process itself. It is important to provide clear and easily understood directions since people may be traumatized and/or suffering from exposures **(Figure 10.26)**.

- Provide privacy whenever possible (including from overhead vantage points, for example, from circling news helicopters or upper stories of nearby buildings).

- Provide warm water for washing, if possible. If water is cold, allow victims to gradually get wet in order to acclimate to the temperature and avoid cold shock.

- Document and preserve (safeguard the condition of) belongings of victims decontaminated for future identification of victims and forensic examination.

- Provide victims and responders with clean alternative clothing to maintain their privacy and protect them from the weather.

What This Means To You

In mass decon situations, you will wear whatever PPE you have available when you arrive on the scene. In most situations, fire service personnel will wear structural firefighter protective clothing and SCBA **(Figure 10.27)**. As more information about the hazardous material involved is gathered, adjustments can be made accordingly.

In technical decon situations where the hazardous material has been identified, use NIOSH guides and manufacturer's recommendations to determine appropriate chemical protective clothing and respiratory protection. Often, those conducting decon are dressed in an ensemble classified one level below that of the entry team **(Figure 10.28, p. 508)**. Thus, if the entry team is dressed in an Environmental Protection Agency (EPA) Level A ensemble, the decon team is dressed in Level B. In some cases, if you are the first person working in the decon line, you may need to be dressed at the same level as the entry team. In either case, chemical gloves are necessary; fire-fighting gloves should not be used in decontamination procedures **(Figure 10.29, p. 508)**. Because there is a possibility that you may become contaminated while working decontamination operations, you will need to pass through decontamination before leaving the corridor.

Decon supervisors should always have a plan for responders who may run low on air involving SCBA during decon operations. Consult your SOP/Gs and emergency response plans for additional information.

Figure 10.27 In most mass decon situations, fire service responders will initially be wearing firefighter protective clothing and SCBA.

Figure 10.28 In technical decon scenarios, those conducting decon are often dressed in an ensemble classified one level below that of the entry team. *Courtesy of FEMA News Photos, photo by Win Henderson.*

Figure 10.29 Individuals conducting decon operations should *not* wear leather gloves because the gloves can absorb contaminants.

Decontamination Implementation

Factors considered when implementing decontamination include:

- Site Selection
- Decontamination Corridor Layout
- Decontamination Security Considerations
- Cold Weather Decontamination
- Evidence Collection and Decontamination
- Evaluating Effectiveness of Decontamination Operations

Site Selection

The following factors are considered when choosing a decontamination site:

- **Wind Direction** — The decontamination site needs to be upwind of the hot zone to help prevent the spread of airborne contaminants into clean areas. If the decontamination site is improperly located downwind, wind currents will blow mists, vapors, powders, and dusts toward responders and victims. During long term operations, the local weather service can provide assistance in predicting changes in the wind direction and weather.

- **Weather** — Ideally, during cold weather, the site should be protected from blowing winds, especially near the end of the corridor. Victims should be shielded from cold winds when they are removing clothing.

- **Accessibility** — The site must be away from the hazards, but adjacent to the hot zone so that persons exiting the hot zone can step directly into the decontamination corridor. An adjacent site eliminates the chance of contaminating clean areas. It also puts the decontamination site as close as possible to the actual incident.

- **Time** — Time is a major consideration in the selection of a site. The less time it takes personnel to get to and from the hot zone, the longer personnel can work. Four crucial time periods are as follows:

  — Travel time in the hot zone

  — Time allotted to work in the hot zone

  — Travel time back to the decontamination site

  — Decontamination time

- **Terrain and surface material** — The decontamination site ideally is flat or slopes toward the hot zone; thus, anything that may accidentally get released in the decontamination corridor would drain toward or into the contaminated hot zone and persons leaving the decon corridor would enter into a clean area. If the site slopes away from the hot zone, contaminants could flow into a clean area and spread contamination. Finding the perfect topography is not always possible, and first responders may have to place some type of barrier to ensure confinement of an unintentional release. Details:

  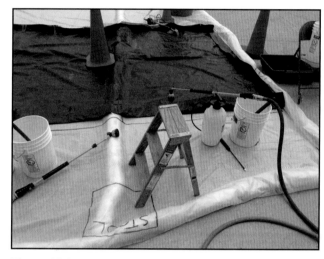

  Figure 10.30 Tarps, plastic sheeting, or salvage covers should be used for flooring, even when the decon corridor is set up on hard surfaces such as concrete or asphalt.

  — Diking around the site prevents accidental contamination escaping.

  — It is best if the site has a hard, nonporous surface to prevent ground contamination.

  — When a hard-surfaced driveway, parking lot, or street is not accessible, some type of impervious covering may be used to cover the ground. Salvage covers or plastic sheeting will prevent contaminated water from soaking into the earth.

  — Covers or sheeting should be used to form the technical decontamination corridor regardless of whether the surface is porous **(Figure 10.30)**.

- **Lighting (and electrical supply)** — The decontamination corridor should have adequate lighting to help reduce the potential for injury to personnel in the area. Selecting a decontamination site illuminated by streetlights, floodlights, or other type of permanent lighting reduces the need for portable lighting. If permanent lighting is unavailable or inadequate, portable lighting will be required. Ideally, the decontamination site will have a ready source of electricity for portable lighting (as well as heaters, water heaters, and other needs). However, if such a source is not available, portable generators will be needed.

- **Drains and waterways** — A decontamination site should NOT be located near storm and sewer drains, creeks, ponds, ditches, and other waterways (unless the sewer system is approved for use as a contained system that can be managed and neutralized). If this situation is not possible, a dike can be constructed to protect the storm drain opening, or a dike may be constructed between the site and a nearby waterway. Protect all environmentally sensitive areas if possible but never delay decon to protect the environment if the delay will increase injury to those affected by the event.

- **Water supply** — Water must be available at the decontamination site if wet decon is used.

Preplans should include pre-designated areas for mass decon at locations likely to be target by terrorists such as government buildings and stadiums. Hospitals must also have plans to decon potentially large numbers of victims who self-present at emergency rooms.

Decontamination Corridor Layout

Responders should establish the decontamination corridor before performing any work in the hot zone. First responders are often involved with setting up and working in the decontamination corridor. The types of decontamination corridors vary as to the numbers of sections or steps used in the decontamination process. Corridors can be straightforward and require only a few steps, or they can be more complex and require a handful of sections and a dozen or more steps. Emergency responders must understand the process and be trained in setting up the type of decontamination required by different materials. Some factors to consider include:

- **Ensure privacy** — Decon tents or decon trailers allow more privacy for individuals going through the decon corridor. Decon officers and ICs need to be particularly sensitive to the needs of women being asked to remove their clothing in front of men (regardless of whether they are victims or other emergency responders). Lawsuits have resulted from situations in which women have felt uncomfortable or even humiliated while going through decon. Providing a private, restricted area such as a tent or trailer in which to conduct decon may prevent similar litigation **(Figure 10.31)**. Use female responders to assist whenever possible when decontaminating women. Do not separate children from their parent(s) or guardian(s).

Figure 10.31 It is important to allow men and women to shower separately if possible. However, family units and others with personal ties (such as children with a babysitter, or an elderly person with a caregiver) should not be separated if they wish to stay together. *Courtesy of New South Wales Fire Brigades.*

- **Bag and tag contaminated clothing/effects** — Various methods can be used, but an accountability system should be implemented. Place clothing and/or personal effects in bags and label the bags with the person's name or other identifier whenever possible. Separate personal effects when possible (wallets, rings, watches, identification cards) into clear plastic bags clearly marked with the person's name or a unique identifying number (triage tag or ticket) **(Figure 10.32)**. These items may need to be decontaminated before being returned. Have some sort of system in place to label or mark all personal effects so that they can be returned to their proper owners after the incident without confusion. All bags that contain contaminated clothing should remain in the warm zone on the dirty side of the decon line. Commercial tagging systems may be used for this purpose or multiple-part plastic hospital identification bracelets for example.

Personal Belongings

Be aware that many individuals will find it very stressful to be separated from their personal belongings. For this reason, responders assisting with disrobing should be sympathetic and empathic about the process to help alleviate anxiety.

To the degree possible, keep track of the status of people and personal belongings. This is important for public information (victim relatives will want to know who went where and why) as well as for crime scene investigation at incidents involving terrorism.

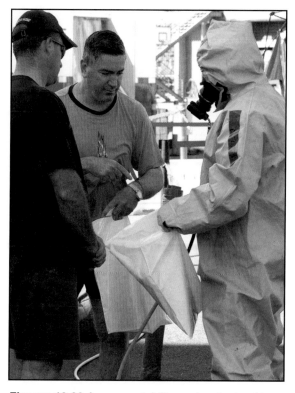

Figures 10.32 An accountability system for tracking contaminated clothing and personal effects must be implemented.

The decontamination corridor may be identified with barrier tape, safety cones, or other items that are visually recognizable **(Figure 10.33)**. Coverings such as a salvage covers or plastic sheeting may also be used to form the corridor. Aside from delineating the corridor and providing privacy, protective covering ensures against environmental harm if contaminated rinse water splashes from a containment basin. Containment basins can be constructed of salvage covers and fire hose or ladders. Some organizations use wading pools or portable drafting tanks as containment basins. Also needed at the site are recovery drums or other types of containers and plastic bags for stowing contaminated tools and PPE.

Decontamination Security Considerations

Law enforcement and military personnel leaving the hot zone must undergo decon **(Figure 10.34, p. 512)**. Performing decon operations on law enforcement and military personnel offers a unique challenge to incident

Figure 10.33 Use barrier tape, safety cones, or similar indicators to mark decon corridors.

Figure 10.34 Everyone leaving the hot zone, including law enforcement and military personnel, must undergo decon. *Courtesy of the U.S. Air Force, photo by Staff Sgt. C. Todd Lopez.*

Figure 10.35 Local policies must establish procedures for managing potentially contaminated weapons, ammunition, and other equipment that could be damaged by exposure to liquid decon

operations. These personnel often carry weapons and will not release them to civilian personnel during decon operations. It may be necessary to include a hazmat trained law enforcement officer involved in the decon operations with the sole responsibility of decontaminating weapons and ensuring their security as the operation proceeds. Special consideration must be given to decon of weapons, ammunition, and other equipment that could be damaged by exposure to liquid decon solutions or water **(Figure 10.35)**. Decon plans must take this equipment into consideration in accordance with local policies and procedures.

Another decontamination corridor may be established for armed emergency services personnel who are leaving the hot zone. Weapons are placed in a hazmat recovery bin supervised by a law enforcement officer wearing the correct level of PPE, as law enforcement personnel disarm and go through decon.

Emergency response personnel should take precautions (such as putting protective booties on dogs' feet) before taking canines into the hot zone of hazmat/WMD incidents and should have their own procedures to decontaminate their animals. However, these animals will be processed through the decontamination corridor, and fire department personnel may have to assist in the decontamination of animals **(Figure 10.36)**.

Criminal suspects may need to be decontaminated. The suspect must be supervised by law enforcement throughout this process. If conducting technical decon, the suspect will go through the same decon steps established for responders and other victims **(Figure 10.37)**. Consideration must be given to whether or not handcuffs must be removed and decontaminated. Follow departmental procedures for decontaminating criminal suspects.

Figure 10.36 Service dogs leaving the hot zone will need to undergo decon. *Courtesy of FEMA News Photos, photo by Jocelyn Augustino.*

Figure 10.37 Criminal suspects will go through the same decon steps established for responders and other victims. Provisions must be made to ensure careful supervision during this process.

NOTE: At hazmat/WMD incidents, there may be requests to decontaminate animals and pets. Contingency plans should include guidelines for decontamination of animals and pets.

Cold Weather Decontamination

Conducting wet decon operations in freezing weather can be difficult to execute safely. Even if showers utilize warm water, run-off water can quickly turn to ice, creating a dangerous slip and fall hazard for both victims and responders. If warm water is not available, susceptible individuals (elderly, very young, individuals with chemical injuries or pre-existing health conditions such as diabetes) can suffer cold shock or hypothermia.

Consideration should be given to protecting victims from exposure to cold temperatures, which can cause hypothermia. Answering the following questions will provide information on how best to protect victims:

- Are wet methods necessary, or can disrobing and dry methods accomplish effective decon?

- Is wind chill a factor?

- Is shelter available for victims during and after decon?

- Is it possible to conduct decon indoors using sprinkler systems, indoor swimming pools, and locker room showers?

- If decon will be conducted indoors (at preplanned facilities, for example), how will victims be transported?

- If decon must be conducted outside in freezing temperatures, how will icy conditions be managed (ie. sand, sawdust, salt)?

WARNING!

Individuals who have been exposed to chemical agents should undergo emergency decon immediately, regardless of ambient temperatures.

Individuals who have been exposed to chemical agents should undergo emergency decon immediately, regardless of ambient temperatures. They should disrobe and thoroughly shower. Dry clothing and warm shelter should be provided as soon as possible after showering.

Evidence Collection and Decontamination

Collection, preservation, and sampling of evidence will be performed under the direction of law enforcement, per established procedures. Decontamination issues associated with these activities will also be determined in conjunction with law enforcement.

Evidence collected on the scene by law enforcement personnel must be appropriately packaged (for example, in approved bags or other evidence containers). Only the exterior of the packaging will be decontaminated as it passes from the hot zone to the cold zone **(Figure 10.38)**. When evidence passes through the decon corridor, chain of custody must be documented per AHJ procedures.

Evaluating Effectiveness of Decontamination Operations

Evaluating the effectiveness of decon operations may be done visually or through the use of monitoring and detection devices or other equipment **(Figure 10.39)**. Generally, the hazardous materials involved will determine the technology or device needed to perform the operations.

If large numbers of people are involved, individuals should be briefly checked after they have gone through the decon process, otherwise check each individual more carefully. These checks should be done as they exit the decon corridor. If contamination is detected, individuals must be redirected through the decon process.

Figure 10.38 Decontamination of evidence will need to be conducted under the direction and supervision of qualified law enforcement officials.

Victims still complaining of symptoms or effects should be checked (or rechecked) for contaminants. If the effectiveness of decon is called into question, victims should go through decon again before transport.

Tools and equipment will normally need to be stored in the decon area until the emergency phase of the operation is completed. After being decontaminated they will need to be checked to ensure all contamination has been removed before being placed back in service. Apparatus will also need to undergo decon if they have been exposed or potentially exposed to hazardous materials. The same monitoring and detection equipment used to determine effectiveness of decon on victims and responders may be used on equipment, tools, and apparatus.

Termination of Decontamination Activities

After concluding decon activities, a debriefing needs to be held as soon as is practical for those involved in the incident. Provide exposed victims with as much information as possible about the delayed health effects of the hazardous materials involved in the incident.

Figure 10.39 The effectiveness of decon operations should be verified using appropriate monitoring and detection equipment. *Courtesy of New South Wales Fire Brigades.*

In some cases, return of personal items may be a law enforcement function because of the evidentiary issues involved. There may be circumstances in which personal effects are immediately returned to the persons undergoing decon.

Additional reports and supporting technical documentation such as incident reports, after action reports, and regulatory citations may be required by emergency response plans and/or SOPs. Exposure records may also need to be filled out and filed.

Exposure records are required for all first responders who have been exposed or potentially exposed to hazardous materials. Follow agency SOP/Gs for filling out exposure records. Information recorded on the exposure report can include:

- Activities performed
- Product involved
- Reason for being there
- Equipment failures
- Malfunction of PPE
- Hazards associated with the product
- Symptoms experienced
- Monitoring levels in use
- Circumstances of exposure

Follow-up examinations should be scheduled with medical personnel, if necessary. The individual, the individual's personal physician, and the individual's employer need to keep copies of these exposure records for future reference.

An activity log must be maintained during the incident or put together afterwards, as appropriate. At a minimum, information for the activity log should be captured during the incident debrief. The activity log may be preformatted, and must document the chronology of the events and activities that occurred during the incident and decon procedure.

In the U.S., OSHA standard 29 *CFR* 1910.1020 (Access to Employee Exposure and Medical Records) should be followed as a guide for requirements involving medical records and maintaining exposure reports. SOPs should spell out additional requirements for local recordkeeping and reports.

Chapter Review

Answer the following questions to review the information provided in this chapter.

1. What is the purpose of decontamination?
2. Give examples of the wet, dry, physical, and chemical methods of decontamination.
3. In what situations should gross decon be performed?
4. What are the advantages of emergency decon?
5. What are the differences between technical decon for ambulatory versus nonambulatory victims?
6. What are the differences between mass decon for ambulatory and nonambulatory victims?
7. What are some considerations when handling deceased victims?
8. List some general guidelines for decontamination operations.
9. What factors influence the implementation of decontamination?
10. What post incident paperwork is unique to hazmat incidents?

On-Scene

Step 1: Isolate contaminated tools and equipment according to SOP/Gs.

Step 2: Conduct decontamination of tools and equipment according to SOP/Gs.

Step 3: Wash and/or isolate PPE according to SOP/Gs, doffing as appropriate.

Step 4: Use hygienic wipes to wipe potential contaminates from face, head, neck, and hands.

In Quarters

Step 5: Shower thoroughly using soap and water.

Step 6: Clean PPE according to SOP/Gs.

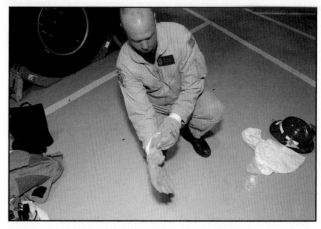

Step 7: Inspect and maintain PPE according to manufacturer's recommendations.

Step 8: Store PPE outside of living and sleeping quarters.

Step 9: Complete required reports and supporting documentation.

Step 1: Ensure that all responders involved in decontamination operations are wearing appropriate PPE for performing emergency decontamination operations.

Step 2: Remove the victim from the contaminated area.

Step 3: Ensure emergency decontamination is set up in a safe area.

Step 4: Wash immediately any contaminated PPE/clothing or exposed body parts with flooding quantities of water.

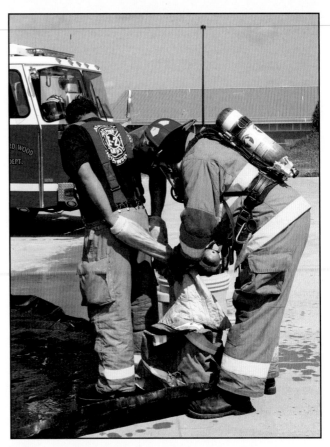

Step 5: Remove mannequin PPE/clothing rapidly, minimizing the spread of contaminants.

Step 6: Perform a quick cycle of head-to-toe rinse, wash, and rinse.

Step 7: Transfer the victim to treatment personnel for assessment, first aid, and medical treatment.

Step 8: Ensure that ambulance and hospital personnel are told about the contaminant involved.

Step 9: Complete required reports and supporting documentation.

Step 1: Ensure proper decontamination method has been chosen to minimize hazards.

Step 2: Ensure that all responders are wearing appropriate PPE for performing technical decontamination operations.

Step 3: Establish technical decontamination corridor for ambulatory decontamination according to the AHJ's SOPs.

Step 4: Ensure the decontamination corridor provides privacy.

Step 5: Establish an initial triage point to evacuate and direct persons.

Step 6: Perform lifesaving intervention if needed.

Step 7: If not an emergency responder, instruct victim to remove potentially contaminated clothing and jewelry, ensuring he/she does not come in further contact with contaminants. Emergency responders may drop tools and or equipment.

Step 8: Instruct person to undergo gross decontamination.

Step 9: Instruct person to undergo secondary decontamination wash.

NOTE: Emergency responders undergoing technical decon will doff PPE per SOPs after the secondary decontamination wash.

Step 10: Instruct person to enter the privacy station, remove undergarments, and shower and wash thoroughly from the top down.

NOTE: Do NOT ask members of the public to remove their clothes to shower unless complete privacy is provided.

Step 11: Provide a clean garment for the person to wear after showering.

Step 12: Monitor for additional contamination using the appropriate detection device.

NOTE: If contamination is detected, repeat the decontamination sequence and/or change the decontamination method, as appropriate.

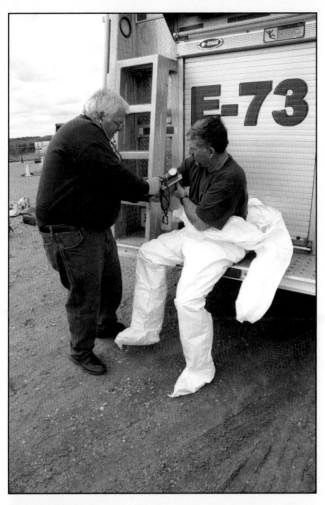

Step 13: Direct the person to the medical evaluation station.

Step 14: Ensure personnel, tools, and equipment are decontaminated.

Step 15: Terminate decontamination operations according to AHJ's policies and procedures.

Step 16: Complete required reports and supporting documentation.

Step 1: Ensure proper decontamination method has been chosen to minimize hazards.

Step 2: Ensure that all responders are wearing appropriate PPE for performing technical decontamination operations.

Step 3: Establish technical decontamination corridor for nonambulatory decontamination according to the AHJ's SOPs.

Step 4: Ensure the decontamination corridor provides privacy for victims.

Step 5: Establish an initial triage point to evaluate and direct persons.

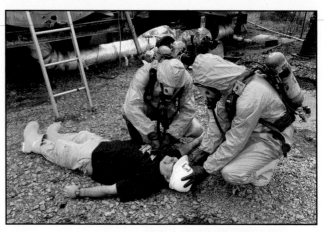

Step 6: Perform lifesaving intervention if needed.

Step 7: Transfer the victim to the nonambulatory wash area of the decontamination station on an appropriate backboard/litter device.

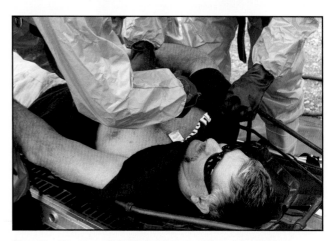

Step 8: Remove all PPE/clothing, jewelry, and personal belongings, and place in appropriate containers. Decontaminate items as required, and safeguard. Use plastic bags with labels for identification.

Step 9: Carefully undress nonambulatory persons, and avoid spreading the contamination when undressing. Do not touch the outside of the clothing to the skin.

NOTE: If biological agents are suspected, a fine water mist can be applied to trap the agent in the clothing and prevent the spread of contamination.

Step 10: Completely wash the victim's entire body using handheld hoses, sponges, and/or brushes, and then rinse.

NOTE: Clean the victim's genital area, armpits, folds in the skin, and nails with special attention. If conscious, instruct the victim to close his/her mouth and eyes during wash and rinse procedures.

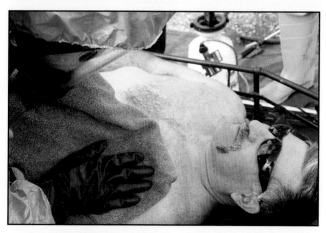

Step 11: Transfer the victim from the wash and rinse stations to a drying station after completing the decontamination process. Ensure that the victim is completely dry.

Step 12: Monitor for additional contamination using the appropriate detection device.

NOTE: If contamination is detected, repeat decontamination wash and/or change decontamination method, as appropriate.

Step 13: Have on-scene medical personnel reevaluate the victim's injuries.

Step 14: Ensure personnel, tools, and equipment are decontaminated.

Step 15: Terminate decontamination operations according to AHJ's policies and procedures.

Step 16: Complete required reports and supporting documentation.

Step 1: Ensure proper decontamination method has been chosen to minimize hazards.

Step 2: Ensure that all responders are wearing appropriate PPE for performing mass decontamination operations.

Step 3: Ensure decontamination operations are set up in a safe area.

Step 4: Ensure the decontamination corridor provides privacy for victims.

Step 5: Prepare fire apparatus for use during mass decontamination.

Step 8: Instruct victims to remove contaminated clothing, ensuring that victims do not come into further contact with any contaminants.

Step 6: Set fire nozzle to fog pattern.

Step 7: Instruct all victims to go through mass decontamination.

Step 9: Instruct victims to keep arms raised as they proceed slowly through the wash area.

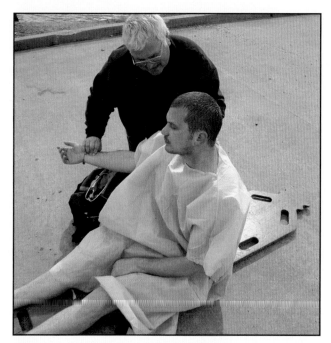

Step 12: Send victims for medical treatment.

Step 13: Inform EMS personnel of contaminant involved and its hazards, if known.

Step 14: Ensure personnel, tools, and equipment are decontaminated.

Step 15: Terminate decontamination operations according to AHJ's policies and procedures.

Step 16: Complete required reports and supporting documentation.

Step 10: Monitor for additional contamination using the appropriate detection device.

NOTE: If contamination is found, instruct victim to go through wash again, as appropriate.

Step 11: Instruct victims to move to a clean area to dry off.

Step 1: Ensure proper decontamination method has been chosen to minimize hazards.

Step 2: Ensure that all responders are wearing appropriate PPE for performing mass decontamination operations.

Step 3: Establish mass decontamination corridor for nonambulatory decontamination in a safe location and according to the AHJ's SOPs.

Step 4: Ensure the decontamination corridor provides privacy for victims.

Step 5: Establish an initial triage point to evaluate and direct persons.

Step 6: Perform lifesaving intervention if necessary.

Step 7: Transfer the victim to the nonambulatory wash area of the decontamination station on an appropriate backboard/litter device.

Step 8: Remove all clothing, jewelry, and personal belongings, and place in appropriate containers. Decontaminate as required, and safeguard personal belongings and items. Use plastic bags with labels for identification.

Step 9: Carefully undress nonambulatory persons, and avoid spreading the contamination when undressing. Do not touch the outside of the clothing to the skin.

NOTE: If biological agents are suspected, a fine water mist can be applied to trap the agent in the clothing and prevent the spread of contamination.

 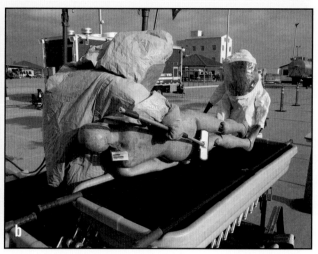

Step 10: Completely wash the victim's entire body using handheld hoses, sponges, and/or brushes and then rinse following AHJ's SOPs for safety.

NOTE: Clean the victim's genital area, armpits, folds in the skin, and nails with special attention. If conscious, instruct the victim to close his/her mouth and eyes during wash and rinse procedures.

Step 11: Transfer the victim from the wash and rinse stations to a drying station after completing the decontamination process. Ensure that the victim is completely dry.

Step 12: Monitor for additional contamination using the appropriate detection device.

NOTE: If contamination is detected, repeat decontamination wash and/or change decontamination method as appropriate.

Step 13: Have on-scene medical personnel reevaluate the victim's injuries.

Step 14: Ensure personnel, tools, and equipment are decontaminated.

Step 15: Terminate decontamination operations according to AHJ's policies and procedures.

Step 16: Complete required reports and supporting documentation.

Chapter Contents

Key Terms

NFPA Job Performance Requirements

This chapter provides information that addresses the following job performance requirements of NFPA 1072, *Standard for Hazardous Materials/Weapons of Mass Destruction Emergency Response Personnel Professional Qualifications (2017)*.

6.7.1

Implementing the Response: Mission-Specific Detection, Monitoring, and Sampling

Learning Objectives

After reading this chapter, students will be able to:

1. Identify uses of concentration, dose, and exposure limits in hazardous materials incidents. (6.7.1)

2. Explain the capabilities and limitations of detection, monitoring, and sampling procedures. (6.7.1)

3. Describe processes for selecting and maintaining detection, monitoring, and sampling devices. (6.7.1)

4. Describe types of monitoring, sampling, and detection devices. (6.7.1)

5. Demonstrate proper use of pH paper to identify hazards. (6.7.1; Skill Sheet 11-1)

6. Demonstrate proper use of pH meters to identify hazards. (6.7.1; Skill Sheet 11-2)

7. Demonstrate proper use of reagent test paper to identify hazards. (6.7.1; Skill Sheet 11-3)

8. Demonstrate the use of a multi-gas meter (carbon monoxide, oxygen, combustible gases, multi-gas, and others) to identify hazards. (6.7.1; Skill Sheet 11-4)

9. Demonstrate proper use of radiation detection instruments to identify hazards. (6.7.1; Skill Sheet 11-5)

10. Demonstrate proper use of dosimeters to identify received dose. (6.7.1; Skill Sheet 11-6)

11. Demonstrate proper use of photoionization detectors to identify hazards. (6.7.1; Skill Sheet 11-7)

12. Demonstrate proper use of colorimetric tubes to identify hazards. (6.7.1; Skill Sheet 11-8)

Chapter 11
Implementing the Response: Mission-Specific Detection, Monitoring, and Sampling

The following topics are explained in this chapter:

- Concentrations and exposures limits
- Detection, monitoring, and sampling basics
- Selection and maintenance of detection, monitoring, and sampling devices
- Hazard-detection equipment

Concentration, Dose, and Exposure Limits

Responders use detection and monitoring devices to detect, identify, and measure hazardous materials **(Figure 11.1, p. 532)**. It is beyond the scope of this manual to thoroughly cover all the detection and monitoring devices available to responders. However, this chapter will address many of the devices that agencies commonly use.

Numerous detection and monitoring devices, such as the following, determine and display the concentration or doses of some materials **(Figure 11.2. p. 532)**:

- Devices that measure *concentrations* — Measure materials that responders might inhale.

- Devices that measure *dosage* — Measure materials that may enter responders' bodies via a means other than inhalation.

Once responders know the actual concentration or dose of the materials on scene, they can determine whether the materials on scene have sufficient concentrations or doses to make them hazardous to responders. Responders may make this determination by referencing the exposure limits of the present materials. **Exposure limits** refer to values expressing the maximum dose or **concentration** to which individuals should be exposed given a specific time frame.

Sources will often state the concentration of a substance in the following terms:

- **Milligrams per cubic meter (mg/m³)** — Expresses concentrations of dusts, gases, or mists in air.

- **Grams per kilogram (g/kg)** — Denotes grams of a substance dosed per kilogram of animal body weight. Commonly used as an expression of a dose in oral and dermal toxicology testing.

> **Exposure Limit** —
> Maximum length of time an individual can be exposed to an airborne substance before injury, illness, or death occurs.

> **Concentration** —
> Quantity of a material in relation to a larger volume of gas or liquid.

Figure 11.1 Monitoring, detection, and sampling are important activities to ensure safety at hazmat incidents.

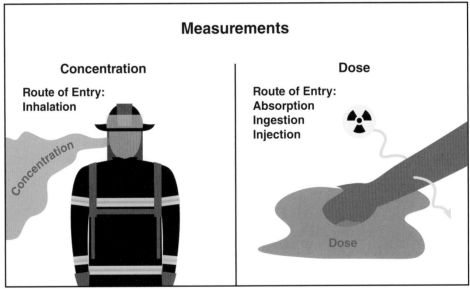

Figure 11.2 Inhaled hazardous materials are measured in concentrations, while materials that enter the body via ingestion, injection, and/or absorption are measured in dosages.

- **Milligrams per kilogram (mg/kg)** — Toxicological dose that denotes milligrams of a substance dosed per kilogram of animal body weight. This unit describes much smaller dosages than g/kg.

- **Micrograms of material per liter of air (µg/L)** — Describes concentrations of chemicals in air.

- **Milligrams per liter (mg/L)** — Expresses concentrations of chemicals in water.

- **Parts per million (ppm)** — May describe the concentration of a gas or vapor in air or the concentration of a specific material in a liquid or solid. This notation describes the relative abundance of a specific material; for example, ppm may describe the number of molecules of gas found within a million air molecules. Or, it may describe the number of molecules of a hazardous material found in a million molecules of a liquid. One microgram is equivalent to 2-3 grains of table salt.

- **Parts per billion (ppb)** — May describe the concentration of a gas or vapor in air or the concentration of a specific material in a liquid or solid. This notation describes the relative abundance of a specific material. Sources will use ppb to describe concentrations much lower than those that can be expressed in ppm. The term ppb is usually used to express extremely low concentrations of unusually toxic gases or vapors.

The NIOSH *Pocket Guide to Hazardous Materials* and other resources provide exposure limits for many materials. Terms describing exposure limits may vary by the source (**Table 11.1, p. 534**). Important terms include:

- **Threshold limit value (TLV®)** — Concentration that can be tolerated during 8-hour workdays.

- **Short-term exposure limit (STEL)** — Concentration that can be tolerated for 15 minutes at a time, provided an appropriate interval between exposures.

- **Threshold limit value-ceiling (TLV®-C)** — Concentration that should not be exceeded during 8-hour workdays.

- **Permissible exposure limit (PEL)** — Concentration at which most people are not adversely affected.

- **Immediately dangerous to life or health (IDHL)** — Concentrations high enough to kill or cause serious injury or illness.

Rather than providing concentrations of specific materials, some meters will instead provide the percentage of a material in the atmosphere, for example, the percentage of oxygen in the air, or percentage of the lower flammable (explosive) limit (LFL) for flammable gases and vapors. As with toxic materials, responders must know precisely what the meter measures, what the displayed numbers or percentages indicate, and what this information means in terms of their safety.

CAUTION
Do not allow gas meters to draw liquids into the probe.

Parts Per Million (ppm) — Method of expressing the concentration of very dilute solutions of one substance in another, normally a liquid or gas, based on volume; expressed as a ratio of the volume of contaminants (parts) compared to the volume of air (million parts). The common unit of measure is equivalent to 1 microgram [1 µg] per liter of water or kg of solid, or a micro liter [1 µL] volume of gas in one liter of air.

Parts Per Billion (ppb) — Method of expressing the concentration of very dilute solutions of one substance in another, normally a liquid or gas, based on volume; expressed as a ratio of the volume of contaminants (parts) compared to the volume of air (billion parts).

Threshold Limit Value (TLV®) — Maximum concentration of a given material in parts per million (ppm) that may be tolerated for an 8-hour exposure during a regular workweek without ill effects.

Short-Term Exposure Limit (STEL) — Fifteen-minute time-weighted average that should not be exceeded at any time during a workday; exposures should not last longer than 15 minutes and should not be repeated more than four times per day with at least 60 minutes between exposures.

Threshold Limit Value/Ceiling (TLV®/C) — Maximum concentration of a given material in parts per million (ppm) that should not be exceeded, even instantaneously.

Permissible Exposure Limit (PEL) — Maximum time-weighted concentration at which 95 percent of exposed, healthy adults suffer no adverse effects over a 40 hour work week; an 8-hour time-weighted average unless otherwise noted. PELs are expressed in either parts per million (ppm) or milligrams per cubic meter (mg/m^3). They are commonly used by OSHA and are found in the NIOSH *Pocket Guide to Chemical Hazards*.

Table 11.1
Exposure Limits

Term	Definition	Exposure Period	Organizaion
IDLH Immediately Dangerous to Life or Health	An atmospheric concentration of any toxic, corrosive, or asphyxiating substance that poses an immediate threat to life. It can cause irreversible or delayed adverse health effects and interfere with the individual's ability to escape from a dangerous atmosphere.*	Immediate (This limit represents the maximum concentration from which an unprotected person can expect to escape in a 30-minute period of time without suffering irreversible health effects.)	**NIOSH** National Institute for Occupational Safety and Health
IDLH Immediately Dangerous to Life or Health	An atmosphere that poses an immediate threat to life, would cause irreversible adverse health effects, or would impair an individual's ability to escape from a dangerous atmosphere.	Immediate	**OSHA** Occupational Safety and Health Administration
LOC Levels of Concern	10% of the IDLH		
PEL Permissible Exposure Limit**	A regulatory limit on the amount or concentration of a substance in the air. PELs may also contain a skin designation. The PEL is the maximum concentration to which the majority of healthy adults can be exposed over a 40-hour workweek without suffering adverse effects.	8-hours Time-Weighted Average (TWA)*** (unless otherwise noted)	**OSHA** Occupational Safety and Health Administration
PEL (C) PEL Ceiling Limit	The maximum concentration to which an employee may be exposed at any time, even instantaneously.	Instantaneous	**OSHA** Occupational Safety and Health Administration
STEL Short-Term Exposure Limit	The maximum concentration allowed for a 15-minute exposure period.	15 minutes (TWA)	**OSHA** Occupational Safety and Health Administration
TLV® Threshold Limit Value†	An occupational exposure value recommended by ACGIH® to which it is believed nearly all workers can be exposed day after day for a working lifetime without ill effect.	Lifetime	**ACGIH®** American Conference of Governmental Industrial Hygienists
TLV®-TWA Threshold Limit Value-Time-Weighted Average	The allowable time-weighted average concentration.	8-hour day or 40-hour workweek (TWA)	**ACGIH®** American Conference of Governmental Industrial Hygienists
TLV®-STEL Threshold Limit Value-Short-Term Exposure Limit	The maximum concentration for a continuous 15-minute exposure period (maximum of four such periods per day, with at least 60 minutes between exposure periods, provided the daily TLV®-TWA is not exceeded).	15 minutes (TWA)	**ACGIH®** American Conference of Governmental Industrial Hygienists

Continued

Table 11.1 (continued)

Term	Definition	Exposure Period	Organizaion
TLV®-C Threshold Limit Value-Ceiling	The concentration that should not be exceeded even instantaneously.	Instantaneous	**ACGIH®** American Conference of Governmental Industrial Hygienists
BEIs® Biological Exposure Indices	A guidance value recommended for assessing biological monitoring results.		**ACGIH®** American Conference of Governmental Industrial Hygienists
REL Recommended Exposure Limit	A recommended exposure limit made by NIOSH.	10-hours (TWA) ††	**NIOSH** National Institute for Occupational Safety and Health
AEGL-1 Acute Exposure Guideline Level-1	The airborne concentration of a substance at or above which it is predicted that the general population, including "susceptible" but excluding "hypersusceptlble" individuals, could experience notable discomfort. †††	Multiple exposure periods: 10 minutes 30 minutes 1 hour 4 hours 8 hours	**EPA** Environmental Protection Agency
AEGL-2 Acute Exposure Guideline Level-2	The airborne concentration of a substance at or above which it is predicted that the general population, including "susceptible" but excluding "hypersusceptible" individuals, could experience irreversible or other serious, long-lasting effects or impaired ability to escape. Airborne concentrations below AEGL-2 but at or above AEGL-1 represent exposure levels that may cause notable discomfort.	Multiple exposure periods: 10 minutes 30 minutes 1 hour 4 hours 8 hours	**EPA** Environmental Protection Agency
AEGL-3 Acute Exposure Guideline Level-3	The airborne concentration of a substance at or above which it is predicted that the general population, including "susceptible" but excluding "hypersusceptible" individuals, could experience life-threatening effects or death. Airborne concentrations below AEGL-3 but at or above AEGL-2 represent exposure levels that may cause irreversible or other serious, long-lasting effects or impaired ability to escape.	Multiple exposure periods: 10 minutes 30 minutes 1 hour 4 hours 8 hours	**EPA** Environmental Protection Agency
ERPG-1 Emergency Response Planning Guideline Level 1	The maximum airborne concentration below which it is believed nearly all individuals could be exposed for up to one hour without experiencing other than mild transient adverse health effects or perceiving a clearly defined objectionable odor.	Up to 1 hour	**AIHA** American Industrial Hygiene Association

Continued

Table 11.1 (concluded)

Term	Definition	Exposure Period	Organizaion
ERPG-2 Emergency Response Planning Guideline Level 2	The maximum airborne concentration below which it is believed nearly all individuals could be exposed for up to one hour without experiencing or developing irreversible or other serious health effects or symptoms that could impair an individual's ability to take protective action.	Up to 1 hour	**AIHA** American Industrial Hygiene Association
ERPG-3 Emergency Response Planning Guideline Level 3	The maximum airborne concentration below which it is believed nearly all individuals could be exposed without experiencing or developing life-threatening health effects.	Up to 1 hour	**AIHA** American Industrial Hygiene Association
TEEL-0 Temporary Emergency Exposure Limits Level 0	The threshold concentration below which most people will experience no appreciable risk of health effects.		**DOE** Department of Energy
TEEL-1 Temporary Emergency Exposure Limits Level 1	The maximum concentration in air below which it is believed nearly all individuals could be exposed without experiencing other than mild transient adverse health effects or perceiving a clearly defined objectionable odor.		**DOE** Department of Energy
TEEL-2 Temporary Emergency Exposure Limits Level 2	The maximum concentration in air below which it is believed nearly all individuals could be exposed without experiencing or developing irreversible or other serious health effects or symptoms that could impair their abilities to take protective action.		**DOE** Department of Energy
TEEL-3 Temporary Emergency Exposure Limits Level 3	The maximum concentration in air below which it is believed nearly all individuals could be exposed without experiencing or developing life-threatening health effects.		**DOE** Department of Energy

* It should be noted that the NIOSH definition only addresses airborne concentrations. It does not include direct contact with liquids or other materials.

** PELs are issued in Title 29 *CFR* 1910.1000, particularly Tables Z-1, Z-2, and Z-3, and are enforceable as law.

*** Time-weighted average means that changing concentration levels can be averaged over a given period of time to reach an average level of exposure.

† TLVs® and BEIs® are guidelines for use by industrial hygienists in making decisions regarding safe levels of exposure. They are not considered to be consensus standards by the ACGIH®, and they do not carry the force of law unless they are officially adopted as such by a particular jurisdiction.

†† NIOSH may also list STELs (15-minute TWA) and ceiling limits.

††† Airborne concentrations below AEGL-1 represent exposure levels that could produce mild odor, taste, or other sensory irritation.

Exposure Safety

For your personal safety:

- Wear appropriate PPE for potential and suspected hazards until detection and monitoring determines the actual hazards present.

- Consider that the lower the exposure limit(s) are, the more potentially harmful the substance may be.

- Realize that you should be safe from any toxic effects if your exposure levels never exceed the lowest numbers for that material; that is, if you are not allergic to the material.

- Never enter IDLH atmospheres without appropriate PPE, including self-contained breathing apparatus (SCBA).

- Exit IDLH atmospheres immediately, if you do not have appropiate PPE.

Detection, Monitoring, and Sampling Basics

Detection, monitoring, and sampling assist in the following mitigation tasks:

- Identify hazards (the potentially hazardous materials present and their concentrations) **(Figure 11.3)**.

- Determine appropriate PPE, tools, and equipment.

- Determine perimeters and the scope of the incident (how far the materials traveled, contaminated areas, and/or potentially safe and contamination-free areas).

- Check the effectiveness of defensive operations.

- Ensure the effectiveness of decon operations.

- Detect leaks from containers or piping systems.

- Monitor the contamination levels of decon runoff.

Figure 11.3 Responders may need to take samples in order to identify unknown materials *Courtesy of Sherry Arasim.*

In order to successfully detect and interpret concentration levels of materials, responders must couple their understanding of the behavior of hazardous materials with an understanding of the detection devices used in their jurisdiction. The state of matter of the material being sampled will affect the monitoring and detection techniques and devices used.

Most agencies that use air detection and monitoring devices will have a variety of equipment to detect a number of different materials and hazards. As a responder, you should remember that most gases sink and displace air, while only a few rise and float above air. To properly determine concentrations of different gases, operate monitoring and detection devices at different heights within a room/area and at different grades within a building (**Figure 11.4**). The knowledge, skills and ability of the individual using the instrument determine the effectiveness of that instrument. The instrument and its user must function as a team and must practice and work together.

Responders using detection, monitoring, and sampling devices must:

- Have a good understanding of the capability of each device.
- Use the devices correctly (**Figure 11.5**).
- Understand what is being measured and how the instrument relays the information to the user.
- Interpret accurately the data each device provides.

Figure 11.4 Because vapor densities vary, and air currents can move hazardous gases and vapors in unexpected ways; samples must be taken at different heights.

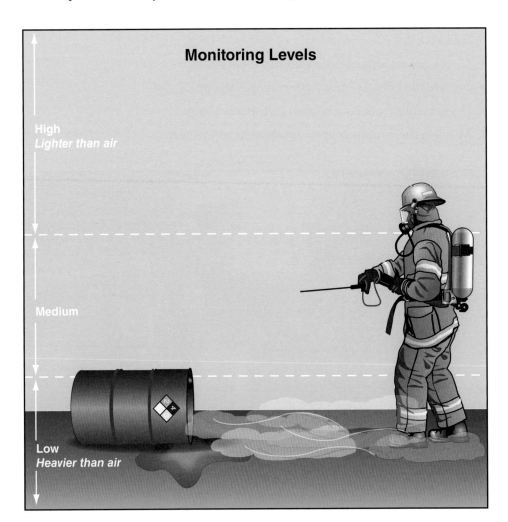

Monitoring Levels

High
Lighter than air

Medium

Low
Heavier than air

- Maintain, field test, and calibrate the devices per manufacturers' instructions.

- Use the devices in accordance with predetermined procedures based on the availability, capabilities, and limitations of personnel, appropriate personal protective equipment (PPE), and other resources available at the incident and in accordance with the IAP.

- Use more than one sampling method and more than one technology to verify monitoring and sampling results, when possible.

WARNING!
Never rely on one type of detection equipment exclusively.

CAUTION
Always operate under the guidance of a hazardous materials technician, an allied professional, an emergency response plan, or standard operating procedures.

Figure 11.5 Responders must have the training to use detection, monitoring, and sampling devices correctly.

A responder who does not understand how to use the devices correctly can easily jeopardize his or her safety and the safety of others. For example, the **instrument response time** may take several seconds **(Figure 11.6, p. 540)**. If responders move too quickly, they may find themselves in a situation where the concentration of the hazardous material is much higher than the meter indicates because they have moved beyond the area where their meter took the sample.

Instrument Response Time — Elapsed time between the movement (drawing in) of an air sample into a monitoring/detection device and the reading (analysis) provided to the user. *Also known as* Instrument Response Time.

WARNING!
To conduct detection, monitoring, and sampling, personnel must have proper training to do so.

WARNING!
All personnel must wear appropriate PPE when operating in potentially hazardous areas.

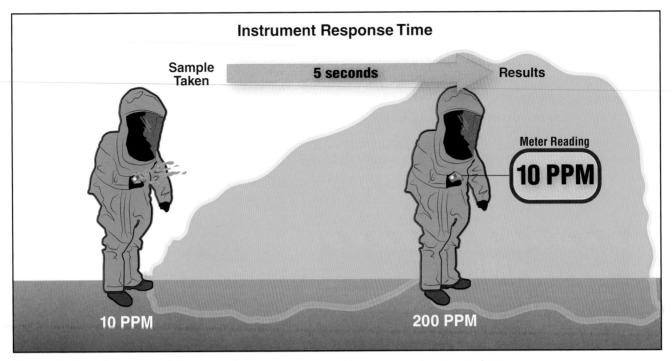

Instrument Response Time

Sample Taken

5 seconds

Results

Meter Reading

10 PPM

10 PPM

200 PPM

Figure 11.6 In order to accurately assess the environment, consider instrument response times.

At WMD or criminal incidents, personnel who conduct sampling activities must follow evidence preservation and chain-of-custody procedures. In order for any recordings to be later admissible as evidence, responders must follow appropriate protocols in regard to chain of custody, packaging, labeling, and transportation of evidence to the testing authority **(Figure 11.7)**. Personnel must record any results of detection and monitoring activities. Responders will need law enforcement assistance in this process.

The hazards present at the incident and the responders' mission will dictate the detection and monitoring strategies and tactics used at the incident as well as the PPE required to perform these strategies and tactics. Strategies and tactics may differ significantly depending on the mission. For instance, responders attempting a rescue will have different objectives than responders attempting the defensive mitigation of a release. Regardless of the mission, hazardous materials incidents will always require size-up and risk assessment. Rescue tactics will be explained in Chapter 12, Implementing the Response: Mission-Specific Victim Rescue and Recovery.

Figure 11.7 When samples are taken at suspected criminal/ WMD incidents, you must follow the chain-of-custody procedures.

OSHA Monitoring

Conduct periodic monitoring as conditions warrant for safe operations, for example, during changing environmental conditions or when moving to a different location. In the U.S., OSHA 29 *CFR* 1910.120(q) requires detection and monitoring before entry into potentially hazardous atmospheres. According to OSHA, responders should conduct detection and monitoring to identify:

- IDLH conditions

- Exposure over permissible exposure limits or published exposure levels

- Exposure over a radioactive material's dose limits or other dangerous condition such as the presence of flammable atmospheres and/or oxygen-deficient environments

Responders may have some idea of what hazardous materials are involved in the incident. If this information is known or suspected, selecting PPE and monitoring and sampling equipment become easier. Responders may consult resources to understand the hazards and properties of the materials involved and to develop an appropriate risk-based response.

Detection Responses

Because detection equipment is designed to detect a specific hazard, the lack of a response does not mean that other hazards are not present. A lack of information on one type of detection device is still information that can be used to determine the type of hazards present. Reasons a meter may not provide a usable result in an environment include:

- Incorrect use

- Calibration

- Low battery

When dealing with unknown materials, take an analytic approach in attempting to identify and characterize the hazards present. This includes monitoring **(Figure 11.8)**:

- Corrosives

- Exothermic reactions

- Flammables

- Oxidizers (and explosives)

- Oxygen levels

- Radiation

- Toxics

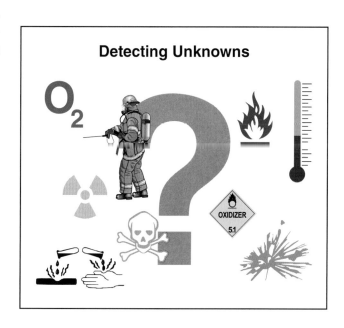

Figure 11.8 When incidents involve unknown substances, responders should monitor for corrosives, exothermic reactions, flammables, oxidizers (and explosives), oxygen levels, radiation, and toxics.

WARNING!
Detection devices do not provide accurate readings if used in an oxygen-deficient atmosphere.

Before beginning detection and monitoring activities, your monitoring plan should address the following questions:

- What purpose does monitoring serve?

- What readings should detectors anticipate? If users detect unanticipated readings, they should consider the possibility of instrument failure, but should not ignore results.

- Which instruments will detect the materials anticipated under the current incident conditions?

- Is more than one hazard present?

- How will current conditions, such as rain, humidity, or temperature, affect monitoring devices?

To stay safe while monitoring, follow the recommended steps:

- Operate under the direction of a hazmat technician, specialist, or allied professional.

- Follow written procedures.

- Understand the limitations of the instruments and detection devices. Follow manufacturers' instructions for **calibration** and usage.

- Remove any damaged device from service immediately. It may no longer be intrinsically safe and may provide false readings **(Figure 11.9)**.

- Be consistent — always wear your appropriate PPE.

- Work with a buddy — always have a back-up team waiting in appropriate PPE.

- Approach the hazard area from upwind.

- Follow local protocols and SOP/Gs in the event an alarm sounds or a hazardous material is detected.

- Pay particular attention to low-lying areas, confined spaces, and containers where vapors and gases will likely concentrate **(Figure 11.10)**.

- Move slowly, making allowances for instruments with significant response times.

- Monitor for vapors and gases at ground level, waist level, and above the head.

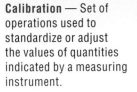

Calibration — Set of operations used to standardize or adjust the values of quantities indicated by a measuring instrument.

WARNING!
Always assume more than one hazard is present.

Potential Causes of Equipment Damage

Contact with
Corrosive Gases

Being Dropped
or Struck

Contact with
Corrosive Liquids

Figure 11.9 Remove damaged devices from service immediately because they may provide false readings.

Figure 11.10 Gases may concentrate in low-lying areas and confined spaces. *Courtesy of MSA.*

Personnel should accurately document monitoring, detection, and sampling results **(Figure 11.11)**:

- Time of the reading
- Location and level of the reading
- Reading obtained
- Instrument used

Record this information in a notebook immediately. Follow AHJ protocols for reporting readings to the Incident Commander.

Action Levels

The AHJ should establish action levels. Action levels can be defined as a response to known or unknown chemicals or products that will trigger some action **(Figure 11.12, p. 544)**. When an action level (or action point) is reached, it may trigger:

- Removal of unprotected or unnecessary personnel
- Additional monitoring
- Alteration or adjustment of PPE
- Total area evacuation

Other factors that may influence action levels include manufacturer's recommendations.

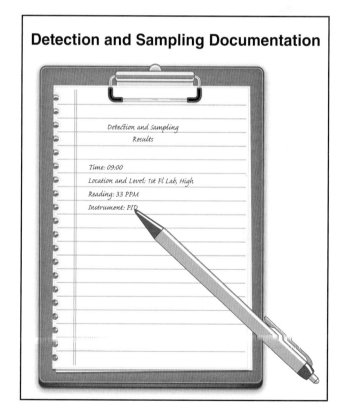

Detection and Sampling Documentation

Detection and Sampling
Results

Time: 09:00
Location and Level: 1st Fl Lab, High
Reading: 33 PPM
Instrument: PID

Figure 11.11 SOPs/SOGs should dictate how to accurately document the results for monitoring, detection, and sampling.

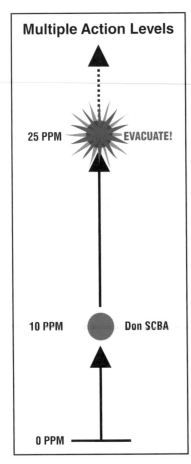

Multiple Action Levels

25 PPM — EVACUATE!

10 PPM — Don SCBA

0 PPM

Figure 11.12 Action levels trigger specific actions such as donning SCBA or evacuating the area. Substances may have multiple action levels.

Determining PPE

If you respond to an incident involving unidentified hazardous materials, you can determine and select personal protective clothing and equipment based on factors described in Chapter 9. Responders should consider the hazards present and determine which protective resources and features will be necessary during detection and monitoring operations. Examples of the thought processes behind selection of each type of equipment are explained in the following paragraphs. At all times, responders must maintain situational awareness and adherence to the instructions of the Incident Commander and the AHJ.

Hazardous solids will usually not travel far unless blown by the wind or dispersed by explosion. Responders at hazmat incidents involving hazardous solids will typically be sufficiently protected when using firefighter protective equipment with SCBA or level C PPE for monitoring and sampling activities.

Hazardous liquids may vaporize and expand rapidly, potentially travel great distances, and fill low-level and confined spaces. Responders at liquid releases may need to don the most protective levels of respiratory and chemical protective equipment before engaging in detection and monitoring activities.

Hazardous gases expand rapidly, filling enclosed spaces and potentially traveling great distances from their source. Corrosive gases can damage and penetrate firefighter protective clothing and SCBA. Toxic gases can kill quickly if inhaled. Flammable gases present an extreme fire hazard while oxidizing gases can cause explosions. Inert gases can displace oxygen levels in enclosed spaces causing an asphyxiation hazard. At gas releases, responders should operate with extreme caution during monitoring and sampling activities **(Figure 11.13)**.

Figure 11.13 Evaluate incidents involving ton cylinders, compressed gas cylinders, and high-pressure vessels carefully to select appropriate PPE. These containers are likely to release gases and may require vapor-protective clothing. *Courtesy of the U.S. Coast Guard, photo by Telfair H. Brown.*

Selection and Maintenance of Detection, Monitoring, and Sampling Devices

When determining what equipment to use for detection and monitoring operations, responders should consider the following:

● **Mission of the operation** — Is the priority of the operation rescue or product control?

● **Suspected hazards involved** — Will responders monitor for radiation at explosive incidents or use pH paper to detect corrosives?

- **Portability and user friendliness** — Some instruments weigh more than others; some are bulky or difficult to use. Given the mission and available PPE, a responder should determine the usability of the devices **(Figure 11.14)**.

- **Instrument reaction time** — Some instruments require a delay ranging from seconds to minutes between the time the instrument detects the material and displays the readings. If the mission necessitates a quick rescue, instruments with long delays may be useless.

- **Sensitivity and selectivity** — Some instruments will detect lower concentrations than others, while others will only detect very specific materials. Responders must consider how well the instrument will detect the desired chemical or chemical family and to what degree.

- **Calibration** — Most instruments need calibration prior to use, per AHJ and manufacturer recommendations. Many factors, such as temperature, humidity, elevation, and atmospheric pressure, can affect this process. Calibrating an instrument in the field can be difficult. Before operating a device, you should do the following:

  — **Calibration Test** (Bump Test) — Ensures that sensors function appropriately for alarms and other functions, but it does not test the accuracy of the sensor **(Figure 11.15)**.

  — **Zeroing** — Resets the memory of an instrument to read at normal (baseline) levels in fresh air. Some instruments will do this function automatically during power-up; some instruments have a dedicated button or menu function.

- **Training** — How difficult is it to learn how to use the instrument? How often does training need to be conducted? Do responders on the scene have adequate training to use the instrument effectively?

> **Calibration Test** — Set of operations used to make sure that an instrument's alerts all work at the recommended levels of hazard detected. *Also known as* Bump Test *and* Field Test.

> **Zeroing** — Resetting an instrument read at normal (baseline) levels in fresh air.

Figure 11.14 Ensure usability before selecting devices.

Figure 11.15 Field tests may include calibration (bump) tests and zeroing.

Before purchasing detection and air monitoring equipment, contact current users of the prospective equipment to get information about the equipment. Current users will offer valuable insight into durability, dependability, weight, ease of use, and many other factors. Only experience using the particular device can provide this information. When considering a purchase, include the cost of filters, probes, internal parts, and calibration; these may add an unforeseen expense. Also consider other factors that may reduce the reliability of instruments, the amount of processing time during each use, and the affect elements (moisture, temperature, and atmosphere) may have on the instrument. Being the first department or team to purchase a new type instrument may not be practical. New concepts and methods of detection make some instruments obsolete very quickly. The cost of the instrument does not determine its effectiveness.

Users must calibrate, maintain, and decontaminate detection and monitoring devices in accordance to manufacturers' directions. Improperly calibrated and maintained devices may give inaccurate and misleading reasons, therefore posing a safety hazard. When performing maintenance and calibration, use the following guidelines:

- Use the manufacturer-recommended calibration gases.

- Calibrate them according to the manufacturer's recommendation and AHJ guidelines **(Figure 11.16)**.

- Store devices in accordance with manufacturers' recommendations with awareness of the expiration dates and shelf-life of some sensors, test strips, and colorimetric tubes.

- Test instruments routinely to ensure proper operation.

Figures 11.16 Use manufacturer-provided calibration gases to ensure proper sensor.

WARNING!
Never rely on one type of detection equipment exclusively.

Zeroing Instruments in the Field

Calibrating an instrument to a calibration gas differs from zeroing an instrument in the field. To truly calibrate an instrument, users adjust the instrument to a known type and concentration of a calibration gas in order to standardize the measurements the instrument will take.

Zeroing in the field adjusts these devices to the existing environment at the site of the hazardous materials release. Avoid zeroing the instrument in locations with potential contaminants. For example, four gas monitors should not be zeroed near running vehicles where carbon monoxide levels may be high.

Hazard Detection Equipment

To determine the appropriate risk-based response for identifying hazards in the field, responders should follow the SOP/Gs of the AHJ. The sections that follow describe detection and monitoring devices for the following hazards, listed alphabetically:

- Corrosives
- Flammables
- Oxidizers
- Oxygen
- Radiation
- Reactives
- Toxics

NOTE: The list of detection equipment in this section is not definitive. Refer to IFSTA's manual **Hazardous Materials Technician** for more information.

Corrosives

A large percentage of hazmat incidents involve corrosive materials. Corrosive gases and vapors can damage detection and monitoring instruments as well as PPE. As a result, at releases involving unknown hazards, a responder's priority is to monitor for pH. The primary equipment used to detect and measure corrosivity are pH meters and pH paper **(Figures 11.17a, b, and c, p. 548)**. Properties of corrosives that can be detected by instruments are described in the next sections.

pH

The concentration of **hydronium** or **hydroxide** ions in a solution determines the solution's pH. The pH scale ranges from 0 to 14. A pH of 7 is neutral; neither acidic nor basic. Acidic substances have excess hydronium ions and a pH less than 7. Basic substances have excess hydroxide ions and a pH greater than 7. As pH levels increase above 7, compounds become more alkaline. Conversely, as pH levels decrease below 7, compounds become increasingly acidic.

Hydronium — Water molecule with an extra hydrogen ion (H_3O^+). Substances/solutions that have more hydronium ions than hydroxide ions have an acidic pH.

Hydroxide — Water molecule missing a hydrogen ion (HO^-). Substances/solutions that have more hydroxide ions than hydronium ions have a basic (alkaline) pH.

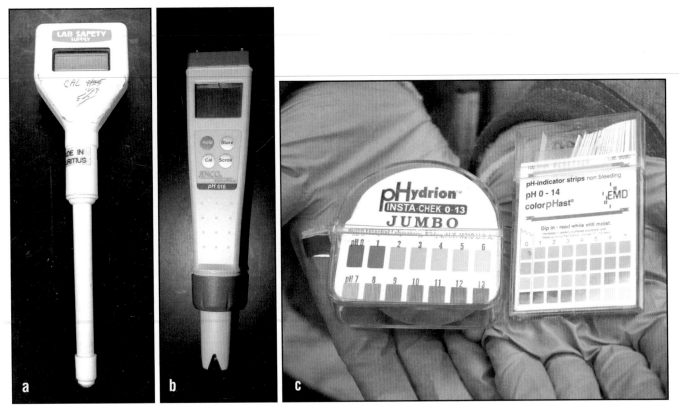

Figures 11.17a, b, and c pH meters **(a and b)** and pH papers **(c)** measure corrosivity.

Concentration

Acidic and basic solutions usually consist of chemicals dissolved in water. The ratio of the amount of chemical to the amount of water determines the solution's concentration. For example, a 95-percent solution of formic acid is composed of 95-percent formic acid and 5-percent water. Generally, the higher the concentration, the more damage the acid or base will do relative to itself. A 98-percent solution of sulfuric acid will burn the skin more rapidly and severely than an equal volume of a 1-percent sulfuric acid solution.

Strength

The number of hydrogen ions or hydronium ions produced, in respect to the original concentration of the acid, determines the strength of an acid or base. Details **(Figure 11.18)**:

- The higher the number of hydronium ions in the solution (concentration), the stronger the acid and the more corrosive it will be relative to other acids of equal concentration. A 98-percent solution of sulfuric acid will dissociate completely and release many more hydronium ions than a 98-percent solution of acetic acid. This property will make the sulfuric acid solution more corrosive than the acetic acid solution.

- The higher the number of hydroxide ions produced in making a basic solution, the more corrosive the base will be relative to other bases of equal concentration. Although they have a similar pH, sodium hydroxide will be far more caustic than sodium bicarbonate because it releases more hydroxide ions in solution.

pH Concentration

Concentration of Hydrogen Ions Compared to Distilled Water	pH
10,000,000	0
1,000,000	1
100,000	2
10,000	3
1,000	4
100	5
10	6
1	7
$1/10$	8
$1/100$	9
$1/1,000$	10
$1/10,000$	11
$1/100,000$	12
$1/1,000,000$	13
$1/10,000,000$	14

Figure 11.18 Acid strength is determined by the number of hydrogen or hydronium ions in the solution. A pH of 1 is 1,000,000 times more acidic than a pH of 7 while a pH of 13 is 1,000,000 times more alkaline.

pH Papers and Meters

pH paper is designed to change color when it comes into contact with corrosive materials; the color of the paper indicates the pH of the material. A standard color system does not exist for pH paper; therefore, different brands may use different colors and configurations. Often, pH paper provides a scale from 0 to 14, but some brands may be more or less specific. If the pH paper has an expiration date, using it after it expires may affect the accuracy of the readout. Responders using pH paper should note the following associations:

- pH 0-3 = especially corrosive acids **(Figure 11.19, p. 550)**
- pH 7 = neutral (water)
- pH 10-14 = especially corrosive bases **(Figure 11.20, p. 550)**
- pH paper stripped or bleached = oxidizers and organic peroxides

The pH paper may not give accurate pH measurements for all materials. For instance, hydrocarbons may appear to give a reading between pH 4-6; however, this is not the true pH of the material **(Figure 11.21, p. 550)**. To test the pH of some hydrocarbons, a responder can wet the pH paper with distilled water and wave the pH paper in the vapor space above the material to test its pH. However, this will only work for materials that have a fairly high vapor pressure.

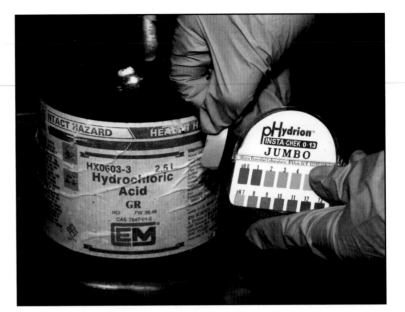

Figure 11.19 A pH of 0-3 indicates a strong acid.

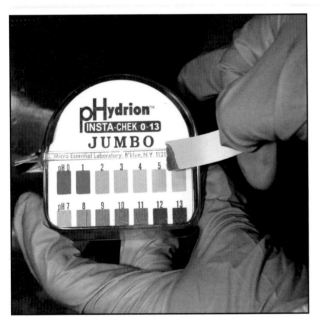

Figure 11.20 A pH of 10-14 indicates an especially corrosive base.

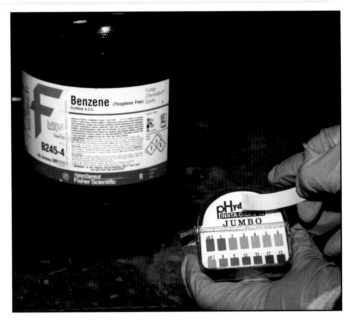

Figure 11.21 Many hydrocarbons may appear to give a reading between 4-6, but this is not a true measure of corrosivity.

pH Paper Limitations

The following factors may limit the usefulness of pH paper:

- Need for close proximity and/or contact with the hazardous material in order to conduct testing

- Inability to detect the concentration of the material

- Difficulty reading the paper if the material sampled is contaminated with oil, mud, or other opaque materials

- Difficulty reading the paper if the material sampled chemically strips the paper or alters it in unexpected ways, such as with highly concentrated acids and bases, certain oxidizers, and hydrocarbons

Responders can attach a strip of pH paper, wetted with distilled water, to their SCBA mask, PPE, a pole, or other instrument (for example, a probe) to ensure the quick detection of corrosive atmospheres. Dry pH paper is slow to react to atmospheric conditions, or it may not react at all.

If pH paper starts to change color, this may indicate the need to evacuate and reevaluate the situation. Evacuation may protect responders wearing only firefighter protective clothing as PPE and may also protect any detection and monitoring devices carried by those responders. **Skill Sheet 11-1** shows how to perform a pH test on an unknown liquid using pH paper.

Responders should be aware that pH meters provide more precise readings than pH paper. However, calibrate pH meters before each use, including rinsing the probe with distilled water before and after calibration. To use a pH meter, insert the probe into the material being tested **(Figure 11.22)**. Temperature, oils, and other contaminates may affect pH meter readings. **Skill Sheet 11-2** shows how to perform a pH test on an unknown liquid using a pH meter.

Figure 11.22 To use a pH meter, dip the probe into the material.

Fluoride Test Paper

Of all the chemical elements, fluorine is the most reactive. Compounds containing fluorine are called *fluorides*. Hydrogen fluoride (HF), a widely used fluoride, is extremely corrosive, toxic, and highly reactive. Responders need the highest level of PPE (Level A) to work with HF because of the extreme effects HF has on health. In situations involving unidentified or unknown hazardous materials, responders should test for fluorides (HF in particular) for their protection.

> # WARNING!
> Responders not wearing Level A PPE should evacuate the area immediately if fluoride test strips change color in air.

Fluoride test papers can determine the presence of fluoride ions and gaseous hydrogen fluoride. In the presence of fluorides, the pinkish-red paper turns yellowish-white. Additionally, chlorates, bromates, and sulphates in significant amounts will cause white discolorations of the paper. Like pH strips, fluoride test paper strips can be attached to PPE or other detection equipment. Before testing an area, follow AHJ regarding whether to wet these strips in tap or distilled water. **Skill Sheet 11-3** shows how to use reagent test paper to test for hazards.

Detection Paper Arrangement Systems

At hazardous materials incidents, hazmat responders may select a number of reagent-indicating papers. Responders may then:

- Arrange them in a unified system to detect a variety of hazards.
- Do this at the same time during initial scene hazard detection and evaluation.

NOTE: These groups of strips are informally known as a *bear claw* or *bear paw* **(Figure 11.23)**.

Depending on the type of incident or specific situation, commonly used reagent or indicating papers that may compose a bear claw include:

- pH paper
- Fluoride paper
- Potassium iodide paper
- M8 and M9 papers for weapons of mass destruction agents
- Spilfyter® paper — 2 or 3 versions
- Water finding — screening to inform results from instrumentation

Manufacturing companies preproduce these strips into groups so that hazardous materials responders do not need to "assemble" these strips into a group at the scene. Examples include:

- **SpilFyter®** is a commercially made product that detects a variety of hazardous materials in a liquid spill **(Figure 11.24)**.
- **HazMat Smart Strip™** is a product that detects a variety of hazardous materials in the air. It warns responders if it detects a hazardous chemical. It identifies eight classes of chemicals with a color-code system.

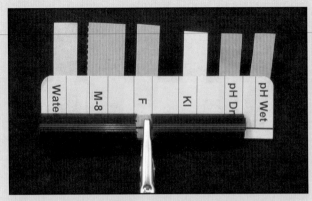

Figure 11.23 Some agencies combine different reagent strips into a *bear claw* or *bear paw. Courtesy of Scott Kerwood.*

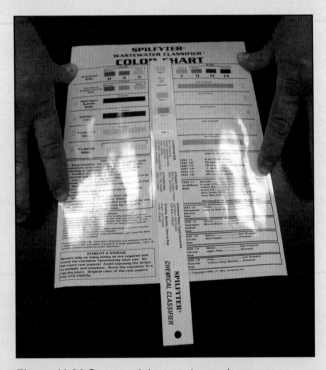

Figure 11.24 Commercial reagent groupings are available.

Combustible Gas Indicator (CGI) — Electronic device that indicates the presence and explosive levels of combustible gases, as relayed from a combustible gas detector.

Flammables

Most fire service responders are already familiar with devices, such as **combustible gas indicators (CGIs),** which measure the amount of flammable vapors and gases in the atmosphere in one of these three ways:

- Percentage of the lower explosive limit (%LEL)
- Parts per million (ppm)
- Percentage of gas per volume of air

Most meters used to measure combustible gases measure the LEL, and for this reason responders may refer to them as *LEL meters* rather than *CGIs*. Typically, LEL meters will sound an alarm at 10% of the LEL of the calibration gas; however, others sound an alarm at different thresholds. Even low percentages of LEL indicate that something is in the air, potentially at dangerously toxic levels.

CGIs have specific calibration issues. Each CGI is calibrated to a specific flammable gas (commonly methane, pentane, propane, or hexane). When responders use a CGI calibrated to one gas (such as methane) to measure other flammable gases/vapors (such as propane), the actual LEL of the gas being measured may differ from the reading the CGI displays **(Table 11.2)**. **Table 11.3** provides examples of conversion factors (also called multipliers or response curves) for various gases. Responders using LEL meters must make allowances for these potential discrepancies in order to correctly interpret LEL readings. Manufacturers provide response curves and conversion factors that are specific to individual meters.

Table 11.3
Sample Conversion Factors

Gas or Vapor	Factor
Hexane	0.68
Hydrogen	0.39
Isopropyl Alcohol	0.73
Methyl Ethyl Ketone	0.90
Methane	0.38
Methanol	0.58
Mineral Spirits	1.58
Nitro Propane	0.95
Octane	1.36
Pentene	0.86
Iso-Pentene	0.86
Isoprene	0.58
Propane	0.56
Styrene	1.27
Vinyl Acetate	0.70
Vinyl Chloride	1.06
O-Xylene	1.36

Table 11.2
Comparison of Actual LEL and Gas Concentrations
with Typical Instrument Readings

Gas Type	Actual % LEL	Actual Gas Concentration	Typical Display Reading (% LEL)
Pentane	50%	0.07%	50%
Methane	50%	2.50%	100%
Propane	50%	1.05%	63%
Styrene	50%	0.55%	26%

Source: *Courtesy of MSA*

Figure 11.25 CGIs are sensitive to oxygen levels. Monitor for oxygen and flammables at the same time. *Courtesy of MSA.*

CGIs may need specific oxygen levels to function properly. Because many CGIs use a combustion chamber to burn the flammable gas, the atmosphere must have enough oxygen to support combustion in order for the instrument to function correctly. Too much oxygen can exaggerate readings or even damage sensors. Therefore, responders should monitor oxygen levels concurrently while using CGIs **(Figure 11.25)**. **Skill Sheet 11-4** provides steps for using a multigas meter.

WARNING!
LEL meters will not provide accurate readings in oxygen deficient or oxygen enriched atmospheres.

Other factors which can influence CGI readings include:

- Catalyst poisons
- Concentrations exceeding 100% of the LEL
- Concentrations exceeding the upper flammable limit (UFL)
- Chlorinated hydrocarbons
- Oxygen-acetylene mixtures

CGI Meter Limitations

Limitations of combustible gas indicator (CGI) meters include:

- As battery power decreases, the meter may lose responsiveness.
- Corrosive gases may damage sensors.
- Extremely cold weather may make meter response sluggish.
- Cell phones, magnetic fields, high voltage lines, radios, and static electricity may interfere with readings.
- Too little or too much oxygen will interfere with accurate readings.

Oxidizers

Determine if any released materials act as oxidizers. Different types of oxidizers may require different types of detection tests. Organic peroxides can initiate explosive polymerization in certain materials, and they are components of improvised explosives such as triacetone triperoxide (TATP) and hexamethylene triperoxide diamine (HMTD). Responders may use peroxide test strips that include a reagent to detect the presence of these materials. A color change to blue after contact for 15 seconds indicates the presence of organic peroxides.

Responders can use potassium iodide (KI) starch paper to test the oxidizing potential of unknown chemicals **(Figure 11.26)**. When contacting oxidizing materials (nitrites and free chlorine), this paper changes color from white to blue/violet, purple, or black. The faster the color changes, the greater the oxidizing potential will be.

One limitation of these test strips is that the responder must be in close proximity to the material in order to use them. If responders detect a peroxide or potential explosive, withdraw immediately and contact EOD/bomb disposal technicians.

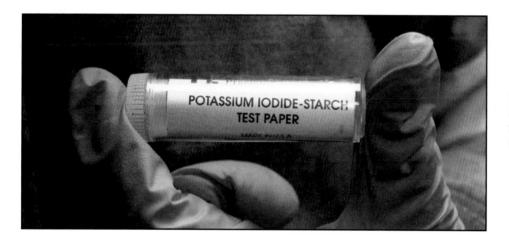

Figure 11.26 Potassium iodide starch paper can be used to test for peroxides and/or potential explosives.

Oxygen

Oxygen meters detect the percentage of oxygen in the air. Below 19.5% oxygen, the atmosphere is considered oxygen deficient and IDLH, requiring use of supplied air such as SCBA. The atmosphere is considered oxygen enriched at concentrations above 23.5%, presenting a potential threat for fire and explosion.

Monitoring for oxygen may seem unnecessary if responders have a nonatmospheric source of air **(Figure 11.27, p. 556)**. However, many detection devices require a certain percentage of atmospheric oxygen to function correctly.

Normal air contains 20.9% oxygen, 78.1% nitrogen, and 1% other gases. Any oxygen readings below 20.9% indicate that a contaminant in the air is displacing the oxygen. This contaminant may potentially exist at toxic or extremely hazardous levels. Because a contaminant will displace air (not just oxygen) proportionally, a one percent drop in oxygen is equivalent to 50,000 ppm of something else in the air **(Figure 11.28, p. 556)**. Even if oxygen levels are not low enough to trigger an alarm, reduced levels of oxygen potentially represent a significant hazard in the form of toxic contaminants. Responders should wear SCBA in these circumstances even if oxygen levels are above 19.5%.

Figure 11.27 Monitor for oxygen even when wearing SCBA. Many devices cannot function properly without sufficient oxygen.

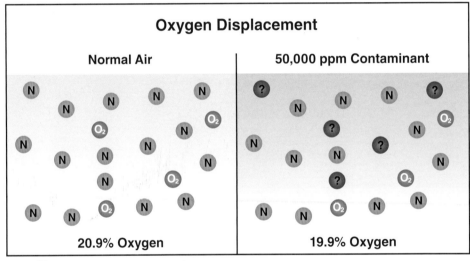

Figure 11.28 A 1% drop in oxygen represents 50,000 ppm of something else in the air. Reduced oxygen levels may represent a potentially significant toxicity hazard.

Oxygen sensors continually degrade, even when not in use. Contact with other types of chemicals, such as other oxidizers and carbon dioxide, can also degrade these sensors. Because of this degradation, responders should replace oxygen sensors frequently.

Humidity, temperature, and elevation can affect the readings from these sensors. Oxygen sensors should be zeroed in clean air at the elevation of the incident (since oxygen levels vary at different elevations).

Limitations of Oxygen Meters

Oxygen meters have several limitations, including the following:

- Corrosive gases can cause rapid sensor failure in some meters.
- Strong oxidizers, such as chlorine, bromine, and fluorine, can cause abnormally high readings (false positives).
- Sensors deteriorate steadily over time and need replacement based on manufacturer's recommendations.
- Changes in temperature (and temperature extremes), humidity, and atmospheric pressure can affect the monitor.

Radiation

Large amounts of radiation exposure are usually expressed in a unit called **Roentgen Equivalent in Man (rem)**; smaller amounts of radiation exposure may be described in **millirem (mrem)**. However, responders should know several terms which express radiation dose and exposure because these units may be used on radiation dose instruments (dosimeters) and radiation survey meters.

Two systems of units are used to measure and express radiation exposure and radiation dose (energy absorbed from the radiation). The U.S. still commonly uses the English System, which has the following units:

- **Roentgen (R)** — Roentgens only measure exposure to gamma and X-ray radiation. Most U.S. dosimeters use *R*. Radiation survey meters use R per hour (R/hr).
- **Radiation absorbed dose (rad)** — Rads express the amount of radiation energy absorbed by a material. This unit applies to any material and all types of radiation, but it does not take into account the potential effect that different types of radiation have on the human body. For example, 1 rad of alpha radiation causes more damage to the human body than 1 rad of gamma radiation.
- **Roentgen equivalent in man (rem)** — Rems express the absorbed dose equivalence as pertaining to a human body. Rem applies to all types of radiation. This unit takes into account the energy absorbed (as measured in rad) and the biological effect on the body due to different types of radiation. Agencies use rem to set dose limits for emergency responders.

The US units of measure can be used together. For gamma and X-ray radiation, apply the following common conversion factor among exposure, absorbed dose, and dose equivalent:

1 R = 1 rad = 1 rem

The SI unit used to measure *absorbed dose* is called **gray (Gy)** whereas the unit for *dose equivalence* is **sievert (Sv)**. Some newer radioactive survey meters and meters outside the U.S use Sievert (**Table 11.4, p. 558**).

Depending on the type of incident, responders may not initially know that the scene may expose them to radiation or contamination. They will not smell, taste, feel, or see radiation. Therefore, responders must have some form of detection instrumentation available to test for the presence of radiation and contamination, particularly at potential terrorist attacks. Emergency responders must always check for radiation at explosive incidents.

Roentgen Equivalent in Man (rem) — English System unit used to express the radiation absorbed dose (rad) equivalence as pertaining to a human body; used to set radiation dose limits for emergency responders. Applied to all types of radiation.

Millirem (mrem) — One thousandth of one Roentgen Equivalent in Man (rem).

Roentgen (R) — English System unit used to measure radiation exposure, applied only to gamma and X-ray radiation; the unit used on most U.S. dosimeters.

Radiation Absorbed Dose (rad) — English System unit used to measure the amount of radiation energy absorbed by a material; its International System equivalent is gray (Gy).

Sievert (Sv) — SI unit of measurement for low levels of ionizing radiation and their health effect in humans.

Gray (Gy) — SI unit of ionizing radiation dose, defined as the absorption of one joule of radiation energy per one kilogram of matter.

Table 11.4 Radiation Unit Equivalents	
100 Rem	1 Sv
1 Rem	10 mSv (millisievert)
1 mrem	10 µSv (microsievert)
1 µrem	.01 µSv

Figure 11.29 Many hand-held radiation survey instruments are affordable and simple to use.

Hand-held portable survey instruments provide first responders the simplest and most affordable option to detect radiation and contamination **(Figure 11.29)**. Responders who train in effective detection methods will help them understand the capabilities, limitations, and operational techniques of radiological survey instruments. Responders should also be reinforced via local ongoing training. If a department does not have instrumentation, it should consult a governmental radiation authority for guidance on lifesaving operations.

Many different models and types of radiological survey instruments are available with a variety of features and controls. Like other instruments used in hazard identification, each has a specific use and each has its limitations. You can divide radiological instruments into three groups:

- Instruments used for measuring radiation exposure
- Instruments used to detect contamination
- Instruments used for dose monitoring and personal dosimetry

Contamination can emit alpha, beta, gamma, or a combination of these types of radiation. Many commonly available survey instruments allow the user the option of changing the detector or probe depending on the intended use of the instrument. Selecting the proper instrument for a specific task depends upon understanding the different types of probes or detectors and how their use affects the operating characteristics of the instrument. Attaching different types of probes to the survey meter can change many radiological survey instruments from radiation detection instruments into contamination detection instruments.

The two general categories of detectors are gas-filled detectors and scintillation detectors. Both of these detector types include radiation survey instruments and contamination survey instruments. The following sections explain these two types of detectors in addition to personal dosimetry devices.

Gas-Filled Detectors

In a gas-filled detector, radiation ionizes the gas inside the detection chamber and the instrument's electronics measure the quantity of ions created. Common examples of gas-filled detectors include ion chambers and **Geiger-Mueller (GM) tubes (Figure 11.30)**.

Ion chambers often use ambient air as the detection gas, which can cause them to be affected by temperature and humidity. Ion chambers often give responses directly proportional to the intensity of the radiation, making them reliable instruments when encountering radiations with varying energies.

Researchers originally developed the **GM detector**, or GM tube, in 1928. Temperature and humidity do not typically affect GM detectors because they are sealed from ambient air. GM tubes with a thin window may detect alpha, beta, and gamma radiation, making them useful for detecting radiological contamination. GM tubes with a sealed metal body are better suited for measuring penetrating gamma radiation that can be an external exposure hazard. The metal case makes this type of probe less suitable for use in detecting radiological contamination. **Skill Sheet 11-5** explains the steps for using a radiation detector to detect radiation.

Geiger-Mueller (GM) Tube — Sensor tube used to detect ionizing radiation. This tube is one element of a Geiger-Mueller detector.

Geiger-Mueller (GM) Detector — Detection device that uses GM tubes to measure ionizing radiation. *Also known as* a Geiger Counter.

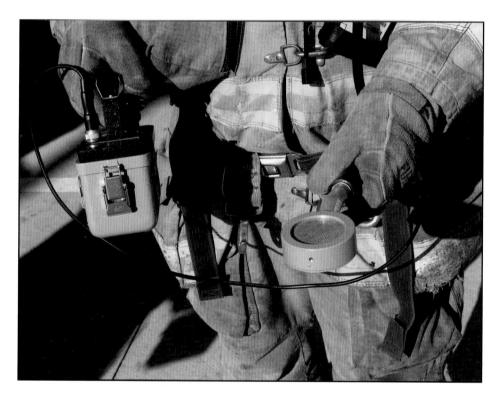

Figure 11.30 A Geiger-Mueller (GM) detector is a gas-filled detector.

Figure 11.31 The crystal in a scintillation detector will produce a small flash of light when interacting with radiation.

Scintillation Detectors

In scintillation detectors, radiation interacts with a crystal, such as sodium iodide, cesium iodide, or zinc sulfide, to produce a small flash of light **(Figure 11.31)**. The electronics of the instrument use a photomultiplier tube to amplify this light pulse thousands of times in order to produce a useful signal. Some scintillation detectors have a thin Mylar® covering over the crystal making them useful for detecting radiological contamination. In general, scintillation detectors help detect small amounts of radiation.

To measure penetrating gamma radiation, some scintillation detectors seal scintillation crystals in a metal body. The need for the photomultiplier tube usually makes scintillation detector probes larger than gas-filled detector probes. They are also susceptible to breakage if not handled properly. Responders who drop the instrument risk shattering the crystal, the photomultiplier tube, or both.

Monitoring and Personal Dosimetry Devices

Dosimetry devices help wearers keep track of their total accumulated radiation dose **(Figures 11.32a and b)**. A dosimeter is like the odometer on a vehicle. Whereas the odometer measures total miles (kilometers) traveled, the dosimeter measures the total amount of dose received. There are several different types of **dosimeters** available. Some commonly used "self-reading" personal dosimeters do not require processing at a lab to retrieve dose information.

A self-reading dosimeter (SRD) measures the radiation dose in Roentgens (R), milliroentgens (mR), sieverts (Sv), or gray (Gy). Generally SRDs only measure gamma and X-ray radiation. SRDs have many names, including:

- Direct reading dosimeter (DRD)
- Pocket ion chamber (PIC)
- Pencil dosimeter

Dosimeter — Detection device used to measure an individual's exposure to an environmental hazard such as radiation or sound.

Figures 11.32a and b Dosimeters are worn to track personal radiation exposure levels.

To read the dosimeter, hold it up to a light source and look through the eyepiece. Responders should always record the SRD reading before they enter a radiation field (hot zone). Read the SRD periodically (at 15- to 30-minute intervals) while working in the hot zone, and read it again upon exit from the hot zone. If the SRD gives a reading that is higher than expected or off the scale, responders should:

- Notify others in the hot zone.
- Tell others to check their SRDs.
- Exit the hot zone immediately.
- Follow local procedures.

Skill Sheet 11-6 provides steps for using a dosimeter.

Protective Action Recommendations, 2008

In August, 2008, U.S. DHS/FEMA published the Planning Guidance for Protection and Recovery Following Radiological Dispersal Device (RDD) and Improvised Nuclear Device (IND) Incidents, in the Federal Register, Vol. 73, No. 149. **Table 11.5** summarizes the recommended public protection guides for radiological and improvised nuclear device incidents. In the early phase of an incident, DHS/FEMA recommends sheltering in place or evacuation if radiation dose levels are predicted to be between 1 and 5 rem (0.01–0.05 Sv).

Table 11.6, p. 562, provides emergency worker guidelines for the early phase of an incident. While the guidelines suggest a 5 rem (0.05 Sv) occupational limit for most situations (with 10 and 25 rem [0.1 and 0.25 Sv] exceptions provided for extraordinary circumstances), these limits may have flexibility. As always, the IC, local SOP/Gs, and the circumstances of the incident determine the appropriate response. Using the ALARA (as low as reasonably achievable) principle should allow responders to conduct many operations, including rescues, below the 5 rem (0.05 Sv) limit.

Table 11.5
Protective Action Guides for RDD and IND Incidents

Phase	Protective Action Recommendation	Protective Action Guide
Early	Sheltering-in-place or evacuation of the Public[a]	1 to 5 *rem* (0.01–0.05 *Sv*) projected dose[b]
	Administration of prophylactic drugs — potassium iodide[c,e] Administration of other prophylactic or decorporation agents[d]	5 *rem* (0.05 *Sv*) projected dose to child thyroid[c, e]
Intermediate	Relocation of the Public	2 *rem* (0.02 *Sv*) projected dose first year. Subsequent years, 0.5 rem/y (0.005 *Sv/y*) projected dose[b]
	Food interdiction	0.5 *rem* (0.005 *Sv*) projected dose, or 5 *rem* (0.05 *Sv*) to any individual organ or tissue in the first year, whichever is limiting
	Drinking water interdiction	0.5 *rem* (0.005 *Sv*) projected dose in the first year

a Should normally begin at 1 *rem* (0.01 *Sv*); take whichever action (or combination of actions) that results in the lowest exposure for the majority of the population. Sheltering may begin at lower levels if advantageous.

b Total Effective Dose Equivalent (TEDE)—the sum of the effective dose equivalent from external radiation exposure and the committed effective dose equivalent from inhaled radioactive material.

c Provides thyroid protection from radioactive iodine only.

d For other information on other radiological prophylactics and medical countermeasures, refer to *http://www.fda.gov/cder/drugprepare/default.htm, http:/www.bt.cdc.gov/radiation, or http://www.orau.gov/reacts*.

e Committed Dose Equivalent (CDE). FDA understands that a KI administration program that sets different projected thyroid radioactive dose thresholds for treatment of different population groups may be logistically impractical to implement during a radiological emergency. If emergency planners reach this conclusion, FDA recommends that KI be administered to both children and adults at the lowest intervention threshold (*i.e.*, >5 *rem* (0.05 *Sv*) projected internal thyroid dose in children) (FDA 2001).

Source: U.S. DHS/Federal Emergency Management Agency's *Planning Guidance for Protection and Recovery Following Radiological Dispersal Device (RDD) and Improvised Nuclear Device (IND) Incidents*, published in the *Federal Register, Vol. 73, No. 149*, Friday, August 1, 2008

Table 11.6
Emergency Worker Guidelines in the Early Phase

Total Effective Dose Equivalent (TEDE)[a] Guideline	Activity	Condition
5 *rem* (0.05 *Sv*)	All occupational exposures	All reasonably achievable actions have been taken to minimize dose.
10 *rem* (0.1 *Sv*)	Protecting valuable property necessary for public welfare (e.g., a power plant)	• All appropriate actions and controls have been implemented; however, exceeding 5 *rem* (0.05 *Sv*) is unavoidable. • Responders have been fully informed of the risks of exposures they may experience. • Dose >5 *rem* (0.05 *Sv*) is on a voluntary basis. • Appropriate respiratory protection and other personal protection is provided and used. • Monitoring available to project or measure dose.
25 *rem* (0.25 *Sv*)[b]....	Lifesaving or protection of large populations. It is highly unlikely that doses would reach this level in an RDD incident; however, worker doses higher than 25 rem (0.25 Sv) are conceivable in a catastrophic incident such as an IND incident.	• All appropriate actions and controls have been implemented; however, exceeding 5 *rem* (0.05 *Sv*) is unavoidable. • Responders have been fully informed of the risks of exposures they may experience. • Dose >5 *rem* (0.05 *Sv*) is on a voluntary basis. • Appropriate respiratory protection and other personal protection is provided and used. • Monitoring available to project or measure dose.

a The projected sum of the effective dose equivalent from external radiation exposure and committed effective dose equivalent from internal radiation exposure.

b EPA's 1992 PAG Manual states that "Situations may also rarely occur in which a dose in excess of 25 rem for emergency exposure would be unavoidable in order to carry out a lifesaving operation or avoid extensive exposure of large populations." Similarly, the NCRP and ICRP raise the possibility that emergency responders might receive an equivalent dose that approaches or exceeds 50 rem (0.5 Sv) to a large portion of the body in a short time (Limitation of Exposure to Ionizing Radiation, National Council on Radiation Protection and Measures, NCRP Report 116 (1993a). If lifesaving emergency responder doses approach or exceed 50 rem (0.5 Sv) emergency responders must be made fully aware of both the acute and the chronic (cancer) risks of such exposure.

Reactives

Responders will want to determine if any released materials have begun to react with themselves or each other. While no current meter or device can detect reactive materials, potentially hazardous chemical reactions will cause a temperature change. For example, if a hazardous material in a container begins to polymerize, it will produce heat. To check for this type of reaction, responders can aim an **infrared thermometer** (or temperature gun) directly at the container **(Figure 11.33)**. This thermometer may detect rising temperatures, indicating a reaction in progress or a temperature increase that may lead to pressure change.

Read the infrared thermometers within the context of the environment and the incident. A metal container sitting on an asphalt pad on a hot day may be quite warm. Responders using infrared thermometers should monitor for rising temperatures or heat readings that are not accounted for in the surrounding environment. In addition to temperature, thermal imagers may also be used to detect liquid levels in containers at hazmat incidents.

Infrared Thermometer — Non-contact measuring device that detects the infrared energy emitted by materials and converts the energy factor into a temperature reading. *Also known as* Temperature Gun.

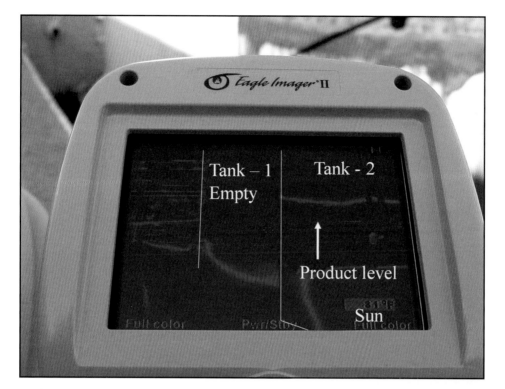

Figure 11.33 Temperature guns and thermal imagers can reveal temperature changes from chemical reactions occurring in containers. They can also show product levels based on temperature differences as shown in this picture. *Courtesy of Barry Lindley.*

Toxics

Many different technologies detect toxic materials. Some instruments (such as carbon monoxide detectors) only detect one chemical, while other instruments identify the presence of large groups of chemicals, such as organic gases and vapors. Some toxic detectors are simple, while others are complex. This section describes the following detection devices:

- Chemical-specific electrochemical cells

- Photoionization detectors (PIDs)

Toxic compounds produce an effect primarily as a function of the dose (amount of a substance ingested or absorbed through skin contact) and the concentration (in this context, the amount of the substance inhaled) of the compound. This principle, termed the **dose-response relationship**, is a key concept in toxicology. Many factors affect the normal dose-response relationship, but typically, as the dose increases, the severity of the toxic response increases **(Figure 11.34, p. 564)**. For example, people exposed to 100 parts per million (ppm) of tetrachloroethylene, a **solvent** commonly used for dry-cleaning fabrics, may experience relatively mild symptoms such as headache and drowsiness. However, people exposed to 200 ppm of tetrachloroethylene may lose motor coordination, and people exposed to 1,500 ppm for 30 minutes may lose consciousness. The severity of the toxic effect also depends on the duration of exposure, a factor that influences the dose of the compound in the body. **Table 11.7. p. 565**, shows factors that influence toxicity.

Toxicity is also a factor of exposure over time. An exposure to 200 ppm of a toxin over the course of 10 minutes will have more exaggerated effects than a cumulative exposure to 200 ppm of a toxin over the course of 48 hours.

Dose-Response Relationship — Comparison of changes within an organism per amount, intensity, or duration of exposure to a stressor over time. This information is used to determine action levels for materials such as drugs, pollutants, and toxins.

Solvent — A substance that dissolves another substance (solute), resulting in a third substance (solution).

Time-Dose/Concentration Relationship

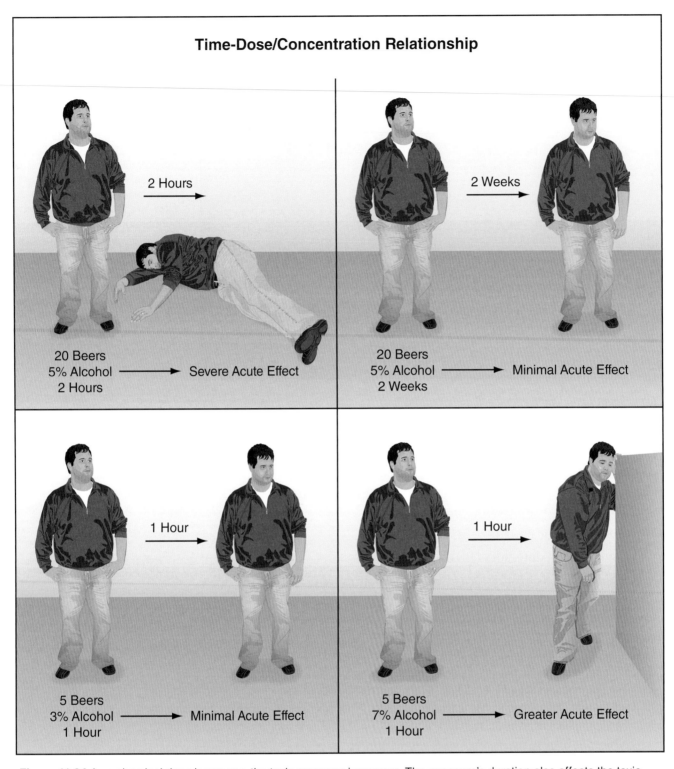

Figure 11.34 As a chemical dose increases, the toxic response increases. The exposure's duration also affects the toxic response. A dose spread over a long period of time may have less effect than the same dose administered over a shorter period.

Table 11.7
Factors Influencing Toxicity

Type of Factor	Examples
Factors related to the chemical	Composition (salt, freebase, etc.), physical characteristics (size, liquid, solid, etc.), physical properties (volatility, solubility, etc.), presence of impurities, breakdown products, carriers
Factors related to exposure	Dose, concentration, route of exposure (inhalation, ingestion, etc.), duration
Factors related to person exposed	Heredity, immunology, nutrition, hormones, age, sex, health status, preceding diseases
Factors related to environment	Media (air, water, soil, etc.), additional chemicals present, temperature, air pressure

Source: U.S. Centers for Disease Control and Prevention (CDC).

Poisons and the measurements of their toxicity are often expressed, on safety data sheets, in terms of **lethal dose (LD)** for amounts ingested and **lethal concentration (LC)** for amounts inhaled. As a general rule, the smaller the value (presented as the LD or LC), the more toxic the substance is **(Figure 11.35)**. The lower the dose or concentration of a substance needed to kill, the more dangerous it is. When establishing these values,

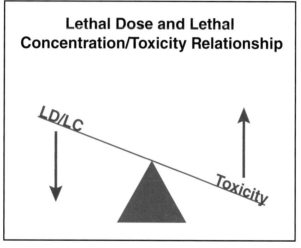

Figure 11.35 The smaller the LD/LC, the greater the toxicity.

researchers most often orally administer the toxin to animals (rats or rabbits) and test the toxin's effects under laboratory conditions over a set period of time.

Lethal Dose

Dose terms are defined as follows:

- **Lethal dose (LD)** — Indicates the minimum amount of solid or liquid that when ingested, absorbed, or injected through the skin will cause death. Sometimes the lethal dose is expressed in conjunction with a percentage such as LD_{50} (most common) or LD_{100}. The number refers to the percentage of an animal test group that the listed dose killed (usually administered orally).

- **Median lethal dose (LD_{50})** — Researchers found that this statistically derived single dose of a substance caused death in 50 percent of animals when administered orally. The LD_{50} value is expressed in terms of weight of test

Lethal Dose (LD) — Concentration of an ingested or injected substance that results in the death of the entire test population. Expressed in milligrams per kilogram (mg/kg); the lower the value, the more toxic the substance.

Lethal Concentration (LC) — Concentration of an inhaled substance that results in the death of the entire test population. Expressed in parts per million (ppm), milligrams per liter (mg/liter), or milligrams per cubic meter (mg/m³); the lower the value, the more toxic the substance.

Median Lethal Dose, 50 Percent Kill (LD_{50}) — Concentration of an ingested or injected substance that results in the death of 50 percent of the test population. LD_{50} is an oral or dermal exposure expressed in milligrams per kilogram (mg/kg); the lower the value, the more toxic the substance.

substance per unit weight of test animal (mg/kg). The term LD_{50} means that half of the test subjects died at that dosage **(Figure 11.36)**. The other half did not die, but the chemical may have made them sick or close to death.

- **Lethal dose low (LDLO or LDL)** — Indicates the lowest administered dose of a material capable of killing a specified test species.

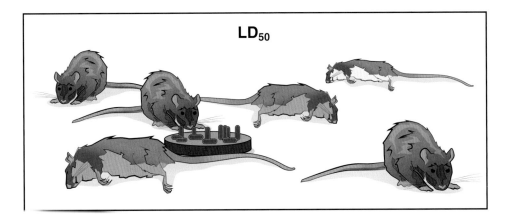

LD$_{50}$

Figure 11.36 The LD$_{50}$ is the dosage at which half the test subjects died.

Median Lethal Concentration, 50 Percent Kill (LC$_{50}$) — Concentration of an inhaled substance that results in the death of 50 percent of the test population. LC$_{50}$ is an inhalation exposure expressed in parts per million (ppm), milligrams per liter (mg/liter), or milligrams per cubic meter (mg/m³); the lower the value, the more toxic the substance.

Lethal Concentration

Concentration terms are defined as follows:

- **Lethal concentration (LC)** — Indicates the minimum concentration of an inhaled substance in the gaseous state that will kill the test group (usually within 1 to 4 hours). Similar to LD, sources may express the **median lethal concentration** as LC_{50}, indicating that concentrations at the listed value killed half of the test group. The 50 percent of the population not killed may suffer effects ranging from no response to severe injury. The following units are often used to quantify LC:

  — Parts per million (ppm)

  — Milligrams per cubic meter (mg/m³)

  — Micrograms of material per liter of air (µg/L)

  — Milligrams per liter (mg/L) (see information box)

- **Lethal concentration low (LCLO or LCL)** — Indicates the lowest concentration of a gas or vapor capable of killing a specified species over a specified time.

Researchers obtain lethal dose and lethal concentration values under laboratory conditions using test animals. Exertion, stress, and individual metabolism or chemical sensitivities (allergies) may make persons more vulnerable to the harmful effects of hazardous materials.

Incapacitating Dose

The *incapacitating dose (ID)* for an organism (such as a human being) indicates the dosage of a chemical or substance required to incapacitate that organism. It is expressed similarly to lethal dose and lethal concentration. Incapacitation can vary from moderate (unable to see, breathless) to severe (convulsions). Chemical warfare agents commonly list IDs. Categories of incapacitating doses include:

- ID_{50} — Dose that incapacitates 50 percent of the population of interest
- ID_{10} — Dose that incapacitates 10 percent of the population of interest

Chemical Specific Detectors

Some chemical monitors use sensors designed to detect a single chemical, such as:

- Carbon monoxide
- Hydrogen sulfide
- Ammonia
- Chlorine
- Hydrazine
- Ethylene oxide
- Hydrogen cyanide
- Phosgene

Some monitors may combine these sensors with a CGI and an oxygen sensor to form two-, three-, or four-gas monitors **(Figure 11.37)**. A typical four-gas monitor will detect LEL, oxygen, carbon monoxide, and hydrogen sulfide.

Usually, these devices sound an alarm when they detect gases at hazardous or potentially hazardous levels. Some devices may combine four-gas monitors with photoionization detectors (PIDs) to create five-gas monitors. The sensors in these devices degrade over time, and temperature and humidity may affect them.

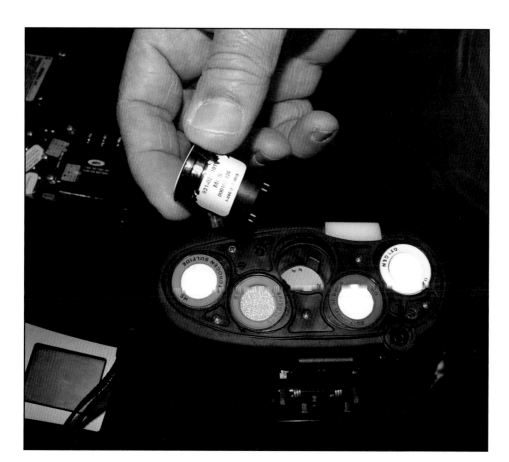

Figure 11.37 Many devices combine sensors in order to detect multiple hazards.

Photoionization Detectors (PIDs)

Photoionization detectors (PIDs) use an ultraviolet lamp to ionize samples of gaseous materials **(Figure 11.38)**. They detect low to very low concentrations of many organic and some inorganic gases and vapors, and they make good general survey instruments, capable of detecting and measuring concentrations in real-time. Although they cannot identify the material(s) present, PIDs

Figure 11.38 Photoionization detectors (PIDs) can detect very low concentrations of organic (and some inorganic) gases and vapors.

can effectively determine that a contaminant is present. Most PIDs use several different lamps (or bulbs) to measure the **ionization potential (IP)** of a material. **Skill Sheet 11-7** demonstrates the steps necessary to use a PID to detect contaminants.

NOTE: Photoionization detectors are highly technical resources and are more commonly used by Hazardous Materials Technicians than by Operations Level personnel.

Responders should use PIDs in the following situations:

- At the edge of a release, where concentrations may be too low to be detected by a CGI
- When responders suspect atmospheric contamination involving either flammable and/or nonflammable atmospheres
- When investigating complaints about odors or strange smells
- When locating low-volume chemical releases
- When evaluating the extent of contamination from a release and assessing risk to the public and environment

---

What This Means To You

Use of Multiple Detectors

At a gasoline spill, a CGI will detect the presence of flammable atmospheres, helping you to avoid explosions. A PID will detect the presence of toxic materials, helping you prevent future illness, such as cancer, from exposure to benzene.

For example, the LEL of gasoline is 1.4% or 10,400 ppm. The *NIOSH Pocket Guide* indicates that the IDLH of benzene (a carcinogen) is 500 ppm and cannot be read by a CGI. A PID, however, will detect benzene and other contaminants that pose a threat to responders at low levels.

Limitations of PIDs

Photoionization detectors have the following limitations:

- Certain models of PID instruments are not intrinsically safe, so they must be used in conjunction with a CGI.

- PIDs cannot identify unidentified/unclassified substances **(Figure 11.39)**.

- Detection of ionization may require several steps, and some materials may require the use of **correction factors**.

- PIDs do not respond to any products with ionization potential (IP) greater than the ultraviolet lamp in the PID.

- Responders should not use PIDs in rain or high humidity environments without the proper filtration attachment.

- Users must periodically clean the lamp window to ensure ionization of the new compounds by the probe (i.e., new air contaminants).

- As with all meters, high winds and humidity may affect readings via dilution of the product.

- Tiny particulates of dust may affect readings.

Correction Factor — Manufacturer-provided number that can be used to convert a specific device's readout to be applicable to another function.

Figure 11.39 PIDs cannot identify the substances detected.

CAUTION

Responders must not allow gas meters to draw liquids into the probe.

Detection Tubes and Chips

Colorimetric indicator tubes consist of a glass tube impregnated with an indicating chemical. The tube is connected to a piston or bellows-type pump **(Figure 11.40, p. 570)**. A known volume of contaminated air is pulled at a predetermined rate through the tube by the pump. The contaminant reacts with the indicator chemical in the tube, producing a change in color where the length is proportional to the contaminant concentration.

Colorimetric Indicator Tube — Small tube filled with a chemical reagent that changes color in a predictable manner when a controlled volume of contaminated air is drawn through it. *Also known as* Detector Tube.

Figure 11.40 A pump is needed to move contaminated air through colorimetric tubes.

Detector tubes are normally chemical-specific **(Figure 11.41)**. Some manufacturers do produce tubes for groups of gases, such as aromatic hydrocarbons or alcohols **(Figure 11.42)**. Concentration ranges on the tubes may be in the ppm or percent range. A preconditioning filter may precede the indicating chemical in order to:

- Remove contaminants (other than the one in question) that may interfere with the measurement
- Remove humidity
- React with a contaminant to change it into a compound that reacts with the indicating chemical

Skill Sheet 11-8 provides steps for using colorimetric tubes to identify hazards.

Figure 11.41 Tubes are chemical specific although some may be designed to react to certain chemical groups, for example, alcohols.

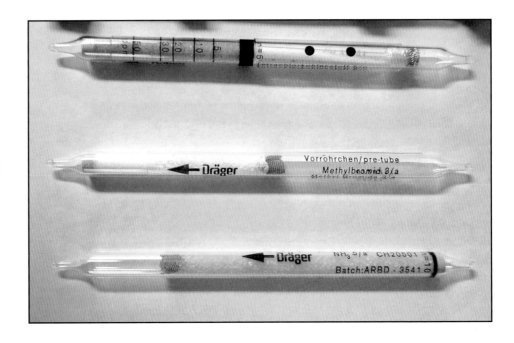

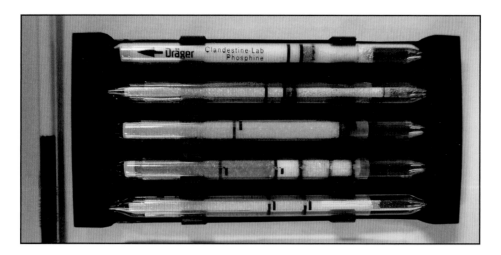

Figure 11.42 Tubes may be packaged for specific uses such as monitoring in illicit labs.

Closely associated with colorimetric tubes technology are colorimetric chips. They may be referred to as a *chip measurement system (CMS)*. CMSs use chemical specific measuring chips with an electronic analyzer. The chemical specific measuring chips have small tubes, sometimes referred to as *capillaries* that are filled with a reagent system for the designated chemical.

Most CMSs are considered direct reading instruments. These electronic analyzing instruments offer a highly reliable measurement for specific gases and vapors in a digital readout format. CMSs tend to offer a fast response, sometimes accurate to within seven percent of measured values for some products. In addition to these features, CMSs are simple to use. Follow manufacturer's instructions and SOPs/SOGs for operating the device.

Chapter Review

Answer the following questions to review the information provided in this chapter.

1. What is the difference between concentration and exposure?

2. Why is it important for first responders to know exposure limits?

3. What are action levels and why are they important to know?

4. What factors should be considered when selecting a monitoring and detection instrument?

5. List the different types of hazard detection equipment and explain their uses at hazmat incidents.

SKILL SHEETS

11-1
Demonstrate proper use of pH paper to identify hazards.

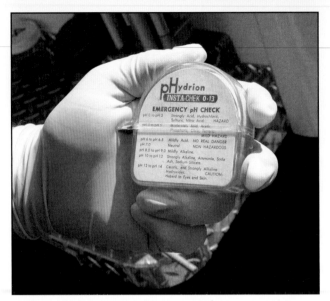

Step 1: Ensure proper detection, monitoring, or sampling method and equipment is chosen.

Step 2: Ensure that all responders are wearing appropriate PPE.

Step 3: Inspect the paper to ensure it has not been exposed or expired.

Step 5: Sample the product.

NOTE: If monitoring vapors or gases, pH paper should be wetted.

Step 4: Remove a piece of appropriate size pH paper from the roll or remove strip from the container and secure the paper.

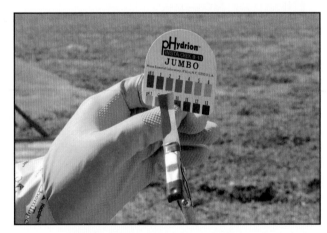

Step 6: Compare results to pH paper color scale to determine if the product is an acid, a base, or neutral. Record results.

NOTE: Confirmation of a corrosive atmosphere will eliminate the use of electronic meters for further testing.

Step 7: Report results according to AHJ's procedures.

Step 8: Dispose of pH papers in accordance with appropriate regulations.

Step 9: Decontaminate equipment and return to operational state per manufacturer's instructions.

Step 10: Complete required reports and supporting documentation.

Step 1: Ensure proper detection, monitoring, or sampling method and equipment is chosen.

Step 2: Ensure that all responders are wearing appropriate PPE.

Step 3: Perform initial inspection to ensure device is serviceable.

Step 4: Turn on the pH meter.

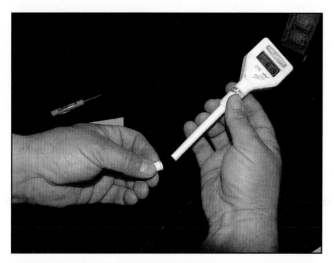

Step 5: Remove the protective cap from the electrode.

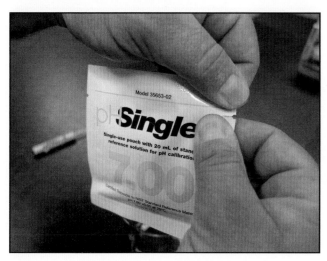

Step 6: Calibrate the pH meter in a test solution with a known pH as per the manufacturer's instructions.

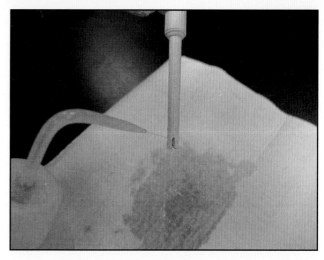

Step 7: Once calibrated, rinse and return the electrode to its operational state as per manufacturer's instructions.

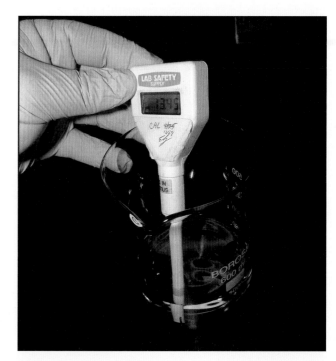

Step 8: Place the electrode in the liquid to be tested and make note of the reading.

Step 9: Report results according to AHJ's procedures.

Step 10: Remove electrode from the liquid, rinse, and return to operational state as per the manufacturer's instructions.

Step 11: Replace the protective cap on the electrode.

Step 12: Turn off the pH meter.

Step 13: Decontaminate equipment and return to operational state per manufacturer's instructions.

Step 14: Complete required reports and supporting documentation.

SKILL SHEETS

11-3

Demonstrate proper use of reagent test paper to identify hazards.

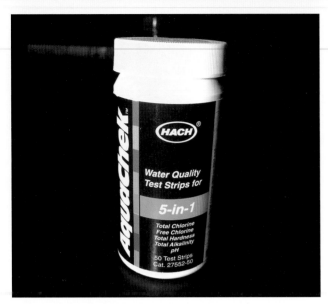

Step 1: Ensure proper detection, monitoring, or sampling method and equipment is chosen.

Step 2: Ensure that all responders are wearing appropriate PPE.

Step 5: Sample the product.

NOTE: Reagent test paper should be wetted as per manufacturer's recommendations.

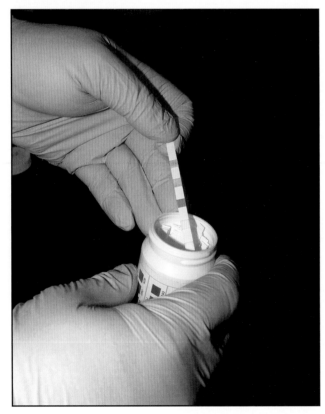

Step 3: Inspect the test paper to ensure it has not been exposed or expired.

Step 4: Remove a piece of appropriate size reagent test paper.

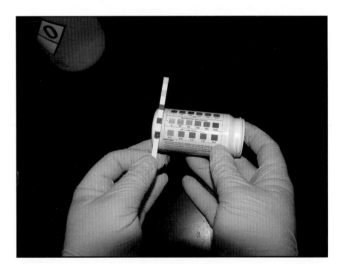

Step 6: Identify any color changes to the reagent test paper and compare them with the provided reference.

Step 7: Report results according to AHJ's procedures.

Step 8: Dispose of pH papers in accordance with appropriate regulations.

Step 9: Decontaminate equipment and return to operational state per manufacturer's instructions.

Step 10: Complete required reports and supporting documentation.

11-4

Demonstrate proper use of a multi-gas meter (carbon monoxide, oxygen, combustible gases, and others) to identify hazards.

SKILL SHEETS

NOTE: Specific procedures will vary depending on the equipment used. Refer to the manufacturer's instructions for complete directions.

Step 1: Ensure proper detection, monitoring, or sampling method and equipment is chosen.

Step 2: Ensure that all responders are wearing appropriate PPE.

Step 3: Perform initial inspection to ensure device is serviceable.

Step 4: Select the monitor and identify the gases it will detect.

Step 6: Perform a "fresh air" calibration of the monitor prior to entry.

Step 5: Perform a bump test to ensure the meter is functioning properly.

Step 7: Properly monitor the area as per AHJ's requirements.

11-4

Demonstrate proper use of a multi-gas meter (carbon monoxide, oxygen, combustible gases, multi-gas, and others) to identify hazards.

SKILL SHEETS

Step 8: Report results according to AHJ's requirements.

Step 9: When monitoring is complete, turn off the instrument.

Step 10: Decontaminate equipment and return to operational state per manufacturer's instructions.

Step 11: Complete required reports and supporting documentation.

NOTE: Specific procedures will vary depending on the equipment used. Refer to the manufacturer's instructions for complete directions.

Step 6: Turn on the meter and test detector against check source.

Step 1: Ensure proper detection, monitoring, or sampling method and equipment is chosen.

Step 2: Ensure that all responders are wearing appropriate PPE.

Step 3: Select the appropriate monitor for the potential hazard(s).

Step 4: Perform initial inspection to ensure device is serviceable.

Step 7: Acquire background radiation levels.

Step 8: Properly monitor the area as per AHJ's requirements.

Step 5: Ensure that the monitor has been maintained and appropriately calibrated according to AHJ's SOPs and manufacturer's instructions.

SKILL SHEETS

11-5

Demonstrate proper use of radiation detection instruments to identify hazards.

Step 9: Determine the presence of ionizing radiation.

Step 10: Compare radiation values to AHJ's SOPs. Record results.

Step 11: Report results according to AHJ's requirements.

Step 12: When monitoring is complete, turn off the instrument.

Step 13: Decontaminate equipment and return to operational state per manufacturer's instructions.

Step 14: Complete required reports and supporting documentation.

11-6

Demonstrate proper use of dosimeters to identify personal dose received.

Step 4: Ensure the dosimeter is properly calibrated.

NOTE: The dosimeter should be logged to you.

Step 5: Ensure the dosimeter reads zero.

Step 1: Ensure proper detection, monitoring, or sampling method and equipment is chosen.

Step 6: Don the dosimeter as per the manufacturer's instructions.

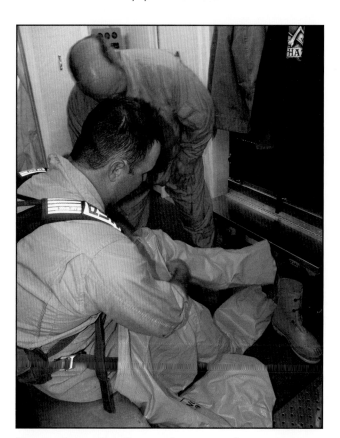

Step 2: Ensure that all responders are wearing appropriate PPE.

Step 3: Perform initial inspection to ensure device is serviceable.

11-6

Demonstrate proper use of dosimeters to identify personal dose received.

Step 7: Perform mission activity.

Step 8: Doff the dosimeter.

Step 9: Follow manufacturer's instructions and AHJ's procedures regarding dosimeter analysis.

Step 10: Report results as per AHJ's procedures.

Step 11: Decontaminate equipment and return to operational state per manufacturer's instructions.

Step 12: Complete required reports and supporting documentation.

NOTE: Specific procedures will vary depending upon the equipment used. Refer to the manufacturer's instructions for complete directions.

Step 1: Ensure proper detection, monitoring, or sampling method and equipment is chosen.

Step 2: Ensure that all responders are wearing appropriate PPE.

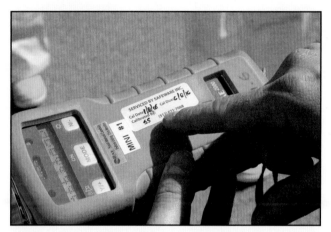

Step 3: Perform initial inspection to ensure device is serviceable.

Step 4: Perform a "fresh air" calibration.

Step 5: Operate the device as per manufacturer's instructions and AHJ's procedures.

Step 6: Properly monitor the area as per AHJ's requirements.

Step 7: Identify conversion factors and apply them as necessary.

Step 8: Report results as per AHJ's procedures.

Step 9: When monitoring is complete, turn off the device.

Step 10: Decontaminate equipment and return to operational state per manufacturer's instructions.

Step 11: Complete required reports and supporting documentation.

SKILL SHEETS

11-8
Demonstrate proper use of colorimetric tubes to identify hazards.

NOTE: Specific procedures will vary depending on the equipment used. Refer to the manufacturer's instructions for complete directions.

Step 5: Reset the counter.

Step 1: Ensure proper detection, monitoring, or sampling method and equipment is chosen.

Step 2: Ensure that all responders are wearing appropriate PPE.

Step 6: Properly break both ends off of the tube(s) using the provided tube cutter.

Step 3: Use the manufacturer's instruction manual to select the proper colorimetric tube for sampling and check expiration dates for the tube.

Step 4: Perform a functional test to check device per the manufacturer's instructions to ensure correct operation.

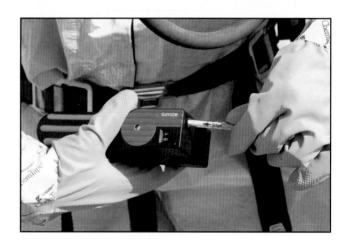

Step 7: Insert the tube into the hand pump in the proper direction.

Step 8: Hold the tip of the tube an appropriate distance away from the product or container opening, taking care not to come into contact with any solid or liquid product.

Step 9: Sample the product based on the manufacturer's instructions.

Step 10: Remove the tube from the pump and read, interpret, and record the results as per the manufacturer's instructions.

Step 11: Dispose of sampling tube in accordance with appropriate regulations.

CAUTION: Used tubes may be a hazardous waste and/or sharps hazard.

Step 12: Decontaminate equipment and return to operational state per manufacturer's instructions.

Step 13: Complete required reports and supporting documentation.

Implementing the Response: Mission-Specific Victim Rescue and Recovery

Chapter Contents

chapter 12

Key Terms

NFPA Job Performance Requirements

This chapter provides information that addresses the following job performance requirements of NFPA 1072, *Standard for Hazardous Materials/Weapons of Mass Destruction Emergency Response Personnel Professional Qualifications (2017)*.

6.8.1

Implementing the Response: Mission-Specific Victim Rescue and Recovery

Learning Objectives

After reading this chapter, students will be able to:

1. Describe considerations for planning and conducting rescue operations. (6.8.1)

2. List rescue equipment. (6.8.1)

3. Describe methods of victim rescue. (6.8.1)

4. Explain recovery operations. (6.8.1)

5. Recognize reports and documentation for victim rescue and recovery. (6.8.1)

6. Perform victim rescue operations at a hazardous materials incident. (6.8.1, Skill Sheet 12-1)

7. Conduct triage. (6.8.1, Skill Sheet 12-2)

Chapter 12
Implementing the Response: Mission-Specific Victim Rescue and Recovery

This chapter will describe the following topics:

- Rescue operations
- Rescue equipment
- Rescue methods
- Recovery operations
- Reports and documentation

Rescue Operations

Operations responders trained to the Mission-Specific Victim Rescue and Recovery Level may be called upon to rescue victims at hazardous materials incidents. Victim rescue and recovery in a hazardous materials/WMD release require various tactics and safety procedures, which depend on the following:

- Incident type
- Number of living victims
- Location of victims
- Whether the victims are ambulatory or nonambulatory

Responders need to evaluate if victim rescue is feasible and evaluate the risks for the responder associated with the rescue.

Developing and implementing a successful rescue and recovery operation requires:

- Training
- Comprehensive understanding of the rescue process
- Information about local capabilities and facilities
- Skills necessary to perform rescues safely and efficiently
- A rescue plan

Knowledge and flexibility are critical elements to ensure responders successfully rescue and recover victims from potentially contaminated environments. When performing rescue and recovery operations, be prepared for the following:

- Direct exposure to the hazards in the hot zone
- Dangers posed by the unstable physical environment of the incident (**Figure 12.1, p. 588**)

Figure 12.1 Rescuers must be prepared for the dangerous and unstable conditions presented by the incident. *Courtesy of the U.S. Marine Corps, photo by Sgt. Christopher D. Reed.*

- Stress from working in protective clothing
- Emotional trauma of a situation involving a high-stress environment
- Decontamination of persons involved in the incident

As with all operations at hazmat incidents, the risks that you identify during the initial size-up will dictate many elements of the response. As responders, you must operate within the structure of the Incident Command System and follow established procedures and local emergency response plan guidelines for conducting rescues **(Figure 12.2)**. Operations Level responders must work under the supervision of a Technician Level responder, an Allied Professional who can continuously assess and observe actions and provide immediate feedback, or under the guidance of SOPs that clearly spell out what actions Operations Level responders can and cannot perform.

Responders operating under SOPs must have a thorough understanding of their responsibilities and the correct protocols for implementing them. Detailed task requirements are outlined in the SOPs. As well, the SOPs provide the tasks that are beyond the Operations Level responder as well as the PPE needed to perform the tasks the responders are assigned.

The Incident Commander, with support from others, will formulate the Incident Action Plan (IAP) at all hazmat incidents, which may include a rescue/recovery plan. After the IAP has been developed, emergency responders will implement the rescue/recovery plan in accordance with local procedures. When assigned rescue and recovery responsibilities, responders must know the plan and follow it. Following the IAP will maximize responders' safety and their ability to save lives.

The response scenario will directly influence initial response operations. If the number of victims is small (one or two victims), emergency responders trained in victim rescue and recovery should be able to handle the incident with a single entry operation. However, if the number of victims is well beyond the number of available responders, emergency responders will be challenged with simultaneously rescuing and removing victims, establishing a safe haven or area of safe refuge until EMS operations can commence and making multiple entries in a short period of time **(Figure 12.3)**. The effects of physical exertion and heat stress will add to the challenges.

Responders conducting rescue missions may come from a variety of emergency responder disciplines and may include firefighters, hazardous material/WMD responders, EMS personnel, law enforcement officers, industrial fire brigade personnel, or a combination of these personnel assembled into a team. In addition to the SOPs of their local agency or AHJ, responders should be familiar with the SOPs for mutual-aid response in their area. **Skill Sheet 12-1** provides steps for performing victim rescue operations at a hazmat incident.

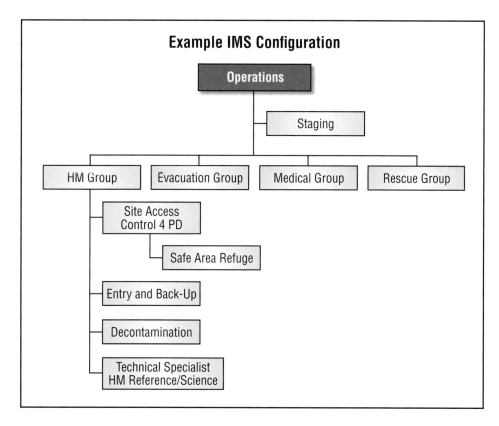

Example IMS Configuration

- Operations
  - Staging
  - HM Group
    - Site Access Control 4 PD
      - Safe Area Refuge
    - Entry and Back-Up
    - Decontamination
    - Technical Specialist HM Reference/Science
  - Evacuation Group
  - Medical Group
  - Rescue Group

Figure 12.2 Logistics section IMS configuration at a hazmat incident with a designated rescue group.

Figure 12.3 Appropriate training will help responders maintain control at mass casualty incidents.

Macdona Chlorine Gas Tank Derailment

In the early morning hours of June, 2004, in Macdona, Texas, one freight train struck another freight train causing the derailment of four locomotives and thirty-eight railcars. As a result of the derailment, a pressure tank car loaded with liquefied chlorine was punctured. Chlorine escaping from the punctured tank immediately vaporized into a cloud of chlorine gas that engulfed the accident area and drifted away from the site.

A nearby resident notified local emergency response authorities via a 9-1-1 call. Due to miscommunication between the resident and the 9-1-1 dispatcher, the dispatcher never heard the caller state that there was a train derailment. Instead, emergency responders were dispatched to assist this resident with difficulty breathing. None of the responders were yet aware that they were responding to a train accident and chlorine gas leak.

When responders approached the accident site, it was still dark so they could not see the wreckage or vapor cloud. The responders began to have difficulty breathing as they became exposed to the chlorine gas. They immediately withdrew from the scene and requested mutual aid from other agencies.

Shortly thereafter, mutual-aid resources, including the local Office of Emergency Management (OEM), began to arrive on scene. OEM established the Unified Incident Command system, activated the emergency operations center, and initiated the emergency management plan. Additional mutual-aid resources were also dispatched to the scene.

While conducting a search of the area, firefighters (wearing protective clothing and breathing apparatus) found the train's engineer stumbling along the roadway. He was in respiratory distress and was transported from the scene for medical attention.

A short time later, responders determined that the derailment wreckage prevented access to nearby residences. Access to the accident site through the wreckage was restricted until hazardous materials responders could properly assess the area.

Once a preliminary technical assessment of the chlorine gas release was completed, a firefighter entry team entered the accident area to attempt a rescue of three people who were reported to be trapped inside their residence. This entry team, however, became disoriented while attempting to advance through the wreckage and inadvertently diverted down the wrong roadway, away from their objective. Along that roadway, the team encountered the body of the train's conductor. Shortly thereafter, one of the entry team firefighters showed signs of dehydration, prompting the dispatch of a second entry team to come to the aid of the first.

A third entry team was dispatched to carry on with the rescue mission. This team successfully advanced through the wreckage and reached the three trapped residents. All three were in considerable respiratory distress. They were transported by helicopter to a local hospital for medical attention.

Lessons Learned:

In the end, three people died as a result of chlorine gas inhalation, including the train conductor and two residents. The train's engineer, twenty-three civilians, and six emergency responders were treated for respiratory distress or other injuries related to the collision and derailment.

Source: National Transportation Safety Board

Determining Feasibility of Rescues

Before implementing any IAP involving victims trapped in a contaminated environment, answer the following basic questions to determine the feasibility of conducting rescue operations:

- Can hazards (including products) be identified?

- What are the other known factors about the incident? Have witnesses provided additional information that might be useful to the decision-making process?

- Are victims within line of sight or is a search needed? Conducting a search for potential victims who are not in sight may increase the risk to responders. Responders conducting searches will have to extend their time in the hot zone to conduct the search, and they may be exposed to hazards that are not detectable from outside the search area.

- Is it a rescue operation or a recovery operation **(Figure 12.4, p. 592)**? Rescue operations are a high priority and may be conducted without complete mitigation of risk. Conduct recovery operations only after the risk to responders has been minimized or eliminated.

- Do on-scene emergency responders have the necessary PPE and training to perform the mission? To make the rescue safely, responders must have the PPE and training necessary to enter the hot zone.

- Do on-scene emergency responders have the necessary equipment to perform the mission **(Figure 12.5, p. 592)**? If needed equipment is not available, responders must wait until it is obtained or arrives on scene.

- Are there enough personnel available to conduct a rescue safely? First-arriving units may need to wait for additional personnel to arrive in order to conduct rescue operations.

- Are there available meters, papers, and instrumentation available that can be used to identify possible hazards and determine what is or is not present? This information may provide additional safety for rescue teams. Many fire departments and industry employers provide basic monitoring and detection devices for their Operations Level employees. If responders are trained in their use (and SOPs allow), use these devices during rescue operations.

- Are other information resources available that could be helpful such as the ERG, NIOSH Pocket Guide, or SDSs?

Rescue Safety

Rescuers must always consider their own safety first. ICs also must consider the hazards to which rescuers may be exposed while conducting search and rescue operations. Safety is the primary concern of rescuers because hurried, unsafe search and rescue operations may have serious consequences for rescuers as well as victims. Safety of responders depends upon rescuers and their officers making a good initial size-up, continuing the size-up throughout the operation, and performing a risk/benefit analysis before each major step in the operation.

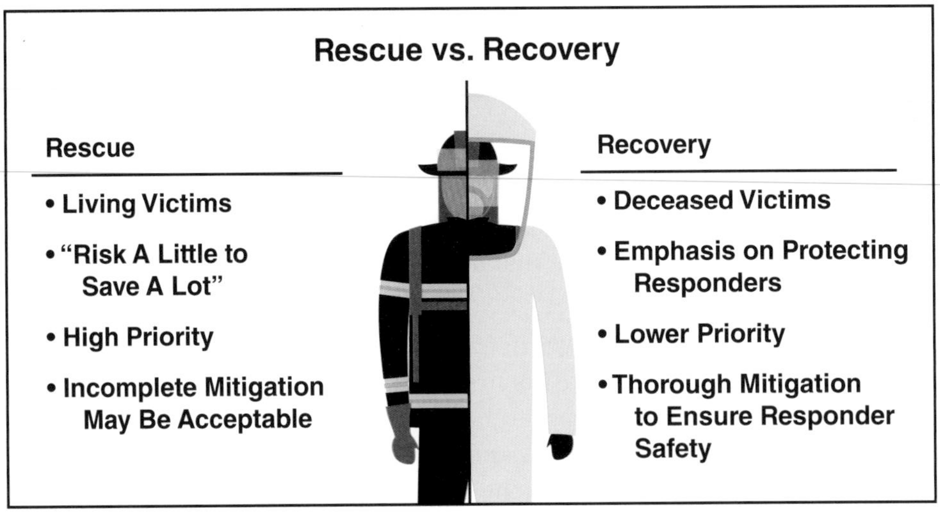

Figure 12.4 Rescue and recovery incidents are managed very differently.

Rescue vs. Recovery

Rescue	Recovery
• Living Victims	• Deceased Victims
• "Risk A Little to Save A Lot"	• Emphasis on Protecting Responders
• High Priority	• Lower Priority
• Incomplete Mitigation May Be Acceptable	• Thorough Mitigation to Ensure Responder Safety

Figure 12.5 Rescuers must have appropriate PPE to enter the hot zone safely. *Courtesy of the U.S. Marine Corps.*

Planning Rescues

When planning a rescue operation, consider all hazards, including hostile human threats, which are or may potentially be present during the incident. Initial rescue and recovery operations directly affect the number of available responders, their level of training, the incident's circumstances, and the PPE available **(Figure 12.6)**.

All unit leaders and supervisors within the ICS organization are responsible for providing proper supervision and overseeing the safety of all entry teams. A Technician Level responder or Allied Professional are typically responsible for risk assessment and the selection of control options.

Because most Operations Level responders are assigned to individual companies or units, the following information is specifically directed towards Operations Level responders assigned in this manner. The IC or the unit leader should establish a hot zone after the initial briefing and situation size-up. Once the hot zone has been established, start an assembly area adjacent to the entry point. The responsibilities of the unit leader include:

- Designate team assignments.
- Brief unit members of the objective and required tasks.
- Maintain both immediate and functional supervision over teams to ensure their safety during the operation.
- Ensure accountability and tracking of personnel.
- Relay any critical or pertinent information received from the teams up the chain of command.
- Ensure a decontamination station is established.
- Ensure a back-up team is in position prior to entry.

The unit leader does not don a chemical protective suit unless the situation indicates that a suit is required to properly supervise the unit's operation. For safety reasons, if possible, add additional responders positioned in areas so that line-of-sight can be maintained. The following situations would require the unit leader to don chemical protective equipment:

- Entry team would not be in line-of-sight.
- Incident requires a complex entry team operation.
- Operation requires several entry teams.

Establish a decontamination station before any PPE-equipped personnel enter into hazardous zones **(Figure 12.7)**. This station's primary function is to provide immediate and adequate decontamination to entry team members so they can be safely removed from PPE with minimum exposure to any material they may have contacted. This station can also be used to decontaminate first responders and/or civilians until shower units arrive on the scene.

The entry team must have at least two trained members in the appropriate level of PPE. Entry team members will perform the actual search, rescue, and removal of victims from the hazard-

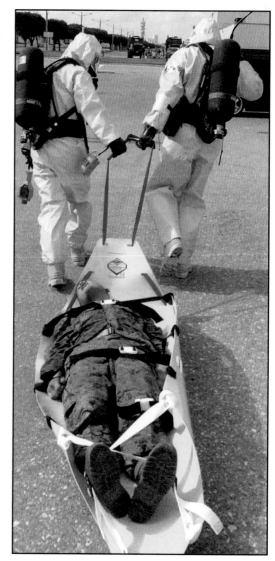

Figure 12.6 Responders must have appropriate PPE, necessary equipment, and enough personnel to conduct a rescue. *Courtesy of the U.S. Marine Corps.*

Figure 12.7 A decontamination station should be established before rescuers enter the hot zone to assist victims.

ous area. While operating in the hot zone, they must stay alert for any clues, signs, or causes that could have contributed to the spill or release. Members must report this information to the unit leader immediately.

Team members must work in close proximity to another (buddy system) at all times and maintain radio contact. If one member must leave the operating zone, the other team member must also leave. If one of the entry team members becomes incapacitated for an unknown reason, the other team member must notify the back-up team that an immediate rescue is needed.

The back-up team consists of two members trained in the appropriate level of PPE, on standby to perform a single task: the removal of a downed responder **(Figure 12.8)**. Their operating time should be less than the entry team's. The unit leader must provide the back-up team functional supervision at all times.

When entering the hot zone, responders must be alert to their surroundings at all times. Follow these basic guidelines:

- Immediately exit any area where chemical contamination is encountered and no living victims are identified.
- Avoid contact with any hazardous or unidentified materials.
- Undergo emergency decontamination immediately upon exiting the hazardous area.
- Immediately obtain medical assistance when needed.
- If conditions in excess of IDLH are detected by monitoring, consider attempting to change the environment (such as ventilation, vapor dispersion/suppression) **(Figure 12.9)**.

Conducting Rescues

Rescue priorities at hazmat incidents may differ from other emergency incidents in which efforts are made to reach the more severely injured victims first in order to save them, if possible. At hazmat incidents, rescuers (with appropriate training and PPE for the hazards present) typically start at the incident's outer edges and work their way in, following these priorities **(Figure 12.10)**:

1. Assist ambulatory casualties to save themselves (direct them to an area of safe refuge to await decontamination).
2. Evacuate nonambulatory casualties showing signs of life.
3. Evacuate nonambulatory casualties showing signs of life from the hot zone.
4. Recover the dead.

NOTE: Always follow procedures established by the authority having jurisdiction when conducting the activities previously listed.

Direct your first efforts at visible victims and then focus on victims out of sight. The following sections will explain the four incident response situations, based upon the status of the victim(s), that affect rescue and recovery at a hazardous materials/WMD release.

Line-of-Sight — Unobstructed, imaginary line between an observer and the object being viewed.

Line-of-Sight with Ambulatory Victims

Address ambulatory victims within the **line-of-sight** first. These victims are generally the farthest away from the release, have experienced the lowest level of exposure and related dose, and require the least amount of time to remove.

Figure 12.8 Backup team members' sole task at the incident is to remain on standby in case an entry team member needs to be rescued.

Figure 12.9 Ventilation may mitigate IDLH atmospheres.

Figure 12.10 Rescue priorities start at the incident's edge and work inward to the hot zone.

If possible, direct these individuals to a safe haven or area of safe refuge within the warm zone until a determination is made regarding the need for decontamination **(Figure 12.11)**. Direct these victims with verbal instructions, signs, hand signals, whistles, or light sources (at night). Responders must also have plans for managing noncooperative victims.

Failure to effectively detect and control contaminated individuals can severely disrupt the best scene management plan. An example of this would be to discover that individuals contaminated with radioactive material were allowed to bypass decon and enter the medical treatment area. Such a discovery would not only extend the hot and warm zone areas unnecessarily but would require completely reestablishing the medical treatment area and possible decontamination or abandonment of medical supplies. First responders should also avoid touching these victims. At crime scenes, they may need to be interviewed by law enforcement as witnesses.

Figure 12.11 Ambulatory victims should be directed to an area of safe refuge or decontamination line.

Line-of-Sight with Nonambulatory Victims

After entry teams have removed the ambulatory victims within the line of sight, they can turn their attention to the nonambulatory victims within line of sight **(Figure 12.12)**. Part of the planning phase is to anticipate what the tools and equipment that will be needed upon entry. Follow local SOP/Gs to determine when, the extent, and the type of triage to be conducted in the hot zone in order to determine which victims should be moved to the decontamination area **(Figure 12.13)**. Typically, responders trained to these mission-specific competencies will move unconscious victims who respond to touch stimulation to the decon area so that medical personnel can treat critical injuries and medically monitor victims as decontamination takes place (or after, as SOP/Gs and training dictate). More triage and treatment will be conducted after victims have been decontaminated.

Figure 12.12 Once ambulatory victims have been moved to safety, rescuers can begin removing visible nonambulatory victims.

Figure 12.13 Follow local SOPs to determine which nonambulatory victims should be moved to the decontamination area.

Non-Line-of-Sight with Ambulatory Victims

Ambulatory victims who are not in the line-of-sight are generally closer to the incident or source of the release and have experienced a greater exposure and related dose. If possible, direct these individuals to a safe haven or area of safe refuge where they can be assessed for decontamination need or medical treatment. Responders may need to enter the hazard area to find these victims, so there is an increased level of risk which must be reflected in planning and safety measures.

Non-Line-of-Sight with Nonambulatory Victims

Nonambulatory victims not within line-of-sight are the last to be rescued from the hot zone. These victims are generally the closest to the hazardous materials/WMD event and have experienced the greatest exposure and related dose. Rescue and removal of these victims typically poses the greatest danger to emergency response personnel and require increased planning and resources to carry out.

Figure 12.14 Thorough triage and medical treatment is provided to nonambulatory victims after they have been decontaminated.

Conducting Triage

Conduct triage of injured victims after they have been moved from the hot zone into an area of refuge in order to determine priority for decon. A more thorough triage will be conducted after victims have gone through the decon process **(Figure 12.14)**. One common system requires responders to quickly assess a victim's status and assign them into four basic categories:

- Priority 1 — Life-threatening injuries and illnesses (highest priority)

- Priority 2 — Serious, but not life-threatening injuries

- Priority 3 — Minor injuries

- Priority 4 — Dead or fatally injured

Responders must be familiar with the triage system used by their agency. **Skill Sheet 12-2** demonstrates the use of one common triage system.

Rescue Equipment

Performing rescue and recovery operations requires specialized tools and equipment. These include:

- Personal protective clothing and equipment suitable for the hazards present
- Triage tags **(Figure 12.15, p. 600)**
- Equipment such as backboards and **Stokes baskets** to quickly package nonambulatory victims for removal
- **SKEDs®**, carts, buggies, and similar devices to move nonambulatory victims **(Figures 12.16a and b, p. 598)**
- Extrication equipment
- Technology used to search for victims such as heat-sensing devices (thermal imagers) and fiber optic cameras

Stokes Basket — Wire or plastic basket-type litter suitable for transporting patients from locations where a standard litter would not be easily secured, such as a pile of rubble, a structural collapse, or the upper floor of a building; may be used with a harness for lifting.

SKED® — Lightweight, compact device for patient packaging; shaped to accommodate a long backboard; may be used with a rope mechanical advantage system.

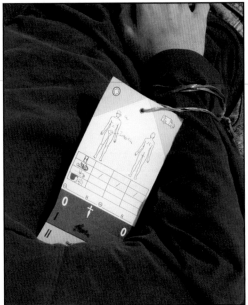

Figure 12.15 Preprinted triage tags can assist in the triage process.

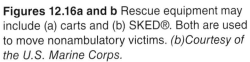

Figures 12.16a and b Rescue equipment may include (a) carts and (b) SKED®. Both are used to move nonambulatory victims. *(b)Courtesy of the U.S. Marine Corps.*

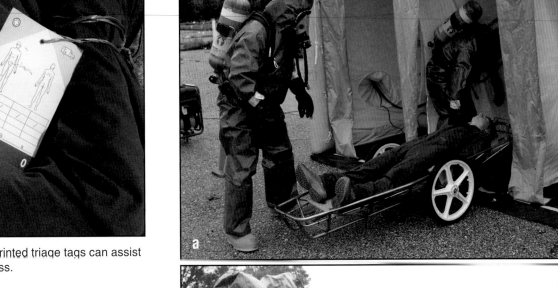

Responders must be trained in the correct use of such equipment, and the equipment must be maintained in good condition. For more information about equipment used for rescue, see IFSTA's **Essentials of Fire Fighting** and **Urban Search and Rescue in Collapsed Structures** manuals.

Rescue Methods

An uninjured victim (or one with minor injuries) may be directed using hand signals or verbal instructions to walk to safety. If physical assistance is needed, one or two rescuers may be needed, depending on how much help is available and the size and condition of the victim.

The chief danger in moving an injured victim quickly from the hot zone is the possibility of aggravating a spinal injury. However, in an extreme emergency, such as an IDLH atmosphere or threat of explosion, the possible spinal injury becomes secondary to the goal of preserving life. In these cases, pull the victim in the direction of the long axis of the body — not sideways.

It is always better to have two or more rescuers when attempting to move an adult. One rescuer can safely carry a small child, but you may need two, three, or even four rescuers to move a large adult. There are a number of lifts, carries, and drags that may be used to move a victim from an area quickly **(Figure 12.17)**.

Realistically, in very hostile environments, victims are removed by whatever means are available at the moment. Sometimes that means grabbing an arm, leg, clothes, belt, hair, or whatever works.

Ideally, rescuers will be able to use some type of litter or SKED® (or other sliding device or material) to remove a victim. These devices include the standard ambulance cot, army litter, scoop stretcher, basket litter, and long backboard **(Figure 12.18)**. For more information on rescue lifts, carries, and drags, as well as using litters and Skeds®, see IFSTA's **Essentials of Fire Fighting** manual.

Figure 12.17 A variety of lifts, carries, and drags can be used to move victims.

Figure 12.18 Victims can be moved on basket litters, backboards, or stretchers. *Courtesy of the U.S. Navy, photo by Mass Communication Specialist 2nd Class Kirk Worley.*

Recovery Operations

All viable victims should be rescued before **recovery** operations begin. Rescue is the removal of both ambulatory and nonambulatory victims who are still living and who have a likelihood of surviving their injuries or exposure. Recovery is removal of the dead.

Recovery operations are a lower priority and should be coordinated by the IC with law enforcement or coroner personnel. Reducing the hazards to create a safer environment to operate in is essential when performing victim recovery. It may be necessary for bodies and human remains to remain in place until law enforcement or investigation efforts are completed. The remains of deceased victims are recovered for body identification.

Recovery — Situation where the victim is determined or presumed to be dead, and the goal of the operation is to recover the body.

Reports and Documentation

The AHJ may have required documents and/or reports that you will need to fill out or assist with completing. Some reports may be situational depending on the incident or factors at the incident. Other types of reports are routine after any type, complexity, or size of event.

Types of reports may include:

- National Fire Incident Reporting System (NFIRS) reports
- Department specific IAPs
- Site safety plans
- NIMS reports such as ICS 208 HM
- After Action Reports
- Exposure reports
- Patient care reports

Chapter Review

Answer the following questions to review the information provided in this chapter.

1. How are rescues prioritized?

2. When is triage conducted?

3. Give examples of the different types of rescue equipment that may be needed at a hazmat/WMD incident.

4. How many rescuers should be available when attempting to move an adult?

5. How does recovery differ from rescue?

6. List the types of documentation that may be required at rescue and recovery operations.

Step 1: Evaluate the situation including the hazards present and the status of victims in order to determine the feasibility of rescue according to AHJ.

Step 2: Ensure proper rescue or recovery method is chosen.

Step 3: Ensure that all responders are wearing appropriate PPE.

Step 4: Visually identify the location and status of victims at the incident.

Step 5: Ask if victims are able to move voluntarily. If they are not able to move, proceed to next step.

Step 6: Establish decontamination appropriate to the incident.

Step 7: Assemble and brief a rapid extraction crew on the task.

Step 8: Communicate intent to begin rapid extraction to Incident Commander.

Step 9: Communicate intent to begin rapid extraction to victims at the incident.

Step 10: Direct rapid extraction crew to begin extraction.

Step 11: Maintain continuous communication during extraction.

Step 12: As extraction crew exits incident area with victims, begin decontamination procedures.

Step 13: Transport decontaminated victims to triage.

Step 14: Complete required reports and supporting documentation.

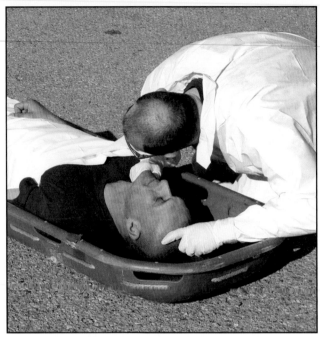

Step 4: Assess patient's respiration.

   a. Attempt to open the airway. If breathing cannot be started by opening the airway, mark Priority 0.

Step 1: Ensure scene safety and proper PPE.

Step 2: Identify patients to be triaged.

Step 3: Assess patient's mobility.

   b. If patient starts breathing, or their respiratory rate is more than 30/minute, mark Priority 1.

   c. If respiratory rate is less than 30/minute, go to next step.

Step 6: Assess patient's level of consciousness
   a. If patient is alert (able to follow simple commands), mark Priority 2.
   b. If any altered mental status, mark Priority 1.

Step 7: Re-triage the Priority 3 "Walking Wounded" patients. Check for any change in medical condition, e.g. shock, mental status, etc.

Step 5: Assess patient's radial pulse.
   a. If patient is breathing, but has no radial pulse, mark Priority 1.
   b. If patient is breathing and has a pulse, go to next step.

Implementing the Response: Mission-Specific Product Control

Chapter Contents

Key Terms

NFPA Job Performance Requirements

This chapter provides information that addresses the following job performance requirements of NFPA 1072, *Standard for Hazardous Materials/Weapons of Mass Destruction Emergency Response Personnel Professional Qualifications (2017)*.

6.6.1

Implementing the Response: Mission-Specific Product Control

Learning Objectives

After reading this chapter, students will be able to:

1. Describe methods of spill control. (6.6.1)

2. Describe methods of leak control. (6.6.1)

3. Describe methods of fire control at a hazardous materials incident. (6.6.1)

4. Perform absorption/adsorption. (6.6.1; Skill Sheet 13-1)

5. Perform damming. (6.6.1; Skill Sheet 13-2)

6. Perform diking operations. (6.6.1; Skill Sheet 13-3)

7. Perform diversion. (6.6.1; Skill Sheet 13-4)

8. Perform retention. (6.6.1; Skill Sheet 13-5)

9. Perform vapor suppression. (6.6.1; Skill Sheet 13-6)

10. Perform vapor dispersion. (6.6.1; Skill Sheet 13-7)

11. Perform dilution. (6.6.1; Skill Sheet 13-8)

12. Perform remote valve shutoff or activate emergency shutoff device. (6.6.1; Skill Sheet 13-9)

Chapter 13
Implementing the Response: Mission-Specific Product Contol

This chapter will explain the following topics that Operations-Level personnel may perform with Mission-Specific training:

- Spill control
- Leak control
- Fire control

Spill Control

Spill-control tactics confine a hazardous material that has been released from its container. These tactics attempt to reduce the amount of contact the product makes with people, property, and the environment, limiting the amount of potential harm the products cause. **Control** actions involving spills are generally defensive in nature **(Figure 13.1)**.

NOTE: Responders should familiarize themselves with their AHJ's policies and procedures for product control as specified in SOPs and emergency response plans.

To prevent further contamination, responders should use spill control to confine the hazardous material after its release. For this reason, spill control is often simply called **confinement**. Some spill-control tactics such as **neutralization** and dispersion minimize the amount of harm that contact with the material causes. Spill control primarily acts as a defensive operation, and responders' safety is a primary consideration.

Spills may involve gases, liquids, or solids. The product involved may be released into the air (as a vapor or gas), into water, and/or onto a surface. The type of release determines the spill-control method needed to control it. For example, in the event of a flammable liquid spill, you must address both the liquid spreading on the ground and the vapors releasing into the air.

To prevent the spread of liquid materials, methods used include building dams or dikes near the source, catching the material in another container, or directing (diverting) the flow to a remote location for collection. Before using equipment to confine spilled materials, ICs need to seek advice from technical sources to determine

Control — To contain, confine, neutralize, or extinguish a hazardous material or its vapor.

Confinement —The process of controlling the flow of a spill and capturing it at some specified location.

Neutralization — Chemical reaction in water in which an acid and base react quantitatively with each other until there are no excess hydrogen or hydroxide ions remaining in the solution.

Figure 13.1 Spill control is typically a defensive action. *Courtesy of Rich Mahaney.*

if the spilled materials will adversely affect the equipment. If the spill involves a corrosive material, it may react with metals or damage other materials **(Figure 13.2)**. Large or rapidly spreading spills may require the use of heavy construction-type equipment, floating confinement booms, or special sewer and storm drain plugs **(Figure 13.3)**.

Spill control is not restricted to controlling liquids. Responders may also need to confine dusts, vapors, and gases with the following:

- Protective covering consisting of a fine spray of water
- A layer of earth
- Plastic sheets
- Salvage covers
- Foam blankets on liquids

Strategically placed water streams can direct gases or allow the water to absorb or move them. Reference sources and training information can provide the proper procedures for confining gases.

Figure 13.2 Ensure that spill control equipment and materials are compatible with the hazardous material. Some corrosives react with metal.

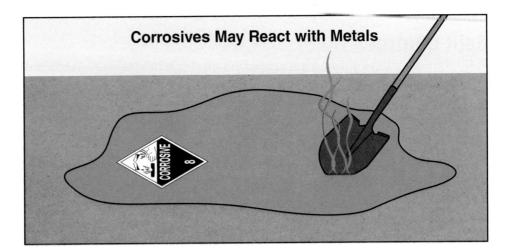

Figure 13.3 Large spills may require floating containment booms. *Courtesy of U.S. EPA.*

The following dictate confinement efforts:

- Material type
- Rate of release
- Speed of spread
- Number of personnel available
- Tools and equipment needed
- Weather
- Topography

CAUTION

Undertake spill-control actions ONLY if you can perform tasks without coming into direct contact with the hazardous material.

Operations level responders take protective actions but do not stop the release unless they can close a remote shutoff valve from a safe location (see Leak Control section). Defensive spill-control tactics that confine hazardous materials include the following:

- Absorption
- Adsorption
- Blanketing/covering
- Dam, dike, diversion, and retention
- Vapor suppression

Rather than attempting to confine the dispersion, some defensive spill-control tactics are diluting the concentration or changing its physical and/or chemical properties aimed at reducing the amount of harm the material causes. These tactics include the following:

- Vapor dispersion
- Ventilation
- Dispersion
- Dilution
- Dissolution
- Neutralization

Neutralization

Many experts consider neutralization a Tech-Level offensive containment tactic **(Figure 13.4)**. However, neutralization aims to reduce or eliminate the chemical hazard of the material rather than physically contain it. Responders use dilution, neutralization, and dissolution tactics infrequently at hazmat incidents and only under specific circumstances due to the other complications that these materials/corrosives can have, including generating large quantities of hazardous waste.

Figure 13.4 Hazmat Technicians usually perform neutralization.

Responders should always operate under the guidance of a hazardous materials technician, an Allied Professional, an emergency response plan, or standard operating procedures. As required by SOPs/SOGs, responders may need to fill out required reports and supporting documentation for product control operations.

Table 13.1 provides a summary of the potential spill control tactics for different types of releases and their resulting dispersions. It also provides an example of a task related to one of the appropriate tactics.

Table 13.1
Spill Control Tactics Used According to Type of Release

Type of Release	Type of Dispersion	Spill Control Tactics	Task Example
Liquid: Airborne Vapor	Hemispheric, cloud, plume, or cone	• Vapor Suppression • Ventilation • Vapor Dispersion • Dissolution	Cover spill with vapor suppressing foam (Vapor Suppression)
Liquid: Surface	Stream	• Diking • Diversion • Retention • Adsorption • Absorption	Dig a ditch to divert a spill away from a stream (Diversion)
Liquid: Surface	Pool	• Absorption • Adsorption (for shallow spills) • Neutralization	Cover spill with an absorbent pillow (Absorption)
Liquid: Surface	Irregular	• Dilution • Absorption • Neutralization	Spray slightly contaminated surfaces with water (Dilution)
Liquid: Waterborne Contamination	Stream or pool	• Damming • Diversion • Retention • Absorption • Dispersion	Place absorbent booms across a river (Absorption)
Solid: Airborne Particles	Hemispheric, cloud, plume, or cone	• Particle Dispersion/ Ventilation • Particle Suppression (wetting material) • Blanketing/Covering	Set up ventilation fans (Particle Dispersion/Ventilation)
Solid: Surface	Pile	• Blanketing/Covering • Vacuuming	Cover spilled material with a tarp or salvage cover (Blanketing/Covering)
Solid: Surface	Irregular	• Blanketing/Covering • Dilution • Dissolution	Spray scattered sprinkles of corrosive powders or dusts with water (Dilution)
Gas: Airborne Gas	Hemispheric, cloud, plume, or cone	• Ventilation • Vapor Dispersion • Dissolution	Spray leaking gas cloud with fog stream (Dissolution)

NIOSH: Iowa Propane Tank Flaming Breach

In Iowa, a call came into the fire department late at night regarding a fire at a turkey farm. Two people riding an off-road vehicle had struck and broken one of two fixed, metal pipes between a propane tank and two vaporizers (devices that receive liquefied petroleum gas in liquid form and add sufficient heat to convert the liquid to a gaseous state). As liquid propane spewed from the pipe, the off-road vehicle operator drove away so that he could call 9-1-1. The propane vapors, which have a vapor density of 1.53 and are heavier than air, spread along the ground and were eventually ignited by the pilot flame at the vaporizers. Burning propane vapors spread throughout the area and began to impinge on the tank, causing the pressure relief valve to activate and send burning propane flames high into the air.

Upon arrival at the fire scene, the Incident Commander made an assessment of the burning tank. Fire had engulfed the propane tank, and it was venting burning propane vapors via two pressure relief vent pipes located on top of the tank. The tank's pressure relief valve was also emitting a loud noise similar to a jet engine.

After seeing the flames and hearing the high-pitched sound being emitted by pressure relief vent pipes located on the top west section of the tank, the Incident Commander decided to allow the tank to burn itself out and to try to save the adjacent buildings. The firefighters positioned themselves in various areas in a semicircle north, northeast, and north-west of the tank. They were about 105 feet (32 m) away from the tank spraying one of the buildings with water when a BLEVE occurred. The BLEVE ripped the tank into four parts, each flying in a different direction. One part of the tank traveled in a north-west direction toward the two firefighters, striking them and killing them instantly **(Figure 13.5)**. Six other firefighters and a deputy sheriff received varying degrees of burns and assorted injuries.

After the incident, investigators discovered that while the tank had two relief valves, the piping connecting the tank to the vaporizers was unprotected and did not have an excess flow valve. At the time of the incident, the tank had a capacity of 18,000 gallons (70 000 L) and held an estimated 10,000 gallons (40 000 L) of liquid propane. The force of the BLEVE illustrates the need for firefighters to consider the type of release and its potential destructive power when establishing the hot zone and deciding on containment measures.

Source: NIOSH

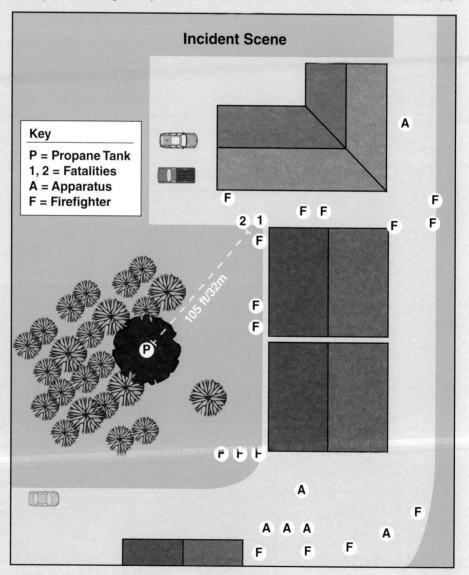

Figure 13.5 This incident scene overview is based on a NIOSH photo.

Absorption

Absorption, like a sponge soaking up water, soaks up or retains a liquid hazardous material in some other material. The bulk of the liquid being absorbed enters the cell structure of the absorbing medium. When choosing an absorbent, it must be chemically compatible with the material being absorbed. Absorbents tend to swell as they absorb the material.

Common absorbents used at hazmat incidents include (**Figure 13.6**):

- Sawdust
- Clays
- Charcoal
- Polyolefin-type fibers
- Specially designed absorbent pads, pillows, booms, and socks

The absorbent is spread directly onto the hazardous material or in a location where the material is expected to flow. After use, responders must treat and dispose of absorbents as hazardous materials because they retain the properties of the materials they absorb.

Responders often use absorption at incidents involving small spills (55 gallons [208 L] or less), such as gasoline or diesel fuel. While some absorbents, such as sawdust, may work best on shallow pools, other types of spills may require different types of absorbents. For example, responders may use absorbent booms for releases involving waterborne spills in streams or pools. For more information about performing absorption, see **Skill Sheet 13-1**.

Adsorption

Adsorption differs from absorption in that the molecules of the liquid hazardous material physically adhere to the adsorbent material rather than being absorbed into its inner spaces. Adsorbents tend not to swell like absorbents. Responders usually use organic-based materials, such as activated charcoal

Figure 13.6 Absorbents are used to soak up hazardous materials.

or carbon, as adsorbents. Adsorbents primarily control shallow liquid spills and increasingly replace soap and water or other decon methods. Make sure that the adsorbent used is compatible with the spilled material in order to avoid potentially dangerous reactions. For more information about performing adsorption, see **Skill Sheet 13-1**.

Blanketing/Covering

Personnel perform blanketing or covering to prevent dispersion of hazardous materials. Operations level responders may not be allowed to perform blanketing/covering actions, depending on the hazards of the material, the nature of the incident, and the distance from which they must operate to ensure their safety. Responders must consider the compatibility between the material being covered and the material covering it **(Figure 13.7)**. For blanketing or covering solids, such as powders and dusts, the following tools are used:

- Tarps
- Plastic sheeting
- Salvage covers
- Other materials (including foam)

Blanketing/covering may also be used as a form of temporary mitigation for radioactive and biological substances, for example, to reduce alpha or beta radiation or prevent the spread of biological materials. Personnel can blanket/cover incidents involving cryogen leaks to cause the released material to auto-refrigerate beneath the tarp or covering **(Figures 13.8a and b, p. 614)**. As a temporary option, responders can cover openings of some liquid containers with plastic sheets or tarps to confine vapors. Blanketing of liquids is essentially the same as vapor suppression (see Vapor Suppression section) because it typically uses an appropriate aqueous (water) foam agent to cover the surface of a spill.

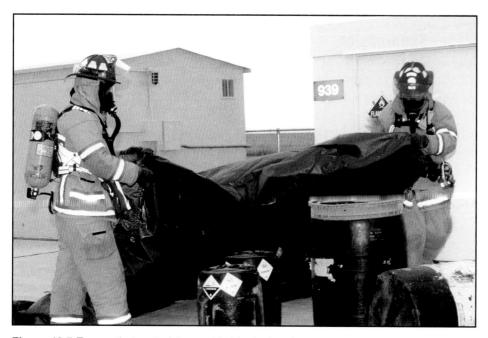

Figure 13.7 Ensure that materials used in blanketing do not react with the hazardous material.

Figures 13.8a and b Covering this anhydrous ammonia release causes it to auto-refrigerate beneath the tarp. *Courtesy of Rich Mahaney.*

Dam — Actions to prevent or limit the flow of a liquid or sludge past a certain area.

Dike — Actions using raised embankments or other barriers to prevent movement of liquids or sludges to another area.

Divert — Actions to direct and control movement of a liquid or sludge to an area that will produce less harm.

Retain — Actions to contain a liquid or sludge in an area where it can be absorbed, neutralized, or removed. Often used as a longer-term solution than other similar product control methods.

Vapor Suppression — Action taken to reduce the emission of vapors at a hazardous materials spill.

Vapor Dispersion — Action taken to direct or influence the course of airborne hazardous materials.

Dam, Dike, Diversion, and Retention

Damming, **diking**, **diverting**, and **retaining** are performed to confine or control a hazardous material **(Figure 13.9)**. These actions control the flow of liquid hazardous materials away from the point of discharge. Responders can use available earthen materials or materials carried on their response vehicles to construct curbs that direct or divert the flow away from gutters, drains, storm sewers, flood-control channels, and outfalls **(Figure 13.10)**. In some cases, it may be desirable to direct the flow into certain locations in order to capture and retain the material for later pickup and disposal. Some dams may permit surface water or runoff to pass over (or under) the dam while holding back the hazardous material **(Figure 13.11)**. Responders must properly dispose of any construction materials that contact the spilled material. See **Skill Sheets 13-2**, **13-3**, **13-4**, and **13-5** for instructions on how to perform damming, diking, diversion, and retention.

Vapor Suppression

Vapor suppression reduces the emission of vapors at a hazmat incident **(Figure 13.12)**. Responders use vapor suppression when they apply fire fighting foam to suppress vapors from flammable and combustible liquids. Other examples of vapor suppression include using water fog from hose streams or chemical vapor suppressants. See **Skill Sheet 13-6** for more information about performing vapor suppression. The Fire Control section later in this chapter addresses the use of fire fighting foam to suppress vapors and extinguish fires.

Vapor Dispersion

Vapor dispersion directs or influences the course of airborne hazardous materials. Pressurized streams of water from hoselines or unattended master streams may help disperse vapors **(Figure 13.13, p. 616)**. These streams create turbulence, which increases the rate of the materials mixing with air and reduces the concentration of the hazardous material. After using water streams for vapor dispersion, responders must confine and analyze runoff water for possible contamination. **Skill Sheet 13-7** provides a set of steps for performing basic vapor dispersion.

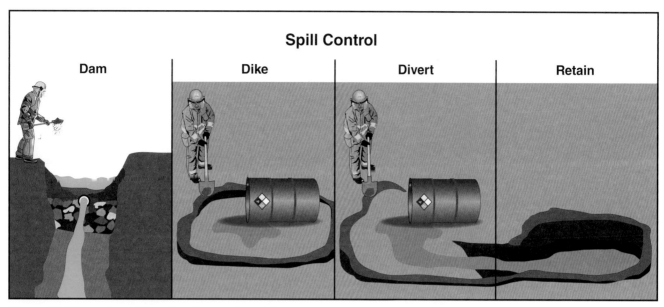

Spill Control

| Dam | Dike | Divert | Retain |

Figure 13.9 Damming, diking, diverting, and retaining are common methods to control liquid spills.

Figure 13.10 Diverting hazardous materials away from storm drains may be necessary to protect the environment.

Figure 13.11 Dams can be constructed to trap materials that are lighter or heavier than water, depending on their specific gravity.

Figure 13.12 Foam is used to suppress flammable liquid vapors.

Figure 13.13 Vapor dispersion uses pressurized water streams from hoselines or unattended master streams.

Ventilation

Ventilation is performed to control air movement using natural or mechanical means. When spills occur inside structures, ventilation can remove and/or disperse harmful airborne particles, vapors, or gases **(Figure 13.14)**. Personnel can apply the same ventilation techniques that they use for smoke removal to hazmat incidents (see IFSTA's **Essentials of Fire Fighting** manual). As with other types of spill control, responders should ensure the compatibility of their ventilation equipment with the hazardous atmosphere. When conducting negative-pressure ventilation, personnel should ensure the fans and other ventilators are compatible with the atmosphere where they are being operated. Equipment must be **intrinsically safe** in a flammable atmosphere. When choosing the type of ventilation to use, remember that positive-pressure ventilation removes atmospheric contaminants more effectively than negative-pressure ventilation.

Dispersion

Dispersion involves breaking up or dispersing a hazardous material that has spilled on a solid or liquid surface. Both chemical and biological agents disperse hazardous materials. Personnel usually use dispersion agents on hydrocarbon spills. Dispersion poses problems of spreading the material over a wide area, and the process itself may cause additional problems. Because of these problems, the use of dispersants may require the approval of government authorities.

Dilution

Dilution is the application of water to a water-soluble material to reduce the hazard. Dilution of liquid materials rarely has practical applications at hazmat incidents in terms of spill control; responders use dilution more frequently during decontamination operations **(Figure 13.15)**. Diluting hazardous water-soluble liquids requires huge volumes of water that may create runoff problems. Responders may use dilution at spills involving small amounts of corrosive material, such as in cases of irregular dispersion or a minor accident in a laboratory. Even then, it is generally considered for use only after spill control methods have been rejected. A simple set of steps for performing dilution are provided in **Skill Sheet 13-8**.

Dilution — Application of water to a water-soluble material to reduce the hazard.

Scope of this List

The first responder is primarily concerned with mitigating hazards from liquid and gas releases. This manual does not cover all techniques. However, this book provides and addresses the tactics for the first responder to aid a Technician during mitigation of an incident.

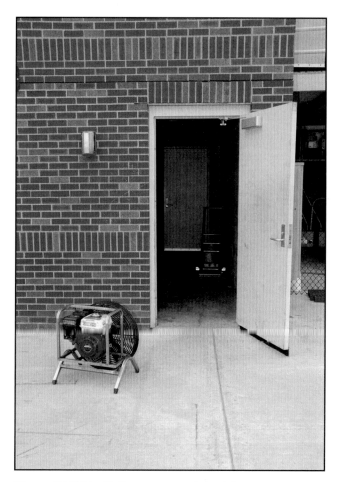

Figure 13.14 Ventilation can move harmful vapors, gases, or other hazardous airborne particles.

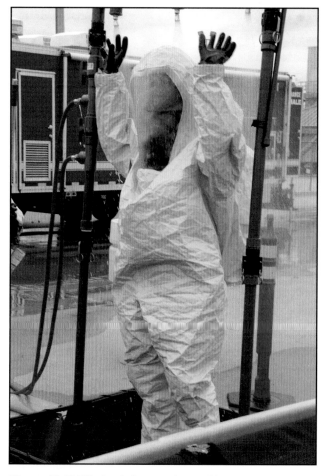

Figure 13.15 Dilution is frequently used in decontamination operations.

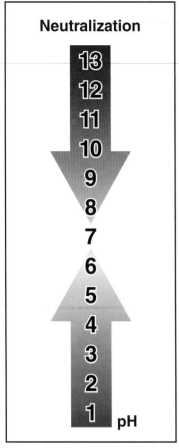

Neutralization

13
12
11
10
9
8
7
6
5
4
3
2
1
pH

Figure 13.16 Neutralization is used to shift pH closer to 7.

Containment — The act of stopping the further release of a material from its container.

Neutralization

Some hazardous materials may be neutralized to minimize the amount of harm that they do upon contact. Usually, neutralization involves raising or lowering the pH of corrosive materials to render them neutral (pH 7) **(Figure 13.16)**. Neutralization can also refer to any chemical reaction that reduces the hazard of the material. Neutralization is a difficult process; for example, adding too much of a neutralizer can cause a pH shift in the opposite direction. These reactions can also release tremendous heat. With few exceptions, responders should only conduct neutralization under the direction of a hazardous materials technician, Allied Professional, or SOP/Gs.

Leak Control

Leak-control tactics are used to contain the product in its original (or another) container, preventing it from escaping. Hazardous materials technicians and specialists perform most leak-control tactics, which are offensive **(Figure 13.17)**.

A leak involves the physical breach in a container through which product escapes. The goal of leak control is to stop or limit the escape or to contain the release either in its original container or by transferring it to a new one. Leak control is often referred to as **containment**. The type of container involved, the type of breach, and properties of the material determine tactics and tasks relating to leak control. Normally, personnel trained below the Technician level do not attempt offensive actions such as leak control. Notable exceptions include situations involving gasoline, diesel, liquefied petroleum gas (LPG), and natural gas fuels. Operations responders can take offensive actions with these fuels provided they have appropriate training, procedures, equipment, and PPE.

Operations level responders may perform leak control by activating emergency shutoff devices on transportation containers and closing shutoff valves at fixed facilities, pipelines, and piping. **Skill Sheet 13-9** covers steps for shutting off a remote valve or operating an emergency shutoff device.

Figure 13.17 Leak control operations are usually performed by Haz Mat Technicians.

Transportation Container Emergency Shutoff Devices

Leak control dictates that personnel enter the hot zone, which puts them at great risk. The IC must remember that the level of training and equipment provided to personnel are limiting factors in performing leak control.

Under safe and acceptable circumstances, Operations responders may operate emergency remote shutoff devices on cargo tank trucks and intermodal containers.

Cargo Tank Truck Shutoff Devices

Most, though not all, cargo tanks have emergency shutoff devices. Device locations may vary, but they are often located behind the driver's side cab **(Figure 13.18)**. Activation of these shutoff devices vary by device, but it is usually as simple as pulling a handle, flipping a switch, or breaking off a fusible device.

By type, cargo tank trucks have the following emergency shutoff device configurations:

Figure 13.18 Cargo tank emergency shutoff device locations vary. Most have one located on the tank, directly behind the driver's side cab. Some will also have one on the rear of the tank.

- **High pressure tanks (MC-331)** — An emergency shutoff device on the left-front corner of the tank (behind the driver). Some will also have one on either the right or the left-rear corner. For example, MC-331s of 3,500 gallon (13 249 L) capacity or larger should have two emergency shutoff devices located remotely from each other — one on the tank behind the driver and the other on the rear of the tank, often on the passenger side **(Figure 13.19)**. These tanks may also have an electronically operated shutdown device that can be activated 150 feet (46 m) from the vehicle. This device may also stop the engine and perform other functions.

- **Nonpressure liquid tanks (MC/DOT-306/406) and low-pressure chemical tanks (MC/DOT-307/407)** — An emergency shutoff device on the left-front corner of the tank (behind the driver) **(Figure 13.20, p. 620)**. Some will also have one on either the right or the left rear corner. Some cargo tanks may have emergency shutoffs in the center of the tank near valves and piping, or built into the valve box **(Figure 13.21, p. 620)**.

- **Corrosive liquid tanks (MC/DOT 312)** — Do NOT typically have emergency shutoff devices.

Figure 13.19 MC-331s will typically have shutoff devices behind the driver's side cab and on the right rear. *Courtesy of Rich Mahaney.*

Figure 13.20 Low-pressure chemical tanks will have shutoffs on the tank behind the driver's side cab. *Courtesy of Rich Mahaney.*

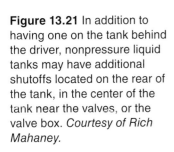

Figure 13.21 In addition to having one on the tank behind the driver, nonpressure liquid tanks may have additional shutoffs located on the rear of the tank, in the center of the tank near the valves, or the valve box. *Courtesy of Rich Mahaney.*

Intermodal Container Emergency Shutoff Device

Gas service (high pressure and cryogenic) intermodal containers will have emergency shutoffs for the bottom internal valve. Other containers may have them, depending on manufacturer or owner. Responders can look for a metal cable running down one side of the frame rail of the intermodal container or from the liquid valve to a fixed point away from the container **(Figures 13.22a-c)**. Pull this cable to activate the emergency shutoff. You may also be able to pull a handle or other device to activate the emergency shutoff device.

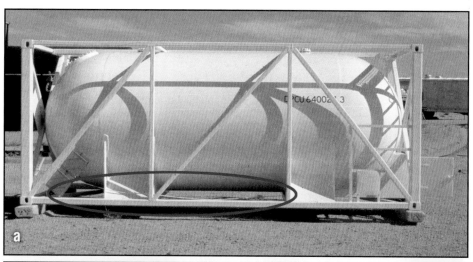

Figures 13.22a-c High pressure and cryogenic intermodal containers will have emergency shutoff devices. Look for a metal cable running down the rail of the container. Pull the cable to activate the device and close the bottom internal valve. *Courtesy of Rich Mahaney.*

Figure 13.23 Remote valves can be closed to stop a material from flowing in pipelines or piping.

Fixed Facility, Pipelines, and Piping Shutoff Valves

Fixed facilities, piping, and pipelines may also have remote shutoff valves. These remote shutoff or control valves can be operated to stop the flow of product to an incident area without entering the hot zone **(Figure 13.23)**. Depending on the diameter and length of piping, a significant amount of product may release for some time before the flow stops.

Responders should NOT shut any valves without direction from facility or pipeline operators **(Figure 13.24)**. In most cases, on-site fixed-facility maintenance personnel or local utility workers know where these valves are located and can be given the authority and responsibility for closing them under the IC's direction. Generally, these personnel will understand the proper procedures and consequences of closing the valve.

Operations level responders who are trained and authorized to operate shutoff valves at their facilities in the event of emergency may do so in accordance with their SOPs.

It may be safe for responders to shut off some natural gas lines, for example, shutting off the gas at the meter to the house or business. Generally, the meter is located outside the structure near the foundation or on the easement near the property line. However, you may find it inside the structure in a basement or mechanical space.

Figure 13.24 Do NOT close valves without direction from facility or pipeline operators. Closing valves without knowledgeable input may cause potentially dangerous consequences. *Courtesy of Texas Commission on Fire Protection.*

The shutoff is an inline valve located on the owner supply side of the meter; that is, between the distribution system and the meter. When the valve is open, the tang (a rectangular bar) is in line with the pipe. To close the valve, use a spanner wrench, pipe wrench, or similar tool to turn the tang until it is 90 degrees to the pipe **(Figure 13.25)**. Contact the local utility company when the gas has been shut off or when any emergency involving natural gas occurs in its service area.

Fire Control

Fire control attempts to minimize the damage, harm, and effect of fire at a hazmat incident. Fire-control tactics aim to **extinguish** fires and prevent ignition of flammable materials. Fire-control tactics may be offensive or defensive, depending on the situation **(Figure 13.26)**.

Responders should consider many factors at hazardous materials incidents where flammable or combustible liquids are present or burning. Responders should consider **(Figure 13.27, p. 624)**:

Figure 13.25 First responders may shut off valves to residential natural gas lines.

- Where vapors may be present or traveling

- Where and what possible ignition sources are present

- Whether to extinguish the fire and how

If the products of combustion present fewer hazards than the leaking chemical, or extinguishment efforts will place firefighters in undue risk, the best course of action may be to protect exposures and let a fire burn until the fuel is consumed. Responders should consider withdrawal as potentially the safest (and best) tactical option due to the following:

Extinguish — To put out a fire completely.

Figure 13.26 Fire-control tactics are used to extinguish fires and prevent ignition of hazardous materials. *Courtesy of Rich Mahaney.*

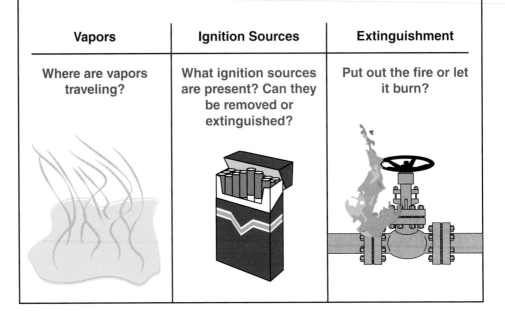

Fire Control Considerations at Flammable/Combustible Liquid Incidents

Vapors	Ignition Sources	Extinguishment
Where are vapors traveling?	What ignition sources are present? Can they be removed or extinguished?	Put out the fire or let it burn?

Figure 13.27 At incidents involving flammable and combustible liquids, responders should always consider where the vapors may be, what ignition sources may be present, and whether to extinguish the fire and how.

- A threat of catastrophic container failure
- Boiling liquid expanding vapor explosion (BLEVE) or other explosion
- The resources needed to control the incident are unavailable

NOTE: The *2016 Emergency Response Guidebook* (*ERG*) provides BLEVE safety precautions on pages 368-369.

WARNING!
Do not assume that relief valves are sufficient to safely relieve excess pressures. Tanks with relief valves may still rupture violently if exposed to heat or flames.

Flammable and Combustible Liquid Spill Control

Most hazmat incidents involve flammable and combustible liquids. Incidents range from spilled fuel at car accidents to major industrial accidents involving bulk containers. The spill control methods used will depend on the incident. Always consider the following:

- Firefighter protective clothing can absorb flammable and combustible liquids, which can later ignite if exposed to an ignition source (**Figure 13.28**). Avoid contact with products and/or contaminated pools, puddles, or streams.

- Vapors from flammable and combustible liquids are usually heavier than air.

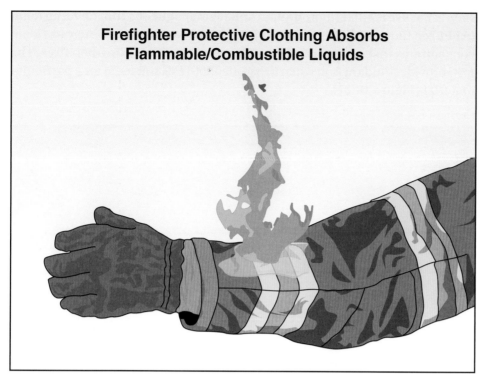

Firefighter Protective Clothing Absorbs Flammable/Combustible Liquids

Figure 13.28 Turnout gear can absorb flammable/combustible liquids, which can later ignite if exposed to an ignition source.

- Flammable and combustible liquids are typically lighter than water and, if so, will float on the surface of water.

- Flammable and combustible liquids are Class B materials; water is an ineffective extinguishing agent.

- Flammable and combustible liquid vapors may be toxic; for example, benzene is a carcinogen.

Controlling vapors is a priority at flammable and combustible liquid spills. Vapor suppression using fire fighting foam can be effective if the foam concentrate is compatible with the hazardous material **(Figure 13.29)**. Before using foam concentrates, responders must proportion (mix with water) and aerate (mix with air) all foam concentrates. Mechanical foam concentrates are divi00ded into two general categories based on the classification of fuels for which they are effective:

- Class A fuel foams (for ordinary combustibles)

- Class B fuel foams (for flammable and combustible liquids)

This section will focus on **Class B foam concentrates** that are used for vapor suppression. There are significant differences in Class B foams. Concentrates designed solely for hydrocarbon fires will not extinguish polar solvent (alcohol-type fuel/liquids that mix) fires regardless of the concentration at which they are used. Water-miscible materials, such as alcohols, esters, and ketones, destroy regular fire fighting foams and require an alcohol-resistant foam agent, therefore responders

Figure 13.29 Fire fighting foam must be compatible with the hazardous material.

Class B Foam Concentrate — Foam fire-suppression agent designed for use on ignited or unignited Class B flammable or combustible liquids. *Also known as* Class B Foam.

should not use regular fluoroprotein and regular **aqueous film forming foam (AFFF)** on those materials (**Figure 13.30**). However, responders may use foam concentrates that are intended for polar solvents on hydrocarbon fires. The *ERG* provides guidance on when to use alcohol-resistant foam for a particular material (**Figure 13.31**).

Figure 13.30 AFFF will not be effective on water-miscible materials such as alcohols, esters, and ketones.

Figure 13.31 The *ERG* provides guidance on the type of foam to use for a material.

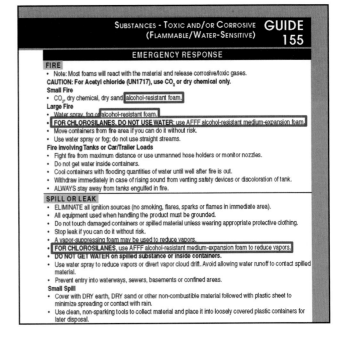

Producing High-Quality Foam

To produce high-quality foam, four elements are necessary: foam concentrate, water, air, and mechanical agitation **(Figure 13.32)**. These four elements must be blended in the correct ratios. Removing one or more elements will result in either no foam or poor-quality foam. Finished foam is produced in two stages.

- Water mixes with foam liquid concentrate to form a foam solution (proportioning stage).

- The foam solution passes through the piping or hose to a foam nozzle or sprinkler that aerates the foam solution to form finished foam (aeration stage).

Aeration produces adequate foam bubbles to form an effective foam blanket. Proper aeration also produces uniform-sized bubbles that form a long-lasting blanket. A good foam blanket maintains an effective cover over a fuel.

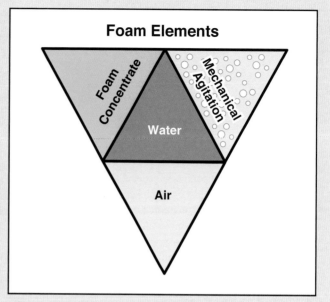

Figure 13.32 Four elements are necessary to produce high-quality foam: foam concentrate, water, air, and mechanical agitation.

WARNING!

When you are performing vapor-suppression tactics, stay upwind from the product and vapors because they may ignite.

What This Means To You

After identifying a flammable or combustible liquid, you can follow the foam recommendations in the orange guide in the *ERG* to determine what type of foam to use. The *ERG* recommends alcohol-resistant foam for polar/water-miscible liquids. It recommends regular foam for nonpolar/water-immiscible liquids.

Various manufacturers use different terms to identify foam concentrates that are effective on polar-solvents. Look for these sets of letters before the name on the foam concentrate container:

- ARC (alcohol-resistant concentrate)
- POL (polar-solvent liquid)
- ATC (alcohol-type concentrate) to use on polar-solvent liquids.

A variety of foams are used in emergency response situations. Refer to **Table 13.2, p. 628**, to review foam types and applications of foams that may be used during hazardous material responses.

NOTE: Refer to **Principles of Foam Fire Fighting** and **Aircraft Rescue and Fire Fighting** manuals for safety and best practices using foam during other types of incidents.

Table 13.2
Foam Concentrate Characteristics/Application Techniques

Type	Characteristics	Application Techniques	Primary Uses
Protein Foam (3% and 6%)	• Protein based • Low expansion • Good reignition (burnback) resistance • Excellent water retention • High heat resistance and stability • Performance can be affected by freezing and thawing • Can freeze protect with antifreeze • Not as mobile or fluid on fuel surface as other low-expansion foams	• Indirect foam stream; do not mix fuel with foam • Avoid agitating fuel during application; static spark ignition of volatile hydrocarbons can result from plunging and turbulence • Use alcohol-resistant type within seconds of proportioning • Not compatible with dry chemical extinguishing agents	• Class B fires involving hydrocarbons • Protecting flammable and combustible liquids where they are stored, transported, and processed
Fluoroprotein Foam (3% and 6%)	• Protein and synthetic based; derived from protein foam • Fuel shedding • Long-term vapor suppression • Good water retention • Excellent, long-lasting heat resistance • Performance not affected by freezing and thawing • Maintains low viscosity at low temperatures • Can freeze protect with antifreeze • Use either freshwater or saltwater • Nontoxic and biodegradable after dilution • Good mobility and fluidity on fuel surface • Premixable for short periods of time	• Direct plunge technique • Subsurface injection • Compatible with simultaneous application of dry chemical extinguishing agents • Deliver through air-aspirating equipment	• Hydrocarbon vapor suppression • Subsurface application to hydrocarbon fuel storage tanks • Extinguishing in-depth crude petroleum or other hydrocarbon fuel fires
Film Forming Fluoroprotein Foam (FFFP) (3% and 6%)	• Protein based; fortified with additional surfactants that reduce the burnback characteristics of other protein-based foams • Fuel shedding • Develops a fast-healing, continuous-floating film on hydrocarbon fuel surfaces • Excellent, long-lasting heat resistance • Good low-temperature viscosity • Fast fire knockdown • Affected by freezing and thawing • Use either freshwater or saltwater • Can store premixed • Can freeze protect with antifreeze • Use alcohol-resistant type on polar solvents at 6% solution and on hydrocarbon fuels at 3% solution • Nontoxic and biodegradable after dilution	• Cover entire fuel surface • May apply with dry chemical agents • May apply with spray nozzles • Subsurface injection • Can plunge into fuel during application	• Suppressing vapors in unignited spills of hazardous liquids • Extinguishing fires in hydrocarbon fuels

Continued

Type	Characteristics	Application Techniques	Primary Uses
Aqueous Film Forming Foam (AFFF) (1%, 3%, and 6%)	• Synthetic based • Good penetrating capabilities • Spreads vapor-sealing film over and floats on hydrocarbon fuels • Can use nonaerating nozzles • Performance may be adversely affected by freezing and storing • Has good low-temperature viscosity • Can freeze protect with antifreeze • Use either freshwater or saltwater • Can premix	• May apply directly onto fuel surface • May apply indirectly by bouncing it off a wall and allowing it to float onto fuel surface • Subsurface injection • May apply with dry chemical agents	• Controlling and extinguishing Class B fires • Handling land or sea crash rescues involving spills • Extinguishing most transportation-related fires • Wetting and penetrating Class A fuels • Securing unignited hydrocarbon spills
Alcohol-Resistant AFFF (3% and 6%)	• Polymer has been added to AFFF concentrate • Multipurpose: Use on both polar solvents and hydrocarbon fuels (use on polar solvents at 6% solution and on hydrocarbon fuels at 3% solution) • Forms a membrane on polar solvent fuels that prevents destruction of the foam blanket • Forms same aqueous film on hydrocarbon fuels as AFFF • Fast flame knockdown • Good burnback resistance on both fuels • Not easily premixed	• Apply directly but gently onto fuel surface • May apply indirectly by bouncing it off a wall and allowing it to float onto fuel surface • Subsurface injection	Fires or spills of both hydrocarbon and polar solvent fuels
High-Expansion Foam	• Synthetic detergent based • Special-purpose, low water content • High air-to-solution ratios: 200:1 to 1,000:1 • Performance not affected by freezing and thawing • Poor heat resistance • Prolonged contact with galvanized or raw steel may attack these surfaces	• Gentle application; do not mix foam with fuel • Cover entire fuel surface • Usually fills entire space in confined space incidents	• Extinguishing Class A and some Class B fires • Flooding confined spaces • Volumetrically displacing vapor, heat, and smoke • Reducing vaporization from liquefied natural gas spills • Extinguishing pesticide fires • Suppressing fuming acid vapors • Suppressing vapors in coal mines and other subterranean spaces and concealed spaces in basements • Extinguishing agent in fixed extinguishing systems • Not recommended for outdoor use

Drainage Time — Amount of time it takes foam to break down or dissolve. *Also known as* Drainage, Drainage Dropout Rate, or Drainage Rate.

Expansion Ratio — 1) Volume of a substance in liquid form compared to the volume of the same number of molecules of that substance in gaseous form. 2) Ratio of the finished foam volume to the volume of the original foam solution. *Also known as* Expansion.

Roll-On Application Method — Method of foam application in which the foam stream is directed at the ground at the front edge of the unignited or ignited liquid fuel spill; foam then spreads across the surface of the liquid. *Also known as* Bounce.

Bank-Down Application Method — Method of foam application that may be employed on an ignited or unignited Class B fuel spill. The foam stream is directed at a vertical surface or object that is next to or within the spill area; foam deflects off the surface or object and flows down onto the surface of the spill to form a foam blanket. *Also known as* Deflection.

Rain-Down Application Method — Foam application method that directs the stream into the air above the unignited or ignited spill or fire, allowing the foam to float gently down onto the surface of the fuel.

Air-Aspirating Foam Nozzle — Foam nozzle designed to provide the aeration required to make the highest quality foam possible; most effective appliance for the generation of low-expansion foam.

Emulsifiers

Emulsifiers are foam concentrates that are used with either Class A or Class B fires. Unlike finished foam that blankets the fuel, an emulsifier mixes with the fuel, breaking it into small droplets and encapsulating them. The resulting emulsion is rendered nonflammable.

Emulsifiers have the following limitations:

- Will not mix thoroughly with deeper fuels. Only use emulsifiers with fuels that are 1-inch (25 mm) deep or less.
- Renders the fuel unsalvageable once it mixes thoroughly with the fuel.
- Does not work effectively with water-soluble or water-miscible fuels because an emulsion cannot be formed between the concentrate and the fuel.
- Can be toxic to fish and aquatic environments, so personnel should consider the effects of run-off.

Foam concentrates vary in their finished-foam quality and, therefore, in their effectiveness. Manufacturers and suppliers will be able to provide information about freeze-protected versions of foams. Foam quality is measured in terms of its 25-percent-drainage time and its **expansion ratio**. **Drainage time** is the time required for one-fourth (25 percent or one-quarter) of the total liquid solution to drain from the foam. Expansion ratio is the volume of finished foam that results from a unit volume of foam solution. In general, the required application rate to control an unignited liquid spill is substantially less than that required to extinguish a spill fire.

Long drainage times result in long-lasting foam blankets. The greater the expansion ratio is, the thicker the foam blanket that can be developed (**Figure 13.33**). All Class B foam concentrates, except the special foams made for acid and alkaline spills, may be used for both fire fighting and vapor suppression. Air-aspirating nozzles produce a larger expansion ratio than water fog nozzles.

All foams have different optimal application methods. Common application methods include:

- The **roll-on application method** involves applying the foam onto the ground at the edge of the spill and rolling it gently onto the material (**Figure 13.34, p. 632**).

- If the spill surrounds some type of obstacle, responders can apply the foam onto the obstacle so the foam will roll down, known as the **bank-down application method** (**Figure 13.35, p. 632**).

- Personnel using the **rain-down method** spray the foam into the air over the target area in a fog pattern (**Figure 13.36, p. 632**). As the foam bubbles burst, the foam melds together to form a film over the fuel. The rain-down method is best used with AFFF.

- Fluoroprotein-type foams ONLY may be plunged directly into the spill.

For vapor suppression, first responders should use **air-aspirating nozzles** rather than water fog nozzles because aerated foam maintains the vapor suppressive blanket longer (**Figures 13.37a and b, p. 633**). For flammable liquid fires, non-aerated AFFF can be effective so water fog nozzles may be used.

Classifications and Expansion Rates of Foam

Classification		Rate

Classification   **Rate**

Low Expansion  Less than 20:1

Mid-Expansion 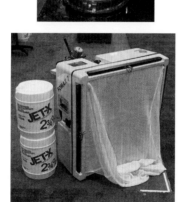 Between 20:1 and 200:1

High Expansion   Greater than 200:1

Figure 13.33 Higher expansion ratios mean thicker foam blankets that take longer to break down.

Adequate vapor suppression relies on selection of the proper foam concentrate. Because finished foam is composed principally of water, you should not use it to cover water-reactive materials. Some fuels destroy foam bubbles; therefore, select a foam concentrate that is compatible with the liquid. Other points to consider when using foam for vapor suppression include:

- Do not use water streams in conjunction with the application of foam. Water destroys and washes away foam blankets.

- Ensure that a material is below its boiling point; foam cannot seal vapors of boiling liquids.

- Do not rely on the film that precedes the foam blanket (such as with AFFF blankets); it is not a reliable vapor suppressant.

- Reapply aerated foam periodically until the foam completely covers the spill.

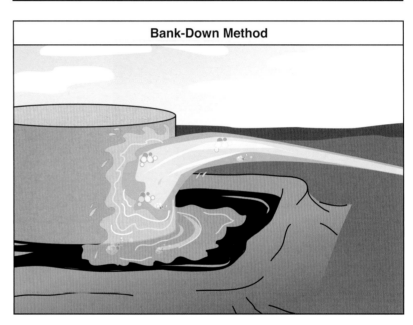

Figure 13.34 The roll-on method of applying foam to extinguish a flammable liquid fire.

Roll-On Method

Bank-Down Method

Figure 13.35 The bank-down method of applying foam to extinguish a flammable liquid fire.

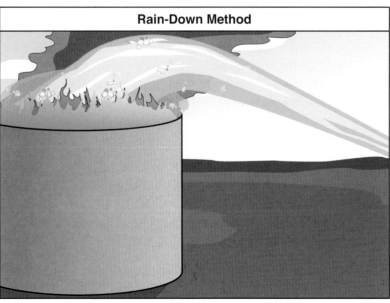

Rain-Down Method

Figure 13.36 The rain-down method of applying foam to extinguish a flammable liquid fire.

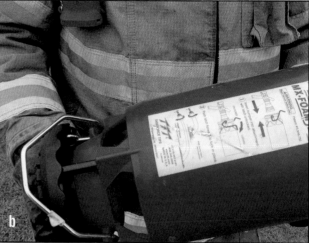

Figures 13.37a and b Air-aspirating nozzles aerate foam better than water fog nozzles, creating a larger expansion ratio.

CAUTION

Do not use the following agents in conjunction with foam: ABC (monoammonium phosphate) dry chemical and some sodium bicarbonate-based BC dry chemical agents will destroy a foam blanket. Other agents, such as potassium-based dry chemical, are compatible with foam.

CAUTION

Before use, check foam compatibility.

Flammable and Combustible Liquid Fire Control

Flammable and combustible liquid fires may be challenging because water is not an effective extinguishing agent. The fire may spread rapidly depending on where the burning product and/or vapors travel. Wind direction and topography may play an important role in fire spread.

NOTE: More information is available in the IFSTA manuals, **Essentials of Fire Fighting** and **Principles of Foam Fire Fighting.**

Proper use of hoselines and extinguishing agents is critical to safely controlling flammable liquid and gas fires. If personnel apply extinguishing agents incorrectly, they may push fuel to unwanted locations, endangering people or exposures. In addition, if you apply the wrong agent, you may place firefighters or personnel in hazardous locations without the possibility of controlling the fire.

WARNING!

If initial water streams are inadequate to cool both the pressurized tank and exposures, give priority to the involved container. Failure to maintain the integrity of the tank will risk the lives of everyone present.

Flammable Liquid Storage Tanks

Cooling a Vessel

Figure 13.38 At flammable liquid storage tank incidents, cooling the tank may be a priority, especially areas of flame impingement and the vapor space.

During incidents involving flammable liquid storage tanks, Operations level responders should maintain a defensible distance. If the Incident Commander decides to engage the fire, personnel should give attention to cooling all areas of the tank located in the vapor space and flame impingement areas — if this is an appropriate tactic for the incident. Responders should deploy water streams to keep the tank's surface wet **(Figure 13.38)**. Fire-control operations may need to assign spotters that can visually confirm that the entire surface of the tank is being cooled. If initial water streams will not cool both the tank and exposures, give priority to the involved container.

NOTE: The amount of water flow needed for each incident will depend on the current situation and the risks/hazards involved.

Using water streams inappropriately on flammable liquid fires can dramatically increase the size and intensity of a fire. Applying extinguishing agents to burning tanks could cause the tanks' contents to overflow and threaten adjacent containers. Responders should confine runoff from water streams applied to hazardous materials until it can be analyzed. Use damming, diking, and retention to confine runoff.

Foam, dry chemical, and water are common extinguishing agents for flammable liquids. Personnel should select foam when they can blanket the fuel, separating it from its air supply. As when using foam for vapor suppression, foam must be compatible with the fuel that is burning, and it must be applied at a sufficient rate to extinguish the fire **(Figure 13.39)**. Protein, fluoroprotein, and aqueous film forming foam (AFFF) have been the mainstay of flammable liquid fire fighting for years. High expansion foam is used for specific fire fighting hazards. Increased production and use of alternative fuels, such as ethanol, has increased the demand for alcohol-resistant foams. The U.S. Pipeline and Hazardous Materials Safety Administration (PHMSA) recommends using *ERG* guide 127 for these materials.

NOTE: Find more detailed information about required flow rate for foam application, necessary duration, and the logistics necessary to support foam fire fighting operations provided in the IFSTA manual, **Principles of Foam Fire Fighting**.

WARNING!

PPE soiled with flammable and combustible liquids may ignite when exposed to heat or an ignition source. Properly decontaminate and inspect the PPE.

Figure 13.39 As with vapor suppression, foam used for fire extinguishment must be compatible with the material burning.

Until the leak is controlled, do not extinguish flammable/combustible liquid fires burning around relief valves or piping. Unburned vapors are usually heavier than air and form pools or pockets of gas in low areas where they may ignite. If you are in a leak area, attempt to control all ignition sources.

An increase in the intensity of sound or fire issuing from a pressure relief device may indicate that the container is overheating and rupture is imminent. Emergency responders should not assume that pressure relief devices will safely relieve excess pressures under severe fire conditions. The rupture of both large and small liquid containers has caused firefighter fatalities.

If a closed pressure container, such as a liquefied gas (LPG) tank, is heated, the liquid inside begins expanding. When the liquid reaches its boiling point, it begins to return to its gaseous state. The change from a liquid to a gas in the confined space increases the internal pressure on the vessel. When too much pressure builds up, the container loses its structural integrity and ruptures, releasing massive amounts of pressure and the flammable contents of the container. The release and subsequent vaporization of these flammable liquids can result in a BLEVE. For a BLEVE to occur, the liquid or liquefied gas must be above its boiling point (at standard temperature and pressure) when the container failure occurs.

Tank failure may occur as a result of mechanical damage to the tank or from direct flame impingement on the vapor space in the tank. The most common cause of a BLEVE is when flames contact the tank shell above the liquid level and the tank shell itself has overheated. When attacking these fires, apply water to the upper portions of the tank, preferably from unattended master stream devices **(Figure 13.40)**.

Figure 13.40 Use unattended master streams when possible, in case of BLEVE.

Personnel most often apply foam to control flammable liquid fires. Responders need these Class B fire fighting techniques in gas utility facilities and accidents involving fuel transport trucks and rail tank cars. Personnel can use water in several forms (cooling agent, mechanical tool, and crew protection) to control Class B fires. The IFSTA manual, **Essentials of Fire Fighting**, provides more information on fighting flammable and combustible fires.

Flammable Gas Fires

In order to prevent a BLEVE, personnel should deploy water streams for maximum effective reach when containers or tanks of flammable gases are exposed to flame impingement. To best achieve this cooling, direct a stream (or streams) at areas on the tank where there is direct flame impingement as well as along the tank's top so that water runs down both sides **(Figure 13.41)**.

Figure 13.41 Cool flammable gas tanks with water running down both sides of the tank to cool the vapor space.

This water stream cools the tank's vapor space. Personnel should also cool the piping and steel supports under tanks to prevent their collapse.

When you use water streams to disperse gas being released under pressure, the mass and velocity of the water streams must exceed the mass and velocity of the escaping gas. Personnel must deliver water streams so that it disperses or disrupts escaping gas **(Figure 13.42)**. Any break in the water stream pattern will allow burning fuel to break through to hoseline crews.

Do not extinguish gas-fed fires burning around relief valves or piping unless turning off the supply can stop the leaking product. An increase in the intensity of sounds or fire issuing from a relief valve indicates pressure within the container is increasing and container failure may be imminent.

Figure 13.42 When using water streams to disperse pressurized gas, the water streams' mass and velocity must exceed the mass and velocity of the escaping gas.

Some gas-fed fires may involve natural gas. The distribution system for natural gas consists of a vast network of surface and sub-surface pipes. The pressure throughout the system will vary based on the material, elevation, use, and many other factors. Natural gas may also be compressed, stored, and shipped in cylinders marked as **compressed natural gas (CNG)**. Natural gas is also shipped and stored as a liquid **(LNG)** and is subject to BLEVE in this form.

NOTE: Refer to the IFSTA manual, **Essentials of Fire Fighting**, for information on controlling a pressurized flammable gas container fire.

Excavation equipment breaking through underground pipes is a common cause of natural gas (CNG) and liquefied petroleum gas (LPG) incidents. **When these breaks occur, contact the utility company immediately.** Even if the gas has not yet ignited, apparatus should approach from and stage on the upwind side, on the side from which the wind is blowing. Firefighters must wear full PPE and prepare for a potential explosion and accompanying fire.

Personnel should first evacuate the area immediately around the break and the area downwind, and they should also eliminate ignition sources. Service connections near the break may have been damaged; therefore, check surrounding buildings for the odor of gas inside. Firefighters should follow their departmental SOPs regarding any kind of interaction with a gas line leak. **If gas is burning, the flame should not be extinguished**. If necessary, use hose streams to protect exposures. Responders should contact the utility company and make an attempt to have the pressurized gas supply stopped. Depending on the AHJ, a hazardous materials response team may or may not be requested.

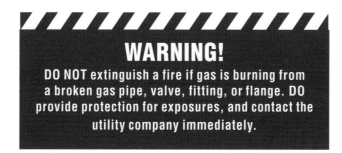

WARNING!
DO NOT extinguish a fire if gas is burning from a broken gas pipe, valve, fitting, or flange. DO provide protection for exposures, and contact the utility company immediately.

Chapter Review

Answer the following questions to review the information provided in this chapter.

1. What is the difference between absorption and adsorption?

2. Under what circumstances might blanketing or covering be used?

3. What types of materials are used in damning, diking, diversion, and retention?

4. What are three common methods of vapor suppression?

5. Explain the difference between negative- and positive-pressure ventilation.

6. What problems are posed by diversion?

7. When is dilution at hazmat incidents most likely to be used?

8. When neutralizing a material, what pH is the goal?

9. Under what circumstances may Operations level responders take offensive actions such as leak control?

10. Under safe circumstances, Operations level responders may operate emergency remote shutoff devices on what types of transportation containers?

11. When should responders operate fixed facility or pipeline remote shutoff valves?

12. When should withdrawal be considered as a tactical option during a hazmat fire?

13. What tactics are used for flammable and combustible liquid spill control?

14. What tactics are used for flammable and combustible liquid fire control?

15. What tactics are used to control flammable gas fires?

Step 1: Ensure proper product control technique is chosen.

Step 2: Ensure that all responders involved in the control function are wearing appropriate PPE for performing absorption/adsorption operations and that appropriate hand tools have been selected.

Step 3: Select a location to efficiently and safely perform the absorption /adsorption operation.

Step 4: Select the most appropriate sorbent/adsorbent.

Step 5: Deploy the sorbent/adsorbent in a manner that most efficiently controls the spill.

Step 6: Upon mitigation of the incident, place any contaminated material, such as clothing, in an approved container for transportation to a disposal location.

Step 7: Seal and label the container and document appropriate information for department records.

Step 8: Decontaminate tools.

Step 9: Advance to decontamination line for decontamination.

Step 10: Complete required reports and supporting documentation.

NOTE: Instructor directions: will need to determine if construction is overflow, underflow, or containment dam.

Step 1: Ensure proper product control technique is chosen.

Step 2: Ensure that all responders involved in the control function are wearing appropriate PPE for performing damming operations and that appropriate hand tools have been selected.

Step 3: Select a location to efficiently and safely perform damming operation.

Step 4: Construct the dam in a manner and location that most effectively controls the spill.

Step 5: Upon mitigation of the incident, place any contaminated material, such as clothing, in an approved container for transportation to a disposal location.

Step 6: Seal and label the container and document appropriate information for department records.

Step 7: Decontaminate tools.

Step 8: Advance to decontamination line for decontamination.

Step 9: Complete required reports and supporting documentation.

Step 1: Ensure proper product control technique is chosen.

Step 2: Ensure that all responders involved in the control function are wearing appropriate PPE for performing diking operations and that appropriate hand tools have been selected.

Step 3: Select a location to efficiently and safely perform the diking operation.

Step 4: Construct the dike in a manner and location that most efficiently controls and directs the spill to a desired location.

Step 5: Upon mitigation of the incident, place any contaminated material, such as clothing, in an approved container for transportation to a disposal location.

Step 6: Seal and label the container and document appropriate information for department records.

Step 7: Decontaminate tools.

Step 8: Advance to decontamination line for decontamination.

Step 9: Complete required reports and supporting documentation.

Step 1: Ensure proper product control technique is chosen.

Step 3: Select a location to efficiently and safely perform the diversion operation.

Step 2: Ensure that all responders involved in the control function are wearing appropriate PPE for performing diversion operations and that appropriate hand tools have been selected.

Step 4: Construct the diversion in a manner and location that most effectively controls and directs the spill to a desired location.

Step 5: Working as a team, use hand tools to break the soil, remove the soil, pile the soil, and pack the soil tightly.

Step 6: Upon mitigation of the incident, place any contaminated material, such as clothing, in an approved container for transportation to a disposal location.

Step 7: Seal and label the container, and document appropriate information for department records.

Step 8: Decontaminate tools.

Step 9: Advance to decontamination line for decontamination.

Step 10: Complete required reports and supporting documentation.

Step 1: Ensure proper product control technique is chosen.

Step 2: Ensure that all responders involved in the control function are wearing appropriate PPE for performing retention operations and that appropriate hand tools have been selected.

Step 3: Select a location to efficiently and safely perform the retention operation.

Step 4: Evaluate the rate of flow of the leak to determine the required capacity of the retention vessel.

Step 8: Decontaminate tools.

Step 9: Advance to decontamination line for decontamination.

Step 10: Complete required reports and supporting documentation.

Step 5: Working as a team, retain the hazardous liquid so that it can no longer flow.

Step 6: Upon mitigation of the incident, place any contaminated material, such as clothing, in an approved container for transportation to a disposal location.

Step 7: Seal and label the container, and document appropriate information for department records.

Step 1: Ensure proper product control technique is chosen.

Step 2: Ensure that all responders involved in the control function are wearing appropriate PPE for performing vapor suppression operations.

Step 3: Select a location to efficiently and safely perform the vapor suppression operation.

Step 4: Evaluate the quantity and surface area of the hazardous material that has leaked.

Step 5: Determine the appropriate type of foam for the type of hazardous material present.

Step 6: Working as a team, deploy the foam eductor and foam, and advance the hoseline and foam nozzle to a position from which to apply the foam.

Step 7: Flow hoseline until finished foam is produced at the nozzle.

Step 8: Apply finished foam in an even layer covering the entire hazardous material spill area.

Step 9: Upon mitigation of the incident, place any contaminated material, such as clothing, in an approved container for transportation to a disposal location.

Step 10: Seal and label the container, and document appropriate information for department records.

Step 11: Decontaminate tools.

Step 12: Advance to decontamination line for decontamination.

Step 13: Complete required reports and supporting documentation.

Step 1: Ensure proper product control technique is chosen.

Step 2: Ensure that all responders involved in the control function are wearing appropriate PPE for vapor dispersion operations.

Step 3: Select a location to efficiently and safely perform the vapor dispersion operation.

Step 4: Working as a team, advance the hoseline to a position to apply agent through vapor cloud to disperse vapors.

Step 5 Constantly monitor the leak concentration, wind direction, exposed personnel, environmental impact, and water stream effectiveness.

Step 6: Upon mitigation of the incident, place any contaminated material, such as clothing, in an approved container for transportation to a disposal location.

Step 7: Seal and label the container, and document appropriate information for department records.

Step 8: Decontaminate tools.

Step 9: Advance to decontamination line for decontamination.

Step 10: Complete required reports and supporting documentation.

Step 1: Ensure proper product control technique is chosen.

Step 2: Ensure that all responders involved in the control function are wearing appropriate PPE for performing dilution operations.

Step 3: Select a location to efficiently and safely perform dilution operation.

Step 4: Evaluate the rate of flow of the leak to determine the required capacity of the retention area and the quantity of water required to dilute the material.

Step 5: Working as a team, monitor and assess the leak, and advance hoselines and tools to retention area.

Step 11: Advance to decontamination line for decontamination.

Step 12: Complete required reports and supporting documentation.

Step 6: Flow water to dilute spilled material.

Step 7: Monitor any diking or dams to ensure integrity of retention area.

Step 8: Upon mitigation of the incident, place any contaminated material, such as clothing, in an approved container for transportation to a disposal location.

Step 9: Seal and label the container, and document appropriate information for department records.

Step 10: Decontaminate tools.

13-9
Perform remote valve shutoff or activate emergency shutoff device.

SKILL SHEETS

Step 1: Ensure proper product control technique is chosen.

Step 2: Ensure that all responders involved in the control function are wearing appropriate PPE for performing remote valve shutoff operations.

Step 3: Identify and locate the emergency remote control valve and/or emergency shutoff device.

Step 6: Notify the Incident Commander of the completed objective.

Step 7: Complete required reports and supporting documentation.

Cargo Tank

Intermodal

Step 4: Operate the remote control valve and/or emergency shutoff device properly.

Step 5: If necessary, decontaminate tools and advance to decontamination line for decontamination.

Implementing the Response: Mission-Specific Evidence Preservation and Public Safety Sampling

Chapter Contents

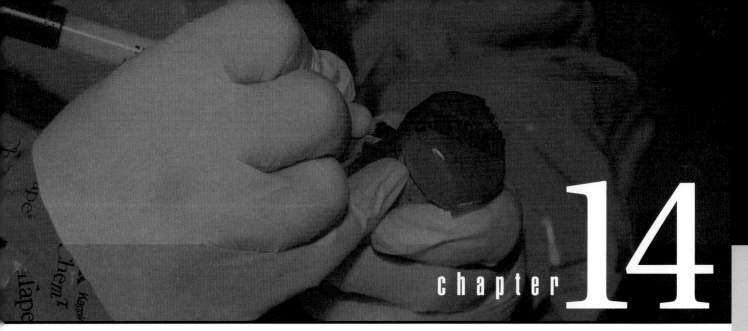

chapter 14

Key Terms

NFPA Job Performance Requirements

This chapter provides information that addresses the following job performance requirements of NFPA 1072, *Standard for Hazardous Materials/Weapons of Mass Destruction Emergency Response Personnel Professional Qualifications (2017)*.

6.5.1

Implementing the Response: Mission-Specific Evidence Preservation and Public Safety Sampling

Learning Objectives

After reading this chapter, students will be able to:

1. Identify hazards at crime scenes involving hazardous materials/WMD. (6.5.1)

2. Recognize agencies and personnel with investigative authority. (6.5.1)

3. List response phases at criminal hazardous materials/WMD incidents. (6.5.1)

4. Describe processes of securing the incident scene. (6.5.1)

5. Explain procedures for identifying, protecting, and preserving potential evidence. (6.5.1)

6. Recognize types of evidence preservation and public safety sampling documentation at hazmat/WMD incidents. (6.5.1)

7. Identify processes for collecting public safety samples.

8. Demonstrate evidence preservation and sampling. (6.5.1; Skill Sheet 14-1)

Chapter 14

Implementing the Response: Mission-Specific Evidence Preservation and Public Safety Sampling

This chapter will explain the following topics:

- Hazards at crime scenes involving hazardous materials/WMD
- Investigative authority
- Response phases at criminal hazardous materials/WMD Incidents
- Securing the scene
- Identifying, protecting, and preserving potential evidence
- Documentation
- Public safety sampling

Hazards at Crime Scenes Involving Hazardous Materials/WMD

When responders observe indicators of criminal activity, they must work with law enforcement agencies to preserve and protect potential evidence. They may also be called to assist law enforcement in gathering evidence for the case against the perpetrator of the crime.

Operations Level responders assigned to crime scene/ WMD incidents must be trained in accordance with the requirements of their jurisdiction **(Figure 14.1)**. They should also operate under the guidance of a hazardous materials technician, an Allied Professional including law enforcement personnel or others with similar authority, an emergency response plan, or standard operating procedures.

Criminal hazardous materials/WMD incidents, environmental crimes, and illicit labs differ greatly in their characterization, location, and associated hazards. Given these differences, responders may need to alter their response actions at each one. However, all operations must be performed within the framework of the Incident Management System (IMS), in accordance with the principles of risk-based response described throughout this manual.

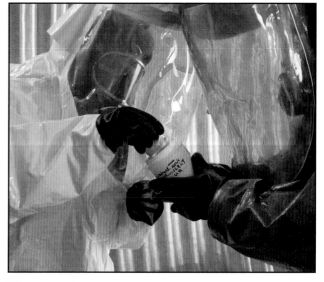

Figure 14.1 Responders who perform evidence sampling tasks must be trained to do so in accordance with the requirements of their jurisdiction.

Some crime scenes may include hazards such as armed individuals, booby traps, or explosives that hazardous device technicians or other specialists must render safe before the hazardous materials team proceeds **(Figure 14.2)**. The risks present at the incident (as determined by assessment of intelligence, warning signs, and detection clues) and the mission being performed will determine the PPE **(Figure 14.3)**.

Decontamination operations must be performed in accordance with SOPs/SOGs and with care being given to preserving evidence. Before response to a potential criminal hazardous materials/WMD incident, responders should plan for responses that follow jurisdictional laws and account for available personnel and equipment. By carrying out duties as planned, responders can achieve the goals of preserving life, stabilizing the incident, and obtaining **forensic evidence.**

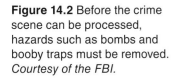

Forensic Evidence — Evidence obtained by scientific methods that is usable in court, for example fingerprints, blood testing, or ballistics.

Figure 14.2 Before the crime scene can be processed, hazards such as bombs and booby traps must be removed. *Courtesy of the FBI.*

Figure 14.3 The risks present and the mission being performed will determine the PPE worn at crime scenes.

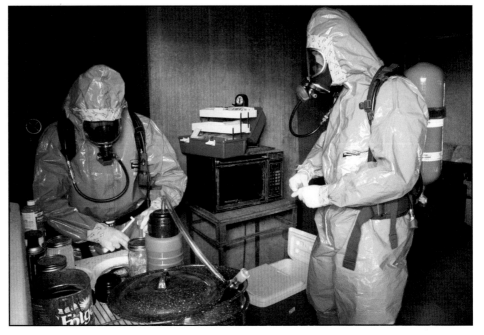

Illicit Laboratories

Subjects who operate illicit drug labs have a vested interest in maintaining their livelihoods as creators and distributors of illegal drugs or chemical, biological, radiological, and explosive materials. As a result, armed guards, attack dogs, and booby traps may be used to protect the labs. All of these must be neutralized before evidence preservation and collection activities begin.

Labs that create substances used in criminal activities or terrorism may have dangerous outer security measures, and all types of labs potentially contain chemical, biological, radiological, and explosive agents that could be accidentally released during a raid **(Figure 14.4)**. It is important that law enforcement tactical teams and hazardous device technicians secure human threats and render explosive hazards safe before beginning other tasks. Chapter 13 provides additional information about operations at illicit labs.

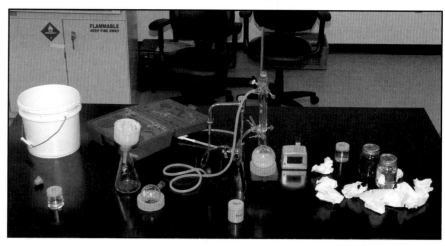

Figure 14.4 Illicit labs may contain a variety of extremely hazardous materials.

Illicit Labs

A deputy sheriff noticed a vehicle at a vacant house in an unincorporated area of his county and stopped to investigate. He noted a strange odor coming from the house and knocked on the door. As he knocked, a man ran from the rear of the house. The deputy sheriff apprehended the man and questioned him. During questioning, the man admitted that he had a meth lab in the house.

The deputy notified the dispatcher, and the county volunteer fire department arrived at the scene. The house was in an isolated area, so the fire department did not enter the house. Instead, they waited for the county hazardous materials team to arrive.

Once the county hazardous materials team arrived, they donned appropriate personal protective equipment (PPE) and entered the house, observing the lab equipment and chemicals in the lab. They made notes, photographed, and sketched the scene. The county law enforcement forensic laboratory technician arrived and assisted the hazmat team in collecting, packaging, and labeling the evidence.

Lessons Learned:

After all the evidence was collected, the proper case report was completed and was submitted to the District Attorney. Later, the District Attorney presented his case to the Grand Jury. The Grand Jury found there was sufficient facts and evidence to indict the suspect, who was tried in a criminal court. The presiding criminal court judge questioned all the evidence, collection methods, and sample analyses and ultimately accepted them for presentation to a jury.

Release or Attack with a WMD Agent

The following items may be used in hazardous materials/WMD attacks:

- Explosives
- Biological toxins
- Toxic industrial chemicals
- Biological pathogens
- Radioactive sources
- Chemical warfare agents
- Improvised nuclear devices

The release/attack site should be cleared or rendered safe of WMD hazards before beginning evidence collection tasks. See Chapter 8 for more information about operations at these incidents.

Environmental Crimes

Environmental crimes involve the illegal use and disposal of hazardous substances and waste which pollute the air, water, or soil and can cause serious injury, chronic illness, or even death. Environmental crime scenes may include additional hazards in the form of armed owners/operators or explosive devices. Law enforcement tactical teams and hazardous device teams must remove or neutralize these hazards before beginning tasks such as emergency operations or mitigation efforts.

Suspicious Letters and Packages

Incidents involving suspicious letters and packages may involve explosives, biological materials, hazardous chemicals, or even radiological materials. Most frequently, suspicious letters and packages involve explosives (mail bombs) and powders **(Figure 14.5)**. A powder incident may be a hoax, but both anthrax and ricin have been sent as a powder through the mail.

If a package is thought to be suspicious, responders should contact law enforcement personnel to investigate further. Follow your AHJ's SOPs for emergency response to these types of incidents. The International Association of Fire Chiefs' (IAFC) *Model Procedures for Responding to a Suspicious Package* provides guidance for responding to these types of incidents.

Investigative Authority

Responders assigned to perform evidence preservation and public safety sampling tasks must be able to identify the investigative authority at hazardous materials/WMD crimes. Typically, hazardous materials crime scenes will initially fall under the jurisdiction of the local law enforcement agency. Depending on the type of crime, location, materials involved, and other factors, this authority may shift to other agencies. Examples:

- Under numerous Federal statutes, if a crime is determined to be a terrorist attack, the investigative authority will shift to the Federal Bureau of Investigation (FBI).

- Depending on the drug type and quantities involved, the Drug Enforcement Administration (DEA) has jurisdiction to investigate crimes involving illegal drugs **(Figure 14.6)**.

- Environmental crimes fall under the jurisdiction of the EPA, depending on the types and quantity of materials involved.

- The Postal Inspection Service will investigate incidents involving suspicious letters or packages.

Multiple agencies may also work together in a Task Force setting when each agency has a vested interest in the prosecution. **Table 14.1, p. 656**, provides a list of U.S. Federal Agencies that may have investigative authority at crimes involving hazardous materials/WMDs.

Figure 14.5 Most letter or package incidents involve white powders.

Figure 14.6 In the U.S., the DEA may have investigative authority at incidents involving illegal drugs, depending on the quantity and type involved. *Courtesy of the DEA.*

CAUTION

All evidence collection and sampling must be conducted in coordination with law enforcement and other members of the criminal justice system. Samples or items that are collected with the intent to become evidence without the direct guidance or supervision from law enforcement may become inadmissible during criminal proceedings.

Table 14.1
U.S. Federal Agencies with Investigative Authority

FBI	The FBI has investigative jurisdiction over violations of more than 200 categories of federal crimes. Top priority has been assigned to five areas: (1) Counterterrorism (2) Drugs/organized crime (3) Foreign counterintelligence (4) Violent crime (5) White-collar crime In addition, the FBI is authorized to investigate matters where prosecution is not contemplated. For example, under the authority of several executive orders, the FBI is responsible for conducting background security checks concerning nominees to sensitive government positions. The FBI also has been directed or authorized by presidential statements or directives to obtain information about activities suspected of jeopardizing the security of the nation.
EPA	EPA agents investigate the most significant and egregious violators of environmental laws which pose significant threats to human health and the environment.
DEA	DEA investigations occur within the context of DEA's responsibilities to enforce the provisions of federal laws and regulations concerning controlled substances, chemical diversion, and drug-trafficking.
ATF	ATF investigations occur within the context of its responsibilities to enforce laws regarding, for example, distilled spirits, beer, wine, tobacco, firearms, explosives, and arson. More specifically, ATF responsibilities include the enforcement of the Gun Control Act of 1968, Title XI of the Organized Crime Control Act of 1970, the National Firearms Act, the Arms Export Control Act, Chapters 51 and 52 of the Internal Revenue Code of 1986, and the Federal Alcohol Administration Act.
Postal Inspection Service	The Postal Inspection Service is responsible for investigating violations of about 200 federal statutes that deal with the integrity and security of mail; the safeguarding of postal employees, property, and the work environment; and the protection of Postal Service revenue and assets.
NPS	The National Park Service (NPS) is responsible for investigating offenses against the United States committed within the national park system in the absence of an investigation by any other federal law enforcement agency. NPS also has authority on and within roads, parks, parkways, and other federal reservations within the District of Columbia. The types of investigations in which NPS is involved include Assimilated Crimes Act investigations, drug enforcement, environmental crimes, crimes against persons, and resource-related crimes, such as plant and wildlife poaching, archaeological site looting, vandalism of historical sites, and simple theft of resources.

Source: U.S. Government Accounting Office

In some cases, more than one agency may be involved in an investigation. For example, at an incident involving a suspicious letter or package, all of the following agencies may be involved in the investigation:

- Postal Inspection Service
- Federal Bureau of Investigation
- Bureau of Alcohol, Tobacco, Firearms, and Explosives
- National Guard Civil support teams
- Public health departments
- Local law enforcement agencies
- Hazardous Device Technicians/military explosive ordnance disposal (EOD) personnel

Local emergency response plans should detail appropriate procedures for requesting assistance from other agencies such as the bomb squad. Local investigative guidelines are complex and dynamic, and local responders must be familiar with their jurisdictional procedures. Consult law enforcement authorities (such as the District or City Attorney) to determine if Operations Level personnel (for example, the local hazmat team) are within their legal authority and have the appropriate training to perform evidence preservation and public safety sampling at the scene.

NOTE: The qualifications, experience, and knowledge of individuals conducting public safety sampling and preservation will be scrutinized closely in criminal cases, especially those involving Federal court.

Response Phases at Criminal Hazardous Materials/WMD Incidents

There are four response phases at criminal hazardous materials/WMD incidents. Understanding these phases can provide a context for first responders to understand how a crime scene involving hazardous materials/WMD is managed and the process under which evidence is collected and sampled in relation to other public safety operations. Establishing Unified Command early at suspected crimes is critical for ensuring that the incident is managed in a fashion that will maximize the goals of all agencies involved.

The four response phases are as follows:

- **Tactical Phase** — Law enforcement removes hostile threats. For example, subjects are arrested, booby traps are neutralized, and explosive devices are removed.

- **Operational Phase** — Life safety objectives are met and the scene is stabilized and secured. While the first priority is life safety, responders should take measures to preserve the scene.

  — Conduct limited public safety sampling to identify hazards.

  — Help make evacuation decisions.

  — Determine proper decontamination methods.

  — Assist in the medical treatment of exposed victims.

  — Follow locally accepted sampling protocols during collection, since the public safety sample may be identified as evidence when the Crime Scene Phase begins **(Figure 14.7)**.

- **Crime Scene Phase** — During this phase, evidence is recovered and packaged for transport, and the crime scene is processed forensically **(Figure 14.8, p. 658)**. These actions must be conducted in accordance with

Figure 14.7 Limited public safety sampling may be conducted during the operational phase.

Figure 14.8 During the crime scene phase, the crime scene is processed forensically.

appropriate law enforcement protocols since materials gathered may be used to prosecute a subject. Law enforcement is responsible for obtaining search warrants and/or consent when necessary. During this phase the critical tasks are:

— Maintaining personal and public safety

— Protecting evidence samples prior to collection

— Preserving evidence samples during and after collection

— Documenting the evidence accurately

— Maintaining the chain of custody during movement to a forensic laboratory

● **Remediation Phase** — In this phase, operations to mitigate any remaining hazards are conducted in order to bring the scene back to a safe condition. Contractors or appropriate Federal authorities typically perform remediation.

Securing the Scene

Upon initial arrival, the safety of personnel is of primary concern. The hazards at the scene and the tactical requirements will determine the necessary PPE for the incident. As soon as the incident is recognized as a potential crime scene, law enforcement must be notified. In the U.S., if you suspect the incident involves WMD or terrorism, contact the FBI WMD Coordinator (usually done through local law enforcement).

Give special care and consideration to the potential existence of forensic evidence present at the scene. If possible, recognize, identify and protect potential **transient evidence**. If applicable warrants are required, law enforcement will obtain and execute them. Take notes to document initial observations.

Transient Evidence — Material that will lose its evidentiary value if it is unpreserved or unprotected; for example, blood in the rain.

Law enforcement responders will establish a security perimeter with an access control point. Once perimeters have been established, law enforcement will document the entry and exit of all people entering and leaving the scene. This documentation is in conjunction with the use of a personnel accountability system, and provides internal checks to ensure that full identification is in place for all personnel in the crime scene.

To aid in securing and protecting the scene, responders should do the following:

- Prevent individuals from altering/destroying physical evidence by restricting their movement, location, and activity while ensuring and maintaining safety at the scene.

- Identify all individuals at the scene who may be suspects or **witnesses** and secure and separate them. Potential witnesses include first responders and will be noted by law enforcement.

- Determine if bystanders are witnesses. If not, remove them from the scene.

- Exclude unauthorized and nonessential personnel (for example, the media) from the scene.

NOTE: Do not take souvenirs, and remember that any photos taken at the scene may become evidence.

Witness — Person called upon to provide factual testimony before a judge or jury.

Identifying, Protecting, and Preserving Potential Evidence

The true value of evidence can be realized only if proper care has been used in observing the simple scientific and legal rules (SOPs). These rules should govern the journey of physical evidence from its discovery to its final appearance as a court exhibit, guided by law enforcement. The steps in this procedure may be described as follows:

1. Security of the scene and the evidence
2. Discovery of the evidence
3. Documentation of the evidence (via photography or sketches)
4. Collection of the object(s) or sample(s)
5. Packaging of the evidence (to include properly sealing the packaging)
6. Submission to laboratories (for forensics and identification)
7. Laboratory examination
8. Custody of the evidence pending trial
9. Transportation to court
10. Exhibition in court

The following sections address topics important to maintaining the viability of evidence:

- Chain of custody
- Identifying potential evidence
- Protecting evidence
- Preserving evidence

Chain of Custody

All evidence must be handled and moved to a law enforcement evidence custodian for documentation into the evidence chain in accordance with the law enforcement AHJ's, **chain of custody** procedures. Chain of custody is the practice of tracking an item of evidence from the time it is found until it is ultimately disposed of or returned. The chain of custody is a written history that must include the name of each person who maintains visual or physical control over the item throughout the process **(Figure 14.9)**. Each person in the chain of custody is a candidate for subpoena to court.

Although the ultimate responsibility for the chain of custody is placed upon the law enforcement AHJ, as a responder taking public health samples, you must be prepared to provide the information required to establish this chain. To ensure that the chain of custody is successfully maintained,

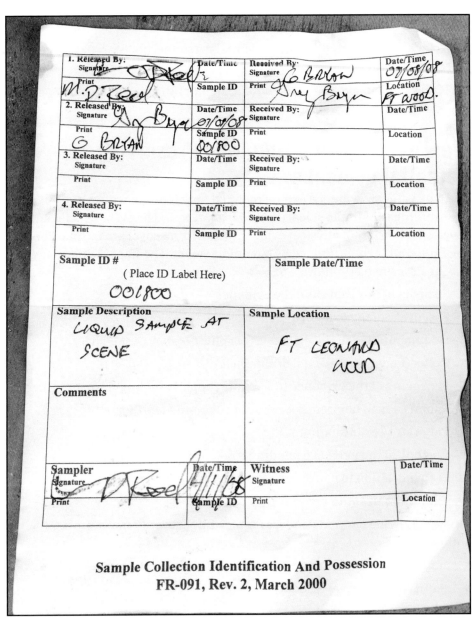

Figure 14.9 Document chain of custody for all evidence samples.

responders should follow SOPs and be prepared to provide the following information:

- Name, identification number, and rank
- Agency
- Date
- Time that evidence changed possession
- Specific location from which the sample was taken
- Incident number or **case identifiers** (if appropriate)
- When and to whom the sample was released (some chain of custody forms ask who evidence was received from as well)

NOTE: If evidence is locked in a vehicle, note this on the evidence form including information as to how long it was there and how secure the vehicle was during that time.

Case Identifier — Alphabetic and/or numeric characters used to identify a case.

Identifying Potential Evidence

Upon arrival, consider everything at the incident scene as potential evidence. Therefore, first responders should try to minimize disturbing the scene as much as possible. Law enforcement personnel will identify relevant items of evidence through investigation, gathering of data, and developing and testing a hypothesis.

Broadly speaking, evidence is any data that may be used to prove or disprove a certain hypothesis. Evidence is usually used to support testimony but can sometimes speak for itself. Three primary classifications of evidence are **(Figure 14.10)**:

- **Direct** — Facts found through the five physical senses to which a person can attest without further support.
- **Circumstantial** — Facts which support presumptions or inferences formed from direct or physical evidence.
- **Physical** — Material objects evaluated during an investigation which tend to prove or disprove facts.

Evidence classifications may overlap. In general, types of evidence can be categorized as something someone can observe (direct evidence), something someone can hold (physical evidence), or something someone can conclude (circumstantial evidence). However, circumstantial evidence can consist of physical evidence.

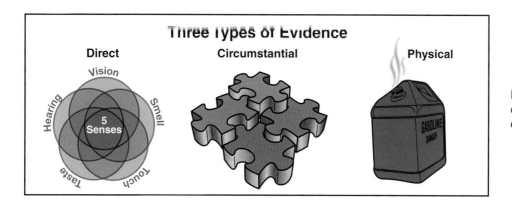

Figure 14.10 The three evidence classifications are direct, circumstantial, and physical.

Protecting Evidence

Securing the scene is an important step in protecting evidence. Keep unnecessary people out of the area. Disturb the scene as little as possible.

Keep suspected evidence in its original position until it is collected, if possible. To protect suspected pieces of evidence from disturbance, use barricades, traffic cones, or other means to alert personnel that potential evidence is present.

Best Practices for Safeguarding Evidence

Do NOT perform any of the following activities at a potential crime scene:

● Smoke or chew tobacco

● Use the telephone or bathroom

● Eat or drink

● Move any items (unless necessary for the safety and well-being of persons at the scene)

● Adjust the thermostat

● Change the power status (whether on or off) of equipment including equipment, lights, appliances, and electrical breakers

● Open doors or windows

● Touch anything unnecessarily

● Reposition moved items

● Deposit litter or spit within the established boundaries of the scene

Figure 14.11 First responders may need to cover or secure physical evidence.

Preserving Evidence

It may be necessary for first responders to protect and preserve evidence when and where it is found until an investigator arrives. While responders should not gather or handle evidence unless it is absolutely necessary, some evidence may be damaged or destroyed if not preserved. In these cases, first responders may need to move, pick up, cover, or otherwise secure physical evidence **(Figure 14.11)**. Documentation of all actions taken is extremely important to maintain evidentiary integrity for an investigation. This includes utilization of proper collection techniques and preservation of the chain of custody. For example, water from hoselines or environmental conditions may wash away footprints, so first responders might need to photograph, cover, and document them before they are destroyed or damaged.

If a first responder handles or collects evidence, he or she then becomes a link in the custody chain. The first responder should accurately document all actions as soon as possible. It may be necessary for this individual to subsequently appear in court.

Documentation

Documentation can be done in many different ways. Local procedures for documentation vary from jurisdiction to jurisdiction. Make sure the response plan follows the procedures or requirements. All scene documentation is **discoverable** in court. Therefore, it needs to be thoroughly and accurately completed. Examples of documentation include:

Discovery — Means by which the plaintiff (one party) obtains information from the opposing party (defendant) to prove its allegation.

- Video
- Photographs **(Figure 14.12)**
- Stills
- Sample logs
- Incident action plan
- Site safety plan

Figure 14.12 Photographs are a good way to document the incident scene as long as local procedures and requirements are followed.

All documentation of actions and observations at the hazardous materials/WMD scene will be passed to crime scene investigators. Individual responders may be called to testify about these observations. Emergency responders must coordinate with the law enforcement agency having jurisdiction in order to avoid problems with inadmissible evidence. Work with the local district attorney to determine what procedures must be followed. Obtain this guidance before the process is started and incorporate it into SOP/Gs.

The law enforcement investigator(s) in charge of the crime scene will compile reports and other documentation pertaining to the crime scene investigation into a case file. This file shall be a record of the actions taken and evidence collected at the scene. This documentation will allow for independent review of the work conducted.

The senior law enforcement investigator will obtain several items for the crime scene case file, such as:

- Initial responding officer(s') documentation
- Fire and hazmat personnel reports
- Emergency medical personnel (including hospital staff) documents

- Entry/exit documentation
- Photographs/videos
- Crime scene sketches/diagrams
- Evidence documentation
- Record of consent form or search warrant
- Site safety plans
- Incident action plans
- Administrative logs
- Responder observations
- Reports such as forensic/technical reports (when they become available)

Public Safety Sampling

Operations Level responders may be required to collect **public safety samples** to determine contaminants or suspected contaminants in support of medical treatment, to determine how to best mitigate the situation, or to determine the type of decontamination required.

When responding to incidents occurring from criminal intent, samples may become evidence **(Figure 14.13)**. This evidence is crucial to law enforcement personnel for criminal investigation and prosecution. In fact, at criminal hazardous materials/Weapons of Mass Destruction (WMD) incidents, evidence preservation and **public safety sampling** is the top priority after life preservation and hazard mitigation. Following the proper procedures will enable responders to collect public safety samples, preserve evidence, and under the direction of law enforcement collect evidence that can aid in the apprehension of suspects and prevent recurrences. Anything at the scene may be considered evidence and may not be recognized for its forensic value until trained evidence technicians arrive and evaluate the situation.

Public safety sampling is performed at suspected criminal hazmat incidents. These incidents often have unidentified, potentially volatile, unusual hazards. For this reason, hazmat response operations differ somewhat from other incidents, for example, the involvement of law enforcement and the need for extreme caution due to the potential of secondary devices.

Public safety sampling is performed to classify potentially hazardous materials at suspicious incidents in order to determine the hazards at the scene and any threats to public safety **(Figure 14.14)**. Public safety samples are used for risk assessments, health, decontamination, and similar actions.

Public safety sampling is used to answer the following questions:

- "Is the building safe to occupy?"
- "When can the employees go back into the _____?" (airport, mayor's office, mall, subway, building, office)
- "Do the employees need to be decontaminated?"
- "Do we need to shut down the _____?" (airport, mayor's office, mall, subway, building, office)
- "Is this white powder dangerous?"
- "What is the size and scope of the hazard?"

Figure 14.13 At criminal incidents, public safety samples may become evidence.

Figure 14.14 Public safety sampling is performed to classify potentially hazardous materials at suspicious incidents.

The steps used during public safety sampling are explained in the following sections:

- Characterizing the scene
- Field screening samples
- Collecting samples
- Evidence laboratories

Skill Sheet 14-1 provides a broad set of steps for evidence preservation and sampling. (**NOTE:** Evidence and samples that are not protected and/or collected in accordance with AHJ procedures and requirements may be inadmissible in court.)

Characterizing the Scene

Hazard characterization is not identification of unknowns. Hazard characterization is the elimination of the *knowns*, which are generally determined through field screening techniques using common field screening equipment such as:

- Alpha, beta, and gamma detectors
- Corrosivity and oxidizer papers
- Fluoride paper
- O_2, volatility, flammability, and chemical warfare agent detectors

Initial monitoring and detection should be performed to **characterize the site and the possible threats.** Hazardous materials teams should be prepared to monitor for any of the following:

> **Site Characterization —** Size-up and evaluation of hazards, problems, and potential solutions of a site.

- Alpha, beta, and gamma radiation
- Corrosives (corrosive atmosphere and corrosive liquids)
- Flammables
- Percentage of oxygen
- Volatile Organic Compounds "VOCs" (which may include toxic industrial chemicals and chemical warfare agents)

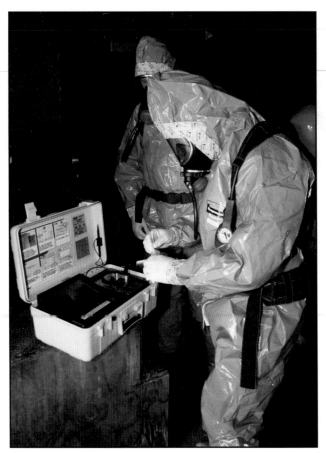

Figure 14.15 If possible, field screen samples before they are collected.

NOTE: For the purposes of law enforcement and evidentiary operations, it is not necessary to *identify* items at the scene, only *classify* them for safe packaging, transportation, and selection of the appropriate receiving forensic laboratory.

Field Screening Samples

Before samples are collected, responders must field screen them to identify specific hazards prior to transportation to a laboratory. These tests are necessary to ensure safety for individuals involved with packing, transporting, and performing lab tests on the samples. A hazardous materials team or other qualified personnel should perform all field screenings. Once the hazards are characterized (field screened) and documented, the materials can be packaged, overpacked, and transported to the appropriate laboratory.

Before field screening, responders should ensure that hazardous device technicians have cleared any potential explosives. Responders should also check for crystallized materials around caps and containers as these are indicators of potentially shock sensitive explosives and reactive chemicals.

To field screen, responders will need to establish a suitable work area that is well-ventilated, potentially outdoors. Ideally, the samples should be field screened in place prior to collection **(Figure 14.15)**. The process for sampling materials is discussed in Chapter 11. **Appendix G** presents a field screening matrix.

Protecting Samples and Evidence

Cross-contamination can happen easily if precautions are not taken to prevent it. If sampling equipment touches a non-sterile surface other than the sample itself, it can no longer be used. Each sampling procedure should be performed with new, sterile equipment, and personnel should change gloves between each separate sample. Law enforcement agencies and/or evidence labs provide protocols that may require control samples to be taken from the contaminated area.

To prevent cross-contamination between samples, act as follows:

● Have two clean sampling team members.

● Label each sample carefully with a sample and seal number.

● Use a commercial hazardous sample packaging system.

● Discard any sampling equipment that touches a non-sterile surface other than the sample itself.

Sampling Methods and Equipment

Samples and evidence can come in many forms. The particular type of material present and the amount of the material will determine sampling method and equipment required. In the case of suspicious letters and packages, the entire letter, envelope, or package (as well as the hazardous materials contained within) should be treated as physical evidence.

Initial monitoring results should classify the general characteristics of the potential threat and type of contaminant that is present (radiological, biological, chemical, or combination). With this information, law enforcement investigators can better determine the correct sampling method.

Collecting Samples

Collect evidence in accordance with the agency's sampling plan while wearing appropriate PPE. Although sample plans will vary (following local or federal protocols), most will include all of the following sampling steps:

1. Prepare evidence containers before entering the exclusion zone, using the system agreed upon by the AHJ and the receiving laboratory.

2. Record the sample location, conditions, and other pertinent information in a field notebook.

3. Confirm that the sample container number agrees with the overpack container number.

4. Wrap the sample container in absorbent material.

5. Place the sample in an overpack container and seal it with tamper-proof tape **(Figure 14.16)**.

6. Compile the chain of custody form.

7. Place the sample and chain of custody form in an approved transport container.

Figure 14.16 Seal overpack containers with tamper-proof tape.

CAUTION
Follow the AHJ's written sampling protocols that outline sampling techniques, types of containers to be used, the sealing process to be used, and other specific procedures!

Include the following protocols in the sample plan:

- Protecting public safety samples and evidence
- Field screening samples prior to laboratory admission
- Labeling and packaging of samples
- Decontaminating samples and evidence

A minimum of two individuals are recommended for a sampling team:

- A primary sampler who takes the samples and handles all sample equipment (sampling tools, container, and sample)
- An assistant to the sampler who handles only clean equipment and provides it to the sampler when needed

A third individual can be added to the sampling team. This individual can provide assistance by documenting, photographing, and monitoring the atmosphere and hazards in the hot zone. Safe external packaging and transportation of evidence is the responsibility of the law enforcement AHJ, in cooperation with the receiving laboratory and the operator of the transport vehicle. Responders must be trained in public safety sampling or evidence collection methods and the equipment used during sampling and collection activities. Many states provide guidance on evidence collection and sampling; agencies should research and identify their local requirements.

Decontaminating Samples and Evidence

Decontamination of evidence will remove contamination from the exterior evidence packaging only; do not open exterior evidence packaging for the purpose of decontaminating interior evidence packaging. Responders should take care during decontamination to preserve the integrity of evidence (such as fingerprints). Many evidence containers will be double-bagged or placed inside multiple containers for protection of the samples and the safety and health of the people handling them. Follow laboratory instructions and procedures for decontamination of evidence packages.

Labeling and Packaging

Law enforcement/or laboratories (for example, laboratories in the Laboratory Response Network, see information box) usually have established labeling and packaging protocols. At a minimum, label all samples with date, time, sample number, sample, and locations of sample site. A seal number is placed on a tamper-proof sample container seal **(Figure 14.17)**. Sample numbers are assigned in the sample log; the log also describes each sample taken.

Depending on requirements, samples may be packaged with commercial hazardous sample packaging systems or noncommercial systems. In the U.S. the commercial systems require Department of Transportation (DOT)

Figure 14.17 Seals must be numbered in accordance with the sample log.

certification. DOT and International Air Transportation Association (IATA) provide standardized commercial packaging equipment including certified safe containers, labels, absorbent, and documentation forms.

When collecting materials, use sterile containers, sterile tools, and certified clean containers should be used. Any container or tool that comes in contact with potential evidence should have a control taken. A *control* is an unused example from each lot of collection containers or tools used. To take a control, place the unused container or unused tool from each lot into an appropriate evidence container. If using a certified clean container, the statement of certification must accompany the control. Controls help establish that the collection containers and tools were not contaminated prior to use.

NOTE: Special containment vessels must be used for highly toxic liquid and pressurized gases.

When shipping samples to the lab, the shipper has to follow the appropriate governmental shipping regulations, for example, DOT in the U.S. or TC in Canada. Take precautions to ensure safe packaging and prepare an emergency action plan.

Laboratory Response Network

In the U.S., the Laboratory Response Network (LRN) was established by the Department of Health and Human Services, Centers for Disease Control and Prevention (CDC) in accordance with Presidential Decision Directive 39, which outlined national antiterrorism policies and assigned specific missions to federal departments and agencies.

Through a collaborative effort involving LRN founding partners, the FBI and the Association of Public Health Laboratories, the LRN became operational in August 1999. Its objective was to ensure an effective laboratory response to bioterrorism by helping to improve the nation's public health laboratory infrastructure, which had limited ability to respond to bioterrorism.

In the years since its creation, the LRN has played an instrumental role in improving the public health infrastructure by helping to boost laboratory capacity. Laboratories are better equipped, their staff levels are increasing, and laboratories are employing advanced technologies.

Public health infrastructure refers to essential public health services, including the people who work in the field of public health, information and communication systems used to collect and disseminate accurate data, and public health organizations at the state and local levels.

Today, the LRN is charged with the task of maintaining an integrated network of state and local public health, federal, military, and international laboratories that can respond to bioterrorism, chemical terrorism, and other public health emergencies. The LRN is a unique asset in the nation's growing preparedness for biological and chemical terrorism. The linking of state and local public health laboratories, veterinary, agriculture, military, and water- and food-testing laboratories is unprecedented.

Evidence Laboratories

Evidence samples are shipped or transported to evidence laboratories for definitive identification. These laboratories are typically selected before the incident, based on a variety of criteria including what types of samples they handle and what information you need or want back from them. Most law enforcement and fire service providers in larger urban areas have already identified laboratories in their area which have specific capabilities and may refer to their predetermined resource list as needed.

Chapter Review

Answer the following questions to review the information provided in this chapter.

1. What types of hazards should the first responder be on the lookout for at potential hazmat crime scenes?

2. Which agencies hold investigative authority in your jurisdiction?

3. During which response phase(s) of a criminal hazmat incident would a non-law enforcement first responder most likely be most active?

4. What actions should you take to secure a possible crime scene?

5. Why is chain of custody so important?

6. List different types of documentation.

7. What actions must be taken to properly protect public safety samples?

NOTE: Bomb squad personnel should conduct field screening for explosives or materials that can cause violent or toxic reactions.

Step 1: Identity incident with potential violation of criminal statues or governmental regulations.

Step 2: Identify law enforcement agency with jurisdiction over evidence and evidence collection.

Step 3: Prepare an evidence collection plan and evidence collection kit for use.

Step 4: Follow all safety procedures to ensure safe entry into hot zone.

Step 5: Ensure that all responders involved in evidence collection are wearing appropriate PPE for performing evidence collection operations in the hot zone.

Step 6: Document all personnel entering hot zone to ensure proper documentation for chain of custody purposes.

Step 7: Enter the scene.

Step 8: Document evidence using photography, sketches, and/or video as determined by the AHJ's SOPs.

Step 9: Collect sample and prepare for field screening of corrosivity, flammability, oxidizers, radioactivity, volatile organic compounds, and fluorides following AHJ's SOPs and protocols for field screening for admission into a forensic laboratory system.

NOTE: Field screening for explosives or materials that can cause violent or toxic reactions should be conducted by bomb squad personnel. If explosives are found, withdraw and follow bomb squad instructions on how to proceed.

Step 10: Seal sample container with tamper-proof seal.

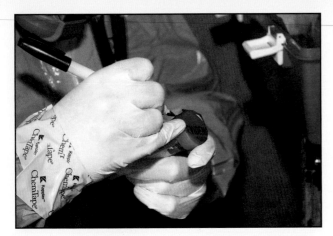

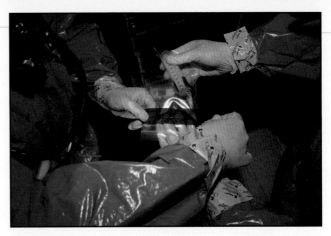

Step 11: Label the seal with date, time, and initials/name of person collecting sample.

Step 12: Document sample location through photograph and/or written documentation.

Step 13: Put sample into secondary container, such as zip-top bag.

Step 14: Label secondary container.

Step 15: Proceed to decontamination line for decontamination.

Step 16: Decontaminate exterior of secondary container while proceeding through decontamination.

NOTE: If any evidence changes custody during decontamination, document on chain of custody form.

Step 17: Follow laboratory instructions for packaging evidence for transportation, ensuring that documentation of chain of custody is performed.

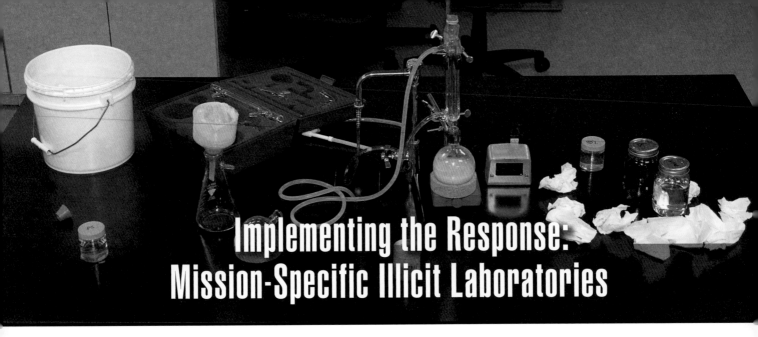

Implementing the Response: Mission-Specific Illicit Laboratories

Chapter Contents

Key Terms

NFPA Job Performance Requirements

This chapter provides information that addresses the following job performance requirements of NFPA 1072, *Standard for Hazardous Materials/Weapons of Mass Destruction Emergency Response Personnel Professional Qualifications (2017)*.

6.9.1

Implementing the Response: Mission-Specific Illicit Laboratories

Learning Objectives

After reading this chapter, students will be able to:

1. Describe general hazards at illicit laboratories. (6.9.1)

2. Identify characteristics of illicit drug labs. (6.9.1)

3. Identify characteristics of chemical agent labs. (6.9.1)

4. Identify characteristics of explosives labs. (6.9.1)

5. Identify characteristics of biological labs. (6.9.1)

6. Identify characteristics of radiological labs. (6.9.1)

7. Describe operations at illicit labs. (6.9.1)

8. Explain remediation of illicit labs. (6.9.1)

9. Identify and avoid booby traps at an illicit laboratory. (6.9.1, Skill Sheet 15-1)

10. Identify and secure an illicit laboratory. (6.9.1, Skill Sheet 15-2)

Chapter 15
Implementing the Response: Mission-Specific Illicit Laboratories

This chapter will describe features of illicit labs, illicit lab types, and typical hazmat responder roles at illicit labs. The following text will include such topics as:

- General hazards at illicit laboratories
- Drug labs
- Chemical agent labs
- Explosives labs
- Biological labs
- Radiological labs
- Operations at illicit labs
- Remediation of illicit labs

General Hazards at Illicit Laboratories

Illicit labs contain many potential dangers for responders. The materials in the lab can be quite hazardous, and responders cannot trust that the people working in the lab followed safe or recognized scientific processes when dealing with those materials **(Figures 15.1a-c)**. Operators of illicit labs may booby trap their labs or attempt to harm responders, rivals, or potential thieves.

In many cases, responders may receive a call for another type of incident (such as a fire), only to discover an illicit lab on scene. Therefore, responders should recognize the indicators for illicit labs and understand the hazards associated with them. Responders must develop and implement a fast and accurate analysis for a successful response to an incident at an illicit lab. A successful response also requires a quickly developed, workable plan that accounts for potential hazards and jurisdictional responsibilities. At illicit labs, responders should always operate under the guidance of a hazardous materials technician, an emergency response plan, standard operating procedures, and Allied Professionals including law enforcement personnel or others with similar authority.

Figures 15.1a-c Illicit labs can be located virtually anywhere. Responders should be wary of booby traps and secondary devices protecting lab locations.

Personnel should identify the kinds of activities taking place in the lab as quickly as possible. Specialized teams exist to assist in this assessment and intelligence process. Examples of these teams in the U.S. include:

- Drug Enforcement Agency (DEA) Clandestine Lab Teams
- Local or State Law Enforcement Lab Teams for the illegal manufacture of drugs
- Federal Bureau of Investigation (FBI) Laboratory Forensic Response Section for the manufacture of WMD materials

These agencies may provide training, insight, or resources that first responders may not have. **Table 15.1** provides a summary of the agencies responsible for various mitigation tactics.

CAUTION
Depending upon the level of sophistication of the individual(s) involved, illicit labs might contain makeshift or on-the-spot equipment or true laboratory grade equipment.

Table 15.1
Illicit Lab Response - Tactical Guidelines

Hazard	Responsibility
Operator present within the laboratory, with access to weapons	Law enforcement tactical teams specifically trained to operate within a hazardous environment
Anti-personnel devices (booby-traps) around and within the laboratory	Bomb Squad personnel trained for these procedures
Hazardous materials/WMD within the illicit laboratory	Technician and Operation Level responders

Illicit lab operators are likely to be hostile and potentially armed. They may also exhibit erratic and disorganized behavior and be unable to provide any clear information to first responders.

First responders should pay close attention to any unusual or atypical behaviors, which may indicate hidden or concealed hazards, including booby traps, narcotics, or weapons.

You should also pay attention to the following:

- Whether the lab operators wear protective equipment
- The presence of animals, such as dogs, poisonous snakes, fire ants, or other venomous insects
- Whether a criminal affiliation appears to be operating the lab

Don a level of protection that reflects your SOP/Gs. You should receive an assessment of the intelligence gathered through investigative processes as well as receive information from:

- Sources (including witness, bystander, or victim observations)
- Suspect interviews
- Confidential informant statements
- Trash/garbage search results
- Discarded personal protective equipment
- Initial first responder intelligence (such as EMS, bomb team, first due personnel)

Additionally, any witness accounts from inside the lab will greatly assist responders in forming a clear picture of the lab, its contents, and its potential hazards. Regardless of the type of lab, watch for instructional manuals or other books, magazines, and internet resources relating to hazardous agents **(Figure 15.2)**.

Try to gather information on the following:

- Materials the lab produces or contains
- Activities of animals in the laboratory
- Layout of the laboratory.

Gathering this information may require responders to interview neighbors or engage in other intelligence gathering information. You may ask potential witnesses the following questions:

- Are you aware of any laboratory glassware or scientific equipment?
- Have you noticed any unusual odors?
- Have you seen any animals that did not appear to be pets?

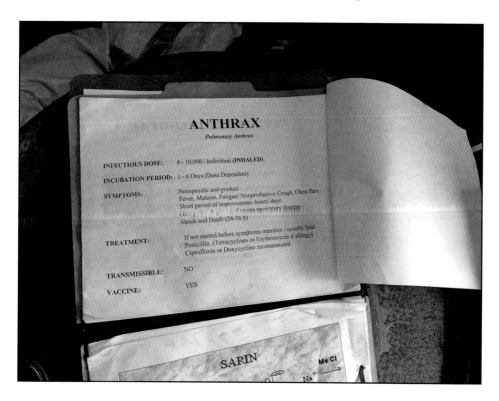

Figure 15.2 Instruction manuals, recipes, internet resources, and other information found in labs may help identify the laboratory type.

- Are there any how-to manuals or "recipes" in the lab?

- Are there any chemical containers? Were they labeled? Did anyone check the refrigerator for chemical containers?

- Did you notice any areas of the building that appeared segregated or draped in plastic?

- Did you see any homemade **gloveboxes (Figure 15.3)**?

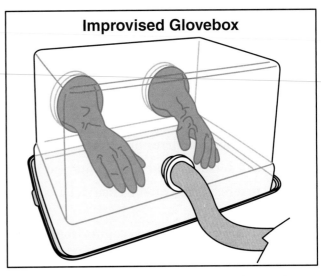

Improvised Glovebox

Figure 15.3 Homemade gloveboxes are simple to make and may indicate a potential biological lab.

- Was there any detection and monitoring equipment present?

- Did you notice any decontamination equipment (bleach, brushes, detergent, and excessive paper towels)?

- Did you look for any discarded personal protective equipment (gloves, gowns, scrubs, coveralls, aprons) or respiratory protection (painters' hoods, dust masks, surgical masks, gas masks)? Did anyone inspect the trash or nearby dumpsters?

The following sections provide more information about hazardous materials, lab operators, booby traps, and other hazards that responders may encounter in illicit labs.

This chapter addresses five primary illicit (illegal) lab types:

- Drug labs

- Chemical agent labs

- Explosives labs

- Biological labs

- Radiological labs

Hazardous Materials

In illicit labs, both the final product and production materials can be harmful. These materials will vary depending on the type of lab, but responders can expect a variety of hazards, such as flammable, volatile, corrosive, toxic, or biological **(Figure 15.4)**. Additional hazards may include active chemical reactions, pressurized materials (liquids, gases), explosive materials, and radioactive materials.

Operators of illicit labs often commit environmental crimes and may do any of the following:

- Possess hazardous substances in residential locations

- Release hazardous vapors into residential areas

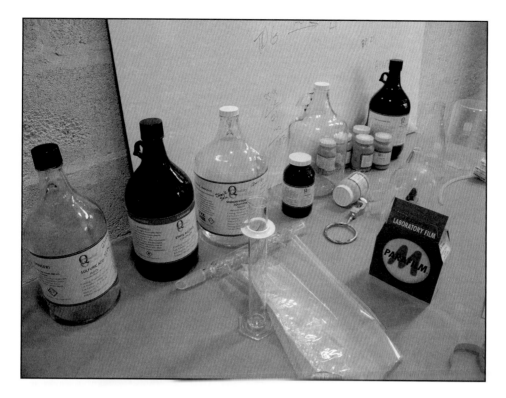

Figure 15.4 At illicit labs, both the final products and the materials used to make them can be very hazardous.

- Dispose of hazardous waste illegally
- Use improper, unapproved processes and locations
- Pour hazardous waste down sanitary sewers (where flammable liquids could ignite)
- Move hazardous materials in improper, unmarked containers
- Mix potentially incompatible materials

NOTE: In some jurisdictions, first responders may be responsible for notifying state or federal environmental protection agencies or initiating hazardous materials cleanup.

Responders must stay alert for these kinds of crimes, as well as other hazards associated with labs. The following sections will describe in greater detail the hazards associated with specific types of labs.

Booby Traps

Responders may encounter booby traps at illicit labs. Tactical teams and bomb technicians should have the skills and training to identify more specific information on potential booby traps. If responders suspect or encounter a potential booby trap, they should rely on specially trained bomb technicians to search for and dismantle any booby traps.

Booby traps can be inside or outside of the lab, and they may include any of the following (**Figure 15.5, p. 682**):

- Explosives (including grenades and dynamite)
- Wires attached to explosives or alerting devices
- Weapons tied to doors

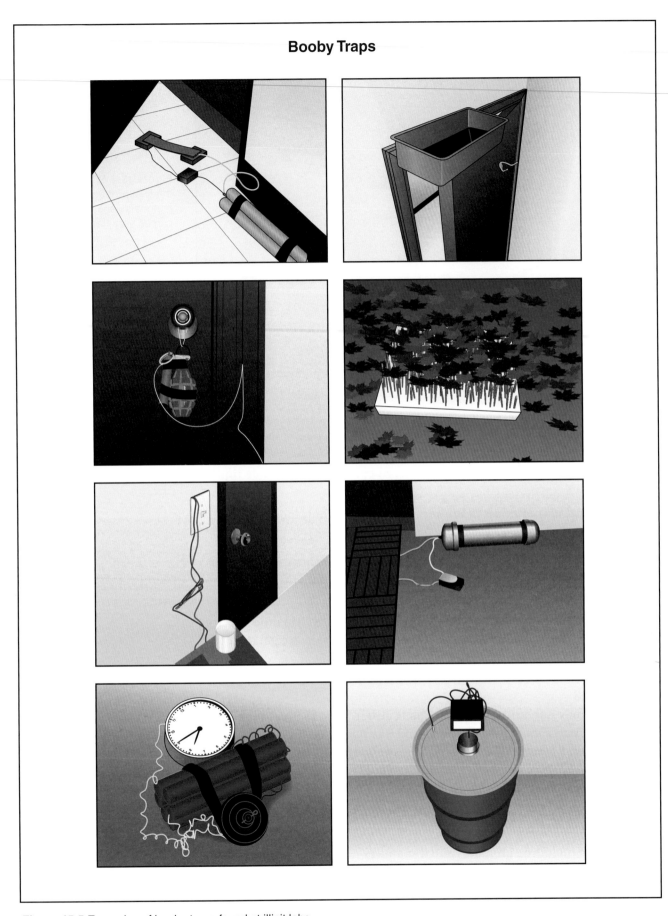

Figure 15.5 Examples of booby traps found at illicit labs.

- Bottles that will break, mixing their chemical contents to produce toxic fumes (includes chemistry glassware designed to mix toxic chemicals when manipulated)
- Reversed on/off switches
- Holes in floors (trap doors)
- Electrified door handles
- Exposed wiring
- Spikes
- Hooks
- Acid

Always maintain good situational awareness and stay alert for booby traps. Properly trained bomb technicians should clear potential anti-personnel devices. Take the following actions to avoid booby traps:

- Limit personnel entering a suspected hazardous area.
- Use intrinsically safe equipment.
- Take aerial photographs for reconnaissance prior to entry.
- Avoid complacency.
- Refrain from handling, touching, or moving items in or around the lab.
- Check doors and openings for wires and/or traps before opening.
- Avoid powering lab equipment on or off.
- Consult with subject matter experts (SME) prior to handling or dismantling unknown or unfamiliar equipment (such as chemistry glassware).

Turning illicit lab equipment on or off may trigger booby traps. In addition, responders should leave on electrical pumps, such as those used in cooling baths in red phosphorous **methamphetamine (meth)** labs, to continue the circulation of cooling water. Interrupting this flow may result in an overheated reaction and ignition of nearby combustibles. Consult a technical expert before powering lab equipment on or off. See **Skill Sheet 15-1** for methods to identify and avoid booby traps at illicit labs.

Methamphetamine (Meth) — Central nervous system stimulant drug that can be produced in small labs. Low dosage medical uses include controlling weight, narcolepsy, and attention deficit hyperactivity disorder. Recreational uses include euphoriant and aphrodisiac qualities. At all dosages, misuse of this drug presents a high risk for personal and social harm.

What This Means To You

Illicit Lab Safety

If you are the first to discover or detect the presence of an illicit lab, use the following guidelines:

- Do not disturb the lab in any way.
- Do not flip switches or turn on or off lights.
- Do not shut off the electricity to the facility.
- Use extreme caution in your movements
- Remain mindful of the potential for booby traps.
- Back out the way you entered.
- Evacuate the surrounding area.
- Request appropriate resources.
- Wait until appropriate personnel or subject matter experts have evaluated the scene.

CAUTION

In an illicit lab, if it is on, leave it on; if it is off, leave it off.

Illicit Drug Labs

Illicit labs can produce a variety of illegal drugs. Some of the most prevalent and hazardous illicit labs found in the U.S. produce methamphetamines. Other labs produce types of illicit drugs, such as:

- Ecstasy (MDMA)
- Phenyl-2-propanone (P2P)
- Phencyclidine (PCP)
- Heroin
- LSD
- Gamma-Hydroxybutyric acid (GHB)

Some labs (also known as *pill mills*) may process or manufacture designer drugs. Because meth labs represent the most common type of clandestine drug lab, this section primarily focuses on the hazards associated with illicit drug labs. Certain types of labs may be more common regionally.

Regulated Drug Labs

Formally regulated drug labs have many of the same hazards as illicit drug labs. Legalizing the development of these materials may not change the hazard profiles of the processes, and some operators may still develop these materials illegally.

Meth is easy to make and uses a variety of ingredients commercially available in local stores. Because of the increasing hazard of meth labs, some U.S. states have placed restrictions on the purchase of items used in making meth.

The process of making meth is called *cooking*, and many different recipes or methods exist. Three of the most common are known as:

- One/Single Bottle method (one pot method)
- Red phosphorous (Red P) method **(Figure 15.6)**
- Nazi/Birch method

The various recipes differ slightly in the process and the chemicals used, but all of them are dangerous because the chemicals are often flammable, corrosive, water reactive, and toxic **(Figures 15.7 a and b). Table 15.2, p. 686,** provides a summary of the products commonly used in cooking meth and the hazards associated with them. Meth labs present a danger to the **meth cook,** the community surrounding the lab, and emergency response personnel who discover the lab.

Meth Cook — 1) Person who generates methamphetamine in an illicit lab. 2) Area with evidence of production of methamphetamine.

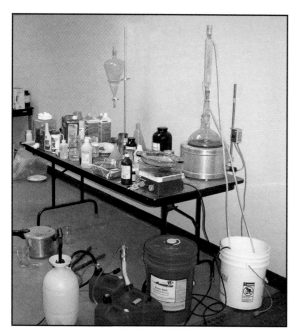

Figure 15.6 An example of a *Red P* meth lab. *Courtesy of MSA.*

CLIA and CLIC

Two national organizations, the Clandestine Laboratory Investigators Association (CLIA) and the Clandestine Laboratory Investigating Chemists Association (CLIC), host regular conferences for personnel interested in learning more about clandestine lab training. CLIA holds an annual one-week training conference taught by hazmat experts. In addition to its conferences, CLIC has published a formal journal that provides data and other information on clandestine labs.

Figures 15.7a and b Products and equipment commonly used in meth labs.

Flammability is perhaps the most serious hazard associated with meth labs. Responders discover many labs only after a fire or explosion occurs. In addition to flammable materials, meth producers also use highly corrosive acids or bases and other extremely toxic materials.

Unique Hazards of Red P

The Red P method of cooking meth produces a byproduct, **phosphine** gas, a highly toxic flammable gas. Some other byproducts are oxidizers. Meth production processes also generate byproducts of hydrogen chloride gas and hydroiodic acid liquid that may vaporize into the atmosphere. Meth lab locations may remain a serious health and environmental hazard for years after the lab is removed unless they undergo an expensive decontamination process

Phosphine — Colorless, flammable, and toxic gas with an odor of garlic or decaying fish; ignites spontaneously on contact with air. Phosphine is a respiratory tract irritant that attacks the cardiovascular and respiratory systems, causing pulmonary edema, peripheral vascular collapse, and cardiac arrest and failure.

Table 15.2
Methamphetamine Sources and Production Hazards

Chemical Name	Common Sources/Uses	Hazards	Production Role
Acetone	• Paint solvent • Nail polish remover	• Highly flammable • Vapor is irritating to eyes and mucous membranes • Inhalation may cause dizziness, narcosis, and coma • Liquid may do damage upon contact • Ingestion may cause gastric irritation, narcosis, and coma	• Pill extraction • Cleaning glassware • Cleaning finished methamphetamine (meth)
Anhydrous Ammonia	• Sold as fertilizer; also used as a refrigerant gas • Stolen from farms and other locations for illegal meth production • Often stored in propane tanks or fire extinguishers at illegal meth labs, which causes the fittings of the tank or extinguisher to turn blue	• Toxic • Corrosive • Flammable • Severe irritant; may cause severe eye damage, skin burns and blisters, chest pain, cessation of breathing, and death	Meth production process
Ethyl Alcohol/ Denatured Alcohol/ Ethanol/ Grain Alcohol	• Sold as solvents • Is the alcohol found in beverages at greatly reduced concentrations	• Highly flammable • Toxic; may cause blindness or death if swallowed • Inhalation may affect central nervous system causing impaired thinking and coordination • Skin and respiratory tract irritant (may be absorbed through the skin) • May affect the liver, blood, kidneys, gastrointestinal tract, and reproductive system	• Used with sulfuric acid to produce ethyl ether (see Ethyl Ether/Ether entry) • Cleaning glassware
Ephedrine	Over-the-counter cold and allergy medications	Harmful if swallowed in large quantities	Primary precursor for meth

Continued

Table 15.2 (continued)

Chemical Name	Common Sources/Uses	Hazards	Production Role
Ethyl Ether/Ether	Starting fluids	• Highly flammable • Oxidizes readily in air to form unstable peroxides that may explode spontaneously • Vapors may cause drowsiness, dizziness, mental confusion, fainting, and unconscious at high concentrations	Separation of the meth base before the *salting-out* process begins, primarily in the *Nazi/Birch* method
Hydrochloric Acid/Muriatic Acid (Other acids can be used as well, including sulfuric acid and phosphoric acid)	Commercial or industrial strength cleaners for driveways, pools, sinks, toilets, etc.	• Toxic; ingestion may cause death • Corrosive; contact with liquid or vapors may cause severe burns • Inhalation may cause coughing, choking, lung damage, pulmonary edema, and possible death • Reacts with metal to form explosive hydrogen gas	Production of water-soluble salts
Hydrogen Peroxide	• Common first aid supply • Used for chemical manufacturing, textile bleaching, food processing, and water purification	• Strong oxidizer • Eye irritant	Extrication of iodine crystals from Tincture of Iodine
Hypophos-phorous Acid	Laboratory Chemical	• Corrosive • Toxic • Generates deadly phosgene during initial reaction	Source of phosphorous in *Red P* method
Iodine	Tincture of iodine	• Toxic • Vapors irritating to respiratory tract and eyes • May irritate eyes and burn skin	• Meth production process • Can be mixed with hydrogen sulfide to make hydriodic acid (strong reducing agent) • Can be mixed with red phosphorus and water to form hydriodic acid
Isopropyl Alcohol	Rubbing Alcohol	• Flammable • Vapors in high concentrations may cause headache and dizziness • Liquid may cause severe eye damage	• Pill extraction • Cleaning finished meth

Continued

Table 15.2 (continued)

Chemical Name	Common Sources/Uses	Hazards	Production Role
Lithium Metal	Lithium batteries	• Flammable solid • Water-reactive (reacts with water to form lithium hydroxide, which can burn the skin and eyes)	Reacts with anhydrous ammonia and ephedrine pseudoephedrine in the *Nazi/Birch* method
Methyl Alcohol	HEET® Gas-Line Antifreeze and Water Remover	• Highly flammable • Vapors may cause headache, nausea, vomiting, and eye irritation • Vapors in high concentrations may cause dizziness, stupor, cramps, and digestive disturbances • Highly toxic when ingested	Pill extraction
Mineral Spirits/ Petroleum Distillate	• Lighter fluid • Paint thinner	• Flammable • Toxic when ingested • Vapors may cause dizziness • May affect central nervous system and kidneys	Separation of meth base before *salting-out* process begins
Naphtha	Camping fuel for stoves and lanterns	• Highly flammable • Toxic when ingested • May affect the central nervous system • May cause irritation to the skin, eyes, and respiratory tract	• Separation of meth base before salting-out process begins • Cleaning preparation
Pseudoephedrine	Over-the-counter cold and allergy medications	Harmful if swallowed in large quantities	Production of meth (same as Ephedrine)
Red Phosphorous	Matches	• Flammable solid • Reacts with oxidizing agents, reducing agents, peroxides, and strong alkalis • When ignited, vapors are irritating to eyes and respiratory tract • Heating in a reaction or cooking process generates deadly phosphine gas • Can convert to white phosphorous (air reactive) when overheated	Mixed with iodine in the *Red P* method; serves as a catalyst by combining with elemental iodine to produce hydriodic acid (HI), which is used to reduce ephedrine or pseudoephedrine to meth
Sodium Hydroxide (Other alkaline materials may also be used such as sodium, calcium oxide, calcium carbonate, and potassium carbonate)	Drain openers	• Very corrosive; burns human skin and eyes • Generates heat when mixed with an acid or dissolved in water	After cooking, an alkaline product such as sodium hydroxide turns the very acidic mixture into a base

Continued

Table 15.2 (concluded)

Chemical Name	Common Sources/Uses	Hazards	Production Role
Sulfuric Acid	Drain openers	• Extremely corrosive • Inhalation of vapors may cause serious lung damage • Contact with eyes may cause blindness • Both ingestion and inhalation may be fatal	Creates the reaction in the salting phase; combines with salt to create hydrogen chloride gas, which is necessary for the *salting-out* phase
Toluene	Solvent often used in automotive fuels	• Flammable • Vapors may cause burns or irritation of the respiratory tract, eyes, and mucous membranes • Inhalation may cause dizziness; severe exposure may cause pulmonary edema • May react with strong oxidizers	Separation of meth base before the *salting-out* process begins
Hydrogen Chloride		• Toxic • Corrosive • Eye irritant • Vapor or aerosol may produce inflammation and may cause ulceration of the nose, throat, and larynx	• Created by adding sulfuric acid to rock salt • Used to salt out meth from base solution
Phosphine Gas		• Very toxic by inhalation • Highly flammable; ignites spontaneously on contact with air and moisture, oxidizers, halogens, chlorine, and acids • May be fatal if inhaled, swallowed, or absorbed through skin • Contact causes burns to skin and eyes	• Byproduct • Produced when red phosphorous and iodine are combined during the cooking process
Hydrogen Iodide/ Hydriodic Acid Gas		• Highly toxic • Attacks mucous membranes and eyes	• Byproduct • Produced when red phosphorous and iodine are combined during the cooking process • Causes the reddish/orange staining commonly found on the walls, ceilings, and other surfaces of meth labs
Hydriodic Acid		• Corrosive • Causes burns if swallowed or comes in contact with skin	• Byproduct • Produced when red phosphorous and iodine are combined during the cooking process

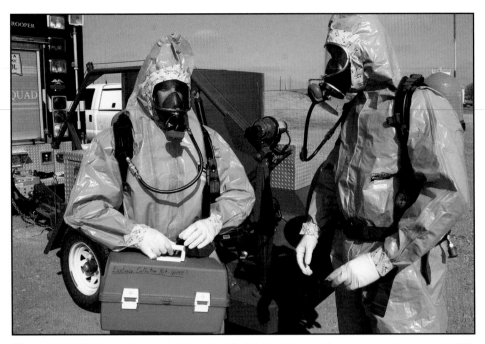

Figure 15.8 PPE may be needed to enter illicit labs. Depending on hazards present, PPE may include CPC with SCBA.

Responders must wear PPE to enter a meth lab **(Figure 15.8)**. If you enter a meth lab before it has been properly decontaminated and ventilated, you may experience the following symptoms:

- Headaches
- Nausea
- Dizziness
- Fatigue
- Shortness of breath
- Coughing
- Chest pain
- Lack of coordination
- Burns
- Death

Risk of injury or toxicity from chemical exposure varies depending on:

- Toxic properties of the chemicals or byproducts
- Quantity and form
- Concentrations
- Duration of exposure
- Route of exposure

Chemicals and products typically found in meth labs include materials that can be categorized by their functions as precursors, solvents, reagents, and catalysts. The following items are commonly found **(Figure 15.9)**:

- Pseudoephedrine (from decongestants such as Sudafed®)
- Red phosphorus

- Iodine crystals
- Elemental sodium, lithium, or potassium (from batteries, pellets, or wire solids)
- Anhydrous ammonia
- White gas, sometimes packaged as Coleman® fuel
- Starting fluid or Ethyl ether
- Sulfuric acid
- Rock salt or table salt
- Hydrochloric acid
- Sodium hydroxide

CAUTION

The presence of sodium, lithium, and other water-reactive substances can complicate fire suppression activities at illicit laboratories because they react with water from hose streams, water-based fire extinguishers, and/or sprinkler systems.

Figure 15.9 Labs may contain a variety of hazards, including highly toxic and flammable materials. *Courtesy of August Vernon.*

In addition to recognizing the types of chemicals typically found in meth labs, first responders should also be familiar with the following types of equipment used to cook meth:

- Condenser tubes — Cools vapors produced during cooking **(Figure 15.10)**
- Filters — Coffee filters, cloth, and cheesecloth
- Funnels/turkey basters — Separates layers of liquids
- Gas containers — Propane cylinders, fire extinguishers, self-contained underwater breathing apparatus (SCUBA) tanks, plastic drink bottles (often attached to some sort of tubing) **(Figure 15.11)**
- Glassware — Particularly Pyrex® or Visions® cookware, mason jars, and other laboratory glassware that can tolerate heating and violent chemical reactions
- Heat sources — Burners, hot plates, microwave ovens, and camp stoves **(Figure 15.12)**
- Grinders — Grinds up ephedrine or pseudoephedrine tablets
- pH papers — Tests the pH levels of the reactions
- Tubing — Glass, plastic, copper, or rubber

Other clues to the presence of meth labs in structures include the following:

- Windows covered with plastic or tinfoil
- Renters who pay landlords in cash

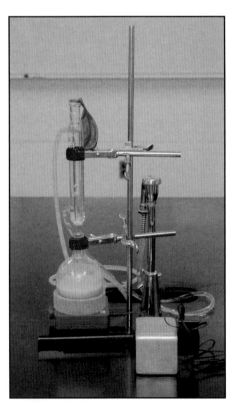

Figure 15.10 Meth labs may use condenser tubes. An example is shown here, connected to the top of a round-bottom flask.

Figure 15.11 Anhydrous ammonia, used in meth production, will turn the brass fittings blue on propane cylinders and other containers.

- Unusual security systems or other devices
- Excessive trash **(Figure 15.13)**
- Increased activity, especially at night
- Unusual structures
- Discoloration of structures, pavement, and soil
- Strong odor of solvents
- Smell of ammonia, starting fluid, or ether
- Iodine- or chemical-stained bathroom or kitchen fixtures

For every pound (0.5 kg) of meth produced, labs generate approximately 6 pounds (3 kg) of hazardous waste. Typically, a mixture of materials make up this waste, presenting unique hazards that may be completely different from the hazards in the illicit lab. To dispose of this waste, operators may:

- Bury the waste
- Dump it in the regular residential trash
- Toss it down the drain to the septic system
- Leave it beside roadways
- Hide it on vacant properties
- Place it in streams or ponds/lakes

Disposal of this waste is expensive, and the cleanup process is potentially dangerous. Many law enforcement departments have contracts with private hazardous materials waste disposal contractors to handle the cleanup and decon of seized illegal meth labs and dumps.

Figure 15.12 Extra heat sources, such as Bunsen burners and hot plates, are often used in meth labs. *Courtesy of Joan Hepler.*

Figure 15.13 Excessive trash may be a lab indicator, particularly if meth product containers are included. *Courtesy of MSA.*

Marijuana Grow Lab and BHO Synthesis Lab Explosions

First responders have encountered an increasing number of hazmat incidents involving the often volatile production of illegal drugs. Drug manufacturers have converted houses and rental properties into marijuana grow labs where manufacturers grow or engineer marijuana plants. Over the past decade, the rise in the designer drug butane hash oil (BHO) has led to explosions and structure fires.

Responders must have the proper equipment and training when they arrive on scene at a grow lab. Responders should coordinate with law enforcement and perhaps other agencies since incidents at these properties could include fires, booby traps, crime scene preservation, and electrical hazards.

While a grow lab might appear like a typical house, several warning signs could help responders identify a structure as a grow lab. Grow labs often have a heavy smell of marijuana that observers can detect from the outside, as well as condensation on the windows from the humidity inside growing rooms. In addition, grow labs might have an unusual amount of security (such as high fences, locked gates, and guard dogs) that could create obstacles for responders.

The production of BHO, a potent form of concentrated marijuana, presents hazards because it uses butane gas. In recent years, individuals attempting to synthesize BHO have suffered severe burns from and died in explosions resembling those from a pipe bomb.

Responders might suspect they have arrived at a scene involving BHO production if they find butane canisters, PVC pipes, and Pyrex dishes near the fire or explosion. Burn victims might need immediate medical care, including the transportation to a burn center.

Chemical Agent Labs

Manufacturers can make chemical warfare agents in illicit laboratories. While the recipes may be easy to find, the actual materials necessary to make chemical warfare agents may not be so accessible. Some ingredients may be common, but access to others is restricted. The following clues may indicate a chemical lab:

Organophosphate Pesticides — Chemicals that kill insects by disrupting their central nervous systems; these chemicals inactivate acetylcholinesterase, an enzyme which is essential to nerve function in insects, humans, and many other animals.

- Military manuals

- Underground "cookbooks"

- Chemicals such as **organophosphate pesticides** that manufacturers would not typically use to make meth or other illegal drugs

- Chemicals such as methyl iodide and phosphorus trichloride (which might indicate attempts to make sarin)

- Lab equipment sophisticated enough to conduct the chemical reactions needed to make chemical agents **(Figures 15.14)**

- Presence of cyanides or acids

Figure 15.14 Chemical agent labs may have very sophisticated laboratory equipment.

Explosives Labs

Some explosives labs do not need to heat or cook any of their materials, and therefore may lack the glassware, tubing, Bunsen burners, chemical bottles, and other paraphernalia and equipment traditionally associated with laboratories. For example, an explosives lab might include a work area in a garage used to assemble custom fireworks or pyrotechnics. However, labs that make explosive chemical mixtures might look more like a traditional industrial or university chemistry lab, and labs that make peroxide-based explosives might look much like a meth or drug lab.

After drug labs, explosives labs are the second most common type of lab encountered. Explosives labs can be mistaken for clandestine drug labs because of the presence of household chemicals. Responders can mistake some improvised explosive materials for narcotics. These labs do not require a lot of equipment or resources, making them easy to establish. Potential manufacturers can easily find recipes on the Internet, in anarchist literature, and in other sources (**Figure 15.15**). Manufacturers can easily incorporate explosive materials, such as black powder or smokeless powder, into an IED.

The materials needed to produce these dangerous explosives can be found in a number of local stores or purchased online. The basic ingredients are a fuel and oxidizer.

Some indicators of a possible explosives lab include:

- Scales and thermometers (**Figure 15.16**)
- Refrigerators/coolers/ice baths (**Figure 15.17**)
- Glassware and laboratory equipment
- Blenders, grinders, mortar and pestle
- Blasting caps/batteries/fuses/switches
- Pipes/end caps/storage containers
- Shrapnel-type materials
- Strong acidic odors
- Explosives, military ordnance

  Other clues to the presence of an explosives lab might include:

- Literature on how to make bombs
- Significant quantities of fireworks
- Hundreds of matchbooks or flares (**Figure 15.18, p. 696**)
- Ammunition such as shotgun shells, black powder, smokeless powder, or blasting caps
- Commercial explosives
- Incendiary materials

Finding these items in conjunction with components that can be used to make IEDs (such as pipes, activation devices, empty fire extinguishers, and propane containers) would give even more evidence of an explosives lab. Also, electronic components such as wires, circuit boards, cellular phones and other items can point towards the possible design of an IED. Use caution when you are inside any type of clandestine laboratory.

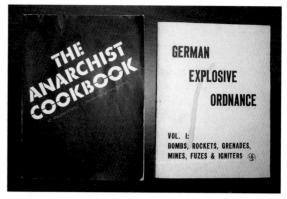

Figure 15.15 Recipes for making explosives are readily available to the public. *Courtesy of August Vernon.*

Figure 15.16 Scales and thermometers may be used in explosives labs.

Figure 15.17 Refrigerators, coolers, and ice baths may be used in explosives labs.

Figure 15.18 Explosives labs may have large numbers of matchbooks. *Courtesy of August Vernon.*

Manufacturers can use any of a number of common chemicals to make homemade explosives (HME). Some common homemade explosives are peroxide-based, but other types of explosives are even simpler. For example, many homemade explosives can be made using pots and pans and no sophisticated equipment. Homemade explosives were introduced in Chapter 8 of this manual.

As of the time of this manual's development, responders have encountered an increasing number of peroxide-based explosives labs. Peroxide-based explosives are sensitive to heat, shock, and friction. Some of the common ingredients that responders may find in a peroxide-based explosives lab include the following:

- Acetone **(Figure 15.19)**
- Ethanol
- Hexamine (solid fuel for camp stoves)
- Hydrogen peroxide
- Strong or weak acids (such as sulfuric or citric acids)

Once manufacturers produce the materials, they can incorporate the materials into a variety of IEDs. Lab operators commonly transport raw materials using a cooling method (such as ice in a cooler). Responders who encounter these raw materials should treat them with caution. If mishandled, the materials in an explosives lab can pose a significant danger.

> ///////////////////
> ## CAUTION
> **First responders are often tasked with monitoring and collecting samples. In an illicit lab environment, be aware that monitoring and sampling activities can trigger explosives or otherwise exacerbate hazards.**

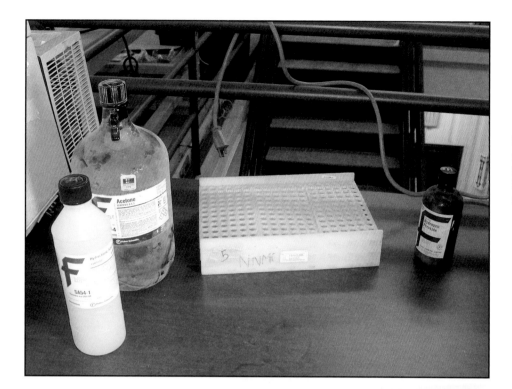

Figure 15.19 The peroxide explosive manufacturing process may use acetone and hydrogen peroxide as well as many acids.

WARNING!
Peroxide-based explosives can look like narcotics (for example, cocaine), but they will react violently with narcotic field-testing kits.

WARNING!
Do not touch white crystals or powder found in any clandestine laboratory!
The crystals or powder may be extremely sensitive to heat, friction, and shock.

WARNING!
Some kinds of energized test equipment may have enough energy to detonate explosive materials.

Biological Labs

Be alert to the indicators of a biological laboratory. Biological labs will have equipment that may include microscopes, growth media, autoclaves, glove boxes, incubators, and refrigerators. Biological labs may contain resources such as acetone, Epson salt, and sodium hydroxide. Botox production may look like a bunch of dirt, rotten food, or garbage in a container.

Biological labs are unlikely to have chemicals such as gasoline, propane, anhydrous ammonia, or other flammable and corrosive liquids. However, they will likely contain chemicals such as acids/bases, alcohols and acetone **(Figure 15.20, p. 698)**.

Autoclave — A device that uses high-pressure steam to sterilize objects.

Virus production requires a different type of lab setup compared to bacterial production. Viral labs may contain:

- T-flasks
- Roller bottles
- Well culture plates
- Incubators
- Culture rolling machines
- Centrifuges **(Figure 15.21)**
- Pipettes
- Disinfection materials
- Live tissue matrices (such as cells, blood, eggs, insects, and live animals)

Figure 15.20 Biological labs may contain acids, bases, acetone, and alcohols.

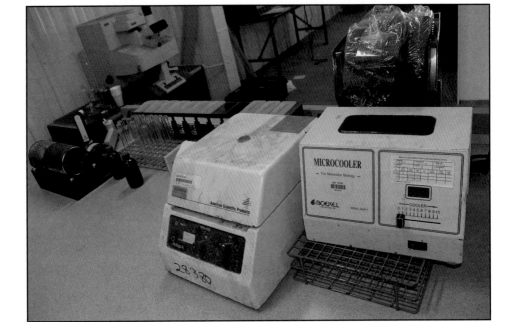

Figure 15.21 Centrifuges may be used in viral labs.

Biological lab indicators include the following:

- The presence of biological materials known to be the source of toxins, such as castor beans, rosary peas, or a botulinum toxin production setup, that may include pressure cookers, mason jars, soil, and a protein such as meat or cat food **(Figure 15.22)**

- The presence of antibiotics and vaccines

- PPE, such as respirators (particularly with HEPA filters), rubber gloves, and masks **(Figure 15.23)**

- Laboratory or test animals and/or related materials, such as cages and food

- Growth containers, such as Petri dishes, glass jars, agar plates, and culture/growth mediums (agar, meat broth, gelatin, meat, or feces) **(Figure 15.24)**

Figure 15.22 Castor beans and rosary peas are sources of biological toxins.

Figure 15.23 PPE, especially respirators with HEPA filters, may be indicators of a biological lab.

Figure 15.24 Agar plates and petri dishes may be used to grow biological cultures.

- Biological safety cabinets or gloveboxes (improvised setups can utilize plastic sheeting, Plexiglas, aquariums, duct tape, and fans) **(Figure 15.25)**

- Laboratory equipment such as incubators or other "equipment" used to control temperature, heat lamps, refrigerators, and fermenters, carboys

- Fewer pieces of specialized glassware; items such as condenser tubes or distillation and reflux setups

- **Cell lysis** and pulverization equipment (ball mills, rock tumblers) **(Figure 15.26)**

- Centrifuge

- Bleach or other sterilization supplies such as antiseptics and autoclaves (or pressure cookers)

- Alterations to building ventilation systems

- Sprayers, nebulizers, or other delivery devices

- Filters / coffee filters, cheesecloth

- Alcohols, acids, and bases

Cell Lysis Equipment — Machinery used to break down the membrane of a cell.

Figure 15.25 Plastic sheeting can be used to make improvised safety cabinets, gloveboxes, or other isolation areas.

Figure 15.26 Cell lysis equipment may include bench-top laboratory roller mills and other pulverization equipment.

WARNING!

Biothreat agent laboratories can yield pathogens that are liquids, powders, and slurries. Powdered agents pose the most serious hazard to responders because they can easily become resuspended, resulting in an increased inhalation and contamination hazard.

Radiological Labs

Radiological laboratories may not look like a traditional laboratory. They may simply be a staging area where assembly takes place. Manufacturers may use radiological materials to create radiological dispersal devices (RDD), radiological exposure devices (RED), or improvised nuclear devices (IND). Examples of radiological materials may include **(Figure 15.27)**:

- Industrial radiography sources **(Figure 15.28)**
- Food, blood, and medical instrument sterilizers
- Radiation treatment machines
- Radiopharmaceuticals
- Soil density and well logging gauges
- Radiochemistry materials
- Nuclear power reactor fuel or related research
- Radioisotope thermoelectric generators (RTG)
- Smoke detectors
- Lantern mantles
- Fiestaware
- Old watches, dials, and gauges
- Rocks, granite, marble, and ceramic tile

Figure 15.27 Examples of radiological materials including smoke detectors, a Fiestaware dish, lantern mantles, and other sources.

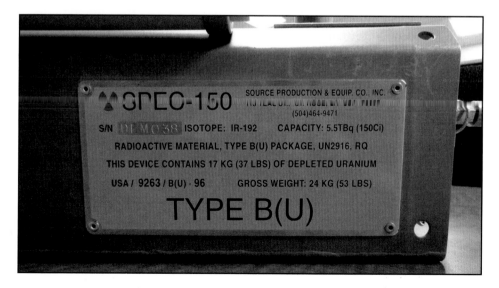

Figure 15.28 An industrial radiography source.

Specially trained personnel and equipment are required to mitigate and detect the hazards in radiological laboratories. This equipment must include a dose rate meter and a contamination meter. These meters measure two different things: gamma radiation for a dose rate and alpha/beta radiation for contamination.

NOTE: Most contamination meters will also detect gamma radiation. The operator must understand the difference between alpha, beta, gamma, and neutron radiation and know how to operate the equipment used in his or her jurisdiction.

Neutron radiation is uncommon and requires specialized detection equipment **(Figure 15.29)**. Radiological material in a clandestine lab may include individual radioactive sources (such as a dish or a rock) or it may be loose, causing contamination (such as dirt, dust, or liquids). Individual sources present less of a hazard internally because they are not airborne. Contamination, by contrast, often presents as an ingestion or inhalation hazard due to the free-moving nature of the material.

Remember to always keep your exposure to radiological materials as low as reasonably achievable (ALARA). To achieve ALARA exposure, you should decrease exposure time to the radiological source; increase your distance from the radiological source; and always use appropriate shielding whenever possible.

Indicators of a radiological laboratory may include:

- Trefoil symbols
- Illness or injury consistent with radiation exposure
- A dose rate reading on a radiation detector or "pager"
- Biological indicators (recent illnesses, hospitalizations, burns, signs of acute radiation sickness)
- Presence of dosimetry (such as TLD, EPD)
- Paperwork, maps, or plans describing the use of radiological materials
- Tongs or tools that provide distance for the user
- Presence of shielding materials (such as lead containers "pigs," lead shot, lead bricks, x-ray aprons)
- Presence of neutron shielding materials (such as paraffin wax, aquarium of oil or water, dense plastics, concrete) **(Figure 15.30)**
- Presence of strong concentrated acids (such as nitric and sulfuric)
- Milling tools or lathes
- Metal grinders
- Presence of medical isotopes, an abnormal number of or disassembled smoke detectors, or industrial radiological sources such as radiography cameras, soil density gauges, or industrial/commercial equipment that is not typical for the occupancy.
- PPE
- Presence of medical countermeasures such as chelating agents or potassium iodide

Figure 15.29 A neutron radiation detector.

Figure 15.30 Dense plastic can be used as shielding from some radiation types.

WARNING!

Special nuclear materials (SNM) emit neutrons which can be reflected and potentiated by materials and objects containing hydrogen (such as water, oil, plastics, wax, and people). Only specialized equipment (neutron detectors) can detect neutrons. The presence of responders near SNM could cause an increase in neutron radiation exposure.

Operations at Illicit Labs

In the U.S., SOP/SOGs for illicit lab responses need to follow rules established by OSHA 1910.120. SOPs should include staffing of positions and activities related to decon, safety, rescue, hazmat, and fire fighting operations. These rules apply to all agencies involved in the response (fire service, law enforcement, environmental contractors, etc).

Responders must preserve evidence as best as they can while mitigating the incident. This attention to potential evidence is particularly important at incidents that are highly likely to be crime scenes.

Chapter 14 details evidence recovery and forensic operations. The processes involved in illicit drug production change frequently; therefore, hazmat/WMD responders must frequently interact with law enforcement drug response teams to learn about new techniques. A responder should also know how to contact his or her state radiation protection office.

State Radiation Protection Office

First responders should know how to contact their state's radiation protection office. If they detect radiation at an incident during a risk-based response, they should immediately contact this office. Each state's radiation protection office will be able to leverage resources at the state and federal level that will enable an effective response and ongoing public protection. The AHJ may have guidance on the information that responders should provide to this office. The FBI headquarters switchboard can assist with contacting the appropriate radiation protection office.

The law enforcement agency having jurisdiction will direct all on-site activities. A search warrant or other protocol will establish parameters for law enforcement authority to seize an illicit laboratory. In order to maintain responders' situational awareness and to enable them to quickly and accurately analyze a situation at an illicit laboratory, law enforcement should provide responders with regular threat briefings on anticipated hazmat or WMD threats.

Because of the many issues and possible outcomes at stake at illicit laboratories, law enforcement jurisdiction, investigative guidelines, and investigative priorities are complicated and ever-changing. Investigative authority at illicit labs may differ based on the type of lab, crime(s) involved, law enforcement jurisdiction, and other factors. Identify specific jurisdictional situations before an illicit laboratory is found. Agencies should review these jurisdictional divisions on a routine basis. In the U.S., jurisdictional confirmation may be needed from:

- Local or state law enforcement authorities
- Federal Bureau of Investigation
- Drug Enforcement Agency
- United States Postal Inspection Service
- Environment Protection Agency

Operations should begin at an illicit laboratory only after a careful risk analysis and an effective Incident Action Plan are developed. Creating this plan can be a complicated process. Most illicit lab responses involve multiple agencies. Each agency has a specific jurisdiction under which that agency will take lead responsibility. Planning a response requires coordination among these agencies. Coordinating agencies may pose a great deal of complications because there are no two illicit labs alike.

Coordination challenges that may comprise more than one entity include:

- Securing and preserving the scene with law enforcement
- Site reconnaissance and hazard identification with bomb squad personnel **(Figure 15.31)**
- Determining atmospheric hazards through air monitoring/detection
- Mitigating immediate hazards while preserving evidence
- Coordinating crime scene operation with the law enforcement agency having investigative authority
- Documenting personnel and scene activities associated with the incident

Figure 15.31 At illicit labs, first responders are likely to conduct site reconnaissance and hazard identification with EOD (bomb squad) personnel.

As explained in Chapter 14, the Operations level responder should become familiar with the local or state and federal procedures for reports and documentation at a crime scene because local procedures vary for crime scene documentation. For example, some jurisdictions do not allow the use of video documentation, while others do not permit digital photographs.

Law enforcement will be responsible for securing and preserving response scenes involving illicit laboratories. **Skill Sheet 15-2** provides basic steps for identifying and securing an illicit lab.

Law enforcement duties may include the following:

- Neutralizing tactical threats
- Rendering safe any explosive ordnance or booby traps
- Taking full accountability and identifying all personnel in the crime scene
- Documenting any items disturbed within the crime scene
- Protecting evidence from potential damage or destruction

Hazmat and bomb squad teams must work together to resolve situations found within illicit drug or WMD laboratories. In many of these illicit labs, hazards will include the following combinations of devices and materials:

- Explosive/chemical
- Explosive/radiological
- Explosive/biological

Teams may also have to work together to clear booby traps and to define the character of explosive ordnance and hazmat hazards. Response agencies should coordinate between bomb squad and hazmat assets in order to prepare for situations involving multiple hazards.

As part of the site risk-assessment process, responders will need to assess atmospheric hazards. Local response plans should ensure the availability and use of proper equipment based on the risks identified during assessment/size-up.

Monitoring and detection equipment should include:

- CGI
- Multi-gas meters
- PID
- Reagent paper
- Radiological monitoring equipment that can detect alpha, beta, and gamma radiation

Responders should coordinate with appropriate law enforcement agencies before conducting reconnaissance operations. Agencies may need information from reconnaissance operations to conduct briefings on intelligence involving the laboratory or site. As in all responses, responses to illicit drug or WMD laboratories should apply standard priorities of scene operations as a part of the response plan. Responders should mitigate immediate hazards while making best efforts to preserve evidence. When you plan a response to potential illicit drug or WMD laboratories, your priority is to coordinate crime scene operation with the law enforcement agency having investigative authority. Look for signs of criminal activity involving chemical, biological, radiological, or explosive materials and devices.

Operations level responders will support law enforcement operations within their scope of training. They should also be familiar with crime scene jurisdictions and procedures as outlined by the AHJ, such as:

- Investigative law enforcement leadership
- Search warrant requirements
- Rules of evidence
- Crime scene documentation
- Photography policies
- Evidence of custodial requirements and chain of custody
- Specific requirements set forth by the prosecuting attorney

NOTE: Ops level responders may be tasked with support of law enforcement officials at an incident.

An effective Incident Action Plan at an illicit lab is based on the following:

- Provides careful analysis of the situation
- Considers the type of laboratory encountered
- Identifies the hazards presented inside and outside of the lab
- Recognizes the jurisdictional parameters of the agencies that will aid in preserving the crime scene
- Provides the safest possible response to the inherently dangerous task that responders face
- Includes considerations for PPE and decontamination
- Provides for the protection and legal disposal of lab equipment and chemicals not needed for evidence

Personal Protective Equipment

At illicit lab responses, responders should assess the following to determine PPE selection:

- The mission and expected hazards
- Intelligence about laboratory operations and contents
- Outward warning signs
- Detection clues, such as any protective clothing used by the operator, activity of animals in the laboratory, and interviews with neighbors

Law enforcement activities may require PPE designed for tactical law enforcement operations. This PPE must be evaluated to ensure it is appropriate for the anticipated hazards as identified during the risk-assessment process. Bomb-squad operations will require the appropriate level of protective garment. Responders may need to augment this garment with incident-appropriate chemical protective clothing. Agencies may develop local procedures to dictate the appropriate PPE for each situation.

Decontamination

Responders should base decontamination procedures upon the results of the risk-assessment process. Tactical entries may require the use of emergency or technical decontamination procedures specifically focused upon the hazards and special needs associated with tactical operations. For example, responders may need to develop procedures for decontaminating weapons, ammunitions, and other specialized equipment **(Figure 15.32, p. 708)**.

What This Means To You

When is Decontamination Necessary?

Even when an incident does not show a clear indication that hazardous materials are in the environment, decontamination may be necessary. For example, if a CBRN SWAT team makes a dynamic entry into an illicit lab and clears an area of suspects, team members may have come into contact with any of the chemicals in that environment. Any people in the area, including any suspects and responders, should be treated as though they were contaminated – and should be appropriately decontaminated. The AHJ and the type of hazard in the environment will determine the decon process, including whether and how to include illicit lab equipment.

Decontamination for tactical scenarios should be based upon a rapid deployment. Agencies should anticipate four potential sources requiring decontamination:

- Uninjured tactical operators and their equipment
- Injured tactical operators
- Uninjured suspects
- Injured suspects

Figure 15.32 Illicit lab SOPs/SOGs must include procedures for decontaminating weapons and ammunition.

Responders must coordinate decontamination procedures with law enforcement tactical teams to resolve potential issues, such as considerations of scene/perimeter, resources (such as law enforcement weapons), equipment, and personnel security, canines, and decontamination procedures. Law enforcement should be engaged with incidents up to the Operations level, as determined by the AHJ.

Remediation of Illicit Labs

Remediation — Fixing or correcting a fault, error, or deficiency.

Responders must familiarize themselves with local, state, and federal agency policies concerning the **remediation** of illicit drug/WMD scenes. Some jurisdictions may hire private contractors to perform remediation activities. Assistance and information should come from:

- Local or state health departments
- Emergency management agencies
- DEA
- EPA
- State/local environmental agencies/departments

As part of an ongoing education and awareness program, agencies should provide regular briefings. Information gathered from these briefings should be considered when developing response remediation plans and incorporated into agencies' local emergency response plans and SOP/Gs, per the AHJ. For example, some jurisdictions require that building deeds, even those of remodeled properties and occupancies, should include the historical presence of an illicit lab. In other cases, the information is not shared to future tenants.

Training and exercising together increases the success and safety of joint operations. Joint training allows agencies to locate the deficiencies in operation, the needed equipment, and/or the assistance needed by other agencies before the actual event occurs. Agencies should write an After Action Report after each incident, with input from all involved, and a copy furnished to all participating agencies/departments.

Chapter Review

Answer the following questions to review the information provided in this chapter.

1. List hazards that may be found at illicit labs.

2. What are clues to the presence of illicit drug labs?

3. What are characteristics of an illegal chemical agent lab?

4. Why might explosives labs lack the presence of laboratory equipment?

5. Explain why an explosives lab might look like an illicit drug lab.

6. What are characteristics of an illegal biological lab?

7. What is ALARA?

8. Give examples of equipment that may be found in a radiological lab.

9. How are PPE needs determined at an illicit lab incident?

10. How are decontamination needs determined at an illicit lab incident?

11. List agencies that can provide assistance and information for remediation of illicit laboratories.

Step 1: Upon suspicion of an illicit lab, Expolsive Ordnance Device (EOD) personnel should be notified of possible response.

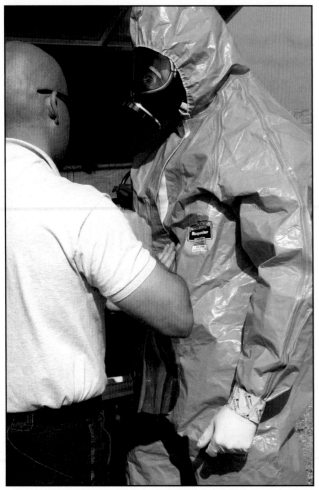

Step 4: Approach the scene carefully, looking for anything suspicious or unusual.

Step 2: Make preparations for safe entry, including appropriate PPE and correct safety procedures.

Step 3: Maintain situational awareness at all times.

15-1
Identify and avoid booby traps at an illicit laboratory.

SKILL SHEETS

Step 5: Before opening doors or windows, examine for any signs of tampering or booby traps. Start low and work upwards, looking for wires, trigger devices, or items that may fall upon opening.

Step 6: If nothing is found, open door slowly and carefully. Proceed cautiously into room.

Step 7: Upon entering, do not touch or change the environment in any way. This includes but is not limited to turning lights or HVAC units on or off or turning electricity to building off.

Step 8: Examine the room in sections; floor to waist, waist to chin, chin to ceiling, and false ceilings if applicable. Look for wires, bottles, pipes, trip wires, or anything out of the ordinary or that arouses your curiosity.

Step 9: If any suspicious items are noticed, back out of the area, retracing your footsteps. Contact the Explosive Ordnance Device (EOD) personnel immediately.

Step 10: Upon their arrival, brief bomb squad personnel on findings.

Step 11: Follow Explosive Ordnance Device (EOD) personnel instructions for proceeding.

Step 1: Work under guidance of hazmat technician, Allied Professional including law enforcement personnel or others with similar authority, an emergency response

plan, or SOPs.

Step 2: Secure the scene.

Step 3: Identify type of laboratory.

Step 4: Identify potential hazards.

Step 5: Select and use appropriate PPE.

Step 6: Protect exposures and personnel.

Step 7: Follow safety procedures.

Step 8: Minimize/avoid hazards.

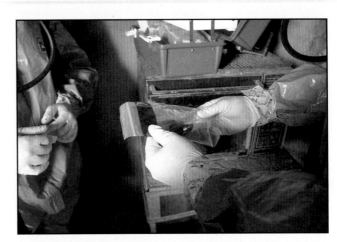

Step 9: Use appropriate control procedures.

Step 10: Identify and preserve evidence.

Step 11: Decontaminate personnel, victims, tools, and equipment.

Step 12: Document and report evidence operations.

Appendices

Courtesy of Rich Mahaney.

Appendix A
Chapter and Page Correlation to NFPA Requirements

NFPA 1072, 2017 JPRs	Chapter References	Page References
4.2.1	1, 2, 3	11-12, 20-28, 45-107, 114-127
4.3.1	1, 2, 3	23-24, 99-104, 114-129
4.4.1	3	113-114
5.2.1	4, 5, 7, 8	137-205, 211-281, 332-334, 365-366, 368-370
5.3.1	1, 4, 6, 8, 9, 10	23-24, 173-175, 203-205, 289-314, 412-413, 421-454, 462-463, 479-483, 487
5.4.1	7, 9, 10	321-354, 421-442, 455-458, 460-467, 486
5.5.1	1, 9, 10	23-24, 460-466, 479-483, 487
5.6.1	7	351-354
6.2.1	7, 9, 10	421-467, 487, 488-496
6.3.1	9, 10	442-454, 460-466, 483-486, 489-494, 496-503, 506-516
6.4.1	9, 10	442-454, 460-466, 483-486, 488-496, 508-516
6.5.1	9, 14	442-454, 460-466, 651-670
6.6.1	9, 13	442-454, 460-466, 607-638
6.7.1	9, 11	442-454, 460-466, 531-572
6.8.1	9, 12	442-454, 460-466, 587-600
6.9.1	9, 15	442-454, 460-466, 677-709

NFPA 472, 2013 Competencies	Chapter References	Page References
4.2.1(1)	1	11
4.2.1(2)	2, 4	76-78, 177-198
4.2.1(3)	4	177-198
4.2.1(4)	1	12
4.2.1(5) and (6)	2	46-72
4.2.1(7)(a) – (f)	2	67, 68, 79-92, 94-95
4.2.1(8)	2	86-88
4.2.1(9)	2	72-81
4.2.1(10)(a) – (g)	2	99-102

NFPA® 472, 2013 Competencies	Chapter References	Page References
4.2.1(11)	2	104-107
4.2.1(12)	2	105
4.2.1(13)	8	365-366
4.2.1(14)	8	365
4.2.1(15)	8	365, 390-392
4.2.1(16)	8	365, 404
4.2.1(17)	8	365, 409
4.2.1(18)	8	412-413
4.2.1(19)	8	365, 371
4.2.1(20)	8	368-370
4.2.2(1)	2	48
4.2.2(2)	2	102
4.2.2(3)	2	103
4.2.3(1)	3	114
4.2.3(2)	3	119
4.4.1(1)	1, 6	16, 289-290
4.4.1(2)	1	17-18
4.4.1(3)(a)	3	124
4.4.1(3)(b)	4	201
4.4.1(3)(c) and (d)	1	20-28
4.4.1(4)(a) – (c)	3	114-127
4.4.1(5)(a) – (d)	3	120
4.4.1(6)(a) – (c)	3	127-129
4.4.1(7)	3	121
4.4.1(8)	3	126
4.4.1(9)(a) and (b)	3	122, 125-127
4.4.1(10)	3	122, 125-127
4.4.1(11) and (12)	3	128-129
4.4.2	3	113-114
5.2.1	5	225-282
5.2.1.1.1(1) – (3)	5	252-259
5.2.1.1.2(1) – (3)(b)	5	264-270
5.2.1.1.3(1) – (7)	5	241-252
5.2.1.1.4(1) – (3)	5	234-240
5.2.1.1.5(1) – (5)	5	273-282

NFPA® 472, 2013 Competencies	Chapter References	Page References
5.2.1.1.6(1) – (2)	5	271-272, 277-278
5.2.1.1.7(1) – (5)	5	273-275
5.2.1.2	5	234-282
5.2.1.2.1(1) – (3)	5	241-252, 261, 271
5.2.1.2.2	2, 5	85-98, 234-240
5.2.1.2.3	2	72-104
5.2.1.3.1(1) – (3)	2	69, 275-277
5.2.1.3.2(1) – (6)	2	93, 96-97
5.2.1.3.3	4	165, 194-196
5.2.1.4	4, 6	199-205, 293-299
5.2.1.5	2, 4	98-104, 199-200
5.2.1.6	8	362-370
5.2.2(1)	4	177-199
5.2.2(2) – 5.2.2(3)(j)	2	101-102
5.2.2(4)(a) – (c)	4	203-205
5.2.2(5)	2, 4, 5	99-102, 203-205, 276-277
5.2.2(6)	7	332-334
5.2.2(7)	7	332-334
5.2.2(8)(a) – (d)	4	163-170
5.2.3(1)(a)(i) – (xv)	4	145-177
5.2.3(1)(b)(i)	10	479
5.2.3(1)(b)(ii)	10	480, 462-463
5.2.3(1)(b)(iii)	1	21-24
5.2.3(1)(b)(iv)	4	176
5.2.3(1)(b)(v) – (vi)	1	21-22
5.2.3(2) – (7)	5	211-225
5.2.3(8)(a) – (j)	4	157-158, 163-173
5.2.3(9)(a) – (g)	8	390
5.2.4(1) – (4)	6	293-299
5.2.4(5)	4	168-170
5.3.1(1)		293-299
5.3.1(2)	6	310-314
5.3.1(3)	7	343
5.3.1(4)	8	368-371
5.3.2(1)	6	310-314

NFPA® 472, 2013 Competencies	Chapter References	Page References
5.3.2(2)	7	342-343
5.3.3(1)(a) – (b)	9	421-431
5.3.3(2)(a) – (b)iii	9	432-440
5.3.4(1) – (5)	10	479-487
5.4.1(1) – (2)	7	335-339
5.4.1(3)(a) and (b)	7	344-347
5.4.1(4)	10	518
5.4.1(5)(a) and (b)	7	323-324
5.4.1(6)	7	344
5.4.2	7	349-351
5.4.3(1)	1	10-20
5.4.3(2)	6	299-303
5.4.3(3)	7	321
5.4.3(4)(a)	7	323-324
5.4.3(4)(b)	7	330-331
5.4.3(5)	7	325
5.4.3(6) and (7)	7	332-334
5.4.4(1) – (3)	7	339-342
5.4.4(4) and (5)	10	455-458
5.4.4(6) and (7)	10	466-467
5.5.1(1) – 5.5.2(1)	7	351-354
5.5.2(2)	7	341
6.2.3.1(1) – 6.2.5.1	9	421-467
6.3.3.1 – 6.4.6.1(4)	10	479-516
6.5.2.1(1)(a) – 6.5.4.2	14	651-670
6.6.3.1(1) – 6.6.4.2	13	607-637
6.7.3.1 – 6.7.4.2	11	531-571
6.8.3.1(1) – 6.8.4.1(5)	12	587-600
6.9. 2.1(1) – 6.9.4.1.5	15	677-709

Appendix B
OSHA Plan States

OSHA State-Plan States and Non-State-Plan States

State-Plan States	Non-State-Plan States
Alaska	Alabama
Arizona	Arkansas
California	Colorado
Connecticut (state and local government employees only)	Delaware
Hawaii	District of Columbia
Indiana	Florida
Iowa	Georgia
Kentucky	Guam
Maryland	Idaho
Michigan	Illinois
Minnesota	Kansas
Nevada	Louisiana
New Mexico	Maine
New York (state and local government employees only)	Massachusetts
North Carolina	Mississippi
Oregon	Missouri
Puerto Rico	Montana
South Carolina	Nebraska
Tennessee	New Hampshire
Utah	New Jersey
Vermont	North Dakota
Virginia	Ohio
Virgin Islands	Oklahoma
Washington	Pennsylvania
Wyoming	Rhode Island
	South Dakota
	Texas
	West Virginia
	Wisconsin

Appendix C
Sample Standard Operating Guideline

TUALATIN VALLEY FIRE AND RESCUE
INCIDENT COMMAND MANUAL

SERIES 300X

OPERATIONAL GUIDELINE
HAZARDOUS MATERIALS RESPONSE

PURPOSE

To provide a standard by which companies trained to the "First Responder Operations" level respond to hazardous materials incidents.

DEFINITIONS

<u>First Responder - Operations</u> - A level of training for first responders to hazardous materials incidents, required by federal and state law; as defined in Oregon Administrative Rule (OAR) 437-01-100(q).

<u>Full Protective Clothing</u> - As it relates to hazardous materials response, full protective clothing means turnouts and SCBA.

<u>On-Scene Commander</u> - A level of training for Incident Commanders on hazardous materials incidents, required by federal and state law; as defined by OAR 437-01-100(q).

<u>Responsible Party</u> - Federal and state regulators assign responsibility for incident clean-up (and costs) to the party who is responsible for the hazardous materials incident (i.e., a fixed facility, transportation agent, etc.).

<u>HMRT</u> - Hazardous Materials Response Team.

<u>Hazardous Materials Group Supervisor (HMRT Leader)</u> - HazMat Group Supervisor reports to the Incident Commander (or Operations Section Chief, if staffed) and is responsible for hazardous materials tactical operations. The HazMat Group Supervisor position is staffed by the Hazardous Materials Response Team Leader.

*   <u>Emergency Response Guidebook</u> - North American Emergency Response Guidebook; formerly "DOT Emergency Response Guidebook."

PROCEDURES

I.   **TRAINING REQUIREMENTS**

    A.   All response personnel must meet the training requirements for "First Responder – Operations" level.

    B.   Incident Commanders on hazardous materials incidents must meet the training requirements for "On-Scene Commander".

II.   INCIDENT COMMANDER

A.   All incidents involving hazardous materials in a spill, release or fire, may require an Incident Commander trained to the "On-Scene Commander" level. All Battalion Chiefs and ICs on the Overhead Team are trained and required to maintain qualifications to the "On-Scene Commander" level.

B.   The Incident Commander may call for a full or partial HMRT response if incident mitigation is beyond the training and capabilities of a company response. The IC may also call for technical assistance from the HMRT without a response to the incident site, if the situation warrants.

III.   COMPANY FUNCTIONS

A.   Companies will respond for the purpose of protecting nearby persons, property or the environment from a hazardous materials release.

B.   Companies will respond in a *defensive* fashion without coming in contact with the release or taking actions to stop a release that would place them in danger of contact.

C.   The primary function of the Operations level responder is to contain the release from a safe distance, keeping it from spreading and protect exposures. The basic functions are:
- isolate the hazard area and control access
- hazard and risk assessment
- basic control, containment and/or confinement procedures appropriate to the level of training and personal protective clothing and equipment.

D.   Companies will not take any actions on hazardous materials incidents that cannot be safely performed in full protective clothing.

IV.   HAZARDOUS MATERIALS RESPONSE AND OPERATIONS

A.   While enroute to the scene:

1.   Contact Fire Comm and obtain available information regarding:
   a.   The nature of the incident, e.g., fixed facility, transportation related. etc.
   b.   The type of product(s) involved, if known.
   c.   The best direction for approaching the scene from upwind, upgrade and upstream.
   d.   Who is on-scene that may have information on the nature of the incident.
   e.   Any information on the incident conditions that may be known and can be provided while enroute to the incident scene.

*   2.   The HazMat Team may be contacted via Fire Comm for technical assistance or response, as appropriate.

3.   Approach the incident scene with caution.
   a.   Approach the incident scene from upwind, upgrade, upstream or at a right angle to the wind direction and/or gradient.
   b.   Consider escape routes. Be aware of situations that require entering areas with egress restrictions, such as fenced compounds.
   c.   Position vehicle/apparatus headed away from the incident scene at a safe distance.

B.   On Arrival

    1.   Establish Command and give size-up.

    2.   Establish a Unified Command if multiple agencies/jurisdictions are involved.

    3.   Ensure a qualified "On-Scene Commander" (i.e., Battalion Chief) is enroute to the scene.

    4.   Continuously evaluate need for HazMat Team technical assistance or response.

C.   Establish Safe Zone and Control Access

    1.   Determine the hazard area and establish the Hot Zone, Warm Zone and Cold Zone boundaries.

        a.   Based on initial observations, identify a safe distance for initial incident isolation to begin. Some recommendations include:
- Single drum, not leaking — minimum 150' in all directions
- Single drum, leaking — 500' in all directions
- Tank car or tank truck with BLEVE potential - half mile in all directions

        b.   Isolate and deny entry to:
- The general public
- Anyone not in proper protective clothing and equipment
- Anyone without a specific assignment

    2.   Establish the Command Post in the Cold Zone.

    3.   Identify and establish the Staging Area location in the Cold Zone.

    4.   Communicate the Zone information, Command Post and Staging Area locations to Fire Comm and incoming units.

    5.   Determine a safe approach for incoming units and direct them to locations at the Safe Zone Perimeter that will facilitate isolation of the incident, i.e., intersections to block and re-direct traffic, etc. All others should be directed to the Staging Area until assigned.

    6.   Request police assistance as needed to:

        a.   Handle Cold Zone Perimeter control to relieve fire units for incident mitigation.

        b.   Handle public evacuations.

        c.   Handle public notification for sheltering in place

    7.   While isolating the incident scene:

        a.   Treat all vapor clouds as being toxic and handle accordingly.

        b.   Do not walk into, through or touch any spilled materials.

        c.   Observe local on-site weather and wind conditions and adjust accordingly.

        d.   Position at a safe distance and utilize your binoculars!

*   D.   Attempt to Identify the Product.

If the product is *known*, proceed to Section V and isolate in accordance with appropriate Emergency Response Guidebook recommendations. Record observations on the hazmat incident worksheet. (Provide the diagram to the incoming Battalion Chief or HazMat Response Team.)

If the product is *unknown,* from a safe distance attempt to gather as much information as possible.

Use Emergency Response Guide #111 isolation recommendations until the material is identified. Record observations on the hazmat incident worksheet.

1.   Life Safety is the number one priority. Do not rush into the scene to effect a rescue without first identifying the hazards.

2.   Attempt to identify outward warning signs that are indicators of the presence of hazardous materials. These include:
   a.   Individuals that have collapsed or are vomiting inside the hazardous area (HMRT response).
   b.   Any evidence of fire, as indicated by smoke, greatly increases all hazards.
   c.   A loud roar of increasing pitch from a container's operating relief valve (HMRT response).
   d.   Evidence of a leak, indicated by a hissing sound.
   e.   Birds and insects falling out of the sky (HMRT response).

AND/OR

3.   Attempt to identify the material(s) involved by using:
   a.   Placards/labels
   b.   Container markings
   c.   Driver/operator provided information including shipping papers.

4.   After determining product:
   a.   Perform rescue, if needed, using safety guidelines related to that product.
   b.   Re-evaluate distances for isolated area.

5.   Communicate your observations to Fire Comm.

6.   Anticipate shifting winds when establishing perimeters; consult with the weather service to obtain accurate forecasts of changes that might impact your incident scene and perimeters.

7.   Eliminate ignition sources if flammable materials are involved. Remember that non-flammable materials, such as anhydrous ammonia are, in fact, flammable, so always identify if the product has a flammable range.

8.   Request additional fire, law enforcement and public works resources, as needed, to secure the incident scene and maintain perimeter control.

*

9.   If large dikes and dams need to be built to control spill, consider requisition for heavy equipment and/or assistance of public works resources.

E.   Conduct a Risk/Benefit Analysis which includes asking the following questions in relation to the incident you are addressing:

1.   What would the outcome be if we did absolutely nothing and allowed the incident to go through natural stabilization?

2.   Once you have identified the outcomes of natural stabilization, the next question you should ask is "Can we change the outcomes of natural stabilization?"

3.   If the answer to this question is "NO", then isolate the hazard area, deny entry, and protect exposures such as people, the environment and adjacent property/equipment.

4. If the answer to this question is "YES", then the next question to ask is "What is the cost of my intervention?"

IF THE INCIDENT COMMANDER DETERMINES DEFENSIVE OPERATIONS CAN STABILIZE/ CONTAIN THE INCIDENT *AND* IT CAN BE DONE IN FULL PROTECTIVE CLOTHING (TURNOUTS AND SCBA), THE IC SHALL CONDUCT OPERATIONS IN ACCORDANCE WITH THE "DEFENSIVE OPERATIONAL GUIDELINES".

V. **DEFENSIVE OPERATIONAL GUIDELINES**
A. Attempt to stop/slow/control leak using defensive techniques (such as turning off a valve, etc.).

B. If the leak cannot be stopped, utilize an appropriate containment procedure to prevent the material from flowing and increasing the exposed surface area (i.e., using dirt or absorbent).

VI. **DECONTAMINATION:** Perform field decontamination as directed by the Incident Commander and/or HazMat Response Team.

NOTE: *ALL CONTAMINATED PATIENTS MUST BE DECONTAMINATED OR PACKAGED FOR TRANSPORT IN A WAY TO PREVENT CONTAMINATION OF TRANSPORT UNITS AND HOSPITALS.*

VII. **CLEAN-UP**
A. If the incident is on a roadway or public access area, the Incident Commander must ensure that a public safety agency (coordinate with law enforcement officials, if available) remains on-scene to continue isolation procedures and standby until the clean-up company arrives.

B. If a responsible party is not on-scene and making arrangements for clean-up and disposal, contact the on-duty HMRT Team Leader for further instructions.

NOTE: *FIRE DEPARTMENT PERSONNEL SHALL NOT ENGAGE IN CLEAN-UP OPERATIONS. THE APPROPRIATE ROLE IS CONTAINMENT/ STABILIZATION. DO NOT TAKE HAZARDOUS MATERIALS FROM AN INCIDENT TO ANY FIRE DISTRICT FACILITY.*

VIII. **CONDUCT TERMINATION PROCEDURES**
A. Prior to the demobilization and release of any equipment from the scene, conduct a debriefing of all response personnel (including cooperating agencies).

B. An effective debriefing should:
1. Inform *all responders* exactly what hazardous materials were involved and the accompanying signs and symptoms of exposure.

2. Provide information for personal exposure records.

3. Identify equipment damage and unsafe conditions requiring immediate attention or isolation for further evaluation.

4. Conduct a post-incident analysis and critique. This may be done at the station.

HAZARDOUS MATERIALS RESPONSE
CHECKLIST

CHECKLIST USE

The checklist should be considered as a minimum requirement for this position. Users of this manual should feel free to augment this list as necessary. Note that some activities are one-time actions and others are on-going or repetitive for the duration of an incident.

____ While enroute to the scene, you may utilize the HazMat Team as a technical resource (contact via Fire Comm).

____ Approach incident cautiously, uphill, upwind, park headed away from incident, consider escape routes.

____ Establish Command.

____ Establish and maintain site access control. Establish initial Zones (Hot: min. of 150'; warm; cold). Establish Command Post and Staging locations.

____ Attempt to identify materials involved by using placards/labels, container markings, shipping papers and driver provided information. Use Guide #111 if spilled product is unknown.

____ Perform rescue only when the rescue operation can be done safely.

____ Request additional fire, law enforcement and public works resources as needed. Consider requisition for heavy equipment.

____ Conduct risk/benefit analysis.

____ If the Incident Commander determines defensive operations can stabilize/contain the incident, conduct defensive operations.
- Attempt to stop/slow/control leak using defensive techniques (such as turning off a valve, etc.).
- If the leak cannot be stopped, utilize an appropriate containment procedure to prevent the material from flowing and increasing the exposed surface area.

____ Perform field decontamination as directed by the Incident Commander.

____ Clean-Up

If the incident is on a roadway or public access area, the Incident Commander must ensure that a public safety agency remains on-scene to continue isolation procedures and standby until the clean-up company arrives. If the responsible party is not on-scene and making arrangements for clean-up and disposal, contact the on-duty HMRT Team Leader for further instructions.

NOTE: Fire Department personnel shall not engage in clean-up operations.

____ Prior to the demobilization and release of any equipment from the scene, conduct a debriefing of all response personnel (including cooperating agencies).

Appendix D
UN Class Placards and Labels

Table D.1 provides the United Nations (UN) placards and labels required for the transportation of dangerous goods.

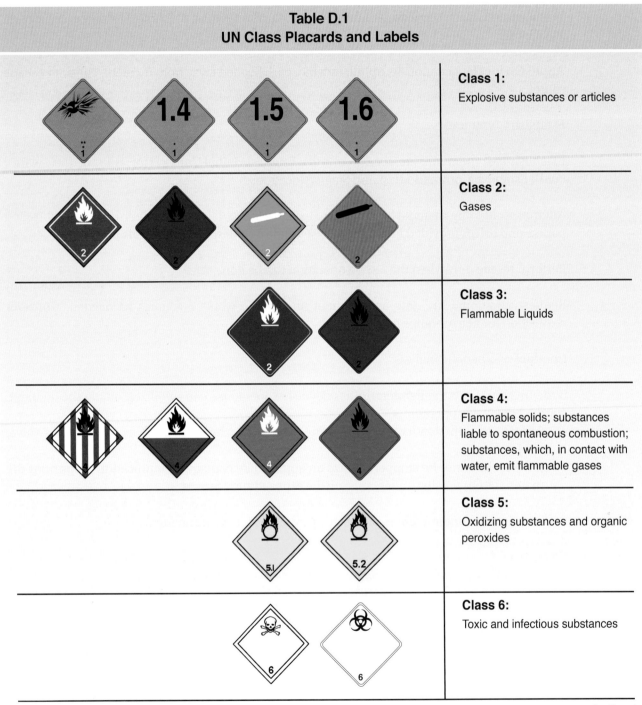

		Class 1: Explosive substances or articles
		Class 2: Gases
		Class 3: Flammable Liquids
		Class 4: Flammable solids; substances liable to spontaneous combustion; substances, which, in contact with water, emit flammable gases
		Class 5: Oxidizing substances and organic peroxides
		Class 6: Toxic and infectious substances

Table D.1
UN Class Placards and Labels

Continued

Class 7:

Radioactive material

Class 8:

Corrosive substances

Class 9:

Miscellaneous dangerous substances and articles

Appendix E
GHS Summary

Global Harmonized System of Classification and Labeling of Chemicals (GHS)

The Globally Harmonized System of Classification and Labeling of Chemicals, or GHS for short, is an effort to create a world-wide, universal chemical hazard communication and container labeling system. This was done because of the large number of hazardous materials and a multitude of systems being used in different countries. For example, in the United States, there are an estimated 650,000 hazardous materials that require communication and labeling. Every manufacturer did it slightly different requiring multiple Material Safety Data Sheets (MSDS) and labels for both domestic and international shipments. Systems in the United States used OSHA requirements and either National Fire Protection (NFPA) diamonds or Hazardous Materials Information System (HMIS) components on the Label.

OSHA has stated unequivocally that HMIS and NFPA ratings, by themselves, are not sufficient for workplace labels since one of the problems with NFPA and HMIS systems is that they do not include a product identifier and do not cover all potential hazards since they are primarily concerned with emergency response. GHS will standardize the system for Safety Data Sheets (SDS) and labels.

GHS provides standardized definitions for chemical hazards, such as flammable and combustible liquids. GHS also addresses classification of chemicals by types of hazard and proposes standardized hazard communication elements, including labels and safety data sheets.

Health, Environmental, and Physical Hazards

Countries with systems have different requirements for hazard definitions as well as information to be included on a label or material safety data sheet. For example, a product may be considered flammable or toxic in one country, but not in another to which it is being shipped. GHS hazard classification criteria were adopted by consensus for physical hazards and key health and environmental classes. Those for health and environmental hazards are:

- Acute toxicity
- Skin corrosive/irritant
- Serious eye damage/eye irritant
- Respiratory or skin sensitization
- Germ cell mutagenicity
- Carcinogenicity
- Reproductive toxin
- Target organ systemic toxicity – single and repeated dose
- Hazardous to the aquatic environment

The categories describing physical hazards include:

- Explosives
- Flammability – gases, aerosols, liquids, solids

- Oxidizers – liquid, solid, gases
- Self-reactive
- Pyrophoric – liquids, solids
- Self-heating
- Organic peroxides
- Corrosive to metals
- Gases under pressure
- Water-activated flammable gases

For each of these hazards standardized label elements were developed including symbols, signal words and hazard statements. A standard format and approach to how GHS information appears on safety data sheets and labels was also developed.

Labels

There are about 35 different types of information that are currently required on labels by different systems. To harmonize, key information elements needed to be identified. Transport pictograms or labels and placards are the standard UN system currently used in most of the world. These new elements or items will be standard on labels.

- Supplier Identifier
- Product Identifier
- Signal Word
- Hazard Statement(s)
- Precautionary Statement(s)
- Pictogram(s)

Supplier Identifier is the Name, Address and Telephone Number of the chemical manufacturer, importer or other responsible party.

Product Identifier is how the hazardous chemical is identified. This can be (but is not limited to) the chemical name, code number or batch number. The manufacturer, importer or distributor can decide the appropriate product identifier. The same product identifier must be both on the label and in section 1 of the SDS.

Signal Words are used to indicate the relative level of severity of the hazard and alert the reader to a potential hazard on the label. There are only two words used as signal words, "Danger" and "Warning." Within a specific hazard class, "Danger" is used for the more severe hazards and "Warning" is used for the less severe hazards. There will only be one signal word on the label no matter how many hazards a chemical may have. If one of the hazards warrants a "Danger" signal word and another warrants the signal word "Warning," then only "Danger" should appear on the label.

Hazard Statements describe the nature of the hazard(s) of a chemical, including, where appropriate, the degree of hazard. For example: "Causes damage to kidneys through prolonged or repeated exposure when absorbed through the skin." All of the applicable hazard statements must appear on the label.

Hazard statements may be combined where appropriate to reduce redundancies and improve readability. The hazard statements are specific to the hazard classification categories, and chemical users should always see the same statement for the same hazards no matter what the chemical is or who produces it.

These are also listed as the "H" codes in Section 2 the SDS. There are 72 individual and 17 combined Hazard statements - these are assigned a unique alphanumerical code which consists of one letter and three numbers as follows:

a) the letter "H" (for "hazard statement");
b) a number designating the type of hazard as follows:
- "2" for physical hazards
- "3" for health hazards
- "4" for environmental hazards
c) two numbers corresponding to the sequential numbering of hazards arising from the intrinsic properties of the substance or mixture, such as explosive properties (codes from 200 to 210), flammability (codes from 220 to 230), etc.

Precautionary Statements describe recommended measures that should be taken to minimize or prevent adverse effects resulting from exposure to the hazardous chemical or improper storage or handling. There are four types of precautionary statements: prevention (to minimize exposure); response (in case of accidental spillage or exposure emergency response, and first-aid); storage; and disposal. For example, a chemical presenting a specific target organ toxicity (repeated exposure) hazard would include the following on the label: "Do not breathe dust. Get attention if you feel unwell. Dispose of container in accordance with local regulations."

Precautionary statements may be combined on the label to save on space and improve readability. For example, "Keep away from heat, spark and open flames," "Store in a well-ventilated place," and "Keep cool" may be combined to read: "Keep away from heat, sparks and open flames and store in a cool, well-ventilated place." Where a chemical is classified for a number of hazards and the precautionary statements are similar, the most stringent statements must be included on the label. In this case, the chemical manufacturer, importer, or distributor may impose an order of precedence where phrases concerning response require rapid action to ensure the health and safety of the exposed person. In the self-reactive hazard category Types C, D, E or F, three of the four precautionary statements for prevention are:

• "Keep away from heat/sparks/open flame/hot surfaces. - No Smoking.";

• "Keep/Store away from clothing/.../ combustible materials";

• "Keep only in original container."

These three precautionary statements could be combined to read: "Keep in original container and away from heat, open flames, combustible materials and hot surfaces. No Smoking."

There are 116 individual and 33 combined Precautionary statements – these are assigned a unique alphanumerical code which consists of one letter and three numbers as follows:

a) the letter "P" (for "precautionary statement");

b) one number designating the type of precautionary statement as follows: - "1" for general precautionary statements

      -   "2" for prevention precautionary statements

      -   "3" for response precautionary statements

      -   "4" for storage precautionary statements

      -   "5" for disposal precautionary statements

c) two numbers (corresponding to the sequential numbering of precautionary statements)

Supplementary Information may provide additional instructions or information that it deems helpful. An example of an item that may be considered supplementary is the personal protective equipment (PPE) pictogram indicating what workers handling the chemical may need to wear to protect them.

Pictograms are graphic symbols used to communicate specific information about the hazards of a chemical. On hazardous chemicals being shipped or transported from a manufacturer, importer or distributor, the required pictograms consist of a red square frame set at a point with a black hazard symbol on a white background, sufficiently wide to be clearly visible. While the GHS uses a total of nine pictograms, OSHA will only enforce the use of eight. The environmental pictogram is not mandatory but may be used to provide additional information.

Health Hazard	Flame	Exclamation Mark
• Carcinogen • Mutagenicity • Reproductive toxicity • Respiratory sensitizer • Target organ toxicity • Aspiration toxicity	• Flammables • Pyrophorics • Self-heating • Emits flammable gas • Self-reactives • Organic peroxides	• Irritant (skin and eye) • Skin sensitizer • Acute toxicity (harmful) • Narcotic effects • Respiratory tract irritant • Hazardous to ozone layer (non-mandatory)
Gas Cylinder	**Corrosion**	**Exploding Bomb**
• Gases under pressure	• Skin Corrosion/ burns • Eye damage • Corrosive to metals	• Explosives • Self-reactives • Organic peroxides
Flame Over Circle	**Environment (Non-mandatory)**	**Skull and Crossbones**
• Oxidizers	• Aquatic toxicity	• Acute toxicity (fatal or toxic)

So a label for a container for methanol would look something like this:

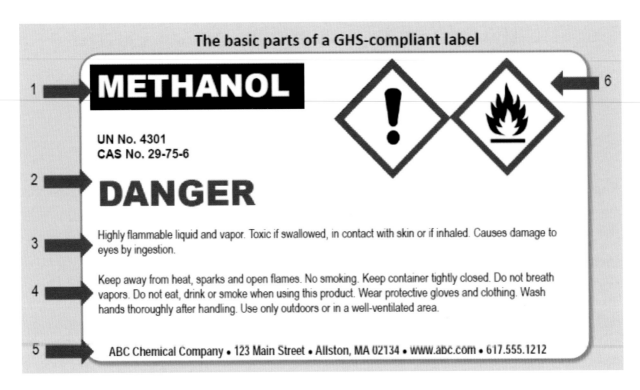

The basic parts of a GHS-compliant label

1. Product identifiers
2. Signal words
3. Hazard statements
4. Precautionary statements
5. Supplier identification
6. Pictograms

Safety Data Sheets

Comparison between SDS and MSDS		
Sections	GHS SDS	OSHA MSDS
1. Product and company identification	• GHS product identifier. • Other means of identification. • Recommended use of the chemical and restrictions on use. • Supplier's details (including name, address, phone number etc). • Emergency phone number.	• Product identity same as on label. • Name address and telephone number of the manufacturer, distributor, employer or other responsible party.
2. Hazards identification	• GHS classification of the substance/mixture and any regional information. • GHS label elements, including precautionary statements. (Hazard symbols may be provided as a graphical reproduction of the symbols in black and white or the name of the symbol, e.g., flame, skull and crossbones.) • Other hazards which do not result in classification (e.g., dust explosion hazard) or are not covered by the GHS.	• health hazards including acute and chronic effects, listing target organs or systems • signs & symptoms of exposure • conditions generally recognized as aggravated by exposure • primary routes of exposure • if listed as a carcinogen by OSHA, IARC, NTP • physical hazards, including the potential for fire, explosion, and reactivity

Figure G.4

3. Composition / information on ingredients	<ul><li>Substance</li><li>Chemical identity</li><li>Common name, synonyms, etc.</li><li>CAS number, EC number, etc.</li><li>Impurities and stabilizing additives which are themselves classified and which contribute to the classification of the substance.</li><li>Mixture</li><li>The chemical identity and concentration or concentration ranges of all ingredients which are hazardous within the meaning of the GHS and are present above their cut-off levels.</li><li>Cut-off level for reproductive toxicity, carcinogenicity and category 1 mutagenicity is ³ 0.1%</li><li>Cut-off level for all other hazard classes is ³ 1%</li><li>Note: For information on ingredients, the competent authority rules for CBI take priority over the rules for product identification</li></ul>	<ul><li>Chemical and common name of ingredients contributing to known hazards</li><li>For untested mixtures, the chemical & common name of ingredients at 1% or more that present a health hazard and those that present a physical hazard in the mixture</li><li>Ingredients at 0.1% or greater, if carcinogens</li></ul>
4. First-aid measures	<ul><li>Description of necessary measures, subdivided according to the different routes of exposure, i.e., inhalation, skin and eye contact and ingestion.</li><li>Most important symptoms/effects, acute and delayed.</li><li>Indication of immediate medical attention and special treatment needed, if necessary.</li></ul>	<ul><li>emergency & first aid procedures</li></ul>
5. Firefighting measures	<ul><li>Suitable (and unsuitable) extinguishing media.</li><li>Specific hazards arising from the chemical (e.g., nature of any hazardous combustion products).</li><li>Special protective equipment and precautions for fire-fighters.</li></ul>	<ul><li>generally applicable control measures</li><li>flammable property information such as flashpoint</li><li>physical hazards including the potential for fire, explosion, and reactivity</li></ul>
6. Accidental release measures	<ul><li>Personal precautions, protective equipment and emergency procedures.</li><li>Environmental precautions.</li><li>Methods and materials for containment and cleaning up.</li></ul>	<ul><li>procedures for clean up of spills and leaks</li></ul>
7. Handling and storage	<ul><li>Precautions for safe handling.</li><li>Conditions for safe storage, including any incompatibilities.</li></ul>	<ul><li>Precautions for safe handling & use, including appropriate hygenic practices.</li></ul>
8. Exposure controls/personal protection	<ul><li>Control parameters (e.g., occupational exposure limit values or biological limit values).</li><li>Appropriate engineering controls.</li><li>Individual protection measures, such as personal protective equipment.</li></ul>	<ul><li>General applicable control measures</li><li>appropriate engineering controls and work practices</li><li>protective measures during maintenance & repair</li><li>personal protective equipment</li><li>Permissible exposure levels, threshold limit values, listed by OSHA, ACGIH, or established company limits.</li></ul>

9. Physical and chemical properties	• Appearance (physical state, color, etc.) • Odor • Odor threshold • pH • melting point/freezing point • initial boiling point and boiling range • flash point: • evaporation rate • flammability (solid, gas) • upper/lower flammability or explosive limits • vapor pressure • vapor density • relative density: • solubility(ies) • partition coefficient: n-octanol/water • auto-ignition temperature • decomposition temperature	• Characteristics of hazardous chemicals such as vapor pressure & density. • Physical hazards including the potential for fire, explosion, and reactivity.
10. Stability and reactivity	• Chemical stability. • Possibility of hazardous reactions. • Conditions to avoid (e.g., static discharge, shock or vibration). • Incompatible materials, • Hazardous decomposition products.	• organic peroxides, pyrophoric, unstable # (reactive), or water-reactive hazards • physical hazards, including reactivity and hazardous polymerization
11. Toxicological information	• Concise but complete and comprehensible description of the various toxicological (health) effects and the available data used to identify those effects, including: • Information on the likely routes of exposure (inhalation, ingestion, skin and eye contact); • Symptoms related to the physical, chemical and toxicological characteristics; • Delayed and immediate effects and also chronic effects from short- and long-term exposure;. • Numerical measures of toxicity (such as acute toxicity estimates).	• See also Section 2 [health hazards Including acute and chronic effects, listing target organs or systems • signs & symptoms of exposure • primary routes of exposure • if listed as a carcinogen by OSHA, IARC, NTP]
12. Ecological information	• Ecotoxicity (aquatic and terrestrial, where available). • Persistence and degradability • Bioaccumulative potential • Mobility in soil • Other adverse effects	• No present requirements.
13. Disposal considerations	• Description of waste residues and information on their safe handling and methods of disposal, including any contaminated packaging.	• No present requirements, • See section 7,

14. Transport information	• UN number. • UN Proper shipping name. • Transport Hazard class(es). • Packing group, if applicable. • Marine pollutant (Y/N). • Special precautions which a user needs to be aware of or needs to comply with in connection with transport or conveyance either within or outside their premises.	• No present requirements,
15. Regulatory information	• Safety, health and environmental regulations specific for the product in question.	• No present requirements.
16. Other information	• Other information including information on preparation and revision of the SDS.	• Date of preparation of MSDS or date of last change

Other changes classification system

The biggest obstacle for emergency responders will be the change from NFPA / HMIS Rating systems and OSHA's Classification / Category System. The GHS system is opposite from NFPA. No symbols or color coding. The change in term to Category will also be different.

NFPA / HMIS	OSHA / GHS Categories
0-4 0-least hazardous 4-most hazardous	1-4 1-most severe hazard 4-least severe hazard • The Hazard category numbers are NOT required to be on labels but are required on SDSs in Section 2. • Numbers are used to CLASSIFY hazards to determine what label information is required.

Global Harmonized System of Classification and Labeling of Chemicals (GHS)

The Globally Harmonized System of Classification and Labeling of Chemicals, or GHS for short, is an effort to create a world-wide, universal chemical hazard communication and container labeling system. This was done because of the large number of hazardous materials and a multitude of systems being used in different countries. For example, in the United States, there are an estimated 650,000 hazardous materials that require communication and labeling. Every manufacturer did it slightly different requiring multiple Material Safety Data Sheets (MSDS) and labels for both domestic and international shipments. Systems in the United States used OSHA requirements and either National Fire Protection (NFPA) diamonds or Hazardous Materials Information System (HMIS) components on the Label.

OSHA has stated unequivocally that HMIS and NFPA ratings, by themselves, are not sufficient for workplace labels since one of the problems with NFPA and HMIS systems is that they do not include a product identifier and do not cover all potential hazards since they are primarily concerned with emergency response. GHS will standardize the system for Safety Data Sheets (SDS) and labels.

GHS provides standardized definitions for chemical hazards, such as flammable and combustible liquids. GHS also addresses classification of chemicals by types of hazard and proposes standardized hazard communication elements, including labels and safety data sheets.

Health, Environmental, and Physical Hazards

Countries with systems have different requirements for hazard definitions as well as information to be included on a label or material safety data sheet. For example, a product may be considered flammable or toxic in one country, but not in another to which it is being shipped. GHS hazard classification criteria were adopted by consensus for physical hazards and key health and environmental classes. Those for health and environmental hazards are:

- Acute toxicity
- Skin corrosive/irritant
- Serious eye damage/eye irritant
- Respiratory or skin sensitization
- Germ cell mutagenicity
- Carcinogenicity
- Reproductive toxin
- Target organ systemic toxicity – single and repeated dose
- Hazardous to the aquatic environment

The categories describing physical hazards include:

- Explosives
- Flammability – gases, aerosols, liquids, solids

Appendix F
Color Codes for Incident Command Positions

From NFPA, Standard on Emergency Services Incident Management System and Command Safety, 2014 Edition, Appendix A

Multi-Agency Events

Command Staff — *White*

Operations Section Chief — *Red*

Planning Section Chief — *Blue*

Logistics Section Chief — *Orange*

Finance/Administration Section Chief — *Green*

HAZARDOUS EVIDENCE COLLECTION AND FIELD SCREENING MATRIX

Location		Case ID		Collect on Team & Initials
Date/Time		Case Agent		
		Scientist		
Radiation det	S/N:	Last Cal.:	Cal. Due:	Alpha Beta Probe
Background	*Rad Dose	**Rad Count		S/N:
5 Gas det	S/N:	Last Cal.:	Cal. Due:	EOD Clear?

Perform fresh air calibration and background in cold zone
* = dose rate expressed as xR/hr using internal GM
** = counts per minute on contact or smear
Record Rad Count for evidence not in containers as inside (no outside)
Record non-zero CO or H₂S MultiRAE readings in comments
pH, H₂O, K and F papers for liquids only

Item #	Location (Floor, Room, Table, etc.)	Description		Suggested Collection Method	Suggested Container Type

*Rad Dose (Ludlum)	**Rad Count (outside)	**Rad Count (inside)	O₂ (%)	Flamm (% LEL)	VOC	pH	H₂O (+/-)	Oxidizer (KI)	Fluor (F)	Comments (Observations Other Screening Methods, Changes to Collection, etc.)
μR/hr mR/hr	cpm kcpm	cpm kcpm			ppb ppm					

Item #	Location (Floor, Room, Table, etc.)	Description		Suggested Collection Method	Suggested Container Type

*Rad Dose (Ludlum)	**Rad Count (outside)	**Rad Count (inside)	O₂ (%)	Flamm (% LEL)	VOC	pH	H₂O (+/-)	Oxidizer (KI)	Fluor (F)	Comments (Observations Other Screening Methods, Changes to Collection, etc.)
μR/hr mR/hr	cpm kcpm	cpm kcpm			ppb ppm					

Item #	Location (Floor, Room, Table, etc.)	Description		Suggested Collection Method	Suggested Container Type

*Rad Dose (Ludlum)	**Rad Count (outside)	**Rad Count (inside)	O₂ (%)	Flamm (% LEL)	VOC	pH	H₂O (+/-)	Oxidizer (KI)	Fluor (F)	Comments (Observations Other Screening Methods, Changes to Collection, etc.)
μR/hr mR/hr	cpm kcpm	cpm kcpm			ppb ppm					

Version February 2015

Page ___ of ___

Continued on back

HAZARDOUS EVIDENCE COLLECTION AND FIELD SCREENING MATRIX

Item #	Location (Floor, Room, Table, etc.)	Description							Suggested Collection Method	Suggested Container Type

*Rad Dose (Ludlum)	**Rad Count (outside)	**Rad Count (inside)	O₂ (%)	Flamm (% LEL)	VOC	pH	H₂O (+/-)	Oxidizer (KI)	Fluor (F)	Comments (Observations, Other Screening Methods, Changes to Collection, etc.)
µR/hr mR/hr	cpm kcpm	cpm kcpm			ppb ppm					

Item #	Location (Floor, Room, Table, etc.)	Description							Suggested Collection Method	Suggested Container Type

*Rad Dose (Ludlum)	**Rad Count (outside)	**Rad Count (inside)	O₂ (%)	Flamm (% LEL)	VOC	pH	H₂O (+/-)	Oxidizer (KI)	Fluor (F)	Comments (Observations, Other Screening Methods, Changes to Collection, etc.)
µR/hr mR/hr	cpm kcpm	cpm kcpm			ppb ppm					

Item #	Location (Floor, Room, Table, etc.)	Description							Suggested Collection Method	Suggested Container Type

*Rad Dose (Ludlum)	**Rad Count (outside)	**Rad Count (inside)	O₂ (%)	Flamm (% LEL)	VOC	pH	H₂O (+/-)	Oxidizer (KI)	Fluor (F)	Comments (Observations, Other Screening Methods, Changes to Collection, etc.)
µR/hr mR/hr	cpm kcpm	cpm kcpm			ppb ppm					

Item #	Location (Floor, Room, Table, etc.)	Description							Suggested Collection Method	Suggested Container Type

*Rad Dose (Ludlum)	**Rad Count (outside)	**Rad Count (inside)	O₂ (%)	Flamm (% LEL)	VOC	pH	H₂O (+/-)	Oxidizer (KI)	Fluor (F)	Comments (Observations, Other Screening Methods, Changes to Collection, etc.)
µR/hr mR/hr	cpm kcpm	cpm kcpm			ppb ppm					

Version February 2015

Continued from front

Page ___ of ___

Courtesy of Barry Lindley.

Glossary

Glossary

A

Absorption — Penetration of one substance into the structure of another, such as the process of picking up a liquid contaminant with an absorbent.

Acetone Peroxide (TATP) — Triacetone triperoxide (TATP) is typically a white crystalline powder with a distinctive acrid (bleach) smell and can range in color from a yellowish to white color. *Similar to* Hexamethylene triperoxide diamine (HMTD).

Acid — Compound containing hydrogen that reacts with water to produce hydrogen ions; a proton donor; a liquid compound with a pH less than 7. Acidic chemicals are corrosive.

Action Options — Specific operations performed in a specific order to accomplish the goals of the response objective.

Activation Energy — Minimum energy that starts a chemical reaction when added to an atomic or molecular system.

Activity — Rate of decay of the isotope in terms of decaying atoms per second. Measured in becquerels (Bq) for small quantities of radiation, and curies (Ci) for large quantities of radiation.

Acute — Characterized by sharpness or severity; having rapid onset and a relatively short duration.

Acute Health Effects — Health effects that occur or develop rapidly after exposure to a hazardous substance.

Adsorb — To collect a liquid or gas on the surface of a solid in a thin layer.

Adsorption — Adherence of a substance in a liquid or gas to a solid. This process occurs on the surface of the adsorbent material.

After-Action Report (AAR) — A concise report that details and analyzes incident operations, provides lessons learned from the incident, and makes recommendations for improvement in future responses.

Agroterrorism — Terrorist attack directed against agriculture, such as food supplies or livestock.

Air-Aspirating Foam Nozzle — Foam nozzle designed to provide the aeration required to make the highest quality foam possible; most effective appliance for the generation of low-expansion foam.

Air Bill — Shipping document prepared from a bill of lading that accompanies each piece or each lot of air cargo. *Similar to* Bill of Lading *and* Waybill.

Air-Purifying Respirator (APR) — Respirator that removes contaminants by passing ambient air through a filter, cartridge, or canister; may have a full or partial facepiece.

Air-Reactive Material — Substance that reacts or ignites when exposed to air at normal temperatures. *Also known as* Pyrophoric.

Alkane — A saturated hydrocarbon, with hydrogen in every possible location. All bonds are single bonds. *Also known as* Paraffin.

Alkene — An unsaturated hydrocarbon with at least one double bond between carbon atoms. *Also known as* Olefin.

Alkyne — An unsaturated hydrocarbon with at least one triple bond. *Also known as* Acetylene.

Allergen — Material that can cause an allergic reaction of the skin or respiratory system. *Also known as* Sensitizer.

Alloy — Substance or mixture composed of two or more metals (or a metal and nonmetallic elements) fused together and dissolved into each other to enhance the properties or usefulness of the base metal.

Ambulatory — People, often responders, who are able to understand directions, talk, and walk unassisted.

American Society of Mechanical Engineers (ASME) — Voluntary standards-setting organization concerned with the development of technical standards, such as those for respiratory protection cylinders.

Ammonium Nitrate and Fuel Oil (ANFO) — High explosive blasting agent made of common fertilizer mixed with diesel fuel or oil; requires a booster to initiate detonation.

Anhydrous — Material containing no water.

Anion — Atom or group of atoms carrying a negative charge.

Antibiotic — Antimicrobial agent made from a mold or a bacterium that kills or slows the growth of bacteria; examples include penicillin and streptomycin. Antibiotics are ineffective against viruses.

Antibody — Specialized protein produced by a body's immune system when it detects antigens (harmful substances). Antibodies can only neutralize or remove the effects of their analogous antigens.

Antidote — Substance that counteracts the effects of a poison or toxin.

Antigen — Toxin or other foreign substance that triggers an immune response in a body.

Aqueous Film Forming Foam (AFFF) — Synthetic foam concentrate that, when combined with water, can form a complete vapor barrier over fuel spills and fires and is a highly effective extinguishing and blanketing agent on hydrocarbon fuels.

Aromatic Hydrocarbon — A hydrocarbon with bonds that form rings. *Also known as* Aromatics, *or* Arene.

Asphyxiant — Any substance that prevents oxygen from combining in sufficient quantities with the blood or from being used by body tissues.

Atom — The smallest complete building block of ordinary matter in any state.

Atomic Number — Number of protons in an atom.

Atomic Stability — Condition where an atom has a filled outer shell and is not seeking electrons. Stable atoms also have the same number of protons and electrons.

Atomic Weight — Physical characteristic relating to the mass of molecules and atoms. A relative scale for atomic weights has been adopted, in which the atomic weight of carbon has been set at 12, although its true atomic weight is 12.01115.

Authority Having Jurisdiction (AHJ) — An organization, office, or individual responsible for enforcing the requirements of a code or standard, or approving equipment, materials, an installation, or a procedure.

Autoclave — A device that uses high-pressure steam to sterilize objects.

Autoignition Temperature — The lowest temperature at which a combustible material ignites in air without a spark or flame. (NFPA 921)

Autoinjector — Spring-loaded syringe filled with a single dose of a lifesaving drug.

Autorefrigeration — Rapid chilling of a liquefied compressed gas as it transitions from a liquid state to a vapor state.

Awareness Level — Lowest level of training established by the National Fire Protection Association for personnel at hazardous materials incidents.

B

Bacteria — Microscopic, single-celled organisms.

Baffle — Partition placed in vehicular or aircraft water tanks to reduce shifting of the water load when starting, stopping, or turning.

Bank-Down Application Method — Method of foam application that may be employed on an ignited or unignited Class B fuel spill. The foam stream is directed at a vertical surface or object that is next to or within the spill area; foam deflects off the surface or object and flows down onto the surface of the spill to form a foam blanket. *Also known as* Deflection.

Bar — Unit of pressure measurement; not part of the SI. Equals 100 000 Pa.

Base — Any alkaline or caustic substance; corrosive water-soluble compound or substance containing group-forming hydroxide ions in water solution that reacts with an acid to form a salt.

Base Metal — In hazardous materials containers, the structural material of a containment vessel itself, independent of welding materials and external supports.

Basic Solution — Solution that has a pH between 7 and 14.

Beam — Structural member subjected to loads, usually vertical loads, perpendicular to its length.

Becquerel (Bq) — International System unit of measurement for radioactivity, indicating the number of nuclear decays/disintegrations a radioactive material undergoes in a certain period of time.

Berm — Temporary or permanent barrier intended to control the flow of water. *Similar to* Dike *and* Dam.

Bill of Lading — Shipping paper used by the trucking industry (and others) indicating origin, destination, route, and product; placed in the cab of every truck tractor. This document establishes the terms of a contract between a shipper and a carrier. It serves as a document of title, contract of carriage, and receipt for goods. *Similar to* Air Bill *and* Waybill.

Binary Explosive — A type of explosive device or material with two components that are explosive when combined but not separately.

Bioassay — Scientific experiment in which live plant or animal tissue or cells are used to determine the biological activity of a substance. *Also known as* Biological Assessment *or* Biological Assay.

Biodegradable — Capable of being broken down into innocuous products by the actions of living things, such as microorganisms.

Biological Agent — Viruses, bacteria, or their toxins which are harmful to people, animals, or crops. When used deliberately to cause harm, may be referred to as a Biological Weapon.

Biological Toxin — Poison produced by living organisms.

Blister Agent — Chemical warfare agent that burns and blisters the skin or any other part of the body it contacts. *Also known as* Vesicant *and* Mustard Agent.

Boiling Liquid Expanding Vapor Explosion (BLEVE) — Rapid vaporization of a liquid stored under pressure upon release to the atmosphere following major failure of its containing vessel. Failure is the result of over-pressurization caused by an external heat source, which causes the vessel to explode into two or more pieces when the temperature of the liquid is well above its boiling point at normal atmospheric pressure.

Boiling Point — Temperature of a substance when the vapor pressure equals atmospheric pressure. At this temperature, the rate of evaporation exceeds the rate of condensation. At this point, more liquid is turning into gas than gas is turning back into a liquid.

Bond Energy — The amount of energy needed to break covalent bonds.

Bonding — Connection of two objects with a metal chain or strap in order to neutralize the static electrical charge between the two. *Similar to* Grounding.

Breach — To make an opening in a structural obstacle (such as a masonry wall) without compromising the overall integrity of the wall to allow access into or out of a structure for rescue, hoseline operations, ventilation, or to perform other functions.

Bung — Cork, plug, or other type of stopper used in a barrel, cask, drum, or keg.

C

Calibrate — Operations to standardize or adjust a measuring instrument.

Calibration — Set of operations used to standardize or adjust the values of quantities indicated by a measuring instrument.

Calibration Test — Set of operations used to make sure that an instrument's alerts all work at the recommended levels of hazard detected. *Also known as* Bump Test *and* Field Test.

Canadian Transportation Emergency Centre (CANUTEC) — Canadian center that provides fire and emergency responders with 24-hour information for incidents involving hazardous materials; operated by Transport Canada, a department of the Canadian government.

Capacity Stencil — Number stenciled on the exterior of a tank car to indicate the volume of the tank. *Also known as* Load Limit Marking.

Carbon Dioxide (CO_2) — Colorless, odorless, heavier than air gas that neither supports combustion nor burns; used in portable fire extinguishers as an extinguishing agent to extinguish Class B or C fires by smothering or displacing the oxygen. CO_2 is a waste product of aerobic metabolism.

Carbon Monoxide (CO) — Colorless, odorless, dangerous gas (both toxic and flammable) formed by the incomplete combustion of carbon. It combines with hemoglobin more than 200 times faster than oxygen does, decreasing the blood's ability to carry oxygen.

Carcinogen — Cancer-producing substance.

CAS® Number — Number assigned by the American Chemical Society's Chemical Abstract Service that uniquely identifies a specific compound.

Case Identifier — Alphabetic and/or numeric characters used to identify a case.

Catalyst — Substance that modifies (usually increases) the rate of a chemical reaction without being consumed in the process.

Cation — Atom or group of atoms carrying a positive charge.

CBRNE — Abbreviation for Chemical, Biological, Radiological, Nuclear, and Explosive. These categories are often used to describe WMDs and other hazardous materials characteristics.

Cell Lysis Equipment — Machinery used to break down the membrane of a cell.

Celsius Scale — International temperature scale on which the freezing point is 0°C (32°F) and the boiling point is 100°C (212°F) at normal atmospheric pressure at sea level. *Also known as* Centigrade Scale.

Chain of Custody — Continuous changes of possession of physical evidence that must be established in court to admit such material into evidence. In order for physical evidence to be admissible in court, there must be an evidence log of accountability that documents each change of possession from the evidence's discovery until it is presented in court.

Chemical Agent — Chemical substance that is intended for use in warfare or terrorist activities to kill, seriously injure, or incapacitate people through its physiological effects. *Also known as* Chemical Warfare Agents.

Chemical Asphyxiant — Substance that reacts to prevent the body from being able to use oxygen. *Also known as* Blood Agent.

Chemical Assessment — An organized approach at quantifying the risks associated with the potential exposure to the chemical.

Chemical Attack — Deliberate release of a toxic gas, liquid, or solid that can poison people and the environment.

Chemical Burn — Injury caused by contact with acids, lye, and vesicants such as tear gas, mustard gas, and phosphorus.

Chemical Degradation — Process that occurs when the characteristics of a material are altered through contact with chemical substances.

Chemical Energy — Potential energy stored in the internal structure of a material that may be released during a chemical reaction or transformation.

Chemical Inventory List (CIL) — Formal tracking document showing details of stored chemicals including location, manufacturer, volume, container type, and health hazards.

Chemical Properties — Relating to the way a substance is able to change into other substances. Chemical properties reflect the ability to burn, react, explode, or produce toxic substances hazardous to people or the environment.

Chemical Protective Clothing (CPC) — Clothing designed to shield or isolate individuals from the chemical, physical, and biological hazards that may be encountered during operations involving hazardous materials.

Chemical Transportation Emergency Center (CHEMTREC®) — Center established by the American Chemistry Council that supplies 24-hour information for incidents involving hazardous materials.

Chemical Warfare Agent — Chemical substance intended for use in warfare or terrorist activity; designed to kill, seriously injure, or seriously incapacitate people through its physiological effects.

Chime — Reinforcement ring at the top (head) of a barrel or drum.

CHLOREP — Program administered and coordinated by The Chlorine Institute to provide an organized and effective system for responding to chlorine emergencies in the United States and Canada, operating 24 hours a day/7 days a week with established phone contacts.

Choking Agent — Chemical warfare agent that attacks the lungs, causing tissue damage.

Chronic — Marked by long duration; recurring over a period of time.

Chronic Health Effect — Long-term health effect resulting from exposure to a hazardous substance.

Class B Foam Concentrate — Foam fire-suppression agent designed for use on ignited or unignited Class B flammable or combustible liquids. *Also known as* Class B Foam.

Class D Fire — Fires of combustible metals such as magnesium, sodium, and titanium.

Cloud — Ball-shaped pattern of an airborne hazardous material where the material has collectively risen above the ground or water at a hazardous materials incident.

Code of Federal Regulations (CFR) — Rules and regulations published by executive agencies of the U.S. federal government. These administrative laws are just as enforceable as statutory laws (known collectively as federal law), which must be passed by Congress.

Coffer Dam — Narrow, empty space (void) between compartments or tanks of a vessel that prevents leakage between them; used to isolate compartments or tanks.

Cold Tapping — Strapping a nozzle or outlet onto a container's tank or piping to assist with the removal or transfer of a product.

Cold Zone — Safe area outside of the warm zone where equipment and personnel are not expected to become contaminated and special protective clothing is not required; the Incident Command Post and other support functions are typically located in this zone. *Also known as* Support Zone.

Colorimetric Indicator Tube — Small tube filled with a chemical reagent that changes color in a predictable manner when a controlled volume of contaminated air is drawn through it. *Also known as* Detector Tube.

Combustible Gas Detector — Device that detects the presence and/or concentration of predefined combustible gases in a defined area. May require additional features to indicate the results to an operator.

Combustible Gas Indicator (CGI) — Electronic device that indicates the presence and explosive levels of combustible gases, as relayed from a combustible gas detector.

Combustible Liquid — Liquid having a flash point at or above 100°F (37.8°C) and below 200°F (93.3°C), per NFPA.

Compatibility Group Letter — Indication on an explosives placard expressed as a letter that categorizes different types of explosive substances and articles for purposes of stowage and segregation.

Compound — Substance consisting of two or more elements that have been united chemically.

Compressed Gas — Gas that, at normal temperature, exists solely as a gas when pressurized in a container, as opposed to a gas that becomes a liquid when stored under pressure.

Computer-Aided Management of Emergency Operations (CAMEO) — A system of software applications that assists emergency responders in the development of safe response plans. It can be used to access, store, and evaluate information critical in emergency response.

Concentration — (1) Percentage (mass or volume) of a material dissolved in water (or other solvent). (2) Quantity of a chemical material inhaled for purposes of measuring toxicity. (3) Quantity of a material in relation to a larger volume of gas or liquid.

Condensation — Process of a gas turning into a liquid state.

Cone — Triangular-shaped pattern of an airborne hazardous material release with a point source at the breach and a wide base downrange.

Confined Space — Space or enclosed area not intended for continuous occupation, having limited (restricted access) openings for entry or exit, providing unfavorable natural ventilation and the potential to have a toxic, explosive, or oxygen-deficient atmosphere.

Confinement — The process of controlling the flow of a spill and capturing it at some specified location.

Consist — Rail shipping paper that contains a list of cars in the train by order; indicates the cars that contain hazardous materials. Some railroads include information on emergency operations for the hazardous materials on board with the consist. *Also known as* Train Consist.

Contagious — Capable of transmission from one person to another through contact or close proximity.

Container — (1) Article of transport equipment that is: (a) of a permanent character and strong enough for repeated use; (b) specifically designed to facilitate the carriage of goods by one or more modes of transport without intermediate reloading; and (c) fitted with devices permitting its ready handling, particularly its transfer from one mode to another. The term "container" does not include vehicles. *Also known as* Cargo Container or Freight Container. (2) Box of standardized size used to transport cargo by truck or railcar when transported over land or by cargo vessels at sea; sizes are usually 8 by 8 by 20 feet or 8 by 8 by 40 feet (2.5 m by 2.5 m by 6 m or 2.5 m by 2.5 m by 12 m).

Containment — The act of stopping the further release of a material from its container.

Contaminant — Foreign substance that compromises the purity of a given substance.

Contamination — Impurity resulting from mixture or contact with a foreign substance.

Continuous Underframe Tank Car — Construction of a rail tank car that includes full support of the tank car. The underframe rests on the truck assembly during transport. *Also known as* Full Sill.

Control — To contain, confine, neutralize, or extinguish a hazardous material or its vapor.

Convulsant — Poison that causes convulsions.

Correction Factor — Manufacturer-provided number that can be used to convert a specific device's read-out to be applicable to another function. *Also known as* Conversion Factor, Multiplier *and* Response Curve.

Corrosive — Capable of causing damage by gradually eroding, rusting, or destroying a material.

Counts per Minute (CPM) — Measure of ionizing radiation in which a detection device registers the rate of returns over time. Primarily used to detect particles, not rays.

Covalent Bond — Chemical bond formed between two or more nonmetals. This chemical bond results in a nonsalt.

Critical Point — The end point of an equilibrium curve. In liquid and vapor response, the conditions under which liquid and its vapor can coexist.

Cross Contamination — Contamination of people, equipment, or the environment outside the hot zone without contacting the primary source of contamination. *Also known as* Secondary Contamination.

Cryogen — Gas that is converted into liquid by being cooled below -130°F (-90°C). *Also known as* Refrigerated Liquid *and* Cryogenic Liquid.

Cryogenic Liquid Storage Tank — Heavily insulated, vacuum-jacketed tanks used to store cryogenic liquids; equipped with safety-relief valves and rupture disks.

Curie (Ci) — English System unit of measurement for radioactivity, indicating the number of nuclear decays/disintegrations a radioactive material undergoes in a certain period of time.

Cyber Terrorism — Premeditated, politically motivated attack against information, computer systems, computer programs, and data which result in violence against noncombatant targets by subnational groups or clandestine agents.

Cylinder — Enclosed container with a circular cross-section used to hold a range of materials. Uses include compressed breathing air, poisons, or radioactive materials. *Also known as* Tank *or* Bottle.

D

Dam — Actions to prevent or limit the flow of a liquid or sludge past a certain area.

Dangerous Goods — (1) Any product, substance, or organism included by its nature or by regulation in any of the nine United Nations classifications of hazardous materials. (2) Alternate term used in Canada and other countries for hazardous materials. (3) Term used in the U.S. and Canada for hazardous materials aboard aircraft.

Datasheet — Document that includes important information regarding a specific utility or resource in a standardized format.

Debriefing — A gathering of information from all personnel that were involved in incident operations.

Decomposition — Chemical change in which a substance breaks down into two or more simpler substances. Result of oxygen acting on a material that results in a change in the material's composition; oxidation occurs slowly, sometimes resulting in the rusting of metals.

Decontamination — Process of removing a hazardous foreign substance from a person, clothing, or area. *Also known as* Decon.

Dedicated Tank Car — Rail tank car that is specked to meet particular parameters unique to the product including pressure relief device, linings, valves, fittings, and attachments. This type of car is often used for a single specified purpose for the life of the car, and may be marked to indicate that exact purpose.

Defending in Place — Taking offensive action to protect persons in immediate danger at hazmat incidents.

Defensive Operations — Operations in which responders seek to confine the emergency to a given area without directly contacting the hazardous materials involved.

Deflagrate — To explode (burn quickly) at a rate of speed slower than the speed of sound.

Demobilization — The process of identifying assets on the scene that are no longer needed and returning them to service.

Density — Mass per unit of volume of a substance; obtained by dividing the mass by the volume.

Detection Limit — The smallest quantity of a material that is identifiable within a stated confidence level.

Detonate — To explode or cause to explode. The level of explosive capability will directly affect the speed of the combustion reaction.

Detonation — Explosion with an energy front that travels faster than the speed of sound.

Detonator — Device used to trigger less sensitive explosives, usually composed of a primary explosive; for example, a blasting cap. Detonators may be initiated mechanically, electrically, or chemically.

Dewar — All-metal container designed for the movement of small quantities of cryogenic liquids within a facility; not designed or intended to meet Department of Transportation (DOT) requirements for the transportation of cryogenic materials.

Diatomic Molecules — Molecules composed of only two atoms that may or may not be the same element.

Dike — Actions using raised embankments or other barriers to prevent movement of liquids or sludges to another area.

Dilution — Application of water to a water-soluble material to reduce the hazard.

Direct-Reading Instrument — A tool that indicates its reading on the tool itself, without requiring additional resources. Each instrument is designed for a specific monitoring purpose.

Discovery — Means by which the plaintiff (one party) obtains information from the opposing party (defendant) to prove its allegation.

Disinfection — Any process that eliminates most biological agents; disinfection techniques may target specific entities. Often uses chemicals.

Dispersion — Act or process of being spread widely.

Dissociation (Chemical) — Process of splitting a molecule or ionic compounds into smaller particles, especially if the process is reversible. *Opposite of* Recombination.

Divert — Actions to direct and control movement of a liquid or sludge to an area that will produce less harm.

Division Number — Subset of a class within an explosives placard that assigns the product's level of explosion hazard.

Dose — Quantity of a chemical material ingested or absorbed through skin contact for purposes of measuring toxicity.

Dose-Response Relationship — Comparison of changes within an organism per amount, intensity, or duration of exposure to a stressor over time. This information is used to determine action levels for materials such as drugs, pollutants, and toxins.

Dosimeter — Detection device used to measure an individual's exposure to an environmental hazard such as radiation or sound.

Drainage Time — Amount of time it takes foam to break down or dissolve. *Also known as* Drainage, Drainage Dropout Rate, *or* Drainage Rate.

Dry Powder — Extinguishing agent suitable for use on combustible metal fires.

Duet Rule — Atoms with only one shell will attempt to maintain two electrons to fill the outer shell at all times, whether by gaining or losing electrons. A complete outer electron shell makes elements very stable.

Dust Explosion — Rapid burning (deflagration), with explosive force, of any combustible dust. Dust explosions generally consist of two explosions: a small explosion or shock wave creates additional dust in an atmosphere, causing the second and larger explosion.

E

Eduction — Process used to mix foam concentrate with water in a nozzle or proportioner; concentrate is drawn into the water stream by the Venturi method. *Also known as* Induction.

Electricity — Form of energy resulting from the presence and flow of charged particles.

Electrochemical Gas Sensor — Device used to measure the concentration of a target gas by oxidizing or reducing the target gas and then measuring the current.

Electron — Subatomic particle with a physical mass and a negative electric charge.

Elevated Temperature Material — Material that when offered for transportation or transported in bulk packaging is (a) in a liquid phase and at temperatures at or above 212°F (100°C), (b) intentionally heated at or above its liquid phase flash points of 100°F (38°C), or (c) in a solid phase and at a temperature at or above 464°F (240°C).

Emergency Breathing Support System (EBSS) — Escape-only respirator that provides sufficient self-contained breathing air to permit the wearer to safely exit the hazardous area; usually integrated into an airline supplied-air respirator system.

Emergency Decontamination — The physical process of immediately reducing contamination of individuals in potentially life-threatening situations, with or without the formal establishment of a decontamination corridor.

Emergency Response Guidebook (ERG) — Manual that aids emergency response and inspection personnel in identifying hazardous materials placards and labels; also gives guidelines for initial actions to be taken at hazardous materials incidents. Developed jointly by Transport Canada (TC), U.S. Department of Transportation (DOT), the Secretariat of Transport and Communications of Mexico (SCT), and with the collaboration of CIQUIME (Centro de Información Química para Emergencias).

Emissivity — Measure of an object's ability to radiate thermal energy.

Encapsulating — Completely enclosed or surrounded, as in a capsule.

End-of-Service-Time Indicator (ESTI) — Warning device that alerts the user that the respiratory protection equipment is about to reach its limit and that it is time to exit the contaminated atmosphere; its alarm may be audible, tactile, visual, or any combination thereof.

Endothermic Reaction — Chemical reaction in which a substance absorbs heat energy.

Engulfment — Dispersion of material as defined in the General Emergency Behavior Model (GEBMO); an engulfing event occurs when matter and/or energy disperses and forms a danger zone.

Evacuation — Controlled process of leaving or being removed from a potentially hazardous location, typically involving relocating people from an area of danger or potential risk to a safer place.

Evaporation — Process of a solid or liquid turning into gas.

Evaporation Rate — Speed at which some material changes from a liquid to a vapor. Materials that change readily to gases are considered volatile.

Evidence — Information collected and analyzed by an investigator.

Excepted Packaging — Container used for transportation of materials that have very limited radioactivity.

Exothermic Reaction — Chemical reaction between two or more materials that changes the materials and produces heat.

Expansion Ratio — 1) Volume of a substance in liquid form compared to the volume of the same number of molecules of that substance in gaseous form. 2) Ratio of the finished foam volume to the volume of the original foam solution. *Also known as* Expansion.

Explosive — Any material or mixture that will undergo an extremely fast self-propagation reaction when subjected to some form of energy.

Explosive Ordnance Disposal (EOD) — Emergency responders specially trained and equipped to handle and dispose of explosive devices. *Also called* Hazardous Devices Units *or* Bomb Squad.

Exposure — (1) Contact with a hazardous material, causing biological damage, typically by swallowing, breathing, or touching (skin or eyes). Exposure may be short-term (acute exposure), of intermediate duration, or long-term (chronic exposure). (2) People, property, systems, or natural features that are or may be exposed to the harmful effects of a hazardous materials emergency.

Exposure Limit — Maximum length of time an individual can be exposed to an airborne substance before injury, illness, or death occurs.

Extinguish — To put out a fire completely.

F

Fahrenheit Scale — Temperature scale on which the freezing point is 32°F (0°C) and the boiling point at sea level is 212°F (100°C) at normal atmospheric pressure.

Ferrous Metal — Metal in which iron is the main constituent element; carbon and other elements are added to the iron to create a variety of metals with various magnetic properties and tensile strengths; varieties include cast and wrought iron, steel and steel alloys; stainless steel, and high-carbon steel.

Fire Point — Temperature at which a liquid fuel produces sufficient vapors to support combustion once the fuel is ignited. Fire point must exceed five seconds of burning duration during the test. The fire point is usually a few degrees above the flash point.

Flame Ionization Detector (FID) — Gas detector that oxidizes all oxidizable materials in a gas stream, and then measures the concentration of the ionized material.

Flame-Resistant (FR) — Material that does not support combustion and is self-extinguishing after removal of an external source of ignition.

Flammability — Fuel's susceptibility to ignition.

Flammable Liquid — Any liquid having a flash point below 100°F (37.8°C) and a vapor pressure not exceeding 40 psi absolute (276 kPa) {2.76 bar}, per NFPA.

Flammable Range — Range between the upper flammable limit and lower flammable limit in which a substance can be ignited. *Also known as* Explosive Range.

Flaring — Controlled release and disposal of flammable gases or liquids through a burning process.

Flash Point — Minimum temperature at which a liquid gives off enough vapors to form an ignitable mixture with air near the surface of the liquid.

Fluorimeter — Device used to detect the fluorescence of a material, especially as pertains to the fluorescent qualities of DNA and RNA.

Forensic Evidence — Evidence obtained by scientific methods that is usable in court, for example finger prints, blood testing, or ballistics.

Fourier Transform Infrared (FT-IR) Spectroscopy — Device that uses a mathematical process to convert detection data onto the infrared spectrum.

Frameless Tank Car — Direct attachment of a rail tank car to the truck assembly. This type of construction transfers all of the stresses of transport from the railcar to the stub sill assembly and the tank itself. *Known as* Stub Sill.

Freezing Point — Temperature at which a liquid becomes a solid at normal atmospheric pressure.

Frostbite — Local tissue damage caused by prolonged exposure to extreme cold.

Fusible Plug — Safety device in pressurized vessels that consists of a threaded metal cylinder with a tapered hole drilled completely through its length; the hole is filled with a metal that has a low, predetermined melting point.

G

Gamma-Ray Spectrometer — Apparatus used to measure the intensity of gamma radiation as compared to the energy of each photon.

Gas — Compressible substance, with no specific volume, that tends to assume the shape of a container. Molecules move about most rapidly in this state.

Gas Chromatograph (GC) — Apparatus used to detect and separate small quantities of volatile liquids or gases via instrument analysis. *Also known as* Gas-Liquid Partition Chromatography (GLPC).

Geiger-Mueller (GM) Detector — Detection device that uses GM tubes to measure ionizing radiation. *Also known as a* Geiger Counter.

Geiger-Mueller (GM) Tube — Sensor tube used to detect ionizing radiation. This tube is one element of a Geiger-Mueller detector.

General Emergency Behavior Model (GEBMO) — Model used to describe how hazardous materials are accidentally released from their containers and how they behave after the release.

Geographic Information Systems (GIS) — Computer software application that relates physical features on the earth to a database to be used for mapping and analysis. The system captures, stores, analyzes, manages, and presents data that refers to or is linked to a location.

Globally Harmonized System of Classification and Labeling of Chemicals (GHS) — International classification and labeling system for chemicals and other hazard communication information, such as safety data sheets.

Glovebox — Sealed container equipped with long-cuff gloves on one facet to allow handling of materials within the container. Commonly used in laboratories and incubators where a vacuum or sterile environment is needed.

Gray (Gy) — SI unit of ionizing radiation dose, defined as the absorption of one joule of radiation energy per one kilogram of matter.

Grounding — Reducing the difference in electrical potential between an object and the ground by the use of various conductors; similar to *Bonding*.

G-Series Agents — Nonpersistent nerve agents initially synthesized by German scientists.

H

Half-Life — The time required for a radioactive material to reduce to half of its initial value.

Halogenated Agent — Chemical compounds (halogenated hydrocarbons) that contain carbon plus one or more elements from the halogen series. Halon 1301 and Halon 1211 are most commonly used as extinguishing agents for Class B and Class C fires. *Also known as* Halogenated Hydrocarbons.

Hazard — Condition, substance, or device that can directly cause injury or loss; the source of a risk.

Hazard and Risk Assessment — Formal review of the hazards and risks that may be encountered by firefighters or emergency responders; used to determine the appropriate level and type of personal and respiratory protection that must be worn. *Also known as* Hazard Assessment.

Hazard Class — Group of materials designated by the Department of Transportation (DOT) that shares a major hazardous property.

Hazard-Control Zones — System of barriers surrounding designated areas at emergency scenes, intended to limit the number of persons exposed to a hazard and to facilitate its mitigation. A major incident has three zones: Restricted (Hot) Zone, Limited Access (Warm) Zone, and Support (Cold) Zone. EPA/OSHA term: Site Work Zones. *Also known as* Control Zones *and* Scene Control Zones.

Hazardous Material — Any substance or material that poses an unreasonable risk to health, safety, property, and/or the environment if it is not properly controlled during handling, storage, manufacture, processing, packaging, use, disposal, or transportation.

Hazardous Materials Profile — A chemical size-up based upon the suspected identity, or not, of a chemical hazard. This is validated with monitoring and detection equipment upon performing a reconnaissance entry. Profiling allows the hazmat technician to predict hazards and validate the actual entry conditions even if the product is not positively identified. *Also known as* Hazard Profile.

Hazardous Materials Technician — Individual trained to use specialized protective clothing and control equipment to control the release of a hazardous material.

Hazardous Waste Operations and Emergency Response (HAZWOPER) — U.S. regulations in Title 29 (Labor) *CFR* 1910.120 for cleanup operations involving hazardous substances and emergency response operations for releases of hazardous substances.

Head Pressure — Pressure exerted by a stationary column of water, directly proportional to the height of the column.

Head Shield — Layer of puncture protection added to the head of tanks. Head shields may or may not be visible, depending on the construction of the tank and the type of protection provided.

Heat — Form of energy associated with the motion of atoms or molecules in solids or liquids that is transferred from one body to another as a result of a temperature difference between the bodies, such as from the sun to the earth. To signify its intensity, it is measured in degrees of temperature.

Heat Cramps — Heat illness resulting from prolonged exposure to high temperatures; characterized by excessive sweating, muscle cramps in the abdomen and legs, faintness, dizziness, and exhaustion.

Heat Exhaustion — Heat illness caused by exposure to excessive heat; symptoms include weakness, cold and clammy skin, heavy perspiration, rapid and shallow breathing, weak pulse, dizziness, and sometimes unconsciousness.

Heat Induced Tear — Rupture of a container caused by overpressure, often along the seam. This type of failure primarily occurs in low-pressure containers transporting flammable/combustible liquids.

Heat Rash — Condition that develops from continuous exposure to heat and humid air; aggravated by clothing that rubs the skin. Reduces the individual's tolerance to heat.

Heat Sink — In thermodynamics, any material or environment that absorbs heat without changing its physical state or appreciably changing temperature.

Heat Stroke — Heat illness in which the body's heat regulating mechanism fails; symptoms include (a) high fever of 105° to 106° F (40.5° to 41.1° C), (b) dry, red, hot skin, (c) rapid, strong pulse, and (d) deep breaths or convulsions. May result in coma or even death. *Also known as* Sunstroke.

Hemispheric Release — Semicircular or dome-shaped pattern of airborne hazardous material that is still partially in contact with the ground or water.

Hexamethylene Triperoxide Diamine (HMTD) — Peroxide-based white powder high explosive organic compound that can be manufactured using nonspecialized equipment. Sensitive to shock and friction during manufacture and handling. *Similar to* acetone peroxide (TATP).

High Explosive — Explosive that decomposes extremely rapidly (almost instantaneously) and has a detonation velocity faster than the speed of sound.

High-Hazard Flammable Trains (HHFT) — Trains that have a continuous block of twenty or more tank cars loaded with a flammable liquid or thirty-five or more cars loaded with a flammable liquid dispersed through a train.

Homemade Explosive (HME) — Explosive material constructed using common household chemicals. The finished product is usually highly unstable.

Hot Tapping — Using welding or cutting to attach a nozzle or outlet onto a container's tank or piping to assist with the removal or transfer of a product.

Hot Zone — Potentially hazardous area immediately surrounding the incident site; requires appropriate protective clothing and equipment and other safety precautions for entry. Typically limited to technician-level personnel. *Also known as* Exclusion Zone.

Hydrocarbon — Organic compound containing only hydrogen and carbon and found primarily in petroleum products and coal.

Hydrogen Cyanide (HCN) — Colorless, toxic, and flammable liquid until it reaches 79° F (26° C). Above that temperature, it becomes a gas with a faint odor similar to bitter almonds; produced by the combustion of nitrogen-bearing substances.

Hydronium — Water molecule with an extra hydrogen ion (H_3O^+). Substances/solutions that have more hydronium ions than hydroxide ions have an acidic pH.

Hydrophilic — Material that is attracted to water. This material may also dissolve or mix in water.

Hydrophobic — Material that is incapable of mixing with water.

Hydrostatic Test — Testing method that uses water under pressure to check the integrity of pressure vessels.

Hydroxide — Water molecule missing a hydrogen ion (HO^-). Substances/solutions that have more hydroxide ions than hydronium ions have a basic (alkaline) pH.

Hypergolic — Substance that ignites when exposed to another substance.

Hypothermia — Abnormally low body temperature.

I

Ignition Temperature — Minimum temperature to which a fuel (other than a liquid) in air must be heated in order to start self-sustained combustion independent of the heating source.

Immediately Dangerous to Life and Health (IDLH) — Description of any atmosphere that poses an immediate hazard to life or produces immediate irreversible, debilitating effects on health; represents concentrations above which respiratory protection should be required. Expressed in parts per million (ppm) or milligrams per cubic meter (mg/m^3); companion measurement to the permissible exposure limit (PEL).

Immiscible — Incapable of being mixed or blended with another substance.

Immunoassay (IA) — Test to measure the concentration of an analyte (material of interest) within a solution.

Improvised Explosive Device (IED) — Any explosive device constructed and deployed in a manner inconsistent with conventional military action.

Incendiary Device — (1) Contrivance designed and used to start a fire. (2) Any mechanical, electrical, or chemical device used intentionally to initiate combustion and start a fire. *Also known as* Explosive Device.

Incident Commander (IC) — Person in charge of the incident command system and responsible for the management of all incident operations during an emergency.

Incident Management System (IMS) — System described in NFPA 1561, *Standard on Emergency Services Incident Management System and Command Safety*, that defines the roles, responsibilities, and standard operating procedures used to manage emergency operations. Such systems may also be referred to as Incident Command Systems (ICS).

Incidental Release — Spill or release of a hazardous material where the substance can be absorbed, neutralized, or otherwise controlled at the time of release by employees in the immediate release area, or by maintenance personnel who are not considered to be emergency responders.

Industrial Packaging — Container used to ship radioactive materials that present limited hazard to the public and the environment, such as smoke detectors.

Inert Gas — Gas that does *not* normally react chemically with another substance or material; any one of six gases: helium, neon, argon, krypton, xenon, and radon.

Infectious — Transmittable; able to infect people.

Infectious Substance — Substance that is known, or reasonably expected, to contain pathogens.

Infrared - Invisible electromagnetic radiant energy at a wavelength in the visible light spectrum greater than the red end but lower than microwaves.

Infrared Thermometer — Non-contact measuring device that detects the infrared energy emitted by materials and converts the energy factor into a temperature reading. *Also known as* Temperature Gun.

Inhalation Hazard — Any material that may cause harm via inhalation. **[DOT]**

Inhibitor — Material that is added to products that easily polymerize in order to control or prevent an undesired reaction. *Also known as* Stabilizer.

Initial Isolation Distance — Distance within which all persons are considered for evacuation in all directions from a hazardous materials incident.

Initial Isolation Zone — Circular zone, with a radius equivalent to the initial isolation distance, within which persons may be exposed to dangerous concentrations upwind of the source and may be exposed to life-threatening concentrations downwind of the source.

Instrument Response Time — Elapsed time between the movement (drawing in) of an air sample into a monitoring/detection device and the reading (analysis) provided to the user. *Also known as* Instrument Reaction Time.

Intermediate Bulk Container (IBC) — Rigid (RIBC) or flexible (FIBC) portable packaging, other than a cylinder or portable tank, that is designed for mechanical handling with a maximum capacity of not more than 3 cubic meters (3,000 L, 793 gal, or 106 ft^3) (49CFR178.700).

Intermodal Container — Freight containers designed and constructed to be used interchangeably in two or more modes of transport. *Also known as* Intermodal Tank, Intermodal Tank Container, *and* Intermodal Freight Container.

International System of Units (SI) — Modern form of the metric system of measurement that standardizes mathematical quantification.

Intrinsically Safe — Describes equipment that is approved for use in flammable atmospheres; must be incapable of releasing enough electrical energy to ignite the flammable atmosphere.

Inverse Square Law — Physical law that states that the amount of radiation present is inversely proportional to the square of the distance from the source of radiation.

Ion — Atom that has lost or gained an electron, thus giving it a positive or negative charge.

Ion Mobility Spectrometry (IMS) — Technique used to separate and identify ionized molecules. The ionize molecules are impeded in travel via a buffer gas chosen for the type of detection intended. Larger ions are slowed more than smaller ions; this difference provides an indication of the ions' size and identity.

Ionic Bond — Chemical bond formed by the transfer of electrons from a metal element to a nonmetal element. This chemical bond results in two oppositely charged ions.

Ionization — Process in which an atom or molecule loses electrons.

Ionization Potential — Energy required to free an electron from its atom or molecule.

Ionize — Process in which an atom or molecule gains a negative or positive charge by gaining or losing electrons.

Ionizing Radiation — Radiation that causes a chemical change in atoms by removing their electrons.

Irritant — Liquid or solid that, upon contact with fire or exposure to air, gives off dangerous or intensely irritating fumes. *Also known as* Irritating Material.

Isolation Perimeter — Outer boundary of an incident that is controlled to prevent entrance by the public or unauthorized persons.

Isotope — Atoms of a chemical element with the usual number of protons in the nucleus, but an unusual number of neutrons; has the same atomic number but a different atomic mass from normal chemical elements.

J

Jubilee Pipe Patch — Modification of a commercial hose clamp; consists of a cylindrical sheet of metal with flanges at each end. The sheet metal can be wrapped over packing around a pipe leak. The flanges can be attached to one another using screws and nuts to form a seal.

L

Label — Four-inch-square diamond-shaped marker required by federal regulations on individual shipping containers that contain hazardous materials, and are smaller than 640 cubic feet (18 m³).

Lethal Concentration (LC) — Concentration of an inhaled substance that results in the death of the entire test population. Expressed in parts per million (ppm), milligrams per liter (mg/liter), or milligrams per cubic meter (mg/m³); the lower the value, the more toxic the substance.

Lethal Dose (LD) — Concentration of an ingested or injected substance that results in the death of the entire test population. Expressed in milligrams per kilogram (mg/kg); the lower the value, the more toxic the substance.

Level A PPE — Highest level of skin, respiratory, and eye protection that can be given by personal protective equipment (PPE), as specified by the U.S. Environmental Protection Agency (EPA); consists of positive-pressure self-contained breathing apparatus, totally encapsulating chemical-protective suit, inner and outer gloves, and chemical-resistant boots.

Level B PPE — Personal protective equipment that affords the highest level of respiratory protection, but a lesser level of skin protection; consists of positive-pressure self-contained breathing apparatus, hooded chemical-protective suit, inner and outer gloves, and chemical-resistant boots.

Level C PPE — Personal protective equipment that affords a lesser level of respiratory and skin protection than levels A or B; consists of full-face or half-mask APR, hooded chemical-resistant suit, inner and outer gloves, and chemical-resistant boots.

Level D PPE — Personal protective equipment that affords the lowest level of respiratory and skin protection; consists of coveralls, gloves, and chemical-resistant boots or shoes.

Limits of Recovery — A container's design strength or ability to hold contents at pressure.

Line-of-Sight — Unobstructed, imaginary line between an observer and the object being viewed.

Liquefied Gas — Confined gas that at normal temperatures exists in both liquid and gaseous states.

Liquefied Natural Gas (LNG) — Natural gas stored under pressure as a liquid.

Liquefied Petroleum Gas (LPG) — Any of several petroleum products, such as propane or butane, stored under pressure as a liquid.

Liquid — Incompressible substance with a constant volume that assumes the shape of its container; molecules flow freely, but substantial cohesion prevents them from expanding as a gas would.

Liquid Splash-Protective Clothing — Chemical-protective clothing designed to protect against liquid splashes per the requirements of NFPA 1992, *Standard on Liquid Splash-Protective Suits for Hazardous Chemical Emergencies*; part of an EPA Level B ensemble.

Local Emergency Planning Committee (LEPC) — Community organization responsible for local emergency response planning. Required by SARA Title III, LEPCs are composed of local officials, citizens, and industry representatives with the task of designing, reviewing, and updating a comprehensive emergency plan for an emergency planning district; plans may address hazardous materials inventories, hazardous material response training, and assessment of local response capabilities.

Local Emergency Response Plan (LERP) — Plan detailing how local emergency response agencies will respond to community emergencies; required by U.S. Environmental Protection Agency (EPA) and prepared by the Local Emergency Planning Committee (LEPC).

Low Explosive — Explosive material that deflagrates, producing a reaction slower than the speed of sound.

Lower Flammable (Explosive) Limit (LFL) — Lower limit at which a flammable gas or vapor will ignite and support combustion; below this limit the gas or vapor is too *lean* or *thin* to burn (too much oxygen and not enough gas, so lacks the proper quantity of fuel). *Also known as* Lower Explosive Limit (LEL).

Low-Pressure Storage Tank — Class of fixed-facility storage tanks that are designed to have an operating pressure ranging from 0.5 to 15 psi (3.45 kPa to 103 kPa) {0.03 bar to 1.03 bar}.

M

Manway — Opening that is large enough to admit a person into a tank trailer or dry bulk trailer. This opening is usually equipped with a removable, lockable cover. *Also known as* Manhole.

Mass Casualty Incident — Incident that results in a large number of casualties within a short time frame, as a result of an attack, natural disaster, aircraft crash, or other cause that is beyond the capabilities of local logistical support.

Mass Decontamination — Process of decontaminating large numbers of people in the fastest possible time to reduce surface contamination to a safe level. It is typically a gross decon process utilizing water or soap and water solutions to reduce the level of contamination, with or without a formal decontamination corridor or line.

Mass Spectrometer — Apparatus used to ionize a chemical and then measure the masses within the sample.

Maximum Allowable Working Pressure (MAWP) — A percentage of a container's test pressure. Can be calculated as the pressure that the weakest component of a vessel or container can safely maintain.

Maximum Safe Storage Temperature (MSST) — Temperature below which the product can be stored safely. This is usually 20-30 degrees cooler than the SADT temperature, but may be much cooler depending on the material.

Mechanical Energy — Energy possessed by objects due to their position or motion, the sum of potential and kinetic energy.

Median Lethal Concentration, 50 Percent Kill (LC$_{50}$) — Concentration of an inhaled substance that results in the death of 50 percent of the test population. LC$_{50}$ is an inhalation exposure expressed in parts per million (ppm), milligrams per liter (mg/liter), or milligrams per cubic meter (mg/m^3); the lower the value, the more toxic the substance.

Median Lethal Dose, 50 Percent Kill (LD$_{50}$) — Concentration of an ingested or injected substance that results in the death of 50 percent of the test population. LD$_{50}$ is an oral or dermal exposure expressed in milligrams per kilogram (mg/kg); the lower the value, the more toxic the substance.

Melting Point — Temperature at which a solid substance changes to a liquid state at normal atmospheric pressure.

Memorandum of Understanding (MOU) — Form of written agreement created by a coalition to make sure that each member is aware of the importance of his or her participation and cooperation.

Mercaptan — A sulfur-containing organic compound often added to natural gas as an odorant. Natural gas is odorless; natural gas treated with mercaptan has a strong odor. *Also known as a* Thiol.

Metadata — Information that provides background and detail about other types of information.

Meth Cook — 1) Person who generates methamphetamine in an illicit lab. 2) Area with evidence of production of methamphetamine.

Methamphetamine (Meth) — Central nervous system stimulant drug that can be produced in small labs. Low dosage medical uses include controlling weight, narcolepsy, and attention deficit hyperactivity disorder. Recreational uses include euphoriant and aphrodisiac qualities. At all dosages, misuse of this drug presents a high risk for personal and social harm.

Micron — Unit of length equal to one-millionth of a meter.

Mild Steel — Class of steel in which a low-level of carbon is the primary alloying agent; available in a variety of formable grades. *Also called* Carbon Steel.

Millimeters of Mercury (mmHg) — Unit of pressure measurement; not part of the SI. Currently defined as a rate rounded to 133 Pascals. Rough equivalent to 1 torr.

Millirem (mrem) — One thousandth of one Roentgen Equivalent in Man (rem).

Miscibility — Two or more liquids' capability to mix together.

Miscible — Materials that are capable of being mixed in all proportions.

Mitigate — (1) To cause to become less harsh or hostile; to make less severe, intense or painful; to alleviate. (2) Third of three steps (locate, isolate, mitigate) in one method of sizing up an emergency situation.

Mixture — Substance containing two or more materials not chemically united.

Mobile Data Terminal (MDT) — Mobile computer that communicates with other computers on a radio system.

Molecular Weight (MW) — Average mass of one molecule. This can be calculated as the sum of the atomic masses of the component atoms.

Monomer — A molecule that may bind chemically to other molecules to form a polymer.

Multiuse Detectors — Device with several types of equipment in one handheld device. Used to detect specific types of materials in an atmosphere.

Munitions — Military reserves of weapons, equipment, and ammunition.

N

National Fire Protection Association (NFPA) — U.S. nonprofit educational and technical association devoted to protecting life and property from fire by developing fire protection standards and educating the public. Located in Quincy, Massachusetts.

National Incident Management System - Incident Command System (NIMS-ICS) — The U.S. mandated incident management system that defines the roles, responsibilities, and standard operating procedures used to manage emergency operations; creates a unified incident response structure for federal, state, and local governments.

National Thread Pipe Taper (NPT) — U.S. standard for pipe threading developed with the intent to create a fluid-tight seal.

Nerve Agent — A class of toxic chemical that works by disrupting the way nerves transfer messages to organs.

Neutralization — Chemical reaction in water in which an acid and base react quantitatively with each other until there are no excess hydrogen or hydroxide ions remaining in the solution.

Neutron — Component of the nucleus of an atom that has a neutral electrical charge yet produces highly penetrating radiation; ultrahigh energy particle that has a physical mass but no electrical charge.

Nondispersive Infrared (NDIR) Sensor — Simple spectroscope that can be used as a gas detector.

Nonflammable — Incapable of combustion under normal circumstances; normally used when referring to liquids or gases.

Nonintervention Operations — Operations in which responders take no direct actions on the actual problem.

Nonionizing Radiation — Series of energy waves composed of oscillating electric and magnetic fields traveling at the speed of light. Examples include ultraviolet radiation, visible light, infrared radiation, microwaves, radio waves, and extremely low frequency radiation.

Nonpersistent Chemical Agent — Chemical agent that generally vaporizes and disperses quickly, usually in less than 10 minutes.

Nucleus — The positively charged central part of an atom, consisting of protons and neutrons.

O

Occupancy — (1) General fire and emergency services term for a building, structure, or residency. (2) Building code classification based on the use to which owners or tenants put buildings or portions of buildings. Regulated by the various building and fire codes. *Also known as* Occupancy Classification.

Octet Rule — Atoms with two or more shells will attempt to maintain eight electrons to fill the outermost shell at all times, whether by gaining or losing electrons. A complete outer electron shell makes elements very stable.

Offensive Operations — Operations in which responders take aggressive, direct action on the material, container, or process equipment involved in an incident.

Olfactory Fatigue — Gradual inability of a person to detect odors after initial exposure; can be extremely rapid with some toxins, such as hydrogen sulfide.

Operations Level — Level of training established by the National Fire Protection Association allowing first responders to take defensive actions at hazardous materials incidents.

Operations Mission-Specific Level — Level of training established by the National Fire Protection Association allowing first responders to take additional defensive tasks and limited offensive actions at hazardous materials incidents.

Organic Peroxide — Any of several organic derivatives of the inorganic compound hydrogen peroxide.

Organophosphate Pesticides — Chemicals that kill insects by disrupting their central nervous systems; these chemicals inactivate acetylcholinesterase, an enzyme which is essential to nerve function in insects, humans, and many other animals.

Other Regulated Material (ORM) — Material, such as a consumer commodity, that does not meet the definition of a hazardous material and is not included in any other hazard class but possesses enough hazardous characteristics that it requires some regulation; presents limited hazard during transportation because of its form, quantity, and packaging.

Overpack — (1) To enclose or secure a container by placing it in a larger container. (2) An outer container designed to enclose or secure an inner container.

Oxidation — Chemical process that occurs when a substance combines with an oxidizer such as oxygen in the air; a common example is the formation of rust on metal.

Oxidation Number — A theoretical number assigned to individual atoms and ions to track whether an oxidation-reduction reaction has taken place. *Also known as* Oxidation Level.

Oxidation-Reduction (Redox) Reaction — Chemical reaction that results in a molecule, ion, or atom gaining or losing an electron. *Also known as* Redox Reaction.

Oxidizer — Any material that readily yields oxygen or other oxidizing gas, or that readily reacts to promote or initiate combustion of combustible materials. (Reproduced with permission from NFPA 400-2010, *Hazardous Materials Code*, Copyright©2010, National Fire Protection Association)

Oxidizing Agent — Substance that oxidizes another substance; can cause other materials to combust more readily or make fires burn more strongly. *Also known as* Oxidizer.

P

Packaging — Shipping containers and their markings, labels, and/or placards.

Pandemic — Epidemic occurring over a very wide area (several countries or continents), usually affecting a large proportion of the population.

Parts Per Billion (ppb) — Method of expressing the concentration of very dilute solutions of one substance in another, normally a liquid or gas, based on volume; expressed as a ratio of the volume of contaminants (parts) compared to the volume of air (billion parts).

Parts Per Million (ppm) — Method of expressing the concentration of very dilute solutions of one substance in another, normally a liquid or gas, based on volume; expressed as a ratio of the volume of contaminants (parts) compared to the volume of air (million parts). The common unit of measure is equivalent to 1 microgram [1 µg] per liter of water or kg of solid, or a micro liter [1 µL] volume of gas in one liter of air.

Pascals (Pa) — SI unit of measure used to indicate internal pressure and stress on a container.

Pathogen — Biological agent that causes disease or illness.

Penetration — Process in which a hazardous material enters an opening or puncture in a protective material.

Periodic Table of Elements — Organizational chart showing chemical elements arranged in order by atomic number, electron configuration, and chemical properties.

Permeation — Process in which a chemical passes through a protective material on a molecular level.

Permissible Exposure Limit (PEL) — Maximum time-weighted concentration at which 95 percent of exposed, healthy adults suffer no adverse effects over a 40-hour work week; an 8-hour time-weighted average unless otherwise noted. PELs are expressed in either parts per million (ppm) or milligrams per cubic meter (mg/m^3). They are commonly used by OSHA and are found in the NIOSH *Pocket Guide to Chemical Hazards*.

Persistence — Length of time a chemical agent remains effective without dispersing.

Persistent Chemical Agent — Chemical agent that remains effective in the open (at the point of dispersion) for a considerable period of time, usually more than 10 minutes.

Person-Borne Improvised Explosives Device (PBIED) — Improvised explosive device carried by a person. This type of IED is often employed by suicide bombers, but may be carried by individuals coerced into carrying the bomb.

pH — Measure of the acidity or alkalinity of a solution.

pH Indicator — Chemical detector for hydronium ions (H_3O^+) or hydrogen ions (H^+). Indicator equipment includes impregnated papers and meters.

Phase — Distinguishable part in a course development or cycle; aspect or part under consideration. In chemistry, a change of phase is marked by a shift in the physical state of a substance caused by a change in heat.

Phosphine — Colorless, flammable, and toxic gas with an odor of garlic or decaying fish; ignites spontaneously on contact with air. Phosphine is a respiratory tract irritant that attacks the cardiovascular and respiratory systems, causing pulmonary edema, peripheral vascular collapse, and cardiac arrest and failure.

Photoionization Detector (PID) — Gas detector that measures volatile compounds in concentrations of parts per million and parts per billion.

Photon — Weightless packet of electromagnetic energy, such as X-rays or visible light.

Physical Properties — Properties that do not involve a change in the chemical identity of the substance, but affect the physical behavior of the material inside and outside the container, which involves the change of the state of the material. Examples include boiling point, specific gravity, vapor density, and water solubility.

Pipeline and Hazardous Materials Safety Administration (PHMSA) — Branch of the U.S. Department of Transportation (DOT) that focuses on pipeline safety and related environmental concerns.

Placard — Diamond-shaped sign that is affixed to each side of a structure or a vehicle transporting hazardous materials to inform responders of fire hazards, life hazards, special hazards, and reactivity potential. The placard indicates the primary class of the material and, in some cases, the exact material being transported; required on containers that are 640 cubic feet (18 m³) or larger.

Plume — Irregularly shaped pattern of an airborne hazardous material where wind and/or topography influence the downrange course from the point of release.

Poison — Any material, excluding gases, that when taken into the body is injurious to health.

Polar Solvent — 1) A material in which the positive and negative charges are permanently separated, resulting in their ability to ionize in solution and create electrical conductivity. Examples include water, alcohol, esters, ketones, amines, and sulfuric acid. 2) Flammable liquids with an attraction for water.

Polarity — Property of some molecules to have discrete areas with negative and positive charges.

Polymer — Large molecule composed of repeating structural units (monomers).

Polymerase Chain Reaction (PCR) — Technique in which DNA is copied to amplify a segment of DNA to diagnose and monitor a disease or to forensically identify an individual.

Polymerization — Chemical reactions in which two or more molecules chemically combine to form larger molecules; this reaction can often be violent.

Positive-Pressure Ventilation (PPV) — Method of ventilating a room or structure by mechanically blowing fresh air through an inlet opening into the space in sufficient volume to create a slight positive pressure within and thereby forcing the contaminated atmosphere out the exit opening.

Postincident Analysis (PIA) — Overview and critique of an incident including feedback from members of all responding agencies. Typically takes place within two weeks of the incident. In the training environment it may be used to evaluate student and instructor performance during a training evolution.

Postincident Critique — Discussion of the incident during the Termination phase of response. Discussion includes responders, stakeholders, and command staff, to determine facets of the response that were successful and areas that can be improved upon.

Powered Air-Purifying Respirator (PAPR) — Motorized respirator that uses a filter to clean surrounding air, then delivers it to the wearer to breathe; typically includes a headpiece, breathing tube, and a blower/battery box that is worn on the belt.

Preincident Survey — Assessment of a facility or location made before an emergency occurs, in order to prepare for an appropriate emergency response. *Also known as* Preplan.

Pressure — Force per unit area exerted by a liquid or gas measured in pounds per square inch (psi) or kilopascals (kPa).

Pressure Relief Device (PRD) — An engineered valve or other device used to control or limit the pressure in a system or vessel, often by venting excess pressure.

Pressure Relief Valve — Pressure control device designed to eliminate hazardous conditions resulting from excessive pressures by allowing this pressure to release in manageable quantities.

Pressure Storage Tank — Class of fixed facility storage tanks divided into two categories: low-pressure storage tanks and pressure vessels.

Pressure Vessel — Fixed-facility storage tanks with operating pressures above 15 psi (103 kPa) {1.03 bar}.

Primary Explosive — High explosive that is easily initiated and highly sensitive to heat; often used as a detonator. *Also known as* Initiation Device.

Protective Action Distance — Downwind distance from a hazardous materials incident within which protective actions should be implemented.

Proton — Subatomic particle with a physical mass and a positive electric charge.

Public Safety Sample — Hazardous materials collected at an incident and used to help inform response and mitigation options.

Public Safety Sampling — Techniques used to collect materials found at a Hazmat/WMD incident that result in a forensically usable and legally defensible sample. Samples are often used when determining response and mitigation options.

Purge — To expel an inert gas through a device's hosing and/or intake system to remove any residual contaminants.

R

Radiation — Energy from a radioactive source emitted in the form of waves or particles, as a result of the decay of an atomic nucleus; process known as *radioactivity*. *Also called* Nuclear Radiation.

Radiation Absorbed Dose (rad) — English System unit used to measure the amount of radiation energy absorbed by a material; its International System equivalent is gray (Gy).

Radiation-Exposure Device (RED) — Powerful gamma-emitting radiation source used as a weapon.

Radioactive Decay — Process in which an unstable radioactive atom loses energy by emitting ionizing radiation and conversion electrons.

Radioactive Material (RAM) — Material with an atomic nucleus that spontaneously decays or disintegrates, emitting radiation as particles or electromagnetic waves at a rate of greater than 0.002 microcuries per gram (Ci/g).

Radioisotope — Unstable atom that releases nuclear energy.

Radiological Dispersal Device (RDD) — Conventional high explosives wrapped with radioactive materials; designed to spread radioactive contamination over a wide area. *Also known as* Dirty Bomb.

Radiological Dispersal Weapons (RDW) — Devices that spread radioactive contamination without using explosives; instead, radioactive contamination is spread using pressurized containers, building ventilation systems, fans, and mechanical devices.

Railcar Initials and Numbers — Combination of letters and numbers stenciled on rail tank cars that may be used to get information about the car's contents from the railroad's computer or the shipper. *Also known as* Reporting Marks.

Rain-Down Application Method — Foam application method that directs the stream into the air above the unignited or ignited spill or fire, allowing the foam to float gently down onto the surface of the fuel.

Raman Spectrometer — Apparatus used to observe the absorption, scattering, and shifts in light when sent through a material. The results are unique to the molecule.

Rapid Intervention Crew or Team (RIC/RIT) — Two or more firefighters designated to perform firefighter rescue; they are stationed outside the hazard and must be standing by throughout the incident. *Previously known as* Rapid Intervention Team (RIT).

Reactive Material — Substance capable of chemically reacting with other substances; for example, material that reacts violently when combined with air or water.

Reactivity — Ability of a substance to chemically react with other materials, and the speed with which that reaction takes place.

Reagent — Chemical that is known to react to another chemical or compound in a specific way, often used to detect or synthesize another chemical.

Recovery — Situation where the victim is determined or presumed to be dead, and the goal of the operation is to recover the body.

Reducing Agent — Fuel that is being oxidized or burned during combustion. *Also known as* Reducer.

Refrigerated Intermodal Container — Cargo container having its own refrigeration unit. *Also known as* Reefer.

Remediation — Fixing or correcting a fault, error, or deficiency.

Resonant Bond — Type of chemical bond in which electrons move freely between the compound atoms. *Also known as* Delocalized Bond.

Response Model — Framework for resolving problems or conflicts using logic, research, and analysis.

Response Objective — Statement based on realistic expectations of what can be accomplished when all allocated resources have been effectively deployed that provide guidance and direction for selecting appropriate strategies and the tactical direction of resources.

Response Option — Specific operations performed in a specific order to accomplish the goals of the response objective.

Retain — Actions to contain a liquid or sludge in an area where it can be absorbed, neutralized, or removed. Often used as a longer-term solution than other similar product control methods.

Rickettsia — Specialized bacteria that live and multiply in the gastrointestinal tract of arthropod carriers, such as ticks and fleas.

Ring Stiffener — Circumferential tank shell stiffener that helps to maintain the tank cross section.

Riot Control Agent — Chemical compound that temporarily makes people unable to function, by causing immediate irritation to the eyes, mouth, throat, lungs, and skin.

Risk-Based Response — Method using hazard and risk assessment to determine an appropriate mitigation effort based on the circumstances of the incident.

Roentgen (R) — English System unit used to measure radiation exposure, applied only to gamma and X-ray radiation; the unit used on most U.S. dosimeters.

Roentgen Equivalent in Man (rem) — English System unit used to express the radiation absorbed dose (rad) equivalence as pertaining to a human body; used to set radiation dose limits for emergency responders. Applied to all types of radiation.

Roll-On Application Method — Method of foam application in which the foam stream is directed at the ground at the front edge of the unignited or ignited liquid fuel spill; foam then spreads across the surface of the liquid. *Also known as* Bounce.

Route of Entry — Pathway via which hazardous materials get into (or affect) the human body.

S

Safety Data Sheet (SDS) — Reference material that provides information on chemicals that are used, produced, or stored at a facility. Form is provided by chemical manufacturers and blenders; contains information about chemical composition, physical and chemical properties, health and safety hazards, emergency response procedures, and waste disposal procedures. *Also known as* Material Safety Data Sheet (MSDS) *or* Product Safety Data Sheet (PSDS).

Safety Officer — Member of the IMS command staff responsible to the Incident Commander for monitoring and assessing hazardous and unsafe conditions and developing measures for assessing personnel safety on an incident. *Also known as* Incident Safety Officer.

Safety Relief Device — Device on cargo tanks with an operating part held in place by a spring; the valve opens at preset pressures to relieve excess pressure and prevent failure of the vessel.

Saponification — Reaction between an alkali and a fatty acid that produces soap.

Saturation — The concentration at which the addition of more solute does not increase the levels of dissolved solute.

Scintillator — Material that glows (luminesces) when exposed to ionizing radiation.

Secondary Device — Bomb or other weapon placed at the scene of an ongoing emergency response that is intended to cause casualties among responders; secondary explosive devices are designed to explode after a primary explosion or other major emergency response event has attracted large numbers of responders to the scene.

Secondary Explosive — High explosive that is designed to detonate only under specific circumstances, including activation from the detonation of a primary explosive. *Also known as* Main Charge Explosive.

Self-Accelerating Decomposition Temperature (SADT) — Lowest temperature at which product in a typical package will undergo a self-accelerating decomposition. The reaction can be violent, usually rupturing the package, dispersing original material, liquid and/or gaseous decomposition products considerable distances.

Self-Contained Breathing Apparatus (SCBA) — Respirator worn by the user that supplies a breathable atmosphere that is either carried in or generated by the apparatus and is independent of the ambient atmosphere. Respiratory protection is worn in all atmospheres that are considered to be Immediately Dangerous to Life and Health (IDLH). *Also known as* Air Mask *or* Air Pack.

Self-Reading Dosimeter (SRD) — Detection device that displays the cumulative reading without requiring additional processing. *Also known as* Direct-Reading Dosimeters (DRDs) *and* Pencil Dosimeters.

Shell — Layer of electrons that orbit the nucleus of an atom. The innermost shell can hold up to two electrons, and each subsequent shell can hold eight. *Also known as* Orbit, Orbital, *and* Ring.

Sheltering in Place — Having occupants remain in a structure or vehicle in order to provide protection from a rapidly approaching hazard, such as a fire or hazardous gas cloud. *Opposite of* evacuation. *Also known as* Protection-in-Place, Sheltering, *and* Taking Refuge.

Short-Term Exposure Limit (STEL) — Fifteen-minute time-weighted average that should not be exceeded at any time during a workday; exposures should not last longer than 15 minutes and should not be repeated more than four times per day with at least 60 minutes between exposures.

Sievert (Sv) — SI unit of measurement for low levels of ionizing radiation and their health effect in humans.

Site Characterization — Size-up and evaluation of hazards, problems, and potential solutions of a site.

Situational Awareness — Perception of the surrounding environment and the ability to anticipate future events.

Size-Up — Ongoing evaluation of influential factors at the scene of an incident.

SKED® — Lightweight, compact device for patient packaging; shaped to accommodate a long backboard; may be used with a rope mechanical advantage system.

Slurry — Suspension formed by a quantity of granulated or powdered solid material that is not completely soluble mixed into a liquid.

Soft Patch — Combination of gasket material and either wooden plugs or wedges inserted into a hole to patch a leak.

Solid — Substance that has a definite shape and size; the molecules of a solid generally have very little mobility.

Solubility — Degree to which a solid, liquid, or gas dissolves in a solvent (usually water).

Soluble — Capable of being dissolved in a liquid (usually water).

Solution — Uniform mixture composed of two or more substances.

Solvent — A substance that dissolves another substance (solute), resulting in a third substance (solution).

Specific Gravity — Mass (weight) of a substance compared to the weight of an equal volume of water at a given temperature. A specific gravity less than one indicates a substance lighter than water; a specific gravity greater than one indicates a substance heavier than water.

Specification Marking — Stencil on the exterior of a tank car indicating the standards to which the tank car was built; may also be found on intermodal containers and cargo tank trucks.

Spectrometer — Apparatus used to measure the intensity of a given sample based on a predefined spectrum such as wavelength or mass.

Spectrophotometer — Apparatus used to measure the intensity of light as an aspect of its color.

Spectroscopy — Study of the results when a material is dispersed into its component spectrum. *Also known as* Spectrography.

Staging Area — Prearranged, temporary strategic location, away from the emergency scene, where units assemble and wait until they are assigned a position on the emergency scene; these resources (personnel, apparatus, tools, and equipment) must then be able to respond within three minutes of being assigned. Staging Area Managers report to the Incident Commander or Operations Section Chief, if one has been established.

Standard Operating Procedure (SOP) — Standard methods or rules in which an organization or fire department operates to carry out a routine function. Usually these procedures are written in a policies and procedures handbook and all firefighters should be well versed in their content.

Standard Transportation Commodity Code (STCC) — Numerical code on the waybill used by the rail industry to identify the commodity. *Also known as* STCC Number.

Static Electricity — Accumulation of electrical charges on opposing surfaces, created by the separation of unlike materials or by the movement of surfaces.

Sterilization — Any process that destroys biological agents and other life forms. Often uses heat.

Stokes Basket — Wire or plastic basket-type litter suitable for transporting patients from locations where a standard litter would not be easily secured, such as a pile of rubble, a structural collapse, or the upper floor of a building; may be used with a harness for lifting.

Street Clothes — Clothing that is anything other than chemical protective clothing or structural firefighters' protective clothing, including work uniforms and ordinary civilian clothing.

Strong Oxidizer — Substance that readily gives off large quantities of oxygen, thereby stimulating combustion; produces a strong reaction by readily accepting electrons from a reducing agent (fuel).

Structural Firefighters' Protective Clothing — General term for the equipment worn by fire and emergency services responders; includes helmets, coats, pants, boots, eye protection, gloves, protective hoods, self-contained breathing apparatus (SCBA), and personal alert safety system (PASS) devices.

Sublimation — Vaporization of a material from the solid to vapor state without passing through the liquid state.

Supervisory Control and Data Acquisition (SCADA) — System that monitors and controls coded signals from preset locations within an infrastructure (pipeline system), industry (manufacturing system), or facility (building system).

Supplied Air Respirator (SAR) — Atmosphere-supplying respirator for which the source of breathing air is not designed to be carried by the user; not certified for fire fighting operations. *Also known as* Airline Respirator System.

Surface Acoustic Wave (SAW) Sensor — Device that senses a physical phenomenon. Electrical signals are transduced to mechanical waves, and then back to electrical signals for analysis.

Surfactant — Chemical that lowers the surface tension of a liquid; allows water to spread more rapidly over the surface of Class A fuels and penetrate organic fuels.

Synergistic Effect — Phenomenon in which the combined properties of substances have an effect greater than their simple arithmetical sum of effects.

Systemic Effect — Damage spread through an entire system; opposite of a local effect, which is limited to a single location.

T

Tapping — Process to attach a nozzle or outlet onto a container's tank or piping to assist with the removal or transfer of a product.

T-Code — Portable tank instruction code used to identify intermodal containers used to transport hazardous materials. This set of codes replace the IMO type listings.

Technical Decontamination — Using chemical or physical methods to thoroughly remove contaminants from responders (primarily entry team personnel) and their equipment; usually conducted within a formal decontamination line or corridor following gross decontamination. *Also known as* Formal Decontamination.

Termination — The phase of an incident in which emergency operations are completed and the scene is turned over to the property owner or other party for recovery operations.

Tertiary Explosive — High explosive that require initiation from a secondary explosive. Tertiary explosives are often categorized with secondary explosives. *Also known as* Blasting Agents.

Thermal Burn — Injury caused by contact with flames, hot objects, and hot fluids; examples include scalds and steam burns.

Thermal Imager — Electronic device that forms images using infrared radiation. *Also known as* Thermal Imaging Camera.

Thermal Insulation — Materials added to decrease heat transfer between objects in proximity to each other.

Threshold Limit Value (TLV®) — Maximum concentration of a given material in parts per million (ppm) that may be tolerated for an 8-hour exposure during a regular workweek without ill effects.

Threshold Limit Value/Ceiling (TLV®/C) — Maximum concentration of a given material in parts per million (ppm) that should not be exceeded, even instantaneously.

Torr — Unit of pressure measurement; not part of the SI. Measured as 1/760 of a standard atmosphere.

Toxic — Poisonous.

Toxic Industrial Material (TIM) — Industrial chemical that is toxic at a certain concentration and is produced in quantities exceeding 30 tons (30 tonnes) per year at any one production facility; readily available and could be used by terrorists to deliberately kill, injure, or incapacitate people. *Also known as* Toxic Industrial Chemical (TIC).

Toxic Inhalation Hazard (TIH) — Volatile liquid or gas known to be a severe hazard to human health during transportation.

Toxicity — Degree to which a substance (toxin or poison) can harm humans or animals. Ability of a substance to do harm within the body.

Toxicology — Study of the adverse effects of chemicals on living organisms.

Toxin — Substance that has the property of being poisonous.

Transient Evidence — Material that will lose its evidentiary value if it is unpreserved or unprotected; for example, blood in the rain.

Transmutation — Conversion of one element or isotope into another form or state.

Transport Index (TI) — Number placed on the label of a package expressing the maximum allowable radiation level in millirem per hour at 1 meter (3.3 feet) from the external surface of the package.

Transportation Mode — Technologies used to move people and/or goods in different environments; for example, rail, motor vehicles, aviation, vessels, and pipelines.

Trench Foot — Foot condition resulting from prolonged exposure to damp conditions or immersion in water; symptoms include tingling and/or itching, pain, swelling, cold and blotchy skin, numbness, and a prickly or heavy feeling in the foot. In severe cases, blisters can form, after which skin and tissue die and fall off.

Triage — System used for sorting and classifying accident casualties to determine the priority for medical treatment and transportation.

Type A Packaging — Container used to ship radioactive materials with relatively high radiation levels.

Type B Packaging — Container used to ship radioactive materials that exceed the limits allowed by Type A packaging, such as materials that would present a radiation hazard to the public or the environment if there were a major release.

Type C Packaging — Container used to ship highly reactive radioactive materials intended for transport via aircraft.

U

UN/NA Number — Four-digit number assigned by the United Nations to identify a specific hazardous chemical. North America (DOT) numbers are identical to UN numbers, unless the UN number is unassigned.

Unified Command (UC) — In the Incident Command System, a shared command role in which all agencies with geographical or functional responsibility establish a common set of incident objectives and strategies. In unified command there is a single incident command post and a single operations chief at any given time.

Unstable Material — Materials that are capable of undergoing chemical changes or that can violently decompose with little or no outside stimulus.

Upper Flammable Limit (UFL) — Upper limit at which a flammable gas or vapor will ignite; above this limit the gas or vapor is too *rich* to burn (lacks the proper quantity of oxygen). *Also known as* Upper Explosive Limit (UEL).

V

Vacuum Relief Valve — Pressure control device designed to introduce outside air into a container during offloading operations.

Valve — Mechanical device with a passageway that controls the flow of a liquid or gas.

Vapor Density — Weight of pure vapor or gas compared to the weight of an equal volume of dry air at the same temperature and pressure. A vapor density less than one indicates a vapor lighter than air; a vapor density greater than one indicates a vapor heavier than air.

Vapor Dispersion — Action taken to direct or influence the course of airborne hazardous materials.

Vapor Explosion — Occurrence when a hot liquid fuel transfers heat energy to a colder, more volatile liquid fuel. As the colder fuel vaporizes, pressure builds in a container and can create shockwaves of kinetic energy.

Vapor Pressure — The pressure at which a vapor is in equilibrium with its liquid phase for a given temperature; liquids that have a greater tendency to evaporate have higher vapor pressures for a given temperature.

Vapor Suppression — Action taken to reduce the emission of vapors at a hazardous materials spill.

Vapor-Protective Clothing — Gas-tight chemical-protective clothing designed to meet NFPA 1991, *Standard on Vapor-Protective Ensembles for Hazardous Materials Emergencies*; part of an EPA Level A ensemble.

Vector — An animate intermediary in the indirect transmission of an agent that carries the agent from a reservoir to a susceptible host.

Vehicle-Borne Improvised Explosives Device (VBIED) — An improvised explosive device placed in a car, truck, or other vehicle. This type of IED typically creates a large explosion.

Ventilation — Systematic removal of heated air, smoke, gases or other airborne contaminants from a structure and replacing them with cooler and/or fresher air to reduce damage and facilitate fire fighting operations.

Virus — Simplest type of microorganism that can only replicate itself in the living cells of its hosts. Viruses are unaffected by antibiotics.

Viscosity — Measure of a liquid's internal friction at a given temperature. This concept is informally expressed as thickness, stickiness, and ability to flow.

Volatility — Ability of a substance to vaporize easily at a relatively low temperature.

W

Warm Zone — Area between the hot and cold zones that usually contains the decontamination corridor; typically requires a lesser degree of personal protective equipment than the Hot Zone. *Also known as* Contamination Reduction Zone *or* Contamination Reduction Corridor.

Water Solubility — Ability of a liquid or solid to mix with or dissolve in water.

Water-Reactive Material — Substance, generally a flammable solid, that reacts when mixed with water or exposed to humid air.

Waybill — Shipping paper used by a railroad to indicate origin, destination, route, and product; a waybill for each car is carried by the conductor. *Similar to* Air Bill *and* Bill of Lading.

Weapon of Mass Destruction (WMD) — Any weapon or device that is intended or has the capability to cause death or serious bodily injury to a significant number of people through the release, dissemination, or impact of toxic or poisonous chemicals or their precursors, a disease organism, or radiation or radioactivity; may include chemical, biological, radiological, nuclear, or explosive (CBRNE) type weapons.

Wet Chemistry — Branch of analysis with a focus on chemicals in their liquid phase.

Wireless Information System for Emergency Responders (WISER) — This electronic resource brings a wide range of information to the hazmat responder such as chemical identification support, characteristics of chemicals and compounds, health hazard information, and containment advice.

Witness — Person called upon to provide factual testimony before a judge or jury.

Z

Zeroing — Resetting an instrument to read at normal (baseline) levels in fresh air.

Index

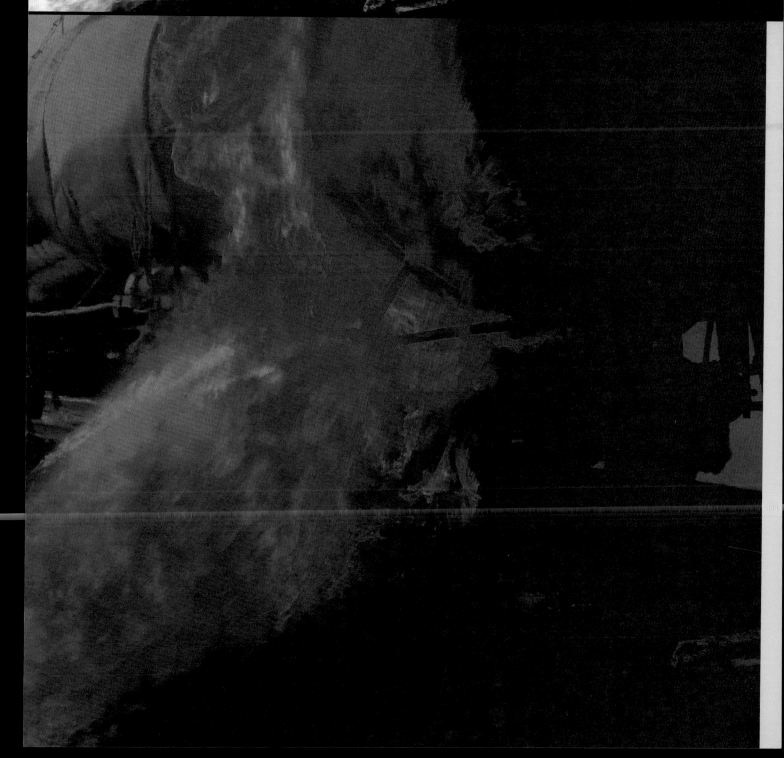

Courtesy of Barry Lindley.

Index

attacks. *See* Explosive/incendiary attacks

binary explosive, 178

black powder, 372, 374

blast-pressure wave (shock wave), 179–180, 372

BLEVE. *See* Boiling liquid expanding vapor explosion (BLEVE)

B-NICE, 366

CBRNE. *See* Chemical, biological, radiological, nuclear, or explosive (CBRNE)

chlorate-based, 379

Class 1 hazard class, 72, 178–181

containers for, 178, 179

defined, 178

devices, 374

dust, 65

dynamite, 65

effects of, 180

fireworks, 383

hazard class, 160

homemade explosives (HME), 375, 377, 696

illicit laboratories, 695–697

incendiary thermal effect, 181

Kansas City, MO, explosion, 30

main charge explosive, 374

M-devices, 383

military explosives, 375, 376

nitrate-based, 379

peroxide-based, 378

placards, 178

pressure containers, 230

seismic effect, 181

shrapnel and fragmentation

definition and description, 181

in grenades, 383

improvised explosive devices, 379

secondary attacks and booby traps, 370

TNT, 372, 374

triacetone triperoxide (TATP), 367, 378, 555

vapor explosion, 183

Explosive/incendiary attacks

anatomy of, 372–373

car and truck bombs, 371

characteristics, 365

commercial/military, 375, 376

high explosives, 373, 375

homemade/improvised materials, 375, 377, 378–379

IED. *See* Improvised Explosive Device (IED)

indicators, 371

low explosives, 374, 375

primary and secondary explosives, 374, 375

response to, 388–389

terrorist incidents, 371–389

tertiary explosives, 374

for WMDs, 367, 368

Export and Import of Hazardous Wastes Regulations (EIHWR), 36

Exposure

defined, 462, 480

GEBMO model, 223–225

environment, 224, 225

people, 224, 225

property, 224, 225

sequence of events, 212

timeframes, 225

identifying, 202

limits, 531, 534–536

radioactive, 166, 168

response objectives and action options, 313

size-up, 211

Extinguish, 623

Extremely hazardous substance, defined, 32

Facility documents, 103

FBI. *See* Federal Bureau of Investigation (FBI)

Federal Bureau of Investigation (FBI)

hazmat incident duties, 34

illicit laboratory investigations, 678

investigative authority, 656

scene control, 658

state radiation protection office, 704

terrorism investigation and prevention, 361

terrorist attack, evidence search perimeter, 336

Federal Emergency Management Agency (FEMA)

evacuating rescuers, 342

incident action planning process, 306

Planning Guidance for Protection and Recovery Following Radiological Dispersal Device (RDD) and Improvised Nuclear Device (IND) Incidents, 561

terrorist attacks, 33

Urban Search and Rescue teams, 334

Federal Hazardous Substances Act (FHSA), 33

Federal Railroad Administration (FRA) regulations, 263

Feed/farm stores, as likely to have hazardous materials, 49

FEMA. *See* Federal Emergency Management Agency (FEMA)

Fertilizer

ammonium nitrate, 13, 28–29, 379

anhydrous ammonia, in hazmat incidents, 39

anhydrous ammonia leak, Shreveport, LA, 30

West, Texas, fertilizer facility fire, 13

FHSA (Federal Hazardous Substances Act), 33

FIBC (flexible intermediate bulk container), 63, 278

Fiber, 139

Field screening samples, 666

Field test, 545

Finance/Administration Section of NIMS-ICS, 326–327

Fire (ordnance) divisions, 94

Fire codes, 48

Fire control

defined, 349

extinguishment, 623

flammable/combustible liquids, 623, 624, 633–636

purpose of, 623

Fire gases, 174

Fire Section of the *ERG*, 123

Fire service ensembles, 452, 453

Fire-entry suit, 435

Fires, response objectives and action options, 2623

Fireworks as explosive devices, 383

First Aid section of the *ERG*, 124

Fixed facility (bulk) pressure container, 59, 64

Fixed facility shutoff valves, 622

Flame retardant, 436

Flame-resistant protective clothing, 436

Flammable liquid and gas

autoignition temperature, 155–156

characteristics, 185

Class 3 hazard class, 72, 183, 185–186

compressed gas cylinders during a fire, 213

defined, 155, 186

divisions, 186

examples, 186

fire control, 636–637

flammability range, 157

flash point, 154–155

highly flammable hazard class, 160

lower flammable (explosive) limit (LFL), 156

placards, 186

specific gravity, 151

upper flammable limit (UFL), 156

units of measurement, 145, 146
Vapor suppression for spill control, 614, 615, 644
Vapor-and-gas-removing filters, 428–429
Vapordome roof tank, 239
Vapor-protective clothing
    cylinder and high-pressure vessel incidents, 544
    defined, 439
    limitations, 440
    not flame resistant, 433, 434
    purpose of, 439
    totally encapsulating chemical-protective (TECP) suits, 439
    to use with positive-pressure SCBA, 439
VBIED (Vehicle-Borne Improvised Explosive Device), 386–388
Vectors for disease transmission, 405, 406
Vehicle-Borne Improvised Explosive Devices (VBIEDs), 386–388
Vehicle-mounted pressure container, 60
Ventilation for spill control, 616, 617
Verbal reports of presence of hazmat, 48
Vesicants, 390, 393–395, 396–397
Vessel cargo carriers, 69–71
Victims
    ambulatory. See Ambulatory victims
    deceased, 505–506
    evacuation of contaminated victims, 346
    identifying, 202
    Mission-Specific Victim Rescue and Recovery Level. See Mission-Specific Victim Rescue and Recovery Level
    nonambulatory. See Nonambulatory victims
    response objectives and action options, 313
Violent rupture, 147
Violent rupture of containers, 218
Viral agents, 403
Virus
    defined, 27, 176
    description, 174
    Ebola virus
        health care provider infection, 177
        outbreak of, 408
        transmission of, 405
        unaffected by antibiotics, 176
    as hazard, 25
    laboratory production, 698
    mechanism of harm, 27–28
    Norwalk Virus, 405
    smallpox, 27, 177, 403, 405
Viscosity, 152–153
Visual detection of hazardous materials, 104–107
Volatility, 393
Vomiting agent, 399
VX nerve agents, 95

W

Warm (limited access) zone, 335, 338
Washing decontamination technique, 491, 494
Water
    availability as decon site selection consideration, 510
    contaminated runoff, 484, 485
    difficulty in containing hazardous materials, 52
    as polar solvent, 149
    rainbow sheen as hazardous material indicator, 105, 106
    solubility of materials, 149–150
Water reactive materials, 186–188
Water reactive materials producing toxic gases, 127
Water-reactive hazard class, 160
Waterways
    as consideration for decon site selection, 509
    as location for hazmat transportation accidents, 50
Waverly overpressurization incident (1978), 215–216

WCB (Workers Compensation Board), 14
Weapon of mass destruction (WMD). See also Terrorism
    biological pathogens, 368
    biological toxins, 367
    defined, 11, 366
    evidence collection procedures, 654
    evidence preservation and chain-of-custody procedures, 540
    expansion of terrorism, 366–367
    explosives, 367, 368, 371
    Improvised Explosive Device (IED), 371. See also Improvised Explosive Device (IED)
    incident resources, 130
    industrial chemicals, 367
    military-grade chemical weapons, 368
    National Medical Response Team-Weapons of Mass Destruction (NMRT-WMD), 333
    nuclear weapons, 368
    protective clothing and equipment for, 432–433
    radiological materials, 368
    ranking of threats, 368
    response phases, 657–658
    toxic industrial materials, 402
    Weapons of Mass Destruction-Civil Support Team (WMD-CST), 332–333
Weather
    cold weather decon, 513–514
    decontamination considerations, 508
    impact on incident progress and mitigation, 202–203
    wind, 121, 459, 508
Well car, 259
Wet decontamination, 483, 484–485
WHMIS (Workplace Hazardous Materials Information System), 37, 89–91
Wildlands, T-card accountability system, 340
Wind
    decontamination considerations, 508
    initial isolation zone and, 121
    wind chill chart, 459
Wireless Information System for Emergency Responders (WISER), 104
WISER (Wireless Information System for Emergency Responders), 104
Witness, 659, 679
WMD. See Weapon of mass destruction (WMD)
WMD-CST (Weapons of Mass Destruction-Civil Support Team), 332–333
Work rotation for heat-exposure prevention, 457
Workers Compensation Board (WCB), 14
Workplace Hazardous Materials Information System (WHMIS), 37, 89–91
Written resources, 98–104
    electronic technical resources, 103–104
    Emergency Response Guidebook (ERG), 102–103. See also Emergency Response Guidebook (ERG)
    facility documents, 103
    safety data sheets, 101–102. See also Safety data sheet (SDS)
    shipping papers, 99–101

X

X-rays, 165
Xylene, 35

Y

Y-cylinders
    characteristics, 273
    other regulated materials (ORM-Ds), 60
    specifications, 273
Yvorra, James G., 307

NOTES